P9-CSA-139

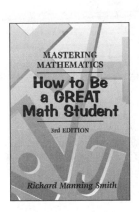

Mastering Mathematics:
How to Be a Great Math Student,
Third Edition
by Richard Manning Smith

A complete guide to improving your learning and performance in math courses!

Providing solid tips for every stage of study, *Mastering Mathematics: How to Be a Great Math Student, Third Edition* stresses the importance of a positive attitude and gives you the tools to succeed in your math course. This practical guide will help you:

- avoid mental blocks during math exams
- identify and improve your areas of weakness
- get the most out of class time
- study more effectively
- overcome a perceived "low math ability"
- be successful on math tests
- get back on track when you are feeling "lost"
  …and much more!

To order a copy of *Mastering Mathematics*, please contact your college store or fill out the form on the back and return with your payment to Brooks/Cole. You can also order on-line at http://www.brookscole.com

ORDER FORM

_____Yes! Send me a copy of *Mastering Mathematics: How to be a Great Math Student*   (ISBN: 0-534-34864-5)

_____Copies x $15.95* =_____

Residents of: AL, AZ, CA, CT, CO, FL, GA, IL, IN, KS, KY, LA, MA, MD, MI, MN, MO, NC, NJ, NY, OH, PA, RI, SC, TN, TX, UT, VA, WA, WI must add appropriate state sales tax.

Subtotal _____
Tax _____
Handling ___$4.00___
Total Due _____

Payment Options

_____ Check or money order enclosed

Bill my   ____VISA ____MasterCard ____American Express

Card Number: _____

Expiration Date: _____

Signature: _____

Note: Credit card billing and shipping addresses must be the same.

Please ship my order to: (Please print)

Name _____

Institution _____

_____

Street Address_____

_____

City _____ State _____ Zip+4_____

Telephone (   )_____ e-mail _____

Your credit card will not be billed until your order is shipped. Prices subject to change without notice. We will refund payment for unshipped out-of-stock titles after 120 days and for not-yet-published titles after 180 days unless an earlier date is requested in writing from you.

Mail to:

Brooks/Cole Publishing Company
Source Code 8BCMA138
511 Forest Lodge Road
Pacific Grove, California 93950-5098
Phone: (408) 373-0728; Fax: (408) 375-6414
e-mail: info@brookscole.com

10/97

* Call after 12/1/97 for current prices: 1-800-487-3575

Elementary
Technical
Mathematics

Elementary
Technical
Mathematics

SEVENTH EDITION

DALE EWEN

Parkland Community College

C. ROBERT NELSON

Champaign Centennial High School

Brooks/Cole Publishing Company

I(T)P® An International Thomson Publishing Company

Pacific Grove • Albany • Belmont • Bonn • Boston • Cincinnati • Detroit • Johannesburg • London
Madrid • Melbourne • Mexico City • New York • Paris • Singapore • Tokyo • Toronto • Washington

Publisher: *Robert W. Pirtle*
Project Developmental Editor: *Elizabeth Rammel*
Marketing Team: *Margaret Parks, Jennifer Huber, and Laura Caldwell*
Editorial Assistant: *Melissa Duge*
Production Coordinator: *Marjorie Z. Sanders*
Production: *Hoyt Publishing Services*
Manuscript Editor: *David Hoyt*
Permissions Editor: *Sue C. Howard*

Interior Design: *Geri Davis*
Cover Design: *Roy R. Neuhaus*
Interior Illustration: *Scientific Illustrators*
Cover Illustration: *Boris Lyubner/The Stock Illustration Source, Inc.*
Art Editor: *Hoyt Publishing Services*
Photo Coordinator: *Sue C. Howard*
Typesetting: *Better Graphics, Inc.*
Printing and Binding: *D. B. Hess Company*

Copyright © 1998 by Brooks/Cole Publishing Company
A division of International Thomson Publishing Inc.

I(T)P   The ITP logo is a registered trademark under license.

For more information, contact:

BROOKS/COLE PUBLISHING COMPANY
511 Forest Lodge Road
Pacific Grove, CA 93950
USA

International Thomson Publishing Europe
Berkshire House 168-173
High Holborn
London WC1V 7AA
England

Thomas Nelson Australia
102 Dodds Street
South Melbourne, 3205
Victoria, Australia

Nelson Canada
1120 Birchmount Road
Scarborough, Ontario
Canada M1K 5G4

International Thomson Editores
Seneca 53
Col Polanco
11560 México, D. F., México

International Thomson Publishing GmbH
Königswinterer Strasse 418
53227 Bonn
Germany

International Thomson Publishing Asia
221 Henderson Road
#05-10 Henderson Building
Singapore 0315

International Thomson Publishing Japan
Hirakawacho Kyowa Building, 3F
2-2-1 Hirakawacho
Chiyoda-ku, Tokyo 102
Japan

All rights reserved. No part of this book may be reproduced, stored in a retrieval system, or transcribed, in any form or by any means—electronic, mechanical, photocopying, recording, or otherwise—without prior written permission of the publisher, Brooks/Cole Publishing Company, Pacific Grove, California 93950.

Printed in the United States of America
10   9   8   7   6   5   4   3   2

Library of Congress Cataloging-in-Publication Data

Ewen, Dale [date]
    Elementary technical mathematics / Dale Ewen, C. Robert Nelson. — 7th ed.
        p.   cm.
    Includes index.
    ISBN 0-534-35127-1 (pbk.)
    1. Mathematics.   I. Nelson, C. Robert.   II. Title.
QA39.2.E9396   1997
510—dc21

97-35750
CIP

Credits continue on page 519.

Preface

Elementary Technical Mathematics, Seventh Edition, is intended for Tech Prep programs and technical, trade, or allied health programs. This book was written for students who have minimal background in mathematics or need considerable review but who plan to learn a technical skill. To become proficient in most technical programs, students must learn basic mathematical skills. To that end, Chapters 1 through 4 cover basic arithmetic operations, fractions, decimals, percent, the metric system, and numbers as measurements. Chapters 5 through 11 present essential algebra needed in technical and trade programs. The essentials of geometry—relationships and formulas for the most common two- and three-dimensional figures—are given in detail in Chapter 12. Chapters 13 and 14 present a short but intensive study of trigonometry that includes right-triangle trigonometry as well as oblique triangles and graphing. The concepts of statistics that are most important to technical fields are discussed in Chapter 15.

We have written this text to match the reading level of most technical students. Visual images engage these readers and stimulate the problem-solving process. We emphasize that these skills are essential for success in technical courses.

The following important text features have been retained from previous editions:

- A large number of applications from a wide variety of technical areas, including auto/diesel mechanics, industrial and construction trades, electronics, agriculture, and allied health.

- Chapter 1 reviews basic concepts in such a way that students or the entire class can easily study only those sections they need to review.

- To reflect the nearly universal use of calculators, basic operations for the scientific calculator are integrated throughout the text. The format used is proven and easy to follow: flowchart, buttons pushed, and the calculator display. The instructor should inform the students when *not* to use a calculator.

- A comprehensive introduction to basic algebra is presented for those students who need it as a prerequisite to more advanced algebra courses. However, the book has been written to allow the omission of selected sections or chapters without loss of continuity, to meet the needs of specific students.

- A chapter review and a chapter test appear at the end of each chapter, as aids for students in preparing for quizzes and exams.
- The text design and second color help make the text more accessible, highlight important concepts, and enhance the art presentation.
- A laminated reference card of useful, frequently referenced information—such as metric system prefixes, English weights and measures, metric and English conversion, and formulas from geometry—is provided, perforated for easy removal and use.

Significant changes in the seventh edition include the following.

- Revised Chapter 1 gives the instructor flexibility in selecting the depth of arithmetic review topics for classes of students and individual students.
- Chapter 2 now focuses on signed numbers and powers of 10.
- Trigonometry is separated into two chapters: Chapter 13, Right Triangle Trigonometry, and Chapter 14, Trigonometry with Any Angle.
- A short introduction to statistical process control has been added to Chapter 15, Basic Statistics.
- The use of a scientific calculator has been integrated throughout the text to reflect its nearly universal use in technical classes and on the job.
- Cumulative reviews are provided at the end of every three chapters to help students review for comprehensive exams.
- Studies show that current students will experience several career changes during their working lives. The chapter-opening pages illustrate various career paths for students to consider as their careers, technology, and the workplace evolve.

Useful ancillaries free to adopters of this text include:

- A comprehensive **Instructor's Manual** with answers to all exercises and chapter tests, two forms of short quizzes, two forms of chapter tests, and a sample comprehensive final exam, all with answers provided.
- **ITP Tools (Mac ISBN: 0-534-35129-8; DOS ISBN: 0-534-35131-X; Win ISBN 0-534-35130-1)** is a suite of programs designed to increase the efficiency and effectiveness of instructors who adopt textbooks published by Brooks/Cole by providing a range of flexible, text-specific testing and tutorial options, as follows. **ITP Test** features text-specific algorithmic test generation, a bank of text-specific test items, and the ability to import your own questions and graphics and to edit questions. ITP Test offers enhanced mathematics text editing and improved algorithm support including trigonometric and logarithmic functions. The computer platforms supported will include Windows 3.1, Macintosh, Windows 95, Windows NT, and PowerMacintosh. **ITP Test On-Line** allows the student to complete an on-line test or practice test. Tests can be delivered on a floppy, local hard disk, LAN volume, or an Internet IP address. **ITP Class Manager** provides the ability to extract and track scores from on-line tests and practice tests created by ITP Test.

We are grateful for the courtesy of the L. S. Starrett Company in allowing us to use photographs of their instruments in Chapter 4. We are also thankful for the contributions of Charles "Chuck" Blesh of Orange Coast College and of Dale Smith of the Ogden-Weber Applied Technology Center. The authors also thank the many faculty members who used earlier editions and who offered suggestions. In particular, we thank the following reviewers:

- Rebecca J. Daggar, Rochester Institute of Technology
- Sandra Dashiell, Thomas Nelson Community College
- John DeCoursey, Vincennes University
- Richard Feeley, Rochester Institute of Technology
- William Galvery, Orange Coast College
- Kenneth Goldberg, New York University
- Kali Kocmoud, New Richmond High School
- Sotos Ktorides, Waukesha County Technical College
- Margaret Ann Leonard, Hudson Valley Community College
- Charles McSurdy, Nashville State Technical Institute
- Linda Nokes, Southwestern Michigan College
- Everett Ralston, California High School
- David Schumann, North Idaho College
- Renald Simmons, Vincennes University
- Deborah Spillius, Waukesha County Technical College
- Mark D. Tenney, Hudson Valley Community College

We thank our Parkland colleagues and students for their valued assistance. Anyone wishing to correspond regarding suggestions or questions should write Dale Ewen at Parkland Community College, 2400 West Bradley Avenue, Champaign, IL 61821, or contact the publisher.

For all their help, we thank our editors, Robert Pirtle and Elizabeth Rammel, and the staff of Brooks/Cole Publishing Company. We also greatly appreciate the diligent efforts of David Hoyt in coordinating production, Better Graphics in typesetting, Carolyn Cain in proofreading, and George Morris of Scientific Illustrators for the outstanding artwork. We are especially grateful to Joyce Ewen for her excellent proofing assistance.

Dale Ewen
C. Robert Nelson

Contents

List of Applications

Basic Concepts

MANUFACTURING
Factory workers often make use of basic mathematical skills. In filling out requisition forms, this supervisor applies many of the concepts discussed in this chapter.

MATHEMATICS AT WORK

Concepts involving basic math are required by almost every business or trade that a person might choose to work in. Daily life often requires working with decimals and percents in one way or another—such as properly balancing a checkbook or making sure bills are correct to avoid being overcharged. Understanding percents helps in making sound decisions about loans and investments that involve interest rates and sales tax.

People who work in shipping and receiving use whole numbers and decimal numbers in computing total weights and shipping charges. Construction trades such as carpentry, roofing, bricklaying, welding, and sheetmetal work use mainly fractions. The blueprints they work with are usually drawn to some fractional scale, such as $\frac{1}{4}$ inch = 1 foot. Machinists, drafters, and workers in automated manufacturing must know fractions and decimals. Fractional measurements are usually converted to decimals before machining parts, especially where tight tolerances are required. Auto mechanics use decimals for shimming parts, boring motors, and setting spark plug gaps. Retail and grocery store clerks must know percent concepts such as sales tax, markup, and markdown. Many retail stores pay an hourly wage plus a commission or percentage of sales.

Unit 1A   REVIEW OF OPERATIONS WITH WHOLE NUMBERS

1.1   Review of Basic Operations

The *positive integers* are the numbers 1, 2, 3, 4, 5, 6, and so on. They can also be written as +1, +2, +3, etc., but usually the *positive* (+) sign is omitted. The *whole numbers* are the numbers 0, 1, 2, 3, 4, 5, 6, and so on. That is, the whole numbers consist of the positive integers and zero.

The value of any digit in a number is determined by its place in the particular number. Each place represents a certain power of ten. By powers of ten, we mean the following:

$10^0 = 1$
$10^1 = 10$
$10^2 = 10 \times 10 = 100$ (the second power of 10)
$10^3 = 10 \times 10 \times 10 = 1000$ (the third power of 10)
$10^4 = 10 \times 10 \times 10 \times 10 = 10,000$ (the fourth power of 10)
etc.

Note: A small superscript number (such as the 2 in 10^2) is called an *exponent*.

The number 2354 means 2 thousands plus 3 hundreds plus 5 tens plus 4 ones.
In the number 236,895,174, each digit has been multiplied by some power of 10, as shown below.

(ten millions)		(hundred thousands)		(thousands)		(tens)	
10^7		10^5		10^3		10^1	
|		|		|		|	
2	3	6,	8	9	5,	1	7	4
|		|		|		|	|
10^8		10^6		10^4		10^2		10^0
(hundred millions)		(millions)		(ten thousands)		(hundreds)		(units)

The "+" (plus) symbol is the sign for addition, as in the expression $5 + 7$. The result of adding the numbers (in this case, 12) is called the *sum*. Integers are added in columns with the digits representing like powers of ten in the same vertical line. (*Vertical* means up and down.)

•   **EXAMPLE 1**   Add: $238 + 15 + 9 + 3564$.

238
15
9
3564
3826

Subtraction is the inverse operation of addition. Therefore, subtraction can be thought of in terms of addition. The "−" (minus) sign is the symbol for subtraction. The quantity 5 − 3 can be thought of as "what number added to 3 gives 5?" The result of subtraction is called the *difference*.

To check a subtraction, add the difference to the second number. If the sum is equal to the first number, the subtraction has been done correctly.

• EXAMPLE 2   Subtract: 2843 − 1928.

SUBTRACT

$$
\begin{array}{r}
2843 \\
-1928 \\
\hline
915
\end{array}
$$

first number
second number
difference

CHECK

$$
\begin{array}{r}
1928 \\
+915 \\
\hline
2843
\end{array}
$$

second number
difference
This sum equals the first number, so 915 is the correct difference.

Next, let's study some applications. In order to communicate about problems in electricity, technicians have developed a "language" of their own. It is a picture language that uses symbols and diagrams. The symbols used most often are listed in Table 2 of Appendix B. The circuit diagram is the most common and useful way to show a circuit. Note how each component (part) of the picture (Figure 1.1a) is represented by its symbol in the circuit diagram (Figure 1.1b) in the same relative position.

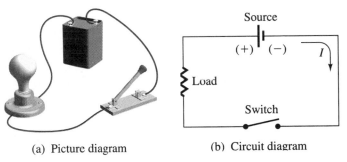

(a) Picture diagram   (b) Circuit diagram

FIGURE 1.1

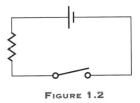

FIGURE 1.2

The light bulb may be represented as a resistance. Then the circuit diagram in Figure 1.1b would appear as in Figure 1.2, where

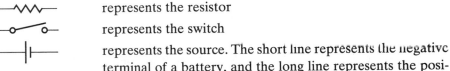

represents the resistor

represents the switch

represents the source. The short line represents the negative terminal of a battery, and the long line represents the positive terminal. The current flows from negative to positive.

There are two basic types of electrical circuits: series and parallel. An electrical circuit with only one path for the current, *I*, to flow is called a *series* circuit (Figure 1.3). An electrical circuit with more than one path for the current to flow is called a *parallel* circuit (Figure 1.4). A circuit breaker or fuse in a house is wired in series with its outlets. The outlets themselves are wired in parallel.

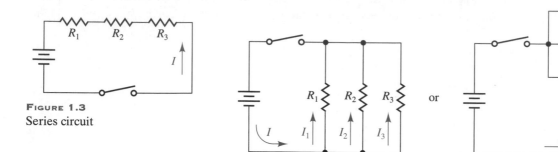

FIGURE 1.3
Series circuit

or

FIGURE 1.4
Parallel circuits

• **EXAMPLE 3** In a series circuit, the total resistance equals the sum of all the resistances in the circuit. Find the total resistance in the series circuit in Figure 1.5. Resistance is measured in ohms, Ω.

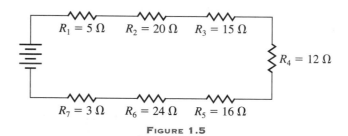

FIGURE 1.5

The total resistance is:

  5 Ω
20 Ω
15 Ω
12 Ω
16 Ω
24 Ω
  3 Ω
95 Ω

• **EXAMPLE 4** Studs are upright wooden or metal pieces in the walls of a building, to which siding, insulation panels, drywall, or decorative paneling are attached. A wall portion with seven studs is shown in Figure 1.6. Studs are normally placed 16 in. on center and are placed double at all internal and external corners of a building. The number of studs needed in a wall can be estimated by finding the number of linear feet (ft) of the wall. How many studs are needed for the exterior walls of the building in Figure 1.7?

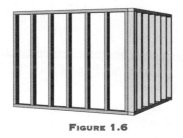

FIGURE 1.6

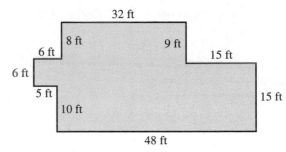

FIGURE 1.7

The outside perimeter of the building is the sum of the lengths of the sides of the building:

48 ft
15 ft
15 ft
9 ft
32 ft
8 ft
6 ft
6 ft
5 ft
10 ft
154 ft

Therefore, approximately 154 studs are needed in the outside wall.

Repeated addition of the same number can be shortened by multiplication. The "×" (times) and the "·" (raised dot) are used to indicate multiplication. When adding the lengths of five pipes, each 7 ft long, we have 7 ft + 7 ft + 7 ft + 7 ft + 7 ft = 35 ft of pipe. In multiplication, this would be 5 × 7 ft = 35 ft. The 5 and 7 are called *factors*, and 35 is called the *product*. Computing areas, volumes, forces, and distances requires skills in multiplication.

• **EXAMPLE 5**   Multiply: 358 × 18.

$$
\begin{array}{r}
358 \\
\times\ 18 \\
\hline
2864 \\
358 \\
\hline
6444
\end{array}
$$

Division is the inverse operation of multiplication. The following are all symbols for division: "÷", "⌐", "/", and "−". The quantity 15 ÷ 5 can also be thought of as "what number times 5 gives 15?" The answer to this question is 3, which is 15 divided by 5. The result, 3, is called the *quotient*. The number to be divided, 15, is called the *dividend*. The number you divide by, 5, is called the *divisor*.

• **EXAMPLE 6**   Divide: 84 ÷ 6.

$$
\begin{array}{r}
14 \quad \leftarrow \text{quotient} \\
6\overline{)84} \quad \leftarrow \text{dividend} \\
\text{divisor} \longrightarrow 6 \\
\hline
24 \\
24 \\
\hline
0 \quad \leftarrow \text{remainder}
\end{array}
$$

• **EXAMPLE 7**   Divide: 115 ÷ 7.

$$
\begin{array}{r}
16 \quad \leftarrow \text{quotient} \\
7\overline{)115} \quad \leftarrow \text{dividend} \\
\text{divisor} \overset{\longrightarrow}{} \quad 7 \\
\hline
45 \\
42 \\
\hline
3 \quad \leftarrow \text{remainder}
\end{array}
$$

The *remainder* (when not 0) is usually written in one of two ways. One way is with an "r" preceding it, as in Example 8a. The other is with the remainder written over the divisor with a fraction bar between them, as in Example 8b. (Fractions are discussed in Unit 1B.)

• **EXAMPLE 8**   Divide: 534 ÷ 24.

$$
\textbf{a.} \quad
\begin{array}{r}
22 \text{ r } 6 \\
24\overline{)534} \\
48 \\
\hline
54 \\
48 \\
\hline
6
\end{array}
\qquad
\textbf{b.} \quad
\begin{array}{r}
22\frac{6}{24} \\
24\overline{)534} \\
48 \\
\hline
54 \\
48 \\
\hline
6
\end{array}
$$

• **EXAMPLE 9**   Find the total piston displacement of an eight-cylinder engine. Each piston displaces a volume of 41 cubic inches (in^3).

$$\text{Total displacement} = 8 \times 41 \text{ in}^3 = 328 \text{ in}^3$$

$$\text{or } 328 \text{ cid (cubic inch displacement)}$$

• **EXAMPLE 10**   Ohm's law states that in a simple electrical circuit, the current I (measured in amps, A) equals the voltage E (measured in volts, V) divided by the resistance R (measured in ohms, Ω). Find the current in the circuit of Figure 1.8.

$$\text{The current } I = \frac{E}{R} = \frac{110}{22} = 5 \text{ A}$$

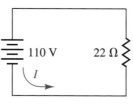

FIGURE 1.8

• **EXAMPLE 11**   A corn planter costs $12,600. It has a 10-year life and a salvage value of $1500. What is the annual depreciation? (Use the straight-line depreciation method.)

The straight-line depreciation method means that the difference between the cost and the salvage value is divided evenly over the life of the item. In this case, the difference between the cost and the salvage value is:

$$
\begin{array}{ll}
\$12,600 & \text{cost} \\
-1,500 & \text{salvage} \\
\hline
\$11,100 & \text{difference}
\end{array}
$$

This difference divided by 10, the life of the item, is $1110. This is the annual depreciation. ————•

Using a Scientific Calculator

Use of a scientific calculator is integrated throughout this text. To demonstrate how to use a calculator, we will (a) use a flowchart to give the directions for each step, (b) show what buttons are pushed and the order in which they are pushed, and (c) show the display at each step. We have chosen to illustrate the most common types of algebraic logic calculators. Yours may differ in the number of digits displayed.

Note: We will always assume that your calculator and its memory are cleared before you begin any calculation.

Use a calculator to add as follows.

• **EXAMPLE 12** Add: 9463
125
9
80

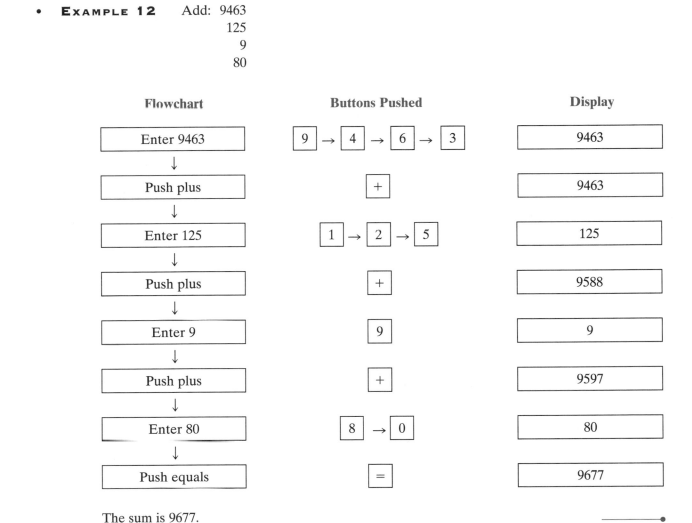

Flowchart	Buttons Pushed	Display
Enter 9463	9 → 4 → 6 → 3	9463
Push plus	+	9463
Enter 125	1 → 2 → 5	125
Push plus	+	9588
Enter 9	9	9
Push plus	+	9597
Enter 80	8 → 0	80
Push equals	=	9677

The sum is 9677. ————•

Use a calculator to subtract as follows.

• **EXAMPLE 13**   Subtract: 3500
                              1628

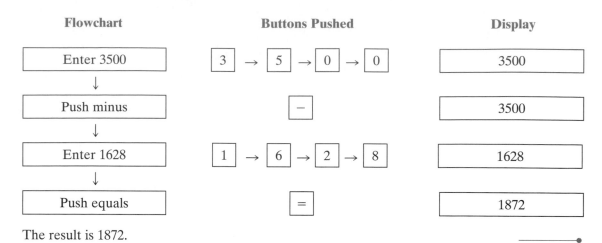

The result is 1872.

• **EXAMPLE 14**   Multiply: 125 × 68.

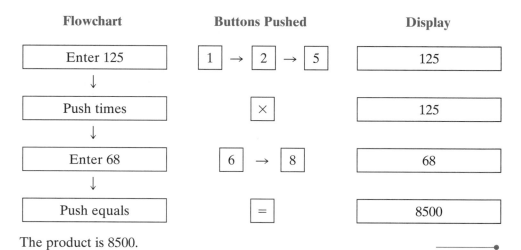

The product is 8500.

To divide numbers using a calculator, follow the steps in Example 15.

• **EXAMPLE 15**   Divide 8700 ÷ 15.

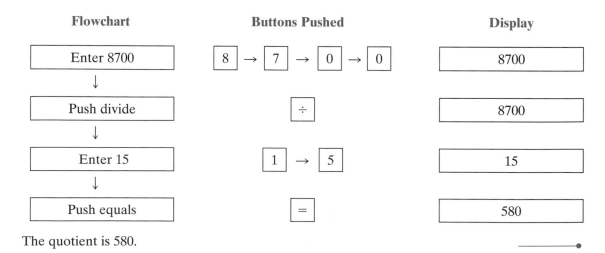

The quotient is 580.

Note: Your instructor will indicate which exercises should be completed using a calculator.

Exercises | 1.1

Add:

1. 832 + 9 + 56 + 2358

2. 64 + 7 + 97 + 543

3. 16 + 2400 + 47 + 3

4. 324 + 973 + 66 + 9430

5. 245
1053

6. 619
297
3142

7. 384
291
147
632

8.   78
107
45
217
9
123

9. 197 + 1072 + 10,877 + 15,532 + 768,098

10. 160,000 + 19,000 + 4,160,000 + 506,000

Subtract and check:

11. 283
152

14. 4000
702

17. 3600
1481

20. 60,000
9,876

12. 603
438

15. 98,405 − 72,397

18. 48,007
27,909

13. 7561
2397

16. 417,286 − 287,156

19. 4000
1180

Find the total resistance in each series circuit:

21.

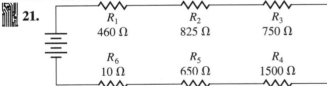

22.

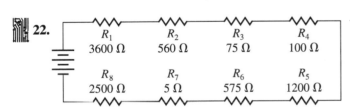

23. Approximately how many studs are needed for the exterior walls in the building shown in Illustration 1? (See Example 4.)

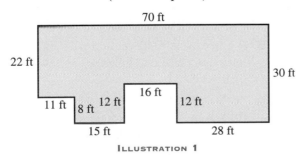

ILLUSTRATION 1

24. A pipe 24 ft long is cut into four pieces, the first 4 ft long, the second 5 ft long, and the third 7 ft long. What is the length of the remaining piece? (Assume no waste from cutting.)

25. Illustration 2 shows the daily cash and credit gasoline receipts at Ralph's Service Station for the week. What are the total cash, total credit, and total receipts for the week?

	Cash	Credit
Sunday	$3375	$1181
Monday	3744	1420
Tuesday	3198	904
Wednesday	3807	893
Thursday	3003	788
Friday	4060	1710
Saturday	3528	2162

ILLUSTRATION 2

26. During the year, a farmer had the income shown in Illustration 3 and the expenses listed in Illustration 4. Find the total income, the total expenses, and the profit.

	Income
Corn sales	$19,282
Soybean sales	51,376
Wheat sales	11,675
Hog sales	189,637
Beef sales	3,154
Custom work	1,593
State gas tax refund	222
Federal gas tax refund	777

ILLUSTRATION 3

	Expenses
Machinery depreciation	$27,937
Machinery repairs	14,924
Gas and oil	10,164
Fertilizer	8,458
Other agricultural chemicals	4,266
Feed	53,387
Building depreciation	2,473
Labor	23,307
Real estate taxes	9,605
Machine hire	710
Veterinary and medicine	8,280
Utilities	1,406
Interest	28,211

ILLUSTRATION 4

27. Total the following input and output (I-O) entries in cubic centimetres (cm^3)* for a patient.
Input: 300 cm^3, 550 cm^3, 150 cm^3, 75 cm^3, 150 cm^3, 450 cm^3, 250 cm^3
Output: 325 cm^3, 150 cm^3, 525 cm^3, 250 cm^3, 175 cm^3

28. A patient is urged by his doctor to drink 3000 cm^3 of fluids each day. Today he had 480 cm^3 for breakfast, 300 cm^3 at midmorning, 650 cm^3 for lunch, 350 cm^3 at midafternoon, and 850 cm^3 for dinner. How much must he drink before bedtime?

Multiply:

29. 567
48

30. 8374
203

31. 3186
257

32. 62,187
234

33. 71,263 × 255

34. 98,447 × 536

35. 37,050 × 716

36. 1520 × 320

37. 6800 × 5200

38. 30,010 × 4080

Divide (use the remainder form with r):

39. 4⟌7236

40. 5⟌308,736

41. 6754 ÷ 6

42. 56,772 ÷ 8

43. 4668 ÷ 12

44. 15,648 ÷ 36

45. 26⟌51,482

46. 108⟌376,544

47. 67,560 ÷ 80

48. $\dfrac{188,000}{120}$

49. A car uses gas at the rate of 31 miles per gallon (mi/gal or mpg) and has a 16-gallon tank. How far can it travel on one tank of gas?

50. A car uses gas at a rate of 12 kilometres per litre (km/L) and has a 65-litre tank. How far can it travel on one tank of gas?

51. Find the total piston displacement of an eight-cylinder engine. Each piston displaces 425 cm^3.

52. An eight-cylinder engine has a total displacement of 400 in^3. Find the displacement of each piston.

53. A four-cylinder engine has a total displacement of 1300 cm^3. Find the displacement of each piston.

54. A car travels 1274 mi and uses 49 gal of gasoline. Calculate its mileage in miles per gallon.

55. A car travels 2340 km and uses 180 L of gasoline. Calculate its "mileage" in kilometres per litre.

56. A garage buys heater hose in 20-metre (m) rolls. Repairing the average car requires 3 m of hose. How many cars can be repaired with one roll?

*Although cm^3 is the "official" metric abbreviation and will be used throughout this book, some readers may be more familiar with the abbreviation "cc," which is still used in some medical and allied health areas.

57. Inventory shows the following lengths of 3-inch steel pipe:

> 5 pieces 18 ft long
>
> 42 pieces 15 ft long
>
> 158 pieces 12 ft long
>
> 105 pieces 10 ft long
>
> 79 pieces 8 ft long
>
> 87 pieces 6 ft long

What is the total linear feet of pipe in inventory?

58. An order of lumber contains 36 boards 12 ft long, 28 boards 10 ft long, 36 boards 8 ft long, and 12 boards 16 ft long. How many boards are contained in the order? How many linear feet of lumber are contained in the order?

59. Two draftpersons operating the same computer plotter work 8 hours each, on a day and night shift basis. One produces 80 drawings per hour; the other produces 120 drawings per hour. What is the difference in their outputs after 30 days?

60. A shipment contains a total of 5232 linear feet of steel pipe. Each piece of pipe is 12 ft long. How many pieces should be expected?

61. How should a window 75 in. wide be placed so that it is centered on a wall 17 ft 5 in. wide?

62. A farmer expects a yield of 165 bushels per acre (bu/acre) from 260 acres of corn. If the corn is stored, how many bushels of storage are needed?

63. A farmer harvests 6864 bushels (bu) of soybeans from 156 acres. What is his yield per acre?

64. A railroad freight car can hold 2035 bu of corn. How many freight cars are needed to haul the expected 12,000,000 bu from a local grain elevator?

65. On a given day, eight steers weighed 856 lb, 754 lb, 1044 lb, 928 lb, 888 lb, 734 lb, 953 lb, and 891 lb. **a.** What is the average weight? **b.** In 36 days, 4320 lb of feed is consumed. What is the average feed consumption per day per steer?

66. What is the weight (in tons) of a stack of hay bales 6 bales wide, 110 bales long, and 15 bales high? The average weight of each bale is 80 lb. (1 ton = 2000 lb.)

67. From a 34-acre field, 92,480 lb of oats are harvested. Find the yield in bushels per acre. (1 bu of oats weighs 32 lb.)

68. A standard bale of cotton weighs approximately 500 lb. How many bales are contained in 15 tons of cotton?

69. A tractor costs $45,000. It has a 10-year life and a salvage value of $3000. What is the annual depreciation? (Use the straight-line depreciation method. See Example 11.)

70. How much pesticide powder would you put in a 400-gal spray tank if 10 gal of spray, containing 2 lb of pesticide, are applied per acre?

Using Ohm's law, find the current I in amps (A) in each electrical circuit (see Example 10):

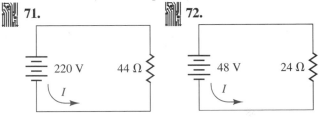

71. 220 V   44 Ω   I

72. 48 V   24 Ω   I

Ohm's law, in another form, states that in a simple circuit the voltage E (measured in volts, V) equals the current I (measured in amps, A) times the resistance R (measured in ohms, Ω). Find the voltage E measured in volts (V) in each electrical circuit:

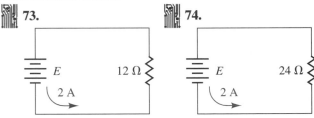

73. E   12 Ω   2 A

74. E   24 Ω   2 A

75. A hospital dietitian determines that each patient needs 4 ounces (oz) of orange juice. How many ounces of orange juice must be prepared for 220 patients?

76. During 24 hours, a patient is given three phenobarbital tablets of 25 mg each. How many milligrams of phenobarbital does the patient receive altogether?

77. To give 800 mg of quinine sulfate from 200-mg tablets, how many tablets would you use?

78. A nurse used two 4-grain potassium permanganate tablets in the preparation of a medication. How much potassium permanganate did she use?

1.2   Order of Operations

The expression 5^3 means to use 5 as a factor 3 times. We say that 5^3 is the third power of 5, where 5 is called the *base* and 3 is called the *exponent*. Here, 5^3 means $5 \times 5 \times 5 = 125$. The expression 2^4 means that 2 is used as a factor 4 times; that is, $2^4 = 2 \times 2 \times 2 \times 2 = 16$. Here, 2^4 is the fourth power of 2.

Just as we use periods, commas, and other punctuation marks to help make sentences more readable, we use *parentheses* "( )", *brackets* "[ ]", and other grouping symbols in mathematics to help clarify the meaning of mathematical expressions. Parentheses not only give an expression a particular meaning, they also specify the order to be followed in evaluating and simplifying expressions.

What is the value of $8 - 3 \cdot 2$? Is it 10? Is it 2? Or is it some other number? It is very important that each mathematical expression have only one value. For this to happen, we all must not only perform the exact *same operations* in a given mathematical expression or problem but also perform them in exactly the *same order*. The following order of operations is followed by all.

ORDER OF OPERATIONS

1. Always do the operations within parentheses or other grouping symbols first.

2. Then evaluate each power, if any. Examples:

$$4 \times 3^2 = 4 \times (3 \times 3) = 4 \times 9 = 36$$
$$5^2 \times 6 = (5 \times 5) \times 6 = 25 \times 6 = 150$$
$$\frac{5^3}{6^2} = \frac{5 \times 5 \times 5}{6 \times 6} = \frac{125}{36}$$

3. Next, perform multiplications and divisions in the order they appear as you read from left to right. For example,

$$
\begin{aligned}
& 60 \times 5 \div 4 \div 3 \times 2 \\
= & \quad 300 \div 4 \div 3 \times 2 \\
= & \qquad 75 \;\; \div 3 \times 2 \\
= & \qquad\qquad 25 \times 2 \\
= & \qquad\qquad\qquad 50
\end{aligned}
$$

4. Finally, perform additions and subtractions in the order they appear as you read from left to right.

Note: If two parentheses or a number and a parenthesis occur next to one another without any sign between them, multiplication is indicated.

By using the above procedure, we find that $8 - 3 \cdot 2 = 8 - 6 = 2$.

• **EXAMPLE 1**   Evaluate: $2 + 5(7 + 6)$.

$$
\begin{aligned}
&= 2 + 5(13) \\
&= 2 + \;\; 65 \\
&= \;\; 67
\end{aligned}
$$

Note: A number next to parentheses indicates multiplication. In the example above, $5(13)$ means 5×13.

• **EXAMPLE 2**   Evaluate: $(9 + 4) \times 16 + 8$.

$$= \quad 13 \quad \times 16 + 8$$
$$= \quad\quad 208 \quad + 8$$
$$= \quad\quad\quad\quad 216$$

• **EXAMPLE 3**   Evaluate: $(6 + 1) \times 3 + (4 + 5)$.

$$= \quad 7 \quad \times 3 + \quad 9$$
$$= \quad\quad\quad 21 + \quad 9$$
$$= \quad\quad\quad\quad 30$$

• **EXAMPLE 4**   Evaluate: $4(16 + 4) + \dfrac{14}{7} - 8$.

$$= 4(\quad 20 \quad) + \dfrac{14}{7} - 8$$
$$= \quad\quad 80 \quad + 2 \quad - 8$$
$$= \quad\quad\quad\quad\quad 74$$

• **EXAMPLE 5**   Evaluate: $7 + (6 - 2)^2$.

$$= 7 + \quad 4^2$$
$$= 7 + \quad 16$$
$$= \quad 23$$

• **EXAMPLE 6**   Evalutc: $25 - 3 \cdot 2^3$.

$$= 25 - 3 \cdot 8$$
$$= 25 - \quad 24$$
$$= \quad 1$$

If pairs of parentheses are nested (parentheses within parentheses, or within brackets), work from the innermost pair of parentheses to the outermost pair. That is, remove the innermost parentheses first, remove the next innermost parentheses second, and so on.

• **EXAMPLE 7**   Evaluate: $6 \times 2 + 3[7 + 4(8 - 6)]$.

$$= 6 \times 2 + 3[7 + 4(\quad 2 \quad)]$$
$$= 6 \times 2 + 3[7 + \quad 8 \quad\quad]$$
$$= 6 \times 2 + 3[\quad\quad 15 \quad\quad]$$
$$= 12 \quad\quad + 45$$
$$= \quad\quad 57$$

Exercises   **1.2**

Evaluate each expression:

1. $8 - 3(4 - 2)$

2. $(8 + 6)4 + 8$

3. $(8 + 6) - (7 - 3)$

4. $4 \times (2 \times 6) + (6 + 2) \div 4$

5. $2(9 + 5) - 6 \times (13 + 2) \div 9$

6. $5(8 \times 9) + (13 + 7) \div 4$

7. $27 + 13 \times (7 - 3)(12 + 6) \div 9$

8. $123 - 3(8 + 9) + 17$

9. $16 + 4(7 + 8) - 3$

10. $(18 + 17)(12 + 9) - (7 \times 16)(4 + 2)$

11. $9 - 2(17 - 15) + 18$

12. $(9 + 7)5 + 13$

13. $(39 - 18) - (23 - 18)$

14. $5(3 \times 7) + (8 + 4) \div 3$

15. $3(8 + 6) - 7(13 + 3) \div 14$

16. $6(4 \times 5) + (15 + 9) \div 6$

17. $42 + 12(9 - 3)(12 + 13) \div 30$

18. $228 - 4 \times (7 + 6) - 8(6 - 2)$

19. $38 + 9 \times (8 + 4) - 3(5 - 2)$

20. $(19 + 8)(4 + 3) \div 21 + (8 \times 15) \div (4 \times 3)$

21. $27 - 2 \times (18 - 9) - 3 + 8(43 - 15)$

22. $6 \times 8 \div 2 \times 9 \div 12 + 6$

23. $12 \times 9 \div 18 \times 64 \div 8 + 7$

24. $18 \div 6 \times 24 \div 4 \div 6$

25. $7 + 6(3 + 2) - 7 - 5(4 + 2)$

26. $5 + 3(7 \times 7) - 6 - 2(4 + 7)$

27. $3 + 17(2 \times 2) - 67$

28. $8 - 3(9 - 2) \div 21 - 7$

29. $28 - 4(2 \times 3) + 4 - (16 \times 8) \div (4 \times 4)$

30. $6 + 4(9 + 6) + 8 - 2(7 + 3) - (3 \times 12) \div 9$

31. $24/(6 - 2) + 4 \times 3 - 15/3$

32. $(36 - 6)/(5 + 10) + (16 - 1)/3$

33. $3 \times 15 \div 9 + (13 - 5)/2 \times 4 - 2$

34. $28/2 \times 7 - (6 + 10)/(6 - 2)$

35. $10 + 4^2$

36. $4 + 2 \cdot 3^2$

37. $\dfrac{20 + (2 \cdot 3)^2}{7 \cdot 2^3}$

38. $\dfrac{(20 - 2 \cdot 5)^2}{3^3 - 2}$

39. $6[3 + 2(2 + 5)]$

40. $5((4 + 6) + 2(5 - 2))$

41. $5 \times 2 + 3[2(5 - 3) + 4(4 + 2) - 3]$

42. $3(10 + 2(1 + 3(2 + 6(4 - 2))))$

1.3　Area and Volume

Area

To measure the length of an object, you must first select a suitable standard unit of length. To measure short lengths, choose a unit such as centimetres or millimetres in the metric system, or inches in the English system. For long distances, choose metres or kilometres in the metric system, or yards or miles in the English system.

To measure the surface area of an object, first select a standard unit of area suitable to the object to be measured. Standard units of area are based on the square and are called square units. For example, a square inch (in^2) is the amount of surface area within a square that measures one inch on a side. A square centimetre (cm^2) is the amount of surface area within a square that is 1 cm on a side. (See Figure 1.9.) The area of any surface is the number of square units contained on that surface.

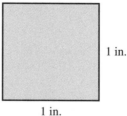

1 in.

1 in.

1 square inch (in^2)

1 cm

1 cm

1 square centimetre (cm^2)

FIGURE 1.9

•   **EXAMPLE 1**   What is the area of a rectangle measuring 4 cm by 3 cm?

Each square in Figure 1.10 represents 1 cm². By simply counting the number of squares, you find that the area of the rectangle is 12 cm².

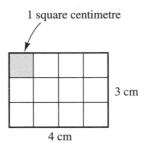

1 square centimetre

3 cm

4 cm

FIGURE 1.10

You can also find the area by multiplying the length times the width:

Area = ℓ × w
 = 4 cm × 3 cm = 12 cm²
 (length)   (width)

Note: cm × cm = cm²   ————————•

•   **EXAMPLE 2**   What is the area of the metal plate represented in Figure 1.11?

Each square represents 1 square inch. By simply counting the number of squares, the area of the metal plate is 42 in².

1 in.   3 in.

6 in.   8 in.

5 in.

9 in.

FIGURE 1.11

Another way to find the area of the figure is to find the areas of two rectangles and then find their difference, as follows:

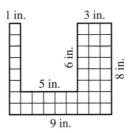

=   8 in.   —   6 in.

9 in.   5 in.

Area of outer rectangle: 9 in. × 8 in. = 72 in²
Area of inner rectangle: 5 in. × 6 in. = 30 in²
Area of metal plate: = 42 in²   ————————•

Volume

In area measurement, the standard units are based on the square and called square units. For *volume* measurement, the standard units are based on the cube and called cubic units. For example, a cubic foot (ft^3) is the amount of space contained in a cube that measures 1 ft (or 12 in.) on each edge, as in Figure 1.12.

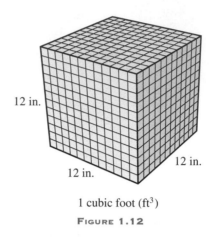

12 in.

12 in.

12 in.

1 cubic foot (ft^3)

FIGURE 1.12

Figure 1.13 shows the relative sizes of a cubic inch and a cubic centimetre. The cubic inch (in^3) is the amount of space contained in a cube that measures 1 in. on each edge. The cubic centimetre (cm^3) is the amount of space contained in a cube that measures 1 cm on each edge. The *volume* of any solid is the number of cubic units contained in the solid.

Figure 1.14 shows that the cubic decimetre (litre) is made up of 10 layers, each containing 100 cm^3, for a total of 1000 cm^3.

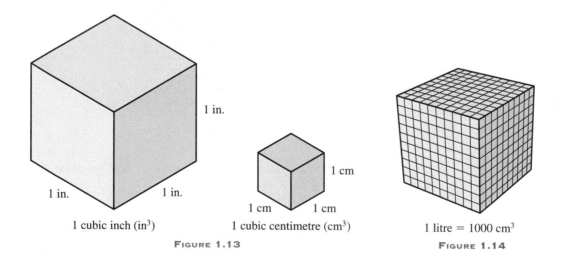

1 in.

1 in.   1 in.

1 cubic inch (in^3)

FIGURE 1.13

1 cm

1 cm   1 cm

1 cubic centimetre (cm^3)

1 litre = 1000 cm^3

FIGURE 1.14

•   **EXAMPLE 3**   Find the volume of a rectangular box 8 cm long, 4 cm wide, and 6 cm high.

Suppose you placed one-centimetre cubes in the box, as in Figure 1.15. On the bottom layer, there would be 8 × 4, or 32, one-cm cubes. In all, there are six such layers, or 6 × 32 = 192 one-cm cubes. Therefore, the volume is 192 cm^3.

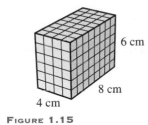

FIGURE 1.15

You can also find the volume by multiplying the length times the width times the height:

$$V = \ell \times w \times h$$
$$= 8 \text{ cm} \times 4 \text{ cm} \times 6 \text{ cm}$$
$$= 192 \text{ cm}^3$$

Note: cm × cm × cm = cm^3

• **EXAMPLE 4** How many cubic inches are in one cubic foot?

The bottom layer of Figure 1.16 contains 12 × 12, or 144, one-inch cubes. There are 12 such layers, or 12 × 144 = 1728 one-inch cubes. Therefore, 1 ft^3 = 1728 in^3.

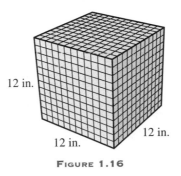

FIGURE 1.16

Exercises | 1.3

1. How many square yards (yd^2) are contained in a rectangle 12 yd long and 8 yd wide?

2. How many square metres (m^2) are contained in a rectangle 12 m long and 8 m wide?

In the following exercises, assume that all corners are square and that like measurements are not repeated because the figures are assumed to have consistent lengths. All three of the following mean that the length of a side is 3 cm:

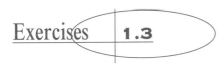

3 cm   3 cm   3 cm

Find the area of each figure.

3.

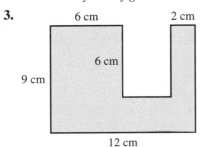

6 cm   2 cm
6 cm
9 cm
12 cm

4.

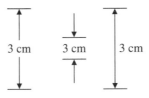

3 in.
8 in.
3 in.
8 in.

5.

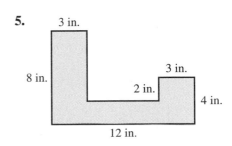

3 in.
8 in.   3 in.
2 in.   4 in.
12 in.

6.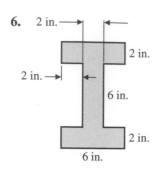

7.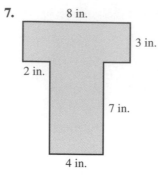

8.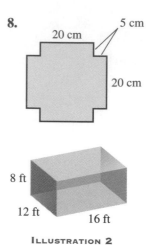

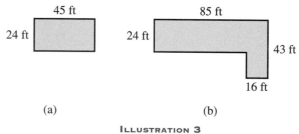

ILLUSTRATION 2

9. How many tiles 4 in. on a side should be used to cover a portion of a wall 48 in. long by 36 in. high? (See Illustration 1.)

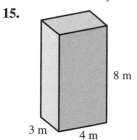

36 in.

48 in.

ILLUSTRATION 1

13. The replacement cost for construction of houses is $38/ft². Determine how much house insurance should be carried on each of the one-story houses in Illustration 3.

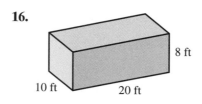

(a)   (b)

ILLUSTRATION 3

10. How many ceiling tiles 2 ft by 4 ft are needed to tile a ceiling that is 24 ft × 26 ft? (Be careful how you arrange the tiles.)

11. How many gallons of paint should be purchased to paint 20 motel rooms as shown in Illustration 2? (Do not paint the floor.) It takes 1 gal to paint 400 square feet (ft²).

12. How many pieces of 4-ft-by-8-ft drywall are needed for the 20 motel rooms in Exercise 11? All four walls and the ceiling in each room are to be drywalled. Assume that the drywall cut out for windows and doors cannot be salvaged and used again.

14. The replacement cost for construction of the building in Illustration 4 is $28/ft². Determine how much insurance should be carried for full replacement.

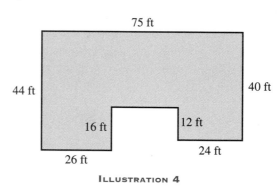

ILLUSTRATION 4

Find the volume of each rectangular solid:

15.

8 m

3 m   4 m

16.

8 ft

10 ft   20 ft

17.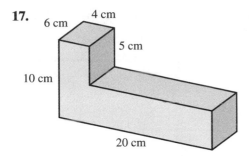

18.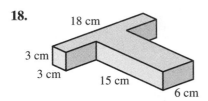

19.
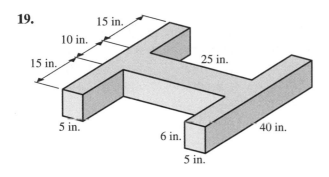

20.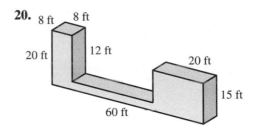

21. Find the volume of a rectangular box 10 cm by 12 cm by 5 cm.

22. Find the volume of a rectangular room 9 ft by 12 ft by 8 ft.

23. Find the weight of a cement floor that is 15 ft by 12 ft by 2 ft if 1 ft^3 of cement weighs 193 lb.

24. A trailer 5 ft by 6 ft by 5 ft is filled with coal. Given that 1 ft^3 of coal weighs 40 lb and 1 ton = 2000 lb, how many tons of coal are in the trailer?

25. A rectangular tank is 8 ft long by 5 ft wide by 6 ft high. Water weighs approximately 62 lb/ft^3. Find the weight of water if the tank is full.

26. A rectangular tank is 9 ft by 6 ft by 4 ft. Gasoline weighs approximately 42 lb/ft^3. Find the weight of gasoline if the tank is full.

27. A building is 100 ft long, 50 ft wide, and 10 ft high. Estimate the cost of heating it at the rate of $25 per 1000 ft^3.

1.4 Formulas

A *formula* is a statement of a rule using letters to represent certain quantities. In physics, one of the basic rules states that *work* equals *force* times *distance*. If a person (Figure 1.17) lifts a 200-lb weight a distance of 3 ft, we say the work done is 200 lb × 3 ft = 600 foot-pounds (ft-lb). The work, *W*, equals the force, *f*, times the distance, *d*, or $W = f \times d$.

FIGURE 1.17

A person pushes against a car weighing 2700 lb but does not move it. The work done is 2700 lb × 0 ft = 0 ft-lb. An automotive technician (Figure 1.18) moves a diesel engine weighing 1100 lb from the floor to a workbench 4 ft high. The work done in moving the engine is 1100 lb × 4 ft = 4400 ft-lb.

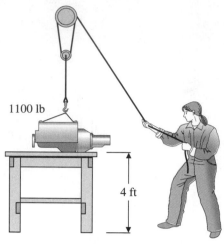

1100 lb

4 ft

FIGURE 1.18

To summarize, if you know the amount of force and the distance the force is applied, the work can be found by simply multiplying the force and distance. The formula $W = f \times d$ is often written $W = f \cdot d$, or simply $W = fd$. Whenever there is no symbol between a number and a letter or between two letters, it is assumed that the operation to be performed is multiplication.

• **EXAMPLE 1**   If $W = fd$, $f = 10$, and $d = 16$, find W.

$$W = fd$$
$$W = (10)(16)$$
$$W = 160$$

• **EXAMPLE 2**   If $I = \dfrac{E}{R}$, $E = 450$, and $R = 15$, find I.

$$I = \frac{E}{R}$$
$$I = \frac{450}{15}$$
$$I = 30$$

• **EXAMPLE 3**   If $P = I^2R$, $I = 3$, and $R = 600$, find P.

$$P = I^2R$$
$$P = (3)^2(600)$$
$$P = (9)(600)$$
$$P = 5400$$

There are many other formulas used in science and technology. Some examples are given here:

a. $d = vt$ **c.** $f = ma$ **e.** $I = \dfrac{E}{R}$

b. $W = IEt$ **d.** $P = IE$ **f.** $P = \dfrac{V^2}{R}$

Formulas from Geometry

The area of a triangle is given by the formula $A = \frac{1}{2}bh$, where b is the length of the base and h is the length of the altitude to the base (Figure 1.19). (An altitude of a triangle is a line from a vertex perpendicular to the opposite side.)

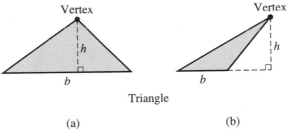

Triangle

(a)   (b)

FIGURE 1.19

• **EXAMPLE 4**   Find the area of a triangle whose base is 18 in. and whose height is 10 in.

$$A = \frac{1}{2}bh$$

$$A = \frac{1}{2}(18 \text{ in.})(10 \text{ in.})$$

$$= 90 \text{ in}^2$$ ————●

The area of a *parallelogram* (a four-sided figure whose opposite sides are parallel) is given by the formula $A = bh$, where b is the length of the base and h is the perpendicular distance between the base and its opposite side (Figure 1.20).

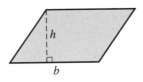

Parallelogram

FIGURE 1.20

• **EXAMPLE 5**   Find the area of a parallelogram with base 24 cm and height 10 cm.

$$A = bh$$
$$A = (24 \text{ cm})(10 \text{ cm})$$
$$= 240 \text{ cm}^2$$ ————●

The area of a *trapezoid* (a four-sided figure with one pair of parallel sides) is given by the formula $A = \left(\frac{a+b}{2}\right)h$, where a and b are the lengths of the parallel sides (called *bases*), and h is the perpendicular distance between the bases (Figure 1.21).

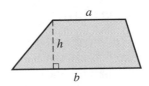

Trapezoid

FIGURE 1.21

• **EXAMPLE 6**   Find the area of the trapezoid in Figure 1.22.

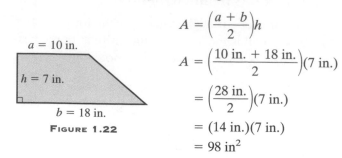

FIGURE 1.22

$$A = \left(\frac{a+b}{2}\right)h$$

$$A = \left(\frac{10 \text{ in.} + 18 \text{ in.}}{2}\right)(7 \text{ in.})$$

$$= \left(\frac{28 \text{ in.}}{2}\right)(7 \text{ in.})$$

$$= (14 \text{ in.})(7 \text{ in.})$$

$$= 98 \text{ in}^2$$

Exercises 1.4

Use the formula W = fd, where f represents a force, and d represents the distance that the force is applied. Find the work done, W:

1. $f = 30, d = 20$ **3.** $f = 1125, d = 10$ **5.** $f = 176, d = 326$

2. $f = 17, d = 9$ **4.** $f = 203, d = 27$ **6.** $f = 2400, d = 120$

*From formulas **a** to **f** on page 20, choose one that contains all the given letters. Then use the formula to find the unknown letter:*

7. If $m = 1600$ and $a = 24$, find f.   **9.** If $E = 120$ and $R = 15$, find I.   **11.** If $I = 29$ and $E = 173$, find P.

8. If $V = 120$ and $R = 24$, find P.   **10.** If $v = 372$ and $t = 18$, find d.   **12.** If $I = 11$, $E = 95$, and $t = 46$, find W.

Find the area of each triangle:

13. $b = 10$ in., $h = 8$ in.   **14.** $b = 36$ cm, $h = 20$ cm   **15.** $b = 54$ ft, $h = 30$ ft   **16.** $b = 188$ m, $h = 220$ ft

Find the area of each rectangle:

17. $b = 8$ m, $h = 7$ m   **18.** $b = 24$ in., $h = 15$ in.   **19.** $b = 36$ ft, $h = 18$ ft   **20.** $b = 250$ cm, $h = 120$ cm

Find the area of each trapezoid:

21. $a = 7$ ft, $b = 9$ ft, $h = 4$ ft   **23.** $a = 96$ cm, $b = 24$ cm, $h = 30$ cm

22. $a = 30$ in., $b = 50$ in., $h = 24$ in.   **24.** $a = 450$ m, $b = 750$ m, $h = 250$ m

25. The volume of a rectangular solid is given by $V = \ell wh$, where ℓ is the length, w is the width, and h is the height of the solid. Find V if $\ell = 25$ cm, $w = 15$ cm, and $h = 12$ cm.

26. Find the volume of a box with dimensions $\ell = 48$ in., $w = 24$ in., and $h = 96$ in.

27. Given $v = v_0 + gt$, $v_0 = 12$, $g = 32$, and $t = 5$. Find v.

28. Given $Q = CV$, $C = 12$, and $V = 2500$, find Q.

29. Given $I = \dfrac{E}{Z}$, $E = 240$, and $Z = 15$, find I.

30. Given $P = I^2R$, $I = 4$, and $R = 2000$, find P.

 **1.5** ## Prime Factorization

Divisibility

A number is *divisible* by a second number if, when you divide the first number by the second number, you get a zero remainder. That is, the second number *divides* the first number.

• **EXAMPLE 1**   12 is divisible by 3, because 3 divides 12.

• **EXAMPLE 2**   124 is not divisible by 7, because 7 does not divide 124. Check with a calculator.

———————•

There are many ways of classifying the positive integers. They can be classified as even or odd, as divisible by 3 or not divisible by 3, as larger than 10 or smaller than 10, and so on. One of the most important classifications involves the concept of a *prime number*: an integer greater than 1 that has no divisors except itself and 1. The first ten prime numbers are 2, 3, 5, 7, 11, 13, 17, 19, 23, and 29.

An integer is *even* if it is divisible by 2; that is, if you divide it by 2, you get a zero remainder. An integer is *odd* if it is not divisible by 2.

In multiplying two or more positive integers, the positive integers are called the *factors* of the product. Thus 2 and 5 are factors of 10, since $2 \times 5 = 10$. The numbers 2, 3, and 5 are factors of 30, since $2 \times 3 \times 5 = 30$. If the factors are prime numbers, they are called *prime factors*. The process of finding the prime factors of an integer is called *prime factorization*. The prime factorization of a given number is the product of its prime factors. That is, each of the factors is prime, and their product equals the given number. One of the most useful applications of prime factorization is in finding the least common denominator (LCD) when adding and subtracting fractions. This application is found in Section 1.7.

• **EXAMPLE 3**   Factor 28 into prime factors.

    a. $28 = 4 \cdot 7$ **b.** $28 = 7 \cdot 4$ **c.** $28 = 2 \cdot 14$

         $= 2 \cdot 2 \cdot 7$ $= 7 \cdot 2 \cdot 2$ $= 2 \cdot 7 \cdot 2$

In each case, you have three prime factors of 28; one factor is 7, the other two are 2's. The factors may be written in any order, but we usually list all the factors in order from smallest to largest. It would not be correct in the examples above to leave $7 \cdot 4$, $4 \cdot 7$, or $2 \cdot 14$ as factors of 28, since 4 and 14 are not prime numbers.

———————•

Short division, a condensed form of long division, is a helpful way to find prime factors. Find a prime number that divides the given number. Divide, using short division. Then find a second prime number that divides the result. Divide, using short division. Keep repeating this process of stacking the quotients and divisors (as shown below) until the final quotient is also prime. The prime factors will be the product of the divisors and the final quotient of the repeated short divisions.

• **EXAMPLE 4**   Find the prime factorization of 12.

    $2\underline{|12}$

    $2\underline{|\ 6}$

       3

The prime factorization of 12 is $2 \cdot 2 \cdot 3$. ———————•

• **EXAMPLE 5**   Find the prime factorization of 56.

    $2\underline{|56}$

    $2\underline{|28}$

    $2\underline{|14}$

       7

The prime factorization of 56 is $2 \cdot 2 \cdot 2 \cdot 7$. ———————•

• **EXAMPLE 6**   Find the prime factorization of 17.

17 has no factors except for itself and 1. Thus, 17 is a prime number. When asked for factors of a prime number, just write "prime" as your answer.

Divisibility Tests

To eliminate some of the guesswork involved in finding prime factors, divisibility tests can be used. Such tests determine whether or not a particular positive integer divides another integer without carrying out the division. Divisibility tests and prime factorization are used to reduce fractions to lowest terms and to find the lowest common denominator. (See Unit 1B.)

The following divisibility tests for certain positive integers are most helpful.

DIVISIBILITY BY 2

If a number ends with an even digit, then the number is divisible by 2.

Note: Zero is even.

• **EXAMPLE 7**   Does 2 divide 4258?

Yes; since 8, the last digit of the number, is even, 4258 is divisible by 2.

Note: Check each example with a calculator.

• **EXAMPLE 8**   Does 2 divide 215,517?

Since 7 (the last digit) is odd, 215,517 is not divisible by 2.

DIVISIBILITY BY 3

If the sum of the digits of a number is divisible by 3, then the number itself is divisible by 3.

• **EXAMPLE 9**   Does 3 divide 531?

The sum of the digits $5 + 3 + 1 = 9$, and 9 is divisible by 3, so 531 is divisible by 3.

• **EXAMPLE 10**   Does 3 divide 551?

The sum of the digits $5 + 5 + 1 = 11$, and 11 is not divisible by 3, so 551 is not divisible by 3.

DIVISIBILITY BY 5

If a number has 0 or 5 as its last digit, then the number is divisible by 5.

• **EXAMPLE 11**   Does 5 divide 2372?

The last digit of 2372 is neither 0 nor 5, so 5 does not divide 2372.

• **EXAMPLE 12**   Does 5 divide 3210?

The last digit of 3210 is 0, so 3210 is divisible by 5.

• **EXAMPLE 13**   Find the prime factorization of 204.

$$\begin{array}{r|l} 2 & 204 \\ 2 & 102 \\ 3 & 51 \\ & 17 \end{array}$$ Last digit is even.
Last digit is even.
Sum of digits is divisible by 3.

The prime factorization of 204 is $2 \cdot 2 \cdot 3 \cdot 17$.

• **EXAMPLE 14**   Find the prime factorization of 630.

$$\begin{array}{r|l} 2 & 630 \\ 3 & 315 \\ 3 & 105 \\ 5 & 35 \\ & 7 \end{array}$$ Last digit is even.
Sum of digits is divisible by 3.
Sum of digits is divisible by 3.
Last digit is 5.

The prime factorization of 630 is $2 \cdot 3 \cdot 3 \cdot 5 \cdot 7$.

Note: As a general rule of thumb:

1. Keep dividing by 2 until the quotient is not even.
2. Keep dividing by 3 until the quotient's sum of digits is not divisible by 3.
3. Keep dividing by 5 until the quotient does not end in 0 or 5.

That is, if you divide out all the factors of 2, 3, and 5, the remaining factors, if any, will be much smaller and easier to work with, and perhaps prime.

Exercises 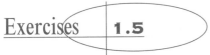 1.5

Which numbers are divisible **a.** *by 3 and* **b.** *by 4?*

1. 15   **2.** 28   **3.** 96   **4.** 172   **5.** 78   **6.** 675

Classify each number as prime or not prime:

7. 53   **9.** 93   **11.** 16   **13.** 39
8. 57   **10.** 121   **12.** 123   **14.** 87

Test for divisibility by 2:

15. 458   **16.** 12,746   **17.** 315,817   **18.** 877,778   **19.** 1367   **20.** 1205

Test for divisibility by 3 and check your results with a calculator:

21. 387   **22.** 1254   **23.** 453,128   **24.** 178,213   **25.** 218,745   **26.** 15,690

Test for divisibility by 5 and check your results with a calculator:

27. 70   **28.** 145   **29.** 366   **30.** 56,665   **31.** 63,227   **32.** 14,601

Test the divisibility of each first number by the second number:

33. 56; 2   **35.** 218; 3   **37.** 528; 5   **39.** 198; 3   **41.** 1,820,670; 2   **43.** 7,215,720; 5

34. 42; 3   **36.** 375; 5   **38.** 2184; 3   **40.** 2236; 3   **42.** 2,817,638; 2   **44.** 5,275,343; 3

Find the prime factorization of each number (use divisibility tests where helpful):

45. 20	**49.** 36	**53.** 51	**57.** 120	**61.** 105
46. 18	**50.** 25	**54.** 56	**58.** 72	**62.** 78
47. 66	**51.** 27	**55.** 42	**59.** 171	**63.** 252
48. 30	**52.** 59	**56.** 45	**60.** 360	**64.** 444

Unit 1A   Review

1. Add: 33 + 104 + 75 + 29   **2.** Subtract: 2301
                                              506

3. Multiply: 3709 × 731   **4.** Divide: 9300 ÷ 15

5. Josh has the following lengths of 3-inch plastic pipe:

   3 pieces 12 ft long

   8 pieces 8 ft long

   9 pieces 10 ft long

   12 pieces 6 ft long

Find the total length of pipe on hand.

6. If one bushel of corn weighs 56 lb, how many bushels are contained in 14,224 lb of corn?

Evaluate each expression:

7. $6 + 2(5 \times 4 - 2)$

8. $3^2 + 12 \div 3 - 2 \times 3$

9. $12 + 2[3(8 - 2) - 2(3 + 1)]$

10. In Illustration 1, find the area.

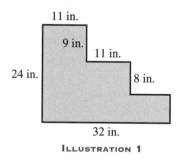

11 in.

9 in.

11 in.

24 in.

8 in.

32 in.

ILLUSTRATION 1

11. In Illustration 2, find the volume.

6 ft

8 ft   15 ft

ILLUSTRATION 2

12. If $d = vt$, $v = 45$, and $t = 4$, find d.

13. If $I = \dfrac{E}{R}$, $E = 120$, and $R = 12$, find I.

14. If $A = \dfrac{1}{2}bh$, $b = 40$, and $h = 15$, find A.

Classify each number as prime or not prime:

15. 51

16. 47

Test for divisibility of each first number by the second number:

17. 195; 3

18. 821; 5

Find the prime factorization of each number:

19. 40

20. 135

Unit 1B   **REVIEW OF OPERATIONS WITH FRACTIONS**

1.6   Introduction to Fractions

The English system of measurement (actually now the U.S. system) is basically a system whose units are expressed as common fractions and mixed numbers. The metric system of measurement is a system whose units are expressed as decimal fractions and powers of ten. As we convert from the English system to the metric system, more computations will be done with decimals which are easier—especially with a calculator. Fewer computations will be done with fractions, which are more difficult. During the transition period, we will need to feel comfortable with both systems. The metric system is developed in Chapter 3.

A *common fraction* may be defined as the ratio or quotient of two integers (say, a and b) in the form $\frac{a}{b}$ (where $b \neq 0$). Examples are $\frac{1}{2}$, $\frac{7}{11}$, $\frac{3}{8}$, and $\frac{37}{22}$. The integer below the line is called the *denominator*. It gives the denomination (size) of equal parts into which the fraction unit is divided. The integer above the line is called the *numerator*. It numerates (counts) the number of times the denominator is used. Look at one inch on a ruler, and then look at the one inch enlarged, as shown in Figures 1.23 and 1.24.

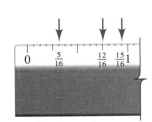

FIGURE 1.23

FIGURE 1.24

$\frac{1}{4}$ in. means 1 of 4 equal parts of an inch.

$\frac{2}{4}$ in. means 2 of 4 equal parts of an inch.

$\frac{3}{4}$ in. means 3 of 4 equal parts of an inch.

$\frac{5}{16}$ in. means 5 of 16 equal parts of an inch.

$\frac{12}{16}$ in. means 12 of 16 equal parts of an inch.

$\frac{15}{16}$ in. means 15 of 16 equal parts of an inch.

Two fractions $\frac{a}{b}$ and $\frac{c}{d}$ are *equal* or *equivalent* if $ad = bc$. That is, $\frac{a}{b} = \frac{c}{d}$ if $ad = bc$ ($b \neq 0$ and $d \neq 0$). For example, $\frac{2}{4}$ and $\frac{8}{16}$ are names for the same fraction, because $(2)(16) = (4)(8)$. There are many other names for this same fraction, such as $\frac{1}{2}, \frac{9}{18}, \frac{10}{20}, \frac{5}{10}, \frac{3}{6}$, and so on.

$\frac{2}{4} = \frac{1}{2}$, because $(2)(2) = (4)(1)$ $\frac{2}{4} = \frac{5}{10}$, because $(2)(10) = (4)(5)$

We have two rules for finding equal (or *equivalent*) fractions.

EQUAL OR EQUIVALENT FRACTIONS

1. The numerator and denominator of any fraction may be *multiplied* by the same number (except zero) without changing the value of the given fraction, thus producing an equivalent fraction. For example, $\frac{4}{9} = \frac{4 \cdot 5}{9 \cdot 5} = \frac{20}{45}$.

2. The numerator and denominator of any fraction may be *divided* by the same number (except zero) without changing the value of the given fraction. For example, $\frac{6}{10} = \frac{6 \div 2}{10 \div 2} = \frac{3}{5}$.

We use the above rules for equivalent fractions (a) to reduce a fraction to lowest terms and (b) to change a fraction to higher terms when adding and subtracting fractions with different denominators.

To *simplify* a fraction means to find an equivalent fraction whose numerator and denominator are *relatively prime*—that is, a fraction whose numerator and denominator have no common divisor. This is also called *reducing a fraction to lowest terms*.

To reduce a fraction to lowest terms, write the prime factorization of the numerator and the denominator. Then divide (cancel) numerator and denominator by any pair of common factors. You may find it helpful to use the divisibility tests in Section 1.5 to write the prime factorizations.

• **EXAMPLE 1**   Simplify: $\dfrac{35}{50}$.

$$\frac{35}{50} = \frac{\cancel{5} \cdot 7}{2 \cdot \cancel{5} \cdot 5} = \frac{7}{10}$$

Note the use of the divisibility test for 5. A last digit of 0 or 5 indicates the number is divisible by 5.

• **EXAMPLE 2**   Simplify: $\dfrac{63}{99}$.

$$\frac{63}{99} = \frac{\cancel{3} \cdot \cancel{3} \cdot 7}{\cancel{3} \cdot \cancel{3} \cdot 11} = \frac{7}{11}$$

Note the use of the divisibility test for 3 twice.

• **EXAMPLE 3**   Simplify: $\dfrac{84}{300}$.

$$\frac{84}{300} = \frac{\cancel{2} \cdot \cancel{2} \cdot \cancel{3} \cdot 7}{\cancel{2} \cdot \cancel{2} \cdot \cancel{3} \cdot 5 \cdot 5} = \frac{7}{25}$$

SIMPLIFYING SPECIAL FRACTIONS

1. Any number (except zero) divided by itself is equal to 1. For example,

$$\frac{3}{3} = 1; \frac{5}{5} = 1; \frac{173}{173} = 1.$$

2. Any number divided by 1 is equal to itself. For example,

$$\frac{5}{1} = 5; \frac{9}{1} = 9; \frac{25}{1} = 25.$$

3. Zero divided by any number (except zero) is equal to zero. For example,

$$\frac{0}{6} = 0; \frac{0}{13} = 0; \frac{0}{8} = 0.$$

4. Any number *divided by zero* is not meaningful and is called *undefined*. For example, $\frac{4}{0}$ is undefined.

A *proper fraction* is a fraction whose numerator is less than its denominator. Examples of proper fractions are $\frac{2}{3}$, $\frac{5}{14}$, and $\frac{3}{8}$. An *improper fraction* is a fraction whose numerator is greater than or equal to its denominator. Examples of improper fractions are $\frac{7}{5}$, $\frac{11}{11}$, and $\frac{9}{4}$.

A *mixed number* is an integer plus a proper fraction. Examples of mixed numbers are $1\frac{3}{4} \left(1 + \frac{3}{4}\right)$, $14\frac{1}{9}$, and $5\frac{2}{15}$.

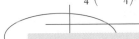

> ### CHANGING AN IMPROPER FRACTION TO A MIXED NUMBER
>
> To change an improper fraction to a mixed number, divide the numerator by the denominator. The quotient is the whole-number part. The remainder over the divisor is the proper fraction part of the mixed number.

• **EXAMPLE 4**   Change $\frac{17}{3}$ to a mixed number.

$$\frac{17}{3} = 17 \div 3 = 3\overline{)17} = 5\frac{2}{3}$$
$$5\ r\ 2$$

• **EXAMPLE 5**   Change $\frac{78}{7}$ to a mixed number.

$$\frac{78}{7} = 78 \div 7 = 7\overline{)78} = 11\frac{1}{7}$$
$$11\ r\ 1$$

If the improper fraction is not reduced to lowest terms, you may find it easier to reduce it before changing it to a mixed number. Of course you may reduce the proper fraction after the division if you prefer.

• **EXAMPLE 6**   Change $\frac{324}{48}$ to a mixed number and simplify.

Method 1: Reduce the improper fraction to lowest terms first.

$$\frac{324}{48} = \frac{\cancel{2} \cdot \cancel{2} \cdot 3 \cdot 3 \cdot 3 \cdot \cancel{3}}{\cancel{2} \cdot \cancel{2} \cdot 2 \cdot 2 \cdot \cancel{3}} = \frac{27}{4} = 4\overline{)27} = 6\frac{3}{4}$$
$$6\ r\ 3$$

Method 2: Change the improper fraction to a mixed number first.

$$\frac{324}{48} = 48\overline{)324}^{\ 6\ r\ 36} = 6\frac{36}{48} = 6\frac{\cancel{2} \cdot \cancel{2} \cdot 3 \cdot \cancel{3}}{\cancel{2} \cdot \cancel{2} \cdot 2 \cdot 2 \cdot \cancel{3}} = 6\frac{3}{4}$$
$$\underline{288}$$
$$36$$

One way to change a mixed number to an improper fraction is to multiply the integer by the denominator of the fraction and then add the numerator of the fraction. Then place this sum over the original denominator.

• **EXAMPLE 7**   Change $2\frac{1}{3}$ to an improper fraction.

$$2\frac{1}{3} = \frac{(2 \times 3) + 1}{3} = \frac{7}{3}$$

● **EXAMPLE 8**   Change $5\frac{3}{8}$ to an improper fraction.

$$5\frac{3}{8} = \frac{(5 \times 8) + 3}{8} = \frac{43}{8}$$

● **EXAMPLE 9**   Change $10\frac{5}{9}$ to an improper fraction.

$$10\frac{5}{9} = \frac{(10 \times 9) + 5}{9} = \frac{95}{9}$$

A number containing an integer and an improper fraction may be simplified as follows.

● **EXAMPLE 10**   Change $3\frac{8}{5}$ **a.** to an improper fraction and then **b.** to a mixed number in simplest form.

a. $3\frac{8}{5} = \frac{(3 \times 5) + 8}{5} = \frac{23}{5}$

b. $\frac{23}{5} = 23 \div 5 = 5\overline{\smash{\big)}23} = 4\frac{3}{5}$
            $4\ r\ 3$

A calculator with a fraction key may be used to simplify fractions as follows. The fraction key often looks similar to $\boxed{a\ b/c}$.

● **EXAMPLE 11**   Reduce $\frac{108}{144}$ to lowest terms.

Flowchart   **Buttons Pushed**   **Display**

Enter $\frac{108}{144}$	$\boxed{1} \rightarrow \boxed{0} \rightarrow \boxed{8} \rightarrow \boxed{a\ b/c} \rightarrow \boxed{1} \rightarrow \boxed{4} \rightarrow \boxed{4}$	$108 \underline{\ \ }/\ 144$
↓		
Push equals	$\boxed{=}$	$3 \underline{\ \ }/\ 4$

Thus, $\frac{108}{144} = \frac{3}{4}$ in lowest terms.

A calculator with a fraction key may be used to change an improper fraction to a mixed number, as follows.

● **EXAMPLE 12**   Change $\frac{25}{6}$ to a mixed number.

Flowchart   **Buttons Pushed**   **Display**

Enter $\frac{25}{6}$	$\boxed{2} \rightarrow \boxed{5} \rightarrow \boxed{a\ b/c} \rightarrow \boxed{6}$	$25 \underline{\ \ }/\ 6$
↓		
Push equals	$\boxed{=}$	$4 \underline{\ \ }/\ 1 \underline{\ \ }/\ 6$

Thus, $\frac{25}{6} = 4\frac{1}{6}$.

Note: The symbol $\underline{\ \ }/$ separates the whole number, the numerator, and the denominator in the display.

A calculator with a fraction key may be used to change a mixed number to an improper fraction as follows. The improper fraction key often looks similar to $\boxed{d/c}$.

• **EXAMPLE 13**   Change $6\frac{3}{5}$ to an improper fraction.

Flowchart	Buttons Pushed	Display
Enter $6\frac{3}{5}$	$\boxed{6} \to \boxed{a\,b/c} \to \boxed{3} \to \boxed{a\,b/c} \to \boxed{5}$	$6 \_\!/ 3 \_\!/ 5$
Push d/c	$\boxed{d/c}$	$33 \_\!/ 5$

Thus, $6\frac{3}{5} = \frac{33}{5}$.

Exercises 1.6

Simplify:

1. $\frac{12}{28}$
2. $\frac{9}{12}$
3. $\frac{36}{42}$
4. $\frac{12}{18}$
5. $\frac{9}{48}$
6. $\frac{8}{10}$
7. $\frac{13}{39}$
8. $\frac{24}{36}$

9. $\frac{48}{60}$
10. $\frac{72}{96}$
11. $\frac{9}{9}$
12. $\frac{15}{1}$
13. $\frac{0}{8}$
14. $\frac{6}{6}$
15. $\frac{9}{0}$
16. $\frac{6}{8}$

17. $\frac{14}{16}$
18. $\frac{7}{28}$
19. $\frac{27}{36}$
20. $\frac{15}{18}$
21. $\frac{12}{16}$
22. $\frac{9}{18}$
23. $\frac{20}{25}$
24. $\frac{12}{36}$

25. $\frac{12}{40}$
26. $\frac{54}{72}$
27. $\frac{32}{48}$
28. $\frac{60}{72}$
29. $\frac{112}{128}$
30. $\frac{330}{360}$
31. $\frac{72}{84}$
32. $\frac{9}{144}$

33. $\frac{126}{210}$
34. $\frac{270}{480}$
35. $\frac{57}{111}$
36. $\frac{30}{162}$
37. $\frac{58}{87}$
38. $\frac{198}{462}$
39. $\frac{112}{144}$
40. $\frac{525}{1155}$

Change each fraction to a mixed number in simplest form:

41. $\frac{78}{5}$
42. $\frac{11}{4}$

43. $\frac{28}{3}$
44. $\frac{21}{3}$

45. $\frac{45}{36}$
46. $\frac{67}{16}$

47. $\frac{57}{6}$
48. $\frac{84}{9}$

49. $5\frac{15}{12}$
50. $2\frac{70}{16}$

Change each mixed number to an improper fraction:

51. $3\frac{5}{6}$
52. $6\frac{3}{4}$

53. $2\frac{1}{8}$
54. $5\frac{2}{3}$

55. $1\frac{7}{16}$
56. $4\frac{1}{2}$

57. $6\frac{7}{8}$
58. $8\frac{1}{5}$

59. $10\frac{3}{5}$
60. $12\frac{5}{6}$

1.7   Addition and Subtraction of Fractions

Technicians must be able to compute fractions accurately, because mistakes on the job can be quite costly. Also, many shop drawing dimensions contain fractions.

ADDING FRACTIONS

$$\frac{a}{c} + \frac{b}{c} = \frac{a+b}{c} \quad (c \neq 0)$$

That is, to add two or more fractions with the same denominator, first add their numerators. Then place this sum over the common denominator.

•   **EXAMPLE 1**   Add: $\frac{1}{8} + \frac{3}{8}$.

$$\frac{1}{8} + \frac{3}{8} = \frac{1+3}{8} = \frac{4}{8} = \frac{1}{2}$$

•   **EXAMPLE 2**   Add: $\frac{2}{16} + \frac{5}{16}$.

$$\frac{2}{16} + \frac{5}{16} = \frac{2+5}{16} = \frac{7}{16}$$

•   **EXAMPLE 3**   Add: $\frac{2}{31} + \frac{7}{31} + \frac{15}{31}$.

$$\frac{2}{31} + \frac{7}{31} + \frac{15}{31} = \frac{2+7+15}{31} = \frac{24}{31}$$

To add fractions with different denominators, we first need to find a common denominator. When reducing fractions to lowest terms, we *divide* both numerator and denominator by the same nonzero number, which does not change the value of the fraction. Similarly, we can *multiply* both numerator and denominator by the same nonzero number without changing the value of the fraction.

How do you determine what number to use for multiplying? It is customary to find the *least common denominator (LCD)* for fractions with unlike denominators. The LCD is the smallest number that has all the denominators as divisors.

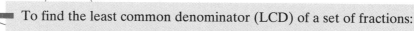

To find the least common denominator (LCD) of a set of fractions:

1. Factor each denominator into its prime factors.
2. Write each prime factor the number of times it appears *most* in any *one* denominator in Step 1. The LCD is the product of these prime factors.

•   **EXAMPLE 4**   Find the LCD of the following fractions: $\frac{1}{6}, \frac{1}{8}$, and $\frac{1}{18}$.

STEP 1 Factor each denominator into prime factors. (Prime factorization may be reviewed in Section 1.5.)

$$6 = 2 \cdot 3$$
$$8 = 2 \cdot 2 \cdot 2$$
$$18 = 2 \cdot 3 \cdot 3$$

STEP 2 Write each prime factor the number of times it appears *most* in any *one* denominator in Step 1. The LCD is the product of these prime factors.

Here, 2 appears once as a factor of 6, three times as a factor of 8, and once as a factor of 18. So 2 appears at most *three* times in any one denominator. Therefore, you have $2 \cdot 2 \cdot 2$ as factors of the LCD. The factor 3 appears at most twice in any one denominator, so you have $3 \cdot 3$ as factors of the LCD. Now 2 and 3 are the only factors of the three given denominators. The LCD for $\frac{1}{6}$, $\frac{1}{8}$, and $\frac{1}{18}$ must be $2 \cdot 2 \cdot 2 \cdot 3 \cdot 3 = 72$. Note that 72 does have divisors 6, 8, and 18.

This procedure is shown in Table 1.1. From the table, we see that the LCD contains the factor 2 three times and the factor 3 two times. Thus, LCD = $2 \cdot 2 \cdot 2 \cdot 3 \cdot 3 = 72$.

TABLE 1.1

		Number of times the prime factor appears	
Prime factor	Denominator	in given denominator	*most* in any one denominator
2	$6 = 2 \cdot 3$	once	
	$8 = 2 \cdot 2 \cdot 2$	three times	three times
	$18 = 2 \cdot 3 \cdot 3$	once	
3	$6 = 2 \cdot 3$	once	
	$8 = 2 \cdot 2 \cdot 2$	none	
	$18 = 2 \cdot 3 \cdot 3$	twice	twice

• **EXAMPLE 5** Find the LCD of $\frac{3}{4}$, $\frac{3}{8}$, and $\frac{3}{16}$.

$$4 = 2 \cdot 2$$
$$8 = 2 \cdot 2 \cdot 2$$
$$16 = 2 \cdot 2 \cdot 2 \cdot 2$$

The factor 2 appears at most *four* times in any one denominator, so the LCD is $2 \cdot 2 \cdot 2 \cdot 2 = 16$. Note that 16 does have divisors 4, 8, and 16.

• **EXAMPLE 6**   Find the LCD of $\dfrac{2}{5}$, $\dfrac{4}{15}$, and $\dfrac{3}{20}$.

$$5 = 5$$
$$15 = 3 \cdot 5$$
$$20 = 2 \cdot 2 \cdot 5$$

The LCD is $2 \cdot 2 \cdot 3 \cdot 5 = 60$.

Of course, if you can find the LCD by inspection, you need not go through the method shown in the examples.

• **EXAMPLE 7**   Find the LCD of $\dfrac{3}{8}$ and $\dfrac{5}{16}$.

By inspection, the LCD is 16, because 16 is the smallest number that has divisors 8 and 16.

After finding the LCD of the fractions you wish to add, change each of the original fractions to a fraction of equal value, with the LCD as its denominator.

• **EXAMPLE 8**   Add $\dfrac{3}{8} + \dfrac{5}{16}$.

First, find the LCD of $\dfrac{3}{8}$ and $\dfrac{5}{16}$. The LCD is 16. Now change $\dfrac{3}{8}$ to a fraction of equal value with a denominator of 16.

Write: $\dfrac{3}{8} = \dfrac{?}{16}$.   Think: $8 \times ? = 16$.

Since $8 \times 2 = 16$, we multiply both the numerator and the denominator by 2. The numerator is 6, and the denominator is 16.

$$\dfrac{3}{8} \times \dfrac{2}{2} = \dfrac{6}{16}$$

Now, using the rule for adding fractions:

$$\dfrac{3}{8} + \dfrac{5}{16} = \dfrac{6}{16} + \dfrac{5}{16} = \dfrac{6+5}{16} = \dfrac{11}{16}$$

Now try adding some fractions for which the LCD is more difficult to find.

• **EXAMPLE 9**   Add: $\dfrac{1}{4} + \dfrac{1}{6} + \dfrac{1}{16} + \dfrac{7}{12}$.

First, find the LCD.

$$4 = 2 \cdot 2$$
$$6 = 2 \cdot 3$$
$$16 = 2 \cdot 2 \cdot 2 \cdot 2$$
$$12 = 2 \cdot 2 \cdot 3$$

Note that 2 is used as a factor at most four times in any one denominator and 3 as a factor at most once. Thus, the LCD $= 2 \cdot 2 \cdot 2 \cdot 2 \cdot 3 = 48$.

Second, change each fraction to an equivalent fraction with 48 as its denominator.

$$\frac{1}{4} = \frac{?}{48} \qquad \frac{1 \times 12}{4 \times 12} = \frac{12}{48}$$

$$\frac{1}{6} = \frac{?}{48} \qquad \frac{1 \times 8}{6 \times 8} = \frac{8}{48}$$

$$\frac{1}{16} = \frac{?}{48} \qquad \frac{1 \times 3}{16 \times 3} = \frac{3}{48}$$

$$\frac{7}{12} = \frac{?}{48} \qquad \frac{7 \times 4}{12 \times 4} = \frac{28}{48}$$

$$\frac{1}{4} + \frac{1}{6} + \frac{1}{16} + \frac{7}{12} = \frac{12}{48} + \frac{8}{48} + \frac{3}{48} + \frac{28}{48}$$

$$= \frac{12 + 8 + 3 + 28}{48}$$

$$= \frac{51}{48}$$

Simplifying, we have

$$\frac{51}{48} = 1\frac{3}{48} = 1\,\frac{\cancel{3} \cdot 1}{\cancel{3} \cdot 16} = 1\frac{1}{16}$$

SUBTRACTING FRACTIONS

$$\frac{a}{c} - \frac{b}{c} = \frac{a - b}{c} \qquad (c \neq 0)$$

To subtract two or more fractions with a common denominator, subtract their numerators and place the difference over the common denominator.

• **EXAMPLE 10**   Subtract: $\dfrac{3}{5} - \dfrac{2}{5}$.

$$\frac{3}{5} - \frac{2}{5} = \frac{3 - 2}{5} = \frac{1}{5}$$

• **EXAMPLE 11**   Subtract: $\dfrac{5}{8} - \dfrac{3}{8}$.

$$\frac{5}{8} - \frac{3}{8} = \frac{5 - 3}{8} = \frac{2}{8} = \frac{\cancel{2} \cdot 1}{\cancel{2} \cdot 4} = \frac{1}{4}$$

To subtract two fractions that have different denominators, first find the LCD. Then express each fraction as an equivalent fraction using the LCD, and subtract the numerators.

• **EXAMPLE 12**   Subtract: $\dfrac{5}{6} - \dfrac{4}{15}$.

$$\frac{5}{6} - \frac{4}{15} = \frac{25}{30} - \frac{8}{30} = \frac{25 - 8}{30} = \frac{17}{30}$$

Addition and Subtraction of Mixed Numbers

To add mixed numbers, find the LCD of the fractions. Add the whole numbers, then add the fractions. Finally, add these two results.

• **EXAMPLE 13**   Add: $2\frac{1}{2}$ and $3\frac{3}{5}$.

$$2\frac{1}{2} = 2\frac{5}{10}$$

$$3\frac{3}{5} = 3\frac{6}{10}$$

$$5\frac{11}{10} = 5 + \frac{11}{10} = 5 + 1\frac{1}{10} = 6\frac{1}{10}$$

To subtract mixed numbers, find the LCD of the fractions. Subtract the fractions, then subtract the whole numbers.

• **EXAMPLE 14**   Subtract: $8\frac{2}{3}$ from $13\frac{3}{4}$.

$$13\frac{3}{4} = 13\frac{9}{12}$$

$$8\frac{2}{3} = 8\frac{8}{12}$$

$$5\frac{1}{12}$$

If the larger of the two mixed numbers does not also have the larger proper fraction, borrow 1 from the whole number. Then add it to the proper fraction before subtracting.

• **EXAMPLE 15**   Subtract: $2\frac{3}{5}$ from $4\frac{1}{2}$.

$$4\frac{1}{2} = 4\frac{5}{10} = 3\frac{15}{10}$$ First change the proper fractions to the LCD, 10. Then

$$2\frac{3}{5} = 2\frac{6}{10} = 2\frac{6}{10}$$ borrow 1 from 4 and add $\frac{10}{10}$ to $\frac{5}{10}$.

$$1\frac{9}{10}$$

• **EXAMPLE 16**   Subtract: $2\frac{3}{7}$ from $8\frac{1}{4}$.

$$8\frac{1}{4} = 8\frac{7}{28} = 7\frac{35}{28}$$ Borrow 1 from 8 and add $\frac{28}{28}$ to $\frac{7}{28}$.

$$2\frac{3}{7} = 2\frac{12}{28} = 2\frac{12}{28}$$

$$5\frac{23}{28}$$

• **EXAMPLE 17**   Subtract: $12 - 4\frac{3}{8}$.

$$12 = 11\frac{8}{8}$$

$$4\frac{3}{8} = 4\frac{3}{8}$$

$$\overline{\phantom{4\frac{3}{8}}} \quad 7\frac{5}{8}$$

Applications Involving Addition and Subtraction of Fractions

An electrical circuit with more than one path for the current to flow is called a *parallel circuit*. See Figure 1.25. The current in a parallel circuit is divided among the branches in the circuit. How it is divided depends on the resistance in each branch. Since the current is divided among the branches, the total current (I_T) of the circuit is the same as the current from the source. This equals the sum of the currents through the individual branches of the circuit. That is, $I_T = I_1 + I_2 + I_3 \cdots$.

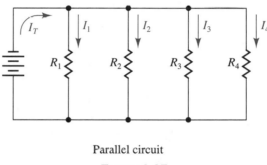

Parallel circuit

FIGURE 1.25

• **EXAMPLE 18**   Find the total current in the parallel circuit in Figure 1.26.

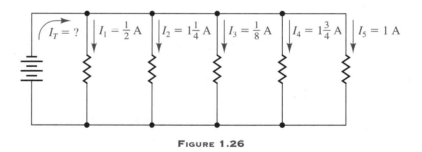

FIGURE 1.26

$$I_T = I_1 + I_2 + I_3 + I_4 + I_5$$

$$\frac{1}{2}\,\text{A} = \frac{4}{8}\,\text{A}$$

$$1\frac{1}{4}\,\text{A} = 1\frac{2}{8}\,\text{A}$$

$$\frac{1}{8}\,\text{A} = \frac{1}{8}\,\text{A}$$

$$1\frac{3}{4}\,\text{A} = 1\frac{6}{8}\,\text{A}$$

$$1\ \ \text{A} = 1\ \ \text{A}$$

$$\overline{} \quad 3\frac{13}{8}\,\text{A} = 4\frac{5}{8}\,\text{A}$$

•   **EXAMPLE 19**   Find the missing dimension in Figure 1.27.

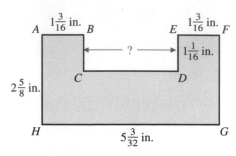

FIGURE 1.27

First, note that the length of side HG equals the sum of the lengths of sides AB, CD, and EF. To find the length of the missing dimension, subtract the sum of side AB and side EF from side HG. That is, first add $AB + EF$:

$AB:$ $\qquad 1\dfrac{3}{16}$ in.

$EF:$ $\qquad 1\dfrac{3}{16}$ in.

$AB + EF:$ $\quad 2\dfrac{6}{16}$ in. or $2\dfrac{3}{8}$ in.

Then subtract $HG - (AB + EF)$:

$HG:$ $\qquad 5\dfrac{3}{32}$ in. $= 5\dfrac{3}{32}$ in. $= 4\dfrac{35}{32}$ in.

$AB + EF:$ $\quad 2\dfrac{3}{8}$ in. $= 2\dfrac{12}{32}$ in. $= 2\dfrac{12}{32}$ in.

$\qquad\qquad\qquad\qquad\qquad\qquad\qquad = 2\dfrac{23}{32}$ in.

Therefore, the missing dimension is $2\dfrac{23}{32}$ in.

•   **EXAMPLE 20**   Find the perimeter (distance around) of Figure 1.27.

The perimeter is the sum of the lengths of all the sides of the figure.

$AB:$ $\quad 1\dfrac{3}{16}$ in. $= 1\dfrac{6}{32}$ in.

$BC:$ $\quad 1\dfrac{1}{16}$ in. $= 1\dfrac{2}{32}$ in.

$CD:$ $\quad 2\dfrac{23}{32}$ in. $= 2\dfrac{23}{32}$ in.

$DE:$ $\quad 1\dfrac{1}{16}$ in. $= 1\dfrac{2}{32}$ in.

$EF:$ $\quad 1\dfrac{3}{16}$ in. $= 1\dfrac{6}{32}$ in.

$FG:$ $\quad 2\dfrac{5}{8}$ in. $= 2\dfrac{20}{32}$ in.

GH: $5\dfrac{3}{32}$ in. $=$ $5\dfrac{3}{32}$ in.

HA: $2\dfrac{5}{8}$ in. $=$ $2\dfrac{20}{32}$ in.

Perimeter: $15\dfrac{82}{32}$ in. $= 15 + 2\dfrac{18}{32}$ in. $= 17\dfrac{9}{16}$ in.

Using a Calculator to Add and Subtract Fractions

• **EXAMPLE 21**   Add: $\dfrac{2}{3} + \dfrac{7}{12}$.

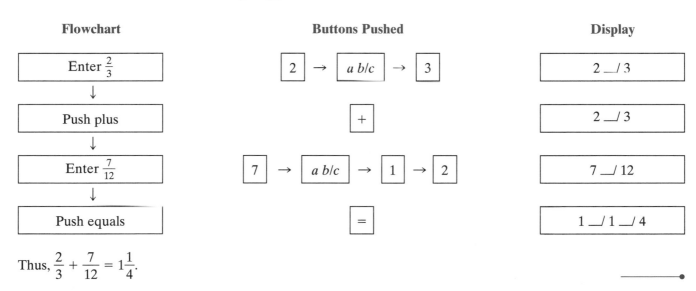

Thus, $\dfrac{2}{3} + \dfrac{7}{12} = 1\dfrac{1}{4}$.

• **EXAMPLE 22**   Add: $7\dfrac{3}{4} + 5\dfrac{7}{12}$.

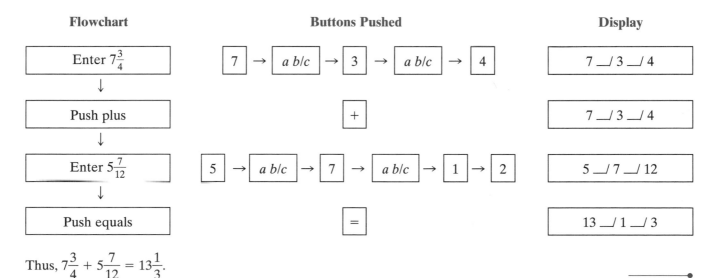

Thus, $7\dfrac{3}{4} + 5\dfrac{7}{12} = 13\dfrac{1}{3}$.

• **EXAMPLE 23** Subtract: $8\frac{1}{4} - 3\frac{5}{12}$.

Flowchart	Buttons Pushed	Display
Enter $8\frac{1}{4}$	$8 \rightarrow \boxed{a\,b/c} \rightarrow 1 \rightarrow \boxed{a\,b/c} \rightarrow 4$	8 _/ 1 _/ 4
↓		
Push minus	$-$	8 _/ 1 _/ 4
↓		
Enter $3\frac{5}{12}$	$3 \rightarrow \boxed{a\,b/c} \rightarrow 5 \rightarrow \boxed{a\,b/c} \rightarrow 1 \rightarrow 2$	3 _/ 5 _/ 12
↓		
Push equals	$=$	4 _/ 5 _/ 6

Thus, $8\frac{1}{4} - 3\frac{5}{12} = 4\frac{5}{6}$.

Exercises 1.7

Find the LCD of each set of fractions:

1. $\frac{1}{2}, \frac{1}{8}, \frac{1}{16}$ **2.** $\frac{1}{3}, \frac{2}{5}, \frac{3}{7}$ **3.** $\frac{1}{6}, \frac{3}{10}, \frac{1}{14}$ **4.** $\frac{1}{9}, \frac{1}{15}, \frac{5}{21}$ **5.** $\frac{1}{3}, \frac{1}{16}, \frac{7}{8}$ **6.** $\frac{1}{5}, \frac{3}{14}, \frac{4}{35}$

Perform the indicated operations and simplify:

7. $\frac{2}{3} + \frac{1}{6}$ **15.** $\frac{1}{5} + \frac{3}{20}$ **23.** $\frac{1}{3} + \frac{1}{6} + \frac{3}{16} + \frac{1}{12}$ **31.** $\frac{7}{8} - \frac{3}{4}$ **39.** $2\frac{1}{2} + 4\frac{3}{4}$ **47.** $3\frac{4}{5} + 9\frac{8}{9}$

8. $\frac{1}{2} + \frac{3}{8}$ **16.** $\frac{3}{4} + \frac{3}{16}$ **24.** $\frac{3}{16} + \frac{1}{8} + \frac{1}{3} + \frac{1}{4}$ **32.** $\frac{9}{64} - \frac{2}{128}$ **40.** $3\frac{5}{8} + 5\frac{3}{4}$ **48.** $4\frac{5}{12} + 6\frac{17}{20}$

9. $\frac{1}{16} + \frac{3}{32}$ **17.** $\frac{4}{5} + \frac{1}{2}$ **25.** $\frac{1}{20} + \frac{1}{30} + \frac{1}{40}$ **33.** $\frac{4}{5} - \frac{3}{10}$ **41.** $3 - \frac{3}{8}$ **49.** $5\frac{6}{7} - 4\frac{11}{12}$

10. $\frac{5}{6} + \frac{1}{18}$ **18.** $\frac{2}{3} + \frac{4}{9}$ **26.** $\frac{1}{14} + \frac{1}{15} + \frac{1}{6}$ **34.** $\frac{7}{16} - \frac{1}{3}$ **42.** $8 - 5\frac{3}{4}$ **50.** $7\frac{2}{9} - 4\frac{5}{6}$

11. $\frac{2}{7} + \frac{3}{28}$ **19.** $\frac{3}{5} + \frac{7}{10}$ **27.** $\frac{3}{10} + \frac{1}{14} + \frac{4}{15}$ **35.** $\frac{9}{14} - \frac{3}{42}$ **43.** $8\frac{3}{16} - 3\frac{7}{16}$ **51.** $3\frac{9}{16} + 4\frac{7}{12} + 3\frac{1}{6}$

12. $\frac{1}{9} + \frac{2}{45}$ **20.** $\frac{1}{8} + \frac{2}{3}$ **28.** $\frac{5}{36} + \frac{11}{72} + \frac{5}{6}$ **36.** $\frac{8}{9} - \frac{5}{24}$ **44.** $5\frac{3}{8} + 2\frac{3}{4}$ **52.** $5\frac{2}{5} + 3\frac{7}{10} + 4\frac{7}{15}$

13. $\frac{3}{8} + \frac{5}{64}$ **21.** $\frac{7}{8} + \frac{5}{16}$ **29.** $\frac{4}{9} + \frac{13}{27} + \frac{2}{3}$ **37.** $\frac{9}{16} - \frac{13}{32} - \frac{1}{8}$ **45.** $7\frac{3}{16} - 4\frac{7}{8}$ **53.** $16\frac{5}{8} - 4\frac{7}{12} - 2\frac{1}{2}$

14. $\frac{3}{10} + \frac{7}{100}$ **22.** $\frac{11}{16} + \frac{17}{32}$ **30.** $\frac{23}{27} + \frac{5}{18} + \frac{5}{36}$ **38.** $\frac{7}{8} - \frac{2}{9} - \frac{1}{12}$ **46.** $8\frac{1}{4} - 4\frac{7}{16}$ **54.** $12\frac{9}{16} - 3\frac{1}{6} + 2\frac{1}{4}$

55. Find the perimeter of the triangular field in Illustration 1.

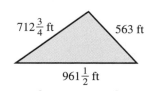

$712\frac{3}{4}$ ft 563 ft

$961\frac{1}{2}$ ft

ILLUSTRATION 1

*Find **a.** the length of the missing dimension and **b.** the perimeter of each figure.*

56.

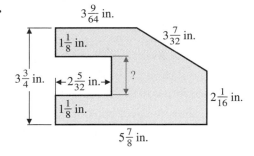

57.

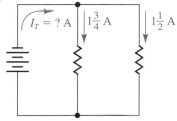

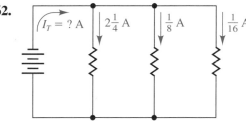

58.

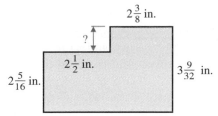

59.

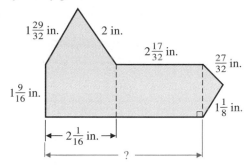

60. The perimeter (sum of the sides) of a triangle is $59\frac{9}{32}$ in. One side is $19\frac{5}{8}$ in., and a second side is $17\frac{13}{16}$ in. How long is the remaining side?

Find the total current in each parallel circuit:

61.

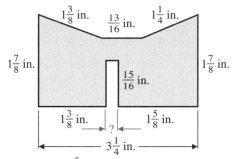

$I_T = ?$ A $1\frac{3}{4}$ A $1\frac{1}{2}$ A

62.

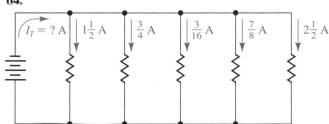

$I_T = ?$ A $2\frac{1}{4}$ A $\frac{1}{8}$ A $\frac{1}{16}$ A

63.

$I_T = ?$ A $\frac{1}{16}$ A $\frac{1}{12}$ A $1\frac{3}{4}$ A

64.

$I_T = ?$ A $1\frac{1}{2}$ A $\frac{3}{4}$ A $\frac{3}{16}$ A $\frac{7}{8}$ A $2\frac{1}{2}$ A

65. Find the length of the shaft in Illustration 2.

$6\frac{3}{4}$ in. $2\frac{7}{8}$ in.

ILLUSTRATION 2

66. Find the distance between the centers of the two endholes of the plate in Illustration 3.

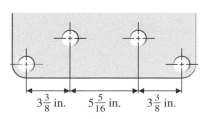

$3\frac{3}{8}$ in. $5\frac{5}{16}$ in. $3\frac{3}{8}$ in.

ILLUSTRATION 3

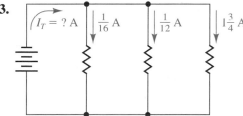

67. In Illustration 4, find **a.** the length of the tool and **b.** the length of diameter A.

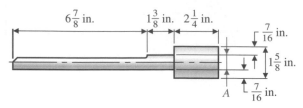

ILLUSTRATION 4

68. A rod $13\frac{13}{16}$ in. long has been cut as shown in Illustration 5. Assume the waste in each cut is $\frac{1}{16}$ in. What is the length of the remaining piece?

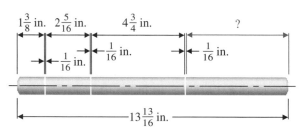

ILLUSTRATION 5

69. Find **a.** the length and **b.** the diameter of the shaft in Illustration 6.

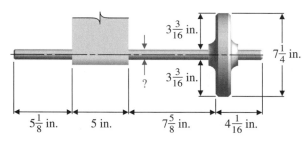

ILLUSTRATION 6

70. Find the missing dimension of the shaft in Illustration 7.

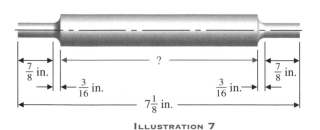

ILLUSTRATION 7

71. Floor joists are spaced 16 in. OC (on center) and are $1\frac{5}{8}$ in. thick. What is the distance between them?

72. If no tap drill chart is available, the correct drill size (*TDS*) can be found by using the formula $TDS = OD - P$. *OD* is the outside diameter and *P* is the pitch of the thread (the distance between successive threads). Find the tap drill size for a $\frac{3}{8}$-in. outside diameter if the pitch is $\frac{1}{16}$ in.

73. How much must the diameter of a $\frac{7}{8}$-in. shaft be reduced so that its diameter will be $\frac{51}{64}$ in.?

74. What is the difference in thickness between a $\frac{7}{16}$-in. steel plate and a $\frac{5}{8}$-in. steel plate?

75. A planer takes a $\frac{3}{32}$-in. cut on a plate that is $1\frac{7}{8}$ in. thick. What is the thickness of the plate after one cut? What is the thickness of the plate after three cuts?

76. A home is built on a $65\frac{3}{4}$-ft-wide lot. The house is $5\frac{5}{12}$ ft from one side of the lot and is $43\frac{5}{6}$ ft wide. (See Illustration 8.) What is the distance from the house to the other side of the lot?

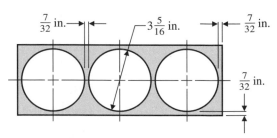

ILLUSTRATION 8

77. See Illustration 9. What width and length steel strip is needed in order to drill three holes of diameter $3\frac{5}{16}$ in.? Allow $\frac{7}{32}$ in. between, and on each side of, the holes.

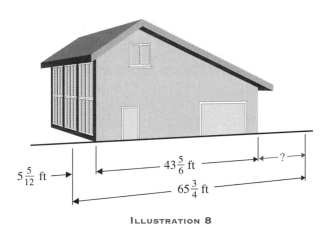

ILLUSTRATION 9

78. Find length x in Illustration 10.

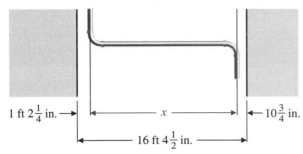

ILLUSTRATION 10

79. Find the total length of the valve in Illustration 11.

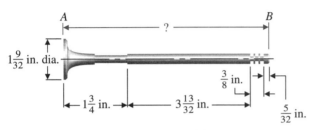

ILLUSTRATION 11

80. Find the total length of the shaft in Illustration 12.

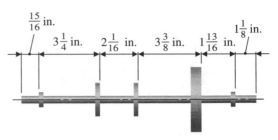

ILLUSTRATION 12

81. A mechanic needs the following lengths of $\frac{3}{8}$-in. copper tubing: $15\frac{3}{8}$ in., $7\frac{3}{4}$ in., $11\frac{1}{2}$ in., $7\frac{7}{32}$ in., and $10\frac{5}{16}$ in. What is the total length of tubing needed?

82. An end view and side view of a shaft are shown in Illustration 13. **a.** Find the diameter of the largest part of the shaft. **b.** Find dimension A of the shaft.

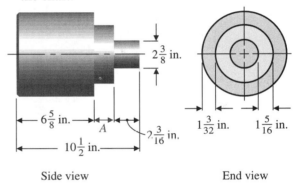

Side view   End view

ILLUSTRATION 13

1.8 Multiplication and Division of Fractions

MULTIPLYING FRACTIONS

$$\frac{a}{b} \times \frac{c}{d} = \frac{a \cdot c}{b \cdot d} \quad (b \neq 0, d \neq 0)$$

To multiply fractions, multiply the numerators and multiply the denominators.

• **EXAMPLE 1** Multiply: $\frac{5}{9} \times \frac{3}{10}$.

$$\frac{5}{9} \times \frac{3}{10} = \frac{5 \cdot 3}{9 \cdot 10} = \frac{15}{90} = \frac{\cancel{15} \cdot 1}{\cancel{15} \cdot 6} = \frac{1}{6}$$

To simplify the work, consider the following:

$$\frac{\overset{1}{\cancel{5}}}{\underset{3}{\cancel{9}}} \times \frac{\overset{1}{\cancel{3}}}{\underset{2}{\cancel{10}}} = \frac{1 \cdot 1}{3 \cdot 2} = \frac{1}{6}$$

This method divides the numerator by 15, or (5×3), and the denominator by 15, or (5×3). It does not change the value of the fraction. ————————•

• **EXAMPLE 2**   Mutliply: $\dfrac{18}{25} \times \dfrac{7}{27}$.

As a shortcut, divide a numerator by 9 and a denominator by 9. Then multiply:

$$\dfrac{\overset{2}{\cancel{18}}}{25} \times \dfrac{7}{\underset{3}{\cancel{27}}} = \dfrac{2 \cdot 7}{25 \cdot 3} = \dfrac{14}{75}$$ ————————•

Any mixed number must be replaced by an equivalent improper fraction before multiplying or dividing two or more fractions.

• **EXAMPLE 3**   Multiply: $8 \times 3\frac{3}{4}$.

$$8 \times 3\dfrac{3}{4} = \dfrac{\overset{2}{\cancel{8}}}{1} \times \dfrac{15}{\underset{1}{\cancel{4}}} = \dfrac{30}{1} = 30$$ ————————•

• **EXAMPLE 4**   Multiply: $\dfrac{9}{16} \times \dfrac{5}{22} \times \dfrac{4}{7} \times 3\frac{2}{3}$.

$$\dfrac{9}{16} \times \dfrac{5}{22} \times \dfrac{4}{7} \times 3\dfrac{2}{3} = \dfrac{\overset{3}{\cancel{9}}}{\underset{4}{\cancel{16}}} \times \dfrac{5}{\underset{2}{\cancel{22}}} \times \dfrac{\overset{1}{\cancel{4}}}{7} \times \dfrac{\overset{1}{\cancel{11}}}{\underset{1}{\cancel{3}}} = \dfrac{15}{56}$$ ————————•

Note: Whenever you multiply several fractions, you may simplify the computation by dividing *any* numerator and *any* denominator by the same number.

DIVIDING FRACTIONS

$$\dfrac{a}{b} \div \dfrac{c}{d} = \dfrac{a}{b} \times \dfrac{d}{c} = \dfrac{a \cdot d}{b \cdot c} \quad (b \neq 0, c \neq 0, d \neq 0)$$

To divide a fraction by a fraction, invert the fraction (interchange numerator and denominator) that follows the division sign ($\div$). Then multiply the resulting fractions.

• **EXAMPLE 5**   Divide: $\dfrac{5}{6} \div \dfrac{2}{3}$.

$$\dfrac{5}{6} \div \dfrac{2}{3} = \dfrac{5}{\underset{2}{\cancel{6}}} \times \dfrac{\overset{1}{\cancel{3}}}{2} = \dfrac{5 \cdot 1}{2 \cdot 2} = \dfrac{5}{4} \text{ or } 1\dfrac{1}{4}$$ ————————•

• **EXAMPLE 6**   Divide: $7 \div \dfrac{2}{5}$.

$$7 \div \dfrac{2}{5} = \dfrac{7}{1} \times \dfrac{5}{2} = \dfrac{35}{2} \text{ or } 17\dfrac{1}{2}$$ ————————•

• **EXAMPLE 7**   Divide: $\dfrac{3}{5} \div 4$.

$$\frac{3}{5} \div 4 = \frac{3}{5} \div \frac{4}{1} = \frac{3}{5} \times \frac{1}{4} = \frac{3}{20}$$

• **EXAMPLE 8**   Divide: $\dfrac{9}{10} \div 5\dfrac{2}{5}$.

$$\frac{9}{10} \div 5\frac{2}{5} = \frac{9}{10} \div \frac{27}{5} = \frac{\overset{1}{\cancel{9}}}{\underset{2}{\cancel{10}}} \times \frac{\overset{1}{\cancel{5}}}{\underset{3}{\cancel{27}}} = \frac{1}{6}$$

When both multiplication and division of fractions occur, invert the first fraction that follows a division sign ($\div$). Then proceed according to the rules for multiplying fractions.

• **EXAMPLE 9**   Perform the indicated operations and simplify: $\dfrac{2}{5} \times \dfrac{1}{3} \div \dfrac{3}{4}$.

$$\frac{2}{5} \times \frac{1}{3} \div \frac{3}{4} = \frac{2}{5} \times \frac{1}{3} \times \frac{4}{3} = \frac{2 \cdot 1 \cdot 4}{5 \cdot 3 \cdot 3} = \frac{8}{45}$$

• **EXAMPLE 10**   Perform the indicated operations and simplify: $7\dfrac{1}{3} \div 4 \times 2$.

$$7\frac{1}{3} \div 4 \times 2 = \frac{\overset{11}{\cancel{22}}}{3} \times \frac{1}{\underset{\underset{1}{\cancel{2}}}{\cancel{4}}} \times \frac{\overset{1}{\cancel{2}}}{1} = \frac{11}{3} \text{ or } 3\frac{2}{3}$$

Applications Involving Multiplication and Division of Fractions

Lumber is usually measured in board feet. One *board foot* is the amount of wood contained in a piece of wood that measures one inch thick and one square foot in area, or its equivalent. (See Figure 1.28.) The number of board feet in lumber may be found by the formula:

$$\text{bd ft} = \frac{\text{number of boards} \times \begin{array}{c}\text{thickness}\\ \text{(in in.)}\end{array} \times \begin{array}{c}\text{width}\\ \text{(in in.)}\end{array} \times \begin{array}{c}\text{length}\\ \text{(in ft)}\end{array}}{12}$$

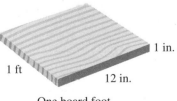

One board foot

FIGURE 1.28

The 12 in the denominator comes from the fact that the simplest form of one board foot can be thought of as a board that is 1 in. thick × 12 in. wide × 1 ft long.

Lumber is either rough or finished. *Rough stock* is lumber that is not planed or dressed; *finished stock* is planed on one or more sides. When measuring lumber, we compute the full size. That is, we compute the measure of the rough stock that is required to make the desired finished piece. When lumber is finished or planed, $\frac{1}{16}$ in. is taken off each side when the lumber is less than $1\frac{1}{2}$ in. thick. If the lumber is $1\frac{1}{2}$ in. or more in thickness, $\frac{1}{8}$ in. is taken off each side. (Note: Lumber for framing houses usually measures $\frac{1}{2}$ in. less than the name that we call the piece. For example, a "two-by-four," a piece 2 in. by 4 in., actually measures $1\frac{1}{2}$ in. by $3\frac{1}{2}$ in.)

•   **EXAMPLE 11**   Find the number of board feet contained in 6 pieces of lumber 2 in. × 8 in. × 16 ft (Figure 1.29).

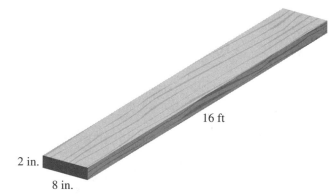

16 ft

2 in.

8 in.

FIGURE 1.29

$$\text{bd ft} = \frac{\begin{array}{c}\text{number of} \\ \text{boards}\end{array} \times \begin{array}{c}\text{thickness} \\ \text{(in in.)}\end{array} \times \begin{array}{c}\text{width} \\ \text{(in in.)}\end{array} \times \begin{array}{c}\text{length} \\ \text{(in ft)}\end{array}}{12}$$

$$= \frac{6 \times 2 \times 8 \times 16}{12} = 128 \text{ bd ft}$$

•   **EXAMPLE 12**   Energy in the form of electrical power is used by industry and consumers alike. Power (in watts, W) equals the voltage (in volts, V) times the current (in amperes, or amps, A). A soldering iron draws a current of $7\frac{1}{2}$ A on a 110-V circuit. Find the wattage, or power, rating of this soldering iron.

$$\text{Power} = (\text{voltage}) \times (\text{current})$$
$$= 110 \quad \times \quad 7\frac{1}{2}$$
$$= 110 \quad \times \quad \frac{15}{2}$$
$$= 825 \text{ W}$$

Power may also be found by computing the product of the square of the current (in amps, A) and the resistance (in ohms, Ω).

• **EXAMPLE 13**   To give $\frac{1}{5}$ grain of Myleran from $\frac{1}{30}$-grain tablets, how many tablets would be given? To find how many tablets would be given, we divide the amount to be given by the amount in each tablet.

$$\frac{1}{5} \div \frac{1}{30} = \frac{1}{\cancel{5}} \times \frac{\overset{6}{\cancel{30}}}{1}$$
$$= 6 \text{ tablets}$$

—————●

• **EXAMPLE 14**   One form of Ohm's law states that the current I (in amps, A) in a simple circuit equals the voltage E (in volts, V) divided by the resistance R (in ohms, Ω). What current is required for a heating element with a resistance of $7\frac{1}{2}$ Ω operating in a 12-V circuit?

$$\text{Current} = (\text{voltage}) \div (\text{resistance})$$
$$= 12 \quad\div\quad 7\frac{1}{2}$$
$$= 12 \quad\div\quad \frac{15}{2}$$
$$= \overset{4}{\cancel{12}} \quad\times\quad \frac{2}{\underset{5}{\cancel{15}}}$$
$$= \frac{8}{5} \text{ or } 1\frac{3}{5} \text{ A}$$

—————●

Using a Calculator to Multiply and Divide Fractions

• **EXAMPLE 15**   Multiply: $2\frac{5}{6} \times 4\frac{1}{2}$.

Flowchart	**Buttons Pushed**	**Display**
Enter $2\frac{5}{6}$	$2 \rightarrow \boxed{a\ b/c} \rightarrow 5 \rightarrow \boxed{a\ b/c} \rightarrow 6$	$2 \_/\ 5 \_/\ 6$
↓		
Push times	$\times$	$2 \_/\ 5 \_/\ 6$
↓		
Enter $4\frac{1}{2}$	$4 \rightarrow \boxed{a\ b/c} \rightarrow 1 \rightarrow \boxed{a\ b/c} \rightarrow 2$	$4 \_/\ 1 \_/\ 2$
↓		
Push equals	$=$	$12 \_/\ 3 \_/\ 4$

Thus, $2\frac{5}{6} \times 4\frac{1}{2} = 12\frac{3}{4}$.

—————●

EXAMPLE 16   Divide: $5\frac{5}{7} \div 8\frac{1}{3}$.

Flowchart	Buttons Pushed	Display
Enter $5\frac{5}{7}$	$5 \rightarrow \boxed{a\,b/c} \rightarrow 5 \rightarrow \boxed{a\,b/c} \rightarrow 7$	$5 \rule{0.5em}{0.4pt}/\,5\,\rule{0.5em}{0.4pt}/\,7$
↓		
Push divide	$\div$	$5 \rule{0.5em}{0.4pt}/\,5\,\rule{0.5em}{0.4pt}/\,7$
↓		
Enter $8\frac{1}{3}$	$8 \rightarrow \boxed{a\,b/c} \rightarrow 1 \rightarrow \boxed{a\,b/c} \rightarrow 3$	$8 \rule{0.5em}{0.4pt}/\,1\,\rule{0.5em}{0.4pt}/\,3$
↓		
Push equals	$=$	$24\,\rule{0.5em}{0.4pt}/\,35$

Thus, $5\frac{5}{7} \div 8\frac{1}{3} = \frac{24}{35}$.

Exercises 1.8

Perform the indicated operations and simplify:

1. $\frac{2}{3} \times 18$

2. $8 \times \frac{1}{2}$

3. $\frac{3}{4} \times 12$

4. $3\frac{1}{2} \times \frac{2}{5}$

5. $1\frac{3}{4} \times \frac{5}{16}$

6. $\frac{1}{3} \times \frac{1}{3} \times \frac{1}{3}$

7. $\frac{16}{21} \times \frac{7}{8}$

8. $\frac{7}{12} \times \frac{45}{56}$

9. $\frac{2}{7} \times 35$

10. $\frac{9}{16} \times \frac{2}{3} \times 1\frac{6}{15}$

11. $\frac{5}{8} \times \frac{7}{10} \times \frac{2}{7}$

12. $\frac{9}{16} \times \frac{5}{9} \times \frac{4}{25}$

13. $2\frac{1}{3} \times \frac{5}{8} \times \frac{6}{7}$

14. $\frac{5}{28} \times \frac{3}{5} \times \frac{2}{3} \times \frac{2}{9}$

15. $\frac{6}{11} \times \frac{26}{35} \times 1\frac{9}{13} \times \frac{7}{12}$

16. $\frac{3}{8} \div \frac{1}{4}$

17. $\frac{3}{5} \div \frac{10}{12}$

18. $\frac{10}{12} \div \frac{3}{5}$

19. $4\frac{1}{2} \div \frac{1}{4}$

20. $18\frac{2}{3} \div 6$

21. $15 \div \frac{3}{8}$

22. $\frac{77}{6} \div 6$

23. $\frac{7}{11} \div \frac{3}{5}$

24. $7 \div 3\frac{1}{8}$

25. $\frac{2}{5} \times 3\frac{2}{3} \div \frac{3}{4}$

26. $\frac{7}{8} \times \frac{1}{2} \div \frac{2}{7}$

27. $\frac{16}{5} \times \frac{3}{2} \times \frac{10}{4} \div 5\frac{1}{3}$

28. $6 \times 6 \times \frac{21}{7} \div 48$

29. $\frac{7}{9} \times \frac{3}{8} \div \frac{28}{81}$

30. $2\frac{1}{3} \times \frac{5}{8} \div \frac{10}{4}$

31. $\frac{2}{7} \times \frac{5}{9} \times \frac{3}{10} \div 6$

32. $\frac{9}{4} \times \frac{9}{4} \times \frac{21}{7} \div 81$

33. $\frac{7}{16} \div \frac{3}{8} \times \frac{1}{2}$

34. $\frac{5}{8} \div \frac{25}{64} \times \frac{5}{6}$

35. A barrel has a capacity of 42 gal. How many gallons does it contain when it is $\frac{3}{4}$ full?

36. Find the area of a rectangle with length $6\frac{1}{3}$ ft and width $3\frac{3}{4}$ ft. (Area = length × width.)

37. Find the perimeter of a rectangle with length $6\frac{1}{3}$ ft and width $3\frac{3}{4}$ ft.

38. Find the perimeter of a rectangle whose length is $2\frac{1}{4}$ times its width and whose width is 6 ft. (See Illustration 1.)

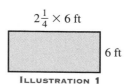

$2\frac{1}{4} \times 6$ ft

6 ft

ILLUSTRATION 1

39. A length of $71\frac{1}{2}$ in. is divided into 16 equal parts. What is the length of each part?

Find the number of board feet in each quantity of lumber:

40. One piece 2 in. × 6 in. × 8 ft

41. 10 pieces 2 in. × 4 in. × 12 ft

42. 24 pieces 4 in. × 4 in. × 16 ft

43. 175 pieces 1 in. × 8 in. × 14 ft

44. Find the total length of eight pieces of steel each $5\frac{3}{4}$ in. long.

45. The outside diameter (OD) of a pipe is $4\frac{9}{32}$ in. The walls are $\frac{7}{32}$ in. thick. Find the inside diameter. (See Illustration 2.)

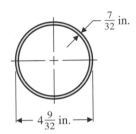

ILLUSTRATION 2

46. A piece of round stock measures $1\frac{3}{8}$ in. in diameter. Specifications call for a finished diameter of $1\frac{11}{32}$ in. What depth of cut is required on a lathe to produce this finished diameter?

47. If a CAD (computer-aided drafting) draftsperson can complete a set of drawings in 16 days, how much of it can the same person complete in $\frac{1}{2}$ day?

48. Two strips of metal are riveted together in a straight line, with nine rivets equally spaced $2\frac{5}{16}$ in. apart. What is the distance between the first and last rivet?

49. Two metal strips are riveted together in a straight line, with 16 equally spaced rivets. The distance between the first and last rivet is $28\frac{1}{8}$ in. Find the distance between any two consecutive rivets.

50. Find length *x*, the distance between centers, in Illustration 3.

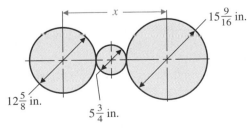

ILLUSTRATION 3

51. From a steel rod 36 in. long, the following pieces are cut:

3 pieces $2\frac{1}{8}$ in. long

2 pieces $5\frac{3}{4}$ in. long

6 pieces $\frac{7}{8}$ in. long

1 piece $3\frac{1}{2}$ in. long

Assume $\frac{1}{16}$ in. waste in each cut. Find the length of the remaining piece.

52. A piece of drill rod 2 ft 6 in. long is to be cut into pins, each $2\frac{1}{2}$ in. long.

a. Assume no loss of material in cutting. How many pins will you get?

b. Assume $\frac{1}{16}$ in. waste in each cut. How many pins will you get?

53. The cutting tool on a lathe moves $\frac{3}{128}$ in. along the piece being turned for each revolution of work. The piece revolves at 45 revolutions per minute. How long will it take to turn a length of $9\frac{9}{64}$-in. stock in one operation?

54. Three vents are equally spaced along a wall 26 ft 6 in. (318 in.) long, as shown in Illustration 4. Find dimension *d*.

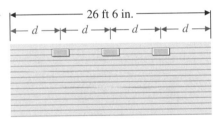

ILLUSTRATION 4

55. A concrete pad for mounting a condensing unit is 4 ft long, $2\frac{2}{3}$ ft wide, and 3 in. $\left(\frac{1}{4}\text{ ft}\right)$ thick. Find its volume in cubic feet.

56. How many lengths of radiator hose, each $5\frac{1}{4}$ in. long, can be cut from a 6-ft roll?

57. Find the load of a circuit that takes $12\frac{1}{2}$ A at 220 V. (See Example 12.)

58. An electric iron requires $4\frac{1}{4}$ A and has a resistance of $24\frac{1}{2}$ Ω. What voltage does it require to operate? ($V = IR$.)

59. An electric hand drill draws $3\frac{3}{4}$ A and has a resistance of $5\frac{1}{3}$ Ω. What power does it use? ($P = I^2R$.)

60. A wiring job requires the following lengths of BX cable:

12 pieces	$8\frac{1}{2}$ ft long	
7 pieces	$18\frac{1}{2}$ ft long	
24 pieces	$1\frac{3}{4}$ ft long	
12 pieces	$6\frac{1}{2}$ ft long	
2 pieces	$34\frac{1}{4}$ ft long	

How much cable is needed to complete the job?

61. What current is required for a heating element with a resistance of $10\frac{1}{2}$ Ω operating in a 24-V circuit? (See Example 14.)

62. How many lengths of wire, each $3\frac{3}{4}$ in. long, can be cut from a 25-ft roll?

63. A total of 19 ceiling outlets are to be equally spaced in a straight line between two points that are $130\frac{1}{2}$ ft apart in a hallway. How far apart will the ceiling outlets be, center to center?

64. If $8\frac{3}{4}$ dozen electrical clips cost $7.35, what did each clip cost?

65. A steer gains $1\frac{3}{4}$ lb a day. How many pounds will it gain in 36 days?

66. Tom needs to apply $1\frac{3}{4}$ gal of herbicide per acre of soybeans. How many gallons of herbicide are needed for 120 acres?

67. An airplane sprayer tank holds 60 gal. If $\frac{3}{4}$ gal of water and $\frac{1}{2}$ lb of pesticide are applied per acre, how much pesticide powder is needed per tankful?

68. If 1 ft³ of cotton weighs $22\frac{1}{2}$ lb, how many cubic feet are contained in a bale of cotton weighing 500 lb? In 15 tons of cotton?

69. A test plot of $\frac{1}{20}$ acre produces 448 lb of shelled corn. Find the yield in bushels per acre. (1 bu of shelled corn weighs 56 lb.)

70. A farmer wishes to concrete his rectangular feed lot, which measures 120 ft by 180 ft. He wants to have a base of 4 in. of gravel covered with $3\frac{1}{2}$ in. of concrete.

a. How many cubic yards of each material must he purchase?

b. What is his total materials cost? Concrete costs $50/yd³ delivered, and gravel costs $5/ton delivered. (1 yd³ of gravel weighs approximately 2500 lb.)

71. A medicine contains $\frac{1}{5}$ alcohol. A bottle holds $2\frac{1}{2}$ oz of this medicine. How many ounces of alcohol does the bottle contain?

72. A patient receives $\frac{1}{4}$ of a $\frac{1}{4}$-grain morphine sulfate tablet. How much of the drug does the patient receive?

73. To give 50 mg of ascorbic acid from 100-mg tablets, how many tablets should be given?

74. To give 50 mg of ascorbic acid from 150-mg tablets, how many tablets should be given?

75. A patient is given $\frac{1}{4}$ of a 5-grain aspirin tablet. How much aspirin does the patient receive?

76. To give 1 grain of digitalis from $1\frac{1}{2}$-grain tablets, how many tablets should be given?

77. A patient is given $\frac{3}{4}$ grain of codeine from $\frac{3}{8}$-grain tablets. How many tablets are given?

78. If you give $\frac{2}{3}$ of a $7\frac{1}{2}$-grain tablet, how many grains does the patient receive?

79. A patient is given $\frac{1}{2}$ grain of Valium from $\frac{1}{6}$-grain tablets. How many tablets are given?

80. The doctor orders 45 mg of prednisone. Each tablet contains 10 mg. How many tablets are given to the patient?

*In a parallel circuit, the total resistance (R_T) is given by the formula**

$$R_T = \cfrac{1}{\cfrac{1}{R_1} + \cfrac{1}{R_2} + \cfrac{1}{R_3} + \cdots}$$

Note: The three dots mean that you should use as many fractions in the denominator as there are resistances in the circuit.

---

*A calculator approach to working with such equations is shown in Section 6.9.

 *Find the total resistance in each parallel circuit:*

81.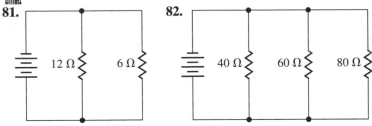

12 Ω   6 Ω

82.

40 Ω   60 Ω   80 Ω

83.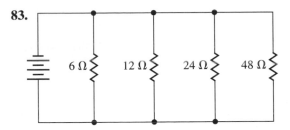

6 Ω   12 Ω   24 Ω   48 Ω

1.9   The English System of Weights and Measures

Centuries ago the thumb, hand, foot, and length from nose to outstretched fingers were used as units of measurement. These methods, of course, were not very satisfactory because people's sizes varied. In the 14th century, King Edward II proclaimed the length of the English inch to be the same as three barleycorn grains laid end to end. (See Figure 1.30) This proclamation helped some, but it did not eliminate disputes over the length of the English inch.

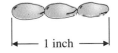

FIGURE 1.30

Each of these methods provides rough estimations of measurements. Actually, *measurement* is the comparison of an *observed* quantity with a *standard unit* quantity. In the estimation above, there is no one standard unit. A standard unit that is constant, accurate, and accepted by all is needed for technical measurements. Today, nations have bureaus to set national standards for all measures.

The English system requires us to understand and be able to use fractions in everyday life. After we have converted to the metric system, the importance of fractional computations will be greatly reduced.

Take a moment to look at the table of English weights and measures on your reference card. Become familiar with most of this table, because you will use it when changing units.

• **EXAMPLE 1**   Change 5 ft 9 in. to inches.

1 ft = 12 in., so 5 ft = 5 × 12 in. = 60 in.

5 ft 9 in. = 60 in. + 9 in. = 69 in.

To change from one unit or set of units to another, we use what is commonly called a *conversion factor*. We know that we can multiply any number or quantity by 1 (one) without changing the value of the original quantity. We also know that any fraction whose numerator and denominator are the same is equal to 1. For example, $\frac{5}{5} = 1$, $\frac{16}{16} = 1$, and $\frac{7 \text{ ft}}{7 \text{ ft}} = 1$. Also, since 12 in. = 1 ft, $\frac{12 \text{ in.}}{1 \text{ ft}} = 1$, and likewise, $\frac{1 \text{ ft}}{12 \text{ in.}} = 1$, because the numerator equals the denominator. We call such names for 1 *conversion factors* (or *unit conversion factors*). The information necessary for forming conversion factors is found in tables, many of which are provided on the reference card included with this book, in case you do not remember them.

CHOOSING CONVERSION FACTORS

The correct choice for a given conversion factor is the one in which the old units are in the numerator of the original expression and in the denominator of the conversion factor, or in the denominator of the original expression and in the numerator of the conversion factor. That is, set up the conversion factor so that the old units cancel each other.

• **EXAMPLE 2**   Change 19 ft to inches.

Since 1 ft = 12 in., the two possible conversion factors are $\frac{1 \text{ ft}}{12 \text{ in.}} = 1$ and $\frac{12 \text{ in.}}{1 \text{ ft}} = 1$. We want to choose the one whose numerator is expressed in the new units (in.) and whose denominator is expressed in the old units (ft); that is, $\frac{12 \text{ in.}}{1 \text{ ft}}$. Therefore,

$$19 \text{ ft} \times \frac{12 \text{ in.}}{1 \text{ ft}} = 19 \times 12 \text{ in.} = 228 \text{ in.}$$

— conversion factor

• **EXAMPLE 3**   Change 8 yd to feet.

3 ft = 1 yd,

$$\text{so } 8 \text{ yd} \times \frac{3 \text{ ft}}{1 \text{ yd}} = 8 \times 3 \text{ ft} = 24 \text{ ft}$$

— conversion factor

• **EXAMPLE 4**   Change 76 oz to pounds.

$$76 \text{ oz} \times \frac{1 \text{ lb}}{16 \text{ oz}} = \frac{76}{16} \text{ lb} = \frac{19}{4} \text{ lb} = 4\frac{3}{4} \text{ lb}$$

— conversion factor

Sometimes it is necessary to use more than one conversion factor.

• **EXAMPLE 5**   Change 6 mi to yards.

In the table on your reference card, there is no expression equating miles with yards. Thus, it is necessary to use two conversion factors.

$$6 \text{ mi} \times \frac{5280 \text{ ft}}{1 \text{ mi}} \times \frac{1 \text{ yd}}{3 \text{ ft}} = \frac{6 \times 5280}{3} \text{ yd} = 10{,}560 \text{ yd}$$

• **EXAMPLE 6**   How could a technician mixing chemicals express 4800 fluid ounces (fl oz) in gallons?

No conversions between fluid ounces and gallons are given in the tables. Use the conversion factors for (a) fluid ounces to pints (pt); (b) pints to quarts (qt); and (c) quarts to gallons.

$$4800 \text{ fl oz} \times \frac{1 \text{ pt}}{16 \text{ fl oz}} \times \frac{1 \text{ qt}}{2 \text{ pt}} \times \frac{1 \text{ gal}}{4 \text{ qt}} = \frac{4800}{16 \times 2 \times 4} \text{ gal} = 37.5 \text{ gal}$$

Conversion factors for   **a.   b.   c.**

The use of a conversion factor is especially helpful for units with which you are unfamiliar, such as changing rods to feet, chains to feet, or fathoms to feet.

• **EXAMPLE 7**   Change 561 ft to rods.

Given 1 rod = 16.5 ft, proceed as follows:

$$561 \text{ ft} \times \frac{1 \text{ rod}}{16.5 \text{ ft}} = 34 \text{ rods}$$

• **EXAMPLE 8**   Change 320 ft/min to ft/h.

Here, choose the conversion factor whose denominator is expressed in the new units (hours) and whose numerator is expressed in the old units (minutes).

$$320 \frac{\text{ft}}{\text{min}} \times \frac{60 \text{ min}}{1 \text{ h}} = 19{,}200 \text{ ft/h}$$

The following example shows how to use multiple conversion factors in more complex units.

• **EXAMPLE 9**   Change 60 mi/h to ft/s.

This requires a series of conversions as follows:

a. from hours to minutes;
b. from minutes to seconds; and
c. from miles to feet.

$$60 \frac{\text{mi}}{\text{h}} \times \frac{1 \text{ h}}{60 \text{ min}} \times \frac{1 \text{ min}}{60 \text{ s}} \times \frac{5280 \text{ ft}}{1 \text{ mi}} = 88 \text{ ft/s}$$

Conversion factors for: **a.   b.   c.**

Exercises   1.9

Fill in each blank:

1. 3 ft 7 in. = _____ in.
2. 6 yd 4 ft = _____ ft
3. 5 lb 3 oz = _____ oz
4. 7 yd 3 ft 6 in. = _____ in.
5. 4 qt 1 pt = _____ pt
6. 6 gal 3 qt = _____ pt
7. 2 bu 2 pecks (pk) = _____ pk
8. 5 bu 3 pk = _____ pk

9. 8 ft = _____ in.
10. 5 yd = _____ ft
11. 3 qt = _____ pt
12. 4 mi = _____ ft
13. 96 in. = _____ ft
14. 72 ft = _____ yd
15. 10 pt = _____ qt
16. 54 in. = _____ ft

17. 88 oz = _____ lb
18. 32 fl oz = _____ pt
19. 14 qt = _____ gal
20. 3 bu = _____ pk
21. 56 fl oz = _____ pt
22. 7040 ft = _____ mi
23. 92 ft = _____ yd
24. 9000 lb = _____ tons

25. 2 mi = _____ yd   **26.** 6000 fl oz = _____ gal   **27.** 500 fl oz = _____ qt   **28.** 3 mi = _____ rods

29. A door is 80 in. in height. Find its height in feet and inches.

30. A plane is flying at 22,000 ft. How many miles high is it?

31. Change the length of a shaft $12\frac{3}{4}$ ft long to inches.

32. A machinist has 15 wrought-iron rods to mill. Each rod weighs 24 oz. What is the total weight of the rods in pounds?

33. The instructions on a carton of chemicals call for mixing 144 fl oz of water, 24 fl oz of chemical No. 1, and 56 fl oz of chemical No. 2. How many quarts are contained in the final mixture?

34. The resistance of 1 ft of No. 32-gauge copper wire is $\frac{4}{25}$ Ω. What is the resistance of 12 yd of this wire?

35. A farmer wishes to wire a shed that is 1 mi from the electricity source in his barn. He uses No. 0-gauge copper wire, which has a resistance of $\frac{1}{10}$ ohm (Ω) per 1000 ft. What is the resistance for the mile of wire?

36. To mix an order of feed, the following quantities of feed are combined: 4200 lb, 600 lb, 5800 lb, 1300 lb, and 2100 lb. How many tons are in the final mixture?

37. Given 1 chain = 66 ft, change 561 ft to chains.

38. Given 1 fathom = 6 ft, change 12 fathoms to feet.

39. Given 1 dram = $27\frac{17}{50}$ grains, change 15 drams to grains.

40. Given 1 ounce = 8 drams, change 96 drams to ounces.

41. Change 4500 ft/h to ft/min.

42. Change 28 ft/s to ft/min.

43. Change $1\frac{1}{5}$ mi/s to mi/min.

44. Change 7200 ft/min to ft/s.

45. Change 40 mi/h to ft/s.

46. Change 64 ft/s to mi/h.

47. Change 24 in./s to ft/min.

48. Change 36 in./s to mi/h.

49. Add: 6 yd 2 ft 11 in.
       2 yd 1 ft 8 in.
       5 yd 2 ft 9 in.
       1 yd   6 in.

50. Subtract: 8 yd 1 ft 3 in.
          2 yd 2 ft 6 in.

Unit 1B   Review

Simplify:

1. $\dfrac{9}{15}$

2. $\dfrac{48}{54}$

Perform the indicated operations and simplify:

5. $\dfrac{5}{6} + \dfrac{2}{3}$

6. $5\dfrac{3}{8} - 2\dfrac{5}{12}$

7. $\dfrac{5}{12} \times \dfrac{16}{25}$

8. $\dfrac{3}{4} \div 1\dfrac{5}{8}$

9. $1\dfrac{2}{3} + 3\dfrac{5}{6} - 2\dfrac{1}{4}$

10. $4\dfrac{2}{3} \div 3\dfrac{1}{2} \times 1\dfrac{1}{2}$

3. Change $\dfrac{27}{6}$ to a mixed number in simplest form.

4. Change $3\dfrac{2}{5}$ to an improper fraction.

11. Find the missing dimension in the figure.

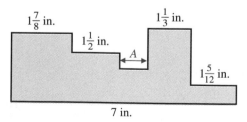

12. A pipe is 72 in. long. Cut three pieces of the following lengths from the pipe: $16\frac{3}{4}$ in., $24\frac{7}{8}$ in., and $12\frac{5}{16}$ in. Assume $\frac{1}{16}$ in. waste in each cut. What length of pipe is left?

13. Find the perimeter of a rectangle with length of $6\frac{1}{4}$ in. and width $2\frac{2}{3}$ in.

14. Find the area of a rectangle with length $6\frac{1}{4}$ in. and width $2\frac{2}{3}$ in.

15. Change 4 ft to inches.

16. Change 24 ft to yards.

17. Change 3 lb to ounces.

18. Change 20 qt to gallons.

19. Change 60 mi/h to ft/s.

20. Subtract 14 ft 4 in.
          8 ft 8 in.

Unit 1C: REVIEW OF OPERATIONS WITH DECIMAL FRACTIONS AND PERCENT

1.10 Addition and Subtraction of Decimal Fractions
Introduction to Decimals

Fractions whose denominators are 10, 100, 1000, or any power of 10 are called *decimal fractions*. Decimal calculations and measuring instruments calibrated in decimals are the basic tools for measurement in the metric system. The common use of the calculator makes a basic understanding of decimal principles necessary.

Recall the place values of the digits of a whole number from Section 1.1. Each digit to the left of the decimal point represents a multiple of a power of 10. Each digit to the right of the decimal point represents a multiple of a power of $\frac{1}{10}$. Study Table 1.2, which shows place values for decimals.

Note that $10^0 = 1$. (See Section 2.5.)

TABLE 1.2   PLACE VALUES FOR DECIMALS

Number	Words	Product form	Exponential form
1,000,000	One million	$10 \times 10 \times 10 \times 10 \times 10 \times 10$	10^6
100,000	One hundred thousand	$10 \times 10 \times 10 \times 10 \times 10$	10^5
10,000	Ten thousand	$10 \times 10 \times 10 \times 10$	10^4
1,000	One thousand	$10 \times 10 \times 10$	10^3
100	One hundred	10×10	10^2
10	Ten	10	10^1
1	One	1	10^0
0.1	One tenth	$\frac{1}{10}$	$\left(\frac{1}{10}\right)^1$ or 10^{-1}
0.01	One hundredth	$\frac{1}{10} \times \frac{1}{10}$	$\left(\frac{1}{10}\right)^2$ or 10^{-2}
0.001	One thousandth	$\frac{1}{10} \times \frac{1}{10} \times \frac{1}{10}$	$\left(\frac{1}{10}\right)^3$ or 10^{-3}
0.0001	One ten-thousandth	$\frac{1}{10} \times \frac{1}{10} \times \frac{1}{10} \times \frac{1}{10}$	$\left(\frac{1}{10}\right)^4$ or 10^{-4}
0.00001	One hundred-thousandth	$\frac{1}{10} \times \frac{1}{10} \times \frac{1}{10} \times \frac{1}{10} \times \frac{1}{10}$	$\left(\frac{1}{10}\right)^5$ or 10^{-5}
0.000001	One millionth	$\frac{1}{10} \times \frac{1}{10} \times \frac{1}{10} \times \frac{1}{10} \times \frac{1}{10} \times \frac{1}{10}$	$\left(\frac{1}{10}\right)^6$ or 10^{-6}

• **EXAMPLE 1**   In the number 123.456, find the place value of each digit and the number it represents.

Digit	Place value	Number represented
1	Hundreds	1×10^2
2	Tens	2×10^1
3	Ones or units	3×10^0 or 3×1
4	Tenths	$4 \times \dfrac{1}{10}$ or 4×10^{-1}
5	Hundredths	$5 \times \left(\dfrac{1}{10}\right)^2$ or 5×10^{-2}
6	Thousandths	$6 \times \left(\dfrac{1}{10}\right)^3$ or 6×10^{-3}

Recall that place values to the left of the decimal point are powers of 10, and place values to the right of the decimal point are powers of $\frac{1}{10}$.

• **EXAMPLE 2**   Write each decimal in words: 0.05; 0.0006; 24.41; 234.001207.

Decimal	Word form
0.05	Five hundredths
0.0006	Six ten-thousandths
24.41	Twenty-four *and* forty-one hundredths
234.001207	Two hundred thirty-four *and* one thousand two hundred seven millionths

Note that the decimal point is read "and."

• **EXAMPLE 3**   Write each number as a decimal and as a common fraction.

Number	Decimal	Common fraction
One hundred four and seventeen hundredths	104.17	$104\dfrac{17}{100}$
Fifty and three thousandths	50.003	$50\dfrac{3}{1000}$
Five hundred eleven hundred-thousandths	0.00511*	$\dfrac{511}{100{,}000}$

*This book follows the common practice of writing a zero before the decimal point in a decimal smaller than 1.

Often, common fractions are easier to use if they are expressed as decimal equivalents. Every common fraction can be expressed as a decimal. A *repeating decimal* is one in which a digit or a group of digits repeats again and again; it may be written as a common fraction.

A bar over a digit or group of digits means that this digit or group of digits is repeated without ending. Each of the following numbers is a repeating decimal:

0.33333 . . . is written 0.3

72.64444 . . . is written 72.64

0.21212121 . . . is written 0.21

6.33120120120 . . . is written 6.33120

A *terminating decimal* is a decimal number with a given number of digits. Examples are 0.75, 12.505, and 0.000612.

> To change a common fraction to a decimal, divide the numerator of the fraction by the denominator.

• **EXAMPLE 4** Change $\dfrac{3}{4}$ to a decimal.

$$
\begin{array}{r}
0.75 \\
4\overline{)3.00} \\
2\,8 \\
\hline
20 \\
20 \\
\end{array}
$$

$\dfrac{3}{4} = 0.75$ (a terminating decimal)

• **EXAMPLE 5** Change $\dfrac{8}{15}$ to a decimal.

$$
\begin{array}{r}
0.5333 \\
15\overline{)8.0000} \\
7\,5 \\
\hline
50 \\
45 \\
\hline
50 \\
45 \\
\hline
50 \\
45 \\
\hline
5 \\
\end{array}
$$

$\dfrac{8}{15} = 0.5333 \ldots = 0.5\overline{3}$ (a repeating decimal)

The result could be written 0.533 or 0.53. It is not necessary to continue the division once it has been established that the quotient has begun to repeat.

Since a decimal fraction can be written as a common fraction with a denominator that is a power of 10, it is easy to change a decimal fraction to a common fraction. Simply use the digits that appear to the right of the decimal point (disregarding beginning zeros) as the numerator. Use the place value of the last digit as the denominator. Any digits to the left of the decimal point will be the whole-number part of the resulting mixed number.

• **EXAMPLE 6** Change each decimal to a common fraction or a mixed number.

Decimal	Common fraction or mixed number
a. 0.3	$\dfrac{3}{10}$
b. 0.17	$\dfrac{17}{100}$
c. 0.25	$\dfrac{25}{100} = \dfrac{1}{4}$
d. 0.125	$\dfrac{125}{1000} = \dfrac{1}{8}$
e. 0.86	$\dfrac{86}{100} = \dfrac{43}{50}$
f. 8.1	$8\dfrac{1}{10}$
g. 13.64	$13\dfrac{64}{100} = 13\dfrac{16}{25}$
h. 5.034	$5\dfrac{34}{1000} = 5\dfrac{17}{500}$

In on-the-job situations, it is often more convenient to add, subtract, multiply, and divide measurements that are in decimal form rather than in fractional form. Except for the placement of the decimal point, the four arithmetic operations are the same for decimal fractions as they are for whole numbers.

• **EXAMPLE 7**   Add 13.2, 8.42, and 120.1.

a. Using decimal fractions:

$$
\begin{array}{r}
13.2 \\
8.42 \\
\underline{120.1} \\
141.72
\end{array}
$$

b. Using common fractions:

$$13\dfrac{2}{10} \;=\; 13\dfrac{20}{100}$$

$$8\dfrac{42}{100} \;=\; 8\dfrac{42}{100}$$

$$120\dfrac{1}{10} \;=\; 120\dfrac{10}{100}$$

$$141\dfrac{72}{100} = 141.72$$

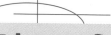

ADDING OR SUBTRACTING DECIMAL FRACTIONS

STEP 1 Write the decimals so that the digits having the same place value are in vertical columns. (Make certain that the decimal points are lined up vertically.)

STEP 2 Add or subtract as with whole numbers.

STEP 3 Place the decimal point between the ones digit and the tenths digit of the sum or the difference. (Be certain the decimal point is in the same vertical line as the other decimal points.)

• **EXAMPLE 8**   Subtract 1.28 from 17.9.

$$
\begin{array}{r}
17.90 \\
\underline{1.28} \\
16.62
\end{array}
$$

Note: Zeros can be supplied after the last digit at the right of the decimal point without changing the value of a number. Therefore 17.9 = 17.90. ———•

• **EXAMPLE 9**   Add 24.1, 26, and 37.02.

$$
\begin{array}{r}
24.10 \\
26.00 \\
\underline{37.02} \\
87.12
\end{array}
$$

Note: A decimal point can be placed at the right of any whole number, and zeros can be supplied without changing the value of the number. ———•

• **EXAMPLE 10**   Perform the indicated operations: 51.6 − 2.45 + 7.3 − 14.92.

$$
\begin{array}{rl}
51.60 & \\
\underline{-\ 2.45} & \\
49.15 & \text{difference} \\
\underline{+\ 7.30} & \\
56.45 & \text{sum} \\
\underline{-14.92} & \\
41.53 & \text{final difference}
\end{array}
$$
———•

• **EXAMPLE 11**   Find the missing dimension in Figure 1.31.

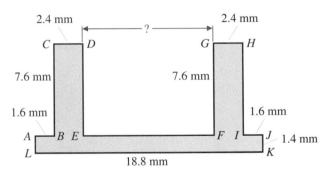

FIGURE 1.31

The missing dimension *EF* equals the sum of the lengths *AB*, *CD*, *GH*, and *IJ* subtracted from the length *LK*. That is, add

AB:	1.6 mm
CD:	2.4 mm
GH:	2.4 mm
IJ:	1.6 mm
	8.0 mm

Then subtract

LK:	18.8 mm
	−8.0 mm
	10.8 mm

That is, length *EF* = 10.8 mm. ———•

• **EXAMPLE 12** As we saw in Unit 1B, the total current in a parallel circuit equals the sum of the currents in the branches of the circuit. Find the total current in the parallel circuit shown in Figure 1.32.

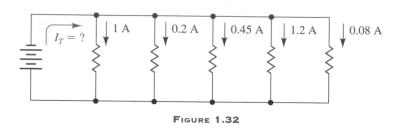

FIGURE 1.32

$$
\begin{array}{ll}
1 & \text{A} \\
0.2 & \text{A} \\
0.45 & \text{A} \\
1.2 & \text{A} \\
\underline{0.08} & \text{A} \\
2.93 & \text{A}
\end{array}
$$

Using a Calculator to Add and Subtract Decimal Fractions

• **EXAMPLE 13** Add: $14.62 + $0.78 + $1.40 + $0.05.

Flowchart	Buttons Pushed	Display
Enter 14.62	1 → 4 → . → 6 → 2	14.62
Push plus	+	14.62
Enter 0.78	. → 7 → 8	0.78
Push plus	+	15.4
Enter 1.40	1 → . → 4 *	1.4
Push plus	+	16.8
Enter 0.05	. → 0 → 5	0.05
Push equals	=	16.85

The sum is $16.85.

*Note: Ending zeros to the right of the decimal point do not have to be entered.

Combinations of addition and subtraction may be performed on a calculator as follows.

- **EXAMPLE 14**   Do as indicated: $74.6 - 8.57 + 5 - 0.0031$.

Flowchart	Buttons Pushed	Display
Enter 74.6	$7 \to 4 \to . \to 6$	74.6
Push minus	$-$	74.6
Enter 8.57	$8 \to . \to 5 \to 7$	8.57
Push plus	$+$	66.03
Enter 5	5	5
Push minus	$-$	71.03
Enter 0.0031	$. \to 0 \to 0 \to 3 \to 1$	0.0031
Push equals	$=$	71.0269

The result is 71.0269.

Exercises   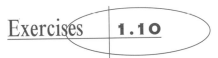 1.10

Write each decimal in words:

1. 0.004   **3.** 0.0005   **5.** 1.00421   **7.** 62.384   **9.** 6.092   **11.** 246.124

2. 0.021   **4.** 7.1   **6.** 14.014023   **8.** 1042.007   **10.** 8.1461   **12.** 12.60048

Write each number both as a decimal and as a common fraction or mixed number.

13. Five and two hundredths

14. One hundred twenty-three and six thousandths

15. Seventy-one and twenty-one ten-thousandths

16. Sixty-five thousandths

17. Sixty-five thousand

18. Three and four ten-thousandths

19. Forty-three and one hundred one ten-thousandths

20. Five hundred sixty-three millionths

Change each common fraction to a decimal:

21. $\dfrac{3}{8}$ **23.** $\dfrac{11}{15}$ **25.** $\dfrac{17}{50}$ **27.** $\dfrac{14}{11}$ **29.** $\dfrac{128}{7}$ **31.** $\dfrac{308}{9}$

22. $\dfrac{16}{25}$ **24.** $\dfrac{2}{5}$ **26.** $\dfrac{11}{9}$ **28.** $\dfrac{128}{25}$ **30.** $\dfrac{603}{24}$ **32.** $\dfrac{230}{6}$

Change each decimal to a common fraction or a mixed number:

33. 0.7   **35.** 0.11   **37.** 0.8425   **39.** 10.76

34. 0.6   **36.** 0.75   **38.** 3.14   **40.** 148.255

Find each sum:

41. 137.64
7.14
0.008
6.1

42. 63
4.7
19.45
120.015

43. 147.49 + 7.31 + 0.004 + 8.4

44. 47 + 6.3 + 20.71 + 170.027

Subtract:

45. 72.4 from 159

46. 3.12 from 4.7

47. 64.718 − 49.41

48. 140 − 16.412

Perform the indicated operations:

49. 18.4 − 13.72 + 4

50. 34.14 − 8.7 − 16.5

51. 0.37 + 4.5 − 0.008

52. 51.7 − 1.11 − 4.6 + 84.1

53. 1.511 + 14.714 − 6.1743

54. 0.0056 + 0.023 − 0.00456 + 0.9005

Use a calculator to do the rest of the exercises:

55. Find the missing dimensions in Illustration 1.

56. Find the perimeter of the figure in Illustration 1.

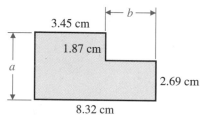

ILLUSTRATION 1

57. Find the length of the shaft shown in Illustration 2.

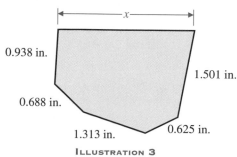

ILLUSTRATION 2

58. The perimeter of the hexagon in Illustration 3 is 6.573 in. Find the length of side *x*.

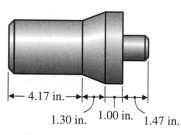

ILLUSTRATION 3

59. Find the perimeter of the figure in Example 11 (Figure 1.31).

60. The weather bureau reported that rainfall for June and July was 5.31 in. and 3.50 in., respectively.
  a. How much less rain fell in July than in June?
  b. What was the total rainfall for the two months?

61. Find the total current in the parallel circuit in Illustration 4.

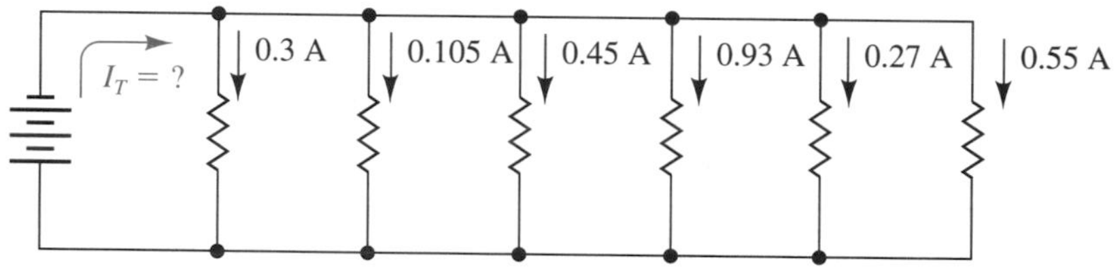

ILLUSTRATION 4

62. As we saw in Unit 1A, the total resistance in a series circuit is equal to the sum of the resistances in the circuit. Find the total resistance in the series circuit in Illustration 5.

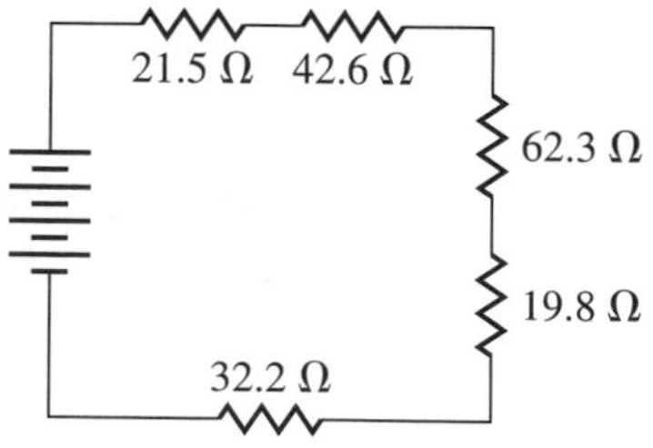

ILLUSTRATION 5

63. Find the total resistance in the series circuit in Illustration 6.

ILLUSTRATION 6

 **64.** In a series circuit, the voltage of the source equals the sum of the separate voltage drops in the circuit. Find the voltage of the source in the circuit in Illustration 7.

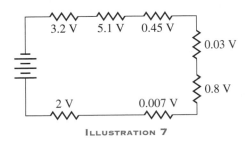

ILLUSTRATION 7

65. Find the difference between the diameters of the circular ends of the taper shown in Illustration 8.

1.625 in. ━━━━━━ 1.093 in

ILLUSTRATION 8

66. Find the missing dimension in each figure in Illustration 9.

67. Find the wall thickness of the pipe in Illustration 10.

68. Find the length, *l*, of the socket in Illustration 11. Also, find the length of diameter *A*.

69. In order to seat a valve properly in an automobile engine, the factory part measuring 1.732 in. must be ground 0.005 in. Find the size of the valve after it is ground.

70. The standard width of a new piston ring is 0.1675 in. The used ring measures 0.1643 in. How much has it worn?

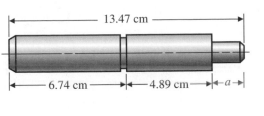

13.47 cm

6.74 cm ━ 4.89 cm ━ *a*

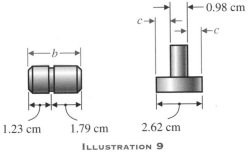

0.98 cm

c

c

b

1.23 cm   1.79 cm   2.62 cm

ILLUSTRATION 9

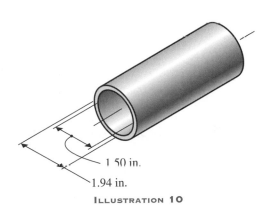

1.50 in.

1.94 in.

ILLUSTRATION 10

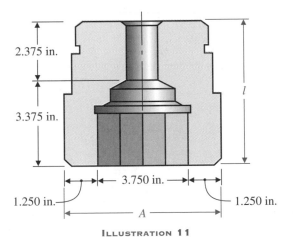

2.375 in.

3.375 in.

l

3.750 in.

1.250 in.

1.250 in.

A

ILLUSTRATION 11

1.11 Rounding Numbers

It is often desirable to make an estimate of a number or a measurement. When a truck driver makes a delivery from one side of a city to another, he or she can only estimate the time it will take to make the trip. An automobile technician must estimate the cost of a repair job and the number of mechanics to assign to that job. On such occasions, estimates are usually *rounded*.

Earlier you found that $\frac{1}{3} = 0.333$. There is no exact decimal value to use in a calculation. You must round 0.333 to a certain number of decimal places, depending on the accuracy needed in a given situation.

There are many rounding procedures in general use today. Some are complicated, and others are simple. We use one of the simplest methods, which will be outlined in the next examples and then stated in the form of a rule.

•   **EXAMPLE 1**   Round 25,348 to the nearest thousand.

Note that 25,348 is more than 25,000 and less than 26,000.

26,000
25,900
25,800
25,700
25,600
25,500
25,400
25,348   As you can see, 25,348 is closer to 25,000 than to 26,000. Therefore,
25,300   25,348 rounded to the nearest thousand is 25,000.
25,200
25,100
25,000

•   **EXAMPLE 2**   Round 2.5271 to the nearest hundredth.

Note that 2.5271 is more than 2.5200 but less than 2.5300.

2.5300 = 2.53
2.5290
2.5280
2.5271   2.5271 is nearer to 2.53 than to 2.52. Therefore, 2.5271 rounded to the
2.5270   nearest hundredth is 2.53.
2.5260
2.5250
2.5240
2.5230
2.5220
2.5210
2.5200 = 2.52

Note: If a number is exactly halfway between two numbers, round up to the larger number.

ROUNDING NUMBERS

To round a number to a particular place value:

1. If the digit in the next place to the right is less than 5, drop that digit and all other following digits. Use zeros to replace any whole-number places dropped.

2. If the digit in the next place to the right is 5 or greater, add 1 to the digit in the place to which you are rounding. Drop all other following digits. Use zeros to replace any whole-number places dropped.

• **EXAMPLE 3** Round each number in the left column to the place indicated in each of the other columns.

Number	Hundred	Ten	Unit	Tenth	Hundredth	Thousandth
a. 158.6147	200	160	159	158.6	158.61	158.615
b. 4,562.7155	4,600	4,560	4,563	4,562.7	4,562.72	4,562.716
c. 7.12579	0	10	7	7.1	7.13	7.126
d. 63,576.15	63,600	63,580	63,576	63,576.2	63,576.15	—
e. 845.9981	800	850	846	846.0	846.00	845.998

Instead of rounding a number to a particular place value, we often need to round a number to a given number of significant digits. *Significant digits* are those digits in a number we are reasonably sure of being able to rely on in a measurement. Here we present a brief introduction to significant digits. (An in-depth discussion of accuracy and significant digits is given in Section 4.1.)

The following digits in a number are significant:

- All nonzero digits (258 has three significant digits)
- All zeros between significant digits (2007 has four significant digits)
- All zeros at the end of a decimal number (2.000 and 0.04500 have four significant digits)

The following digits in a number are not significant:

- All zeros at the beginning of a decimal number less than 1 (0.00775 has three significant digits)
- All zeros at the end of a whole number (3600 has two significant digits)

ROUNDING NUMBERS

To round a number to a given number of significant digits:

1. Count the given number of significant digits from left to right, starting with the first nonzero digit.
2. If the next digit to the right is less than 5, drop that digit and all other following digits. Use zeros to replace any whole-number places dropped.
3. If the next digit to the right is 5 or greater, add 1 to the digit in the place to which you are rounding. Drop all other following digits. Use zeros to replace any whole-number places dropped.

• **EXAMPLE 4** Round each number to three significant digits.

a. 74,123   Count three digits from left to right, which is 1. Since the next digit to the right is less than 5, replace the next two digits with zeros. Thus, 74,123 rounded to three significant digits is 74,100.

b. 0.002976401   Count three digits from left to right, starting with the first nonzero digit, which is 7. Since the next digit to the right is greater than 5, increase the digit by 1 and drop the next four digits. Thus, 0.002976401 rounded to three significant digits is 0.00298.

• **EXAMPLE 5**   Round each number to the given number of significant digits.

Number	Given number of significant digits	Rounded number
a. 2571.88	3	2570
b. 2571.88	4	2572
c. 345,175	2	350,000
d. 345,175	4	345,200
e. 0.0030162	2	0.0030
f. 0.0030162	3	0.00302
g. 24.00055	3	24.0
h. 24.00055	5	24.001

Exercises   1.11

*Round each number to **a.** the nearest hundred and **b.** the nearest ten:*

1. 1652   **2.** 1760   **3.** 3125.4   **4.** 73.82   **5.** 18,675   **6.** 5968

*Round each number to **a.** the nearest tenth, and **b.** the nearest thousandth:*

7. 3.1416   **8.** 0.161616   **9.** 0.05731   **10.** 0.9836   **11.** 0.07046   **12.** 3.7654

Round each number in the left column to the place indicated in each of the other columns:

	Number	Hundred	Ten	Unit	Tenth	Hundredth	Thousandth
13.	636.1825						
14.	1,451.5254						
15.	17,159.1666						
16.	8.171717						
17.	1,543,679						
18.	41,892.1565						
19.	10,649.83						
20.	84.00659						
21.	649.8995						
22.	147.99545						

Round each number to three significant digits:

23. 236,534   **24.** 202.505   **25.** 0.03275

Round each number to two significant digits:

26. 63,914   **27.** 71.613   **28.** 0.03275

Round each number to four significant digits:

29. 1,462,304   **30.** 23.2347   **31.** 0.000337567

Round each number to three significant digits:

32. 20,714   **33.** 1.00782   **34.** 0.00118952

1.12 Multiplication and Division of Decimal Fractions

MULTIPLYING TWO DECIMAL FRACTIONS

1. Multiply the numbers as you would whole numbers.
2. Count the total number of digits to the right of the decimal points in the two numbers being multiplied. Then place the decimal in the product so there is that same total number of digits to the right of the decimal point.

• **EXAMPLE 1**   Multiply: 42.6×1.73.

$$
\begin{array}{r}
42.6 \\
1.73 \\
\hline
12\,78 \\
298\,2 \\
426 \\
\hline
73.698
\end{array}
$$

Note that 42.6 has one digit to the *right* of the decimal point and 1.73 has two digits to the *right* of the decimal point. The product should have three digits to the *right* of the decimal point. ●

• **EXAMPLE 2**   Multiply: 30.6×4200.

$$
\begin{array}{r}
30.6 \\
4200 \\
\hline
61\,200 \\
1224 \\
\hline
128520.0
\end{array}
$$

DIVIDING TWO DECIMAL FRACTIONS

STEP 1 Use the same form as in dividing two whole numbers.
STEP 2 Multiply both the divisor and the dividend (numerator and denominator) by a power of ten that makes the divisor a whole number.
STEP 3 Divide as you did with whole numbers, and place the decimal point in the quotient directly above the decimal point in the dividend.

• **EXAMPLE 3**   Divide 24.32 by 6.4.

$$
\frac{24.32}{6.4} \times \frac{10}{10} = \frac{243.2}{64}
$$

$$
\begin{array}{r}
3.8 \\
64\overline{)243.2} \\
192 \\
\hline
512 \\
512
\end{array}
$$

Or

$$\begin{array}{r} 3.8 \\ 6.4.\overline{)24.3.2} \\ \underline{19\,2} \\ 5\,1\,2 \\ \underline{5\,1\,2} \end{array}$$

Moving the decimal point one place to the right in both the divisor and dividend here is the same as multiplying the numerator and denominator by 10 above.

• **EXAMPLE 4**   Divide 75.1 by 1.62 and round to the nearest hundredth.

To round to the nearest hundredth, you must carry the division out to the thousandths place and then round to hundredths. We show two methods.

Method 1: $\dfrac{75.1}{1.62} \times \dfrac{100}{100} = \dfrac{7510}{162}$

$$\begin{array}{r} 46.358 \\ 162\,\overline{)7510.000} \\ \underline{648} \\ 1030 \\ \underline{972} \\ 580 \\ \underline{486} \\ 940 \\ \underline{810} \\ 1300 \\ \underline{1296} \\ 4 \end{array}$$

Method 2:

$$\begin{array}{r} 46.358 \\ 1.62.\overline{)75.10.000} \\ \underline{64\,8} \\ 10\,30 \\ \underline{9\,72} \\ 58\,0 \\ \underline{48\,6} \\ 9\,40 \\ \underline{8\,10} \\ 1\,300 \\ \underline{1\,296} \\ 4 \end{array}$$

Moving the decimal point two places to the right in both the divisor and dividend here is the same as multiplying both the numerator and denominator by 100 in Method 1.

In both methods, you need to add zeros after the decimal point and carry the division out to the thousandths place. Then round to the nearest hundredth. This gives 46.36 as the result.

• **EXAMPLE 5**   A gasoline station is leased for $1155 per month. How much gasoline must be sold each month to make the cost of the lease equal to 3.5¢ ($0.035) per gallon?

Divide the cost of the lease per gallon into the cost of the lease per month.

$$
0.035. \overline{) \begin{array}{r} 33\ 000. \\ 1155.000. \end{array}}
$$
$$
\begin{array}{r}
105 \\ \hline
105 \\
105
\end{array}
$$

That is, 33,000 gal of gasoline must be sold each month.

• **EXAMPLE 6**

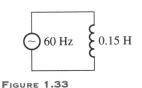

FIGURE 1.33

The inductive reactance (in ohms, Ω) in an ac circuit equals the product of 2π times the frequency (in hertz, Hz, that is, cycles/second) times the inductance (in henries, H). Find the inductive reactance in the ac circuit in Figure 1.33. (Use the π button on your calculator, or use $\pi = 3.14$.)

The inductive reactance is

$$2\pi \times \text{frequency} \times \text{inductance}$$
$$2 \times \pi \times \quad 60 \quad \times \quad 0.15 \quad = 56.5\ \Omega \text{ (rounded to 3 significant digits)}$$

• **EXAMPLE 7**

The effect of both resistance and inductance in a circuit is called *impedance*. Ohm's law for an ac circuit states that the current (in amps, A) equals the voltage (in volts, V) divided by the impedance (in ohms, Ω). Find the current in a 110-V ac circuit that has an impedance of 65 Ω.

$$
\begin{aligned}
\text{current} &= \text{voltage} \div \text{impedance} \\
&= 110 \quad \div \quad 65 \\
&= 1.69\ \text{A (rounded to 3 significant digits)}
\end{aligned}
$$

• **EXAMPLE 8**

A sprayer tank holds 350 gal. Suppose 20 gal of water and 1.25 gal of pesticide are applied to each acre. **a.** How many acres can be treated on one tankful? **b.** How much pesticide is needed per tankful?

a. To find the number of acres treated on one tankful, divide the number of gallons of water *and* pesticide into the number of gallons of a full tank.

$$
21.25. \overline{) \begin{array}{r} 16.4 \\ 350.00.0 \end{array}} \quad \text{or approximately 16 acres/tankful}
$$
$$
\begin{array}{r}
212\ 5 \\ \hline
137\ 50 \\
127\ 50 \\ \hline
10\ 00\ 0 \\
8\ 50\ 0 \\ \hline
1\ 50\ 0
\end{array}
$$

b. To find the amount of pesticide needed per tankful, multiply the number of gallons of pesticide applied per acre times the number of acres treated on one tankful.

$$
\begin{array}{r}
1.25 \\
16 \\ \hline
7\ 50 \\
12\ 5 \\ \hline
20.00
\end{array}
$$
or approximately 20 gal/tankful

Using a Calculator to Multiply and Divide Decimal Fractions

• **EXAMPLE 9**   Multiply: $8.23 \times 65 \times 0.4$.

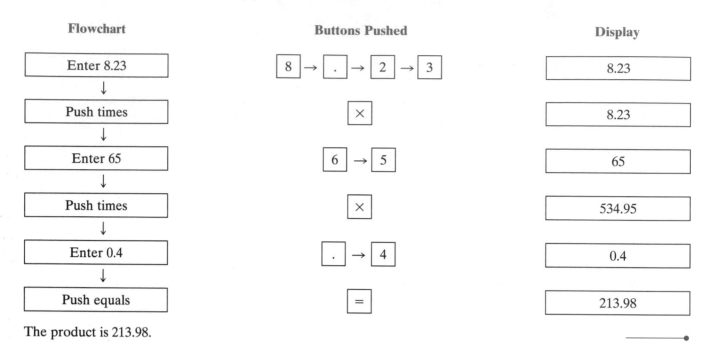

The product is 213.98.

To divide numbers using a calculator, follow the steps in the example below.

• **EXAMPLE 10**   Divide: $3.69 \div 8.2$.

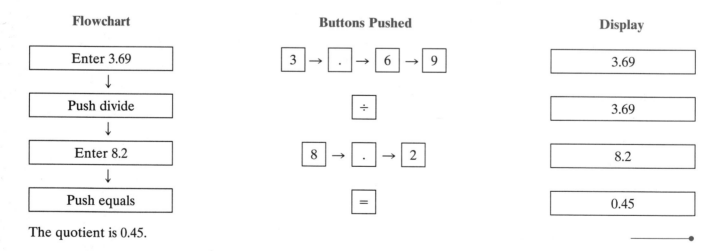

The quotient is 0.45.

• **EXAMPLE 11**   Evaluate: $\dfrac{(18.5)(48)}{(0.25)(240)}$.

Flowchart	**Buttons Pushed**	**Display**
Enter 18.5	$1 \to 8 \to . \to 5$	18.5
Push times	$\times$	18.5
Enter 48	$4 \to 8$	48
Push divide	$\div$	888
Enter 0.25	$. \to 2 \to 5$	0.25
Push divide	$\div$	3552
Enter 240	$2 \to 4 \to 0$	240
Push equals	$=$	14.8

The result is 14.8.

Exercises 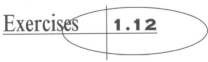 1.12

Multiply:

1.   3.7
   0.15

2. 14.1
   1.7

3. 25.03
   0.42

4. 4.162
   3.14

5. 3050
   5.04

6. 4000
   6.75

7. 5800
   1600

8. 90,000
   0.00705

Divide:

9. $36 \div 1.2$

10. $5.1 \div 1.7$

11. $0.6 \div 0.04$

12. $14.356 \div 0.74$

13. $16.2932 \div 0.11$

14. $328.314 \div 2.1$

Divide and round to the nearest hundredth:

15. $17,500 \div 70.5$

16. $7900 \div 1.52$

17. $75,000 \div 20.4$

18. $1850 \div 0.75$

19. In Illustration 1, find **a.** the perimeter of the outside square and **b.** the length of the center line, ℓ.

20. In Illustration 2, find the perimeter of the octagon, which has eight equal sides.

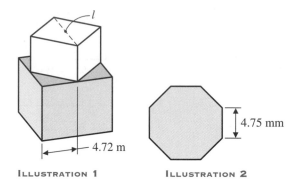

ILLUSTRATION 1   ILLUSTRATION 2

21. The pitch p of a screw is the reciprocal of the number of threads per inch n; that is, $p = \frac{1}{n}$. If the pitch is 0.0125, find the number of threads per inch.

22. A 78-ft cable is to be cut into 3.25-ft lengths. Into how many such lengths can the cable be cut?

23. A steel rod 32.63 in. long is to be cut into 8 pieces. Each piece is 3.56 in. long. Each cut wastes 0.15 in. of rod in shavings. How many inches of the rod are left?

24. How high is a pile of 32 metal sheets if each sheet is 0.045 in. thick?

25. How many metal sheets are in a stack that measures 18 in. high if each sheet is 0.0060 in. thick?

26. A building measures 45 ft 3 in. by 64 ft 6 in. inside. How many square feet of possible floor space does it contain?

27. The cost of excavation is $0.75/yd^3$. Find the cost of excavating a basement 87 feet long, 42 ft wide, and 8 ft deep.

28. Each cut on a lathe is 0.018 in. deep. How many cuts would be needed to turn down 2.640-in. stock to 2.388 in.?

29. Find the total length of the crankshaft shown in Illustration 3.

30. A shop foreman may spend $335 on overtime to complete a job. Overtime pay is $16.75 per hour. How many hours of overtime may he use?

31. Find the total piston displacement of an eight-cylinder engine if each piston displaces a volume of 56.25 in^3.

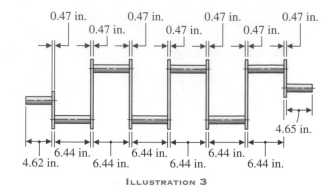

ILLUSTRATION 3

32. Find the total piston displacement of a six-cylinder engine if each piston displaces 0.9 litres (L).

33. A four-cylinder engine has a total displacement of 2.0 L. Find the displacement of each piston.

34. An eight-cylinder engine has a total displacement of 318 in^3. Find the displacement of each piston.

35. The diameter of a new piston is 4.675 in. The average wear per 10,000 mi is 0.007 in. *uniformly over the piston.* **a.** Find the average wear after 80,000 mi. **b.** Find the average diameter of the piston after 100,000 mi.

36. A certain job requires 500 man-hours to complete. How many days will it take for five people working eight hours per day to complete the job?

37. How many gallons of herbicide are needed for 150 acres of soybeans if 1.6 gal/acre are applied?

38. Suppose 10 gal of water and 1.7 lb of pesticide are to be applied per acre. **a.** How much pesticide would you put in a 300-gal spray tank? **b.** How many acres can be covered with one tankful? (Assume the pesticide dissolves in the water and has no volume.)

39. A cattle feeder buys some feeder cattle, which average 550 lb at $55/hundred weight (that is, $55 per hundred pounds, or 55¢/lb). His cost of adding 500 lb of weight gain while the cattle are in his feedlot is $39/hundredweight. The price he receives when he sells them as slaughter cattle is $60/hundredweight. What is his net profit per head?

40. Malathion insecticide is to be applied at a rate of 2 pt/100 gal. of water. How many pints are needed for a tank that holds 20 gal? 60 gal? 150 gal? 350 gal? (Assume the insecticide dissolves in the water and has no volume.)

Find the inductive reactance in each ac circuit (see Example 6):

41.
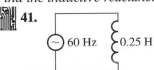

42.
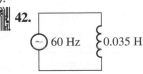

Power (in watts, W) equals voltage times current. Find the power in each circuit:

43.
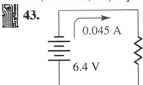

44.
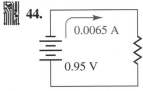

45. Find the current in a 220-V ac circuit that has an impedance of 35.5 Ω. (See Example 7.)

46. A flashlight bulb is connected to a 1.5-V dry cell. If it draws 0.25 A, what is its resistance? (Resistance equals voltage divided by current.)

47. A lamp that requires 0.84 A of current is connected to a 115-V source. What is the lamp's resistance?

48. A heating element operates on a 115-V line. If it has a resistance of 18 Ω, what current does it draw? (Current equals voltage divided by resistance.)

49. A nurse gives three tablets of glyceryl trinitrate, containing 0.150 grains each. How many grains are given?

50. A nurse gives two tablets of ephedrine, containing 0.75 grains each. How many grains are given?

51. An order reads 0.5 mg of digitalis, and each tablet contains 0.1 mg. How many tablets should be given?

52. An order reads 1.25 mg of digoxin, and the tablets on hand are 0.25 mg. How many tablets should be given?

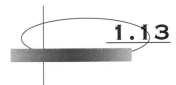

1.13   Percent

Percent is the comparison of any number of parts to 100 parts. The word *percent* means "per hundred." The symbol for percent is %.

You wish to put milk in a pitcher so that it is 25% "full" (Figure 1.34a). First, imagine a line drawn down the side of the pitcher. Then imagine the line divided into 100 equal parts. Each mark shows 1%: that is, each mark shows one out of 100 parts. Finally, count 25 marks from the bottom. The amount of milk below the line is 25% of what the pitcher will hold. Note that 100% is a full, or one whole, pitcher of milk.

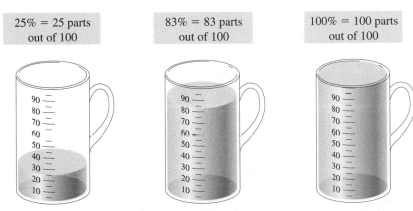

(a) This pitcher is 25% full.   (b) This pitcher is 83% full.   (c) This pitcher is 100% full.

FIGURE 1.34

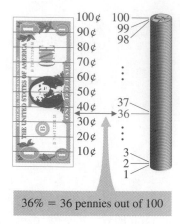

36% = 36 pennies out of 100

FIGURE 1.35

One dollar equals 100 cents or 100 pennies. Then, 36% of one dollar equals 36 of 100 parts, or 36 cents or 36 pennies. (See Figure 1.35.)

To save 10% of your salary, you would have to save $10 out of each $100 earned.

When the United States government spends 22% of its budget on its debt, interest payments are taking $22 out of every $100 the government collects.

A salesperson who earns a commission of 8% receives $8 out of each $100 of goods he or she sells.

A car's radiator holds a mixture that is 25% antifreeze. That is, in each hundred parts of mixture, there are 25 parts of pure antifreeze.

A state charges a 5% sales tax. That is, for each $100 of goods that you buy, a tax of $5 is added to your bill. The $5, a 5% tax, is then paid to the state.

Just remember: *percent* means "per hundred."

Changing a Percent to a Decimal

Percent means the number of parts per 100 parts. Any percent can be written as a fraction with 100 as the denominator.

• **EXAMPLE 1**   Change each percent to a fraction and then to a decimal.

a. $75\% = \dfrac{75}{100} = 0.75$ 75 hundredths

b. $45\% = \dfrac{45}{100} = 0.45$ 45 hundredths

c. $16\% = \dfrac{16}{100} = 0.16$ 16 hundredths

d. $7\% = \dfrac{7}{100} = 0.07$ 7 hundredths

CHANGING A PERCENT TO A DECIMAL

To change a percent to a decimal, move the decimal point two places to the *left* (divide by 100). Then remove the percent sign (%).

• **EXAMPLE 2**   Change each percent to a decimal.

a. $44\% = 0.44$ Move the decimal point two places to the *left* and remove
b. $24\% = 0.24$ the percent sign (%).
c. $115\% = 1.15$
d. $5.7\% = 0.057$
e. $0.25\% = 0.0025$
f. $100\% = 1$

If the percent contains a fraction, write the fraction as a decimal. Then proceed as described above.

• **EXAMPLE 3** Change each percent to a decimal.

a. $12\frac{1}{2}\% = 12.5\% = 0.125$

b. $6\frac{3}{4}\% = 6.75\% = 0.0675$

c. $165\frac{1}{4}\% = 165.25\% = 1.6525$

d. $\frac{3}{5}\% = 0.6\% = 0.006$

When we work problems involving percents, we must use the decimal form of the percent, or its equivalent fractional form.

Changing a Decimal to a Percent

Changing a decimal to a percent is the reverse of what we did in Example 1.

• **EXAMPLE 4** Write 0.75 as a percent.

$$0.75 = \frac{75}{100} \qquad \text{75 hundredths}$$

$$= 75\% \qquad \textit{hundredths} \text{ means percent}$$

CHANGING A DECIMAL TO A PERCENT

To change a decimal to a percent, move the decimal point two places to the *right* (multiply by 100). Write the percent sign (%) after the number.

• **EXAMPLE 5** Change each decimal to a percent.

a. $0.38 = 38\%$ Move the decimal point two places to the *right*. Write
b. $0.42 = 42\%$ the percent sign (%) after the number.
c. $0.08 = 8\%$
d. $0.195 = 19.5\%$
e. $1.25 = 125\%$
f. $2 = 200\%$

Changing a Fraction to a Percent

In some problems, we need to change a fraction to a percent.

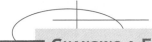

CHANGING A FRACTION TO A PERCENT

1. First, change the fraction to a decimal.
2. Then change this decimal to a percent.

•   **EXAMPLE 6**   Change $\frac{3}{5}$ to a percent.

First, change $\frac{3}{5}$ to a decimal by dividing the numerator by the denominator.

$$\begin{array}{r} 0.6 \\ 5\overline{)3.0} \\ \underline{3\,0} \end{array}$$

Then change 0.6 to a percent by moving the decimal point two places to the right. Write the percent sign (%) after the number.

$$0.6 = 60\%$$

So $\frac{3}{5} = 0.6 = 60\%$.   ————————•

•   **EXAMPLE 7**   Change $\frac{3}{8}$ to a decimal.

$$\begin{array}{r} 0.375 \\ 8\overline{)3.000} \\ \underline{2\,4} \\ 60 \\ \underline{56} \\ 40 \\ \underline{40} \end{array}$$

Then change 0.375 to a percent.

$$0.375 = 37.5\%$$

So $\frac{3}{8} = 0.375 = 37.5\%$.   ————————•

•   **EXAMPLE 8**   Change $\frac{5}{6}$ to a percent.

First, change $\frac{5}{6}$ to a decimal.

$$\begin{array}{r} 0.83 \text{ r } 2 \text{ or } 0.83\frac{2}{6} = 0.83\frac{1}{3} \\ 6\overline{)5.00} \\ \underline{4\,8} \\ 20 \\ \underline{18} \\ 2 \end{array}$$

Note: When the division is carried out to the hundredths place and the remainder is not zero, write the remainder in fraction form, with the remainder over the divisor.

Then change $0.83\frac{1}{3}$ to a percent.

$$0.83\frac{1}{3} = 83\frac{1}{3}\%$$

So $\frac{5}{6} = 0.83\frac{1}{3} = 83\frac{1}{3}\%$.   ————————•

•   **EXAMPLE 9**   Change $1\frac{2}{3}$ to a percent.

First, change $1\frac{2}{3}$ to a decimal.

$$\frac{0.66 \text{ r } 2 \text{ or } 0.66\frac{2}{3}}{3\overline{)2.00}}$$

$$\frac{1\,8}{20}$$

$$\frac{18}{2}$$

That is, $1\frac{2}{3} = 1.66\frac{2}{3}$. Then change $1.66\frac{2}{3}$ to a percent.

$$1.66\frac{2}{3} = 166\frac{2}{3}\%$$

So $1\frac{2}{3} = 1.66\frac{2}{3} = 166\frac{2}{3}\%$.

Changing a Percent to a Fraction

CHANGING A PERCENT TO A FRACTION

1. Change the percent to a decimal.
2. Then change the decimal to a fraction in lowest terms.

• **EXAMPLE 10**   Change 25% to a fraction in lowest terms.

First, change 25% to a decimal by moving the decimal point two places to the *left*. Remove the percent sign (%).

$$25\% = 0.25$$

Then change 0.25 to a fraction. Reduce it to lowest terms.

$$0.25 = \frac{25}{100} = \frac{1}{4}$$

So $25\% = 0.25 = \frac{1}{4}$.

• **EXAMPLE 11**   Change 215% to a mixed number.

First, change 215% to a decimal.

$$215\% = 2.15$$

Then change 2.15 to a mixed number in lowest terms.

$$2.15 = 2\frac{15}{100} = 2\frac{3}{20}$$

So $215\% = 2.15 = 2\frac{3}{20}$.

CHANGING A PERCENT THAT CONTAINS A MIXED NUMBER TO A FRACTION

1. Change the mixed number to an improper fraction.
2. Then multiply this result by $\frac{1}{100}$* and remove the percent sign (%).

_____

*Multiplying by $\frac{1}{100}$ is the same as dividing by 100. This is what we do to change a percent to a decimal.

•   **EXAMPLE 12**   Change $33\frac{1}{3}\%$ to a fraction.

First, change the mixed number to an improper fraction.

$$33\frac{1}{3}\% = \frac{100}{3}\%$$

Then multiply this result by $\frac{1}{100}$ and remove the percent sign (%).

$$\frac{100}{3}\% \times \frac{1}{100} = \frac{\overset{1}{\cancel{100}}}{3} \times \frac{1}{\underset{1}{\cancel{100}}} = \frac{1}{3}$$

So $33\frac{1}{3}\% = \frac{1}{3}$.

•   **EXAMPLE 13**   Change $83\frac{1}{3}\%$ to a fraction.

First, $83\frac{1}{3}\% = \frac{250}{3}\%$.

Then $\frac{250}{3}\% \times \frac{1}{100} = \frac{\overset{5}{\cancel{250}}}{3} \times \frac{1}{\underset{2}{\cancel{100}}} = \frac{5}{6}$.

So $83\frac{1}{3}\% = \frac{5}{6}$.

Exercises   1.13

Change each percent to a decimal:

1. 27%

2. 15%

3. 6%

4. 5%

5. 156%

6. 232%

7. 29.2%

8. 36.2%

9. 8.7%

10. 128.7%

11. 947.8%

12. 68.29%

13. 0.28%

14. 0.78%

15. 0.068%

16. 0.0093%

17. $4\frac{1}{4}\%$

18. $9\frac{1}{2}\%$

19. $\frac{3}{8}\%$

20. $50\frac{1}{3}\%$

Change each decimal to a percent:

21. 0.54

22. 0.25

23. 0.08

24. 0.02

25. 0.62

26. 0.79

27. 2.17

28. 0.345

29. 4.35

30. 0.225

31. 0.185

32. 6.25

33. 0.297

34. 7.11

35. 5.19

36. 0.815

37. 0.0187

38. 0.0342

39. 0.0029

40. 0.00062

Change each fraction to a percent:

41. $\frac{4}{5}$

42. $\frac{3}{4}$

43. $\frac{1}{8}$

44. $\frac{2}{5}$

45. $\frac{1}{6}$

46. $\frac{1}{3}$

47. $\frac{4}{9}$

48. $\frac{3}{7}$

49. $\frac{3}{5}$

50. $\frac{5}{6}$

51. $\frac{13}{40}$

52. $\frac{17}{50}$

53. $\frac{7}{16}$

54. $\frac{15}{16}$

55. $\frac{96}{40}$

56. $\frac{100}{16}$

57. $1\frac{3}{4}$

58. $2\frac{1}{3}$

59. $2\frac{5}{12}$

60. $5\frac{3}{8}$

Change each percent to a fraction or a mixed number in lowest terms:

61. 75%   **65.** 60%   **69.** 275%   **73.** $10\frac{3}{4}\%$ **77.** $17\frac{1}{4}\%$

62. 45%   **66.** 15%   **70.** 325%   **74.** $13\frac{2}{5}\%$ **78.** $6\frac{1}{3}\%$

63. 16%   **67.** 93%   **71.** 125%   **75.** $10\frac{7}{10}\%$ **79.** $16\frac{1}{6}\%$

64. 80%   **68.** 32%   **72.** 150%   **76.** $40\frac{7}{20}\%$ **80.** $72\frac{1}{8}\%$

1.14   Percentage, Base, and Rate

Any percent problem calls for finding one of three things:

    1. the percentage,   **2.** the rate (percent), or   **3.** the base.

Such problems are solved using one of three percent formulas. In these formulas, we let

    P = the percentage
    R = the rate (percent)
    B = the base

The following may help you identify which letter stands for each given number and the unknown in a problem:

    1. The rate, R, usually has either a percent sign (%) or the word *percent* with it.
    2. The base, B, is usually the whole (or entire) amount. The base is often the number that follows the word *of*.
    3. The percentage, P, is usually some fractional part of the base, B. If you identify R and B first, then P will be the number that is not R or B.

Note: The base and the percentage should have the same unit(s) of measure.

• **EXAMPLE 1**   Given: 25% of \$80 is \$20. Identify R, B, and P.

    R is 25%.   25 is the number with a percent sign. Remember to change 25% to the decimal 0.25 for use in a formula.

    B is \$80.   \$80 is the whole amount. It also follows the word *of*.
    P is \$20.   \$20 is the part. It is also the number that is not R or B.   ——————•

• **EXAMPLE 2**   Given: 72% of the 75 students who took this course last year are now working; find how many are now working. Identify R, B, and P.

    R is 72%.   72 is the number with a percent sign.
    B is 75 students.   75 is the whole amount. It also follows the word *of*.
    P is the unknown.   The unknown is the number that is some fractional part of the base. It is also the number that is not R or B.   ——————•

Percent Problems: Finding the Percentage

After you have determined which two numbers are known, you find the third or unknown number by using one of three formulas.

FORMULAS FOR FINDING PERCENTAGE, BASE, AND RATE

1. $P = BR$ Use to find the percentage.

2. $B = \dfrac{P}{R}$ Use to find the base.

3. $R = \dfrac{P}{B}$ Use to find the rate or percent.

Note: After you have studied algebra later in the text, you will need to remember only the first formula.

• **EXAMPLE 3**   Find 75% of 180.

$$R = 75\% = 0.75$$
$$B = 180$$
$$P = \text{the unknown; therefore, use Formula 1.}$$
$$P = BR$$
$$P = (180)(0.75)$$
$$= 135$$

• **EXAMPLE 4** $45 is $9\frac{3}{4}\%$ of what amount?

$$R = 9\frac{3}{4}\% = 9.75\% = 0.0975$$
$$B = \text{the unknown; so use Formula 2.}$$
$$P = \$45$$
$$B = \frac{P}{R}$$
$$B = \frac{\$45}{0.0975}$$
$$= \$461.54$$

• **EXAMPLE 5**   What percent of 20 metres is 5 metres?

$$R = \text{the unknown; use Formula 3.}$$
$$B = 20 \text{ m}$$
$$P = 5 \text{ m}$$
$$R = \frac{P}{B}$$
$$R = \frac{5 \text{ m}}{20 \text{ m}}$$
$$= 0.25 = 25\%$$

• **EXAMPLE 6**   Aluminum is 12% of the mass of a given car. This car has 186 kg of aluminum in it. What is the total mass of the car?

$$R = 12\% = 0.12$$
B = the unknown; use Formula 2.
$$P = 186 \text{ kg}$$
$$B = \frac{P}{R}$$
$$B = \frac{186 \text{ kg}}{0.12}$$
$$= 1550 \text{ kg}$$

• **EXAMPLE 7**   A fuse is a safety device with a core. When too much current flows, the core melts and breaks the circuit. The size of a fuse is the number of amperes of current the fuse can safely carry. A given 50-amp (50 A) fuse blows at 20% overload. What is the maximum current the fuse will carry?

First, find the amount of current overload:

$$R = 20\% = 0.20$$
$$B = 50 \text{ A}$$
P = the unknown; use Formula 1.
$$P = BR$$
$$P = (50 \text{ A})(0.20)$$
$$= 10 \text{ A}$$

The maximum current the fuse will carry is the normal current plus the overload:

$$50 \text{ A} + 10 \text{ A} = 60 \text{ A}$$

• **EXAMPLE 8**   Georgia's salary was $360 per week. Then she was given a raise of $30 per week. What percent raise did she get?

R = the unknown; use Formula 3.
$$B = \$360$$
$$P = \$30$$
$$R = \frac{P}{B}$$
$$R = \frac{\$30}{\$360}$$
$$= 0.08\frac{1}{3} = 8\frac{1}{3}\%$$

• **EXAMPLE 9**   Castings are listed at $9.50 each. A 12% discount is given if 50 or more are bought at one time. We buy 60 castings.

a. What is the discount on one casting?
b. What is the cost of one casting?
c. What is the total cost?

a. Discount equals 12% of $9.50.

$$R = 12\% = 0.12$$
$$B = \$9.50$$
$$P = \text{the unknown (the discount)}$$
$$P = BR$$
$$P = (\$9.50)(0.12)$$
$$= \$1.14 \text{ (the discount on one casting)}$$

b. Cost (of one casting) = list − discount

$$= \$9.50 - \$1.14$$
$$= \$8.36$$

c. Total cost = cost of one casting times the number of castings

$$= (\$8.36)(60)$$
$$= \$501.60$$

You may also need to find the percent increase or decrease in a given quantity.

• **EXAMPLE 10**   Mary's hourly wages changed from $10.40 to $11.05. Find the percent increase in her wages.

First, let's find the change in her wages.

$$\$11.05 - \$10.40 = \$0.65$$

Then, this change is what percent of her original wage?

$$R = \frac{P}{B} = \frac{\$0.65}{\$10.40} = 0.0625 = 6.25\%$$

The process of finding the percent increase or percent decrease may be summarized by the following formula:

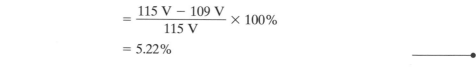

$$\text{percent increase (or percent decrease)} = \frac{\text{the change}}{\text{the original value}} \times 100\%$$

• **EXAMPLE 11**   Normal ac line voltage is 115 volts (V). Find the percent decrease if the line voltage drops to 109 V.

$$\text{percent decrease} = \frac{\text{the change}}{\text{the original value}} \times 100\%$$
$$= \frac{115\text{ V} - 109\text{ V}}{115\text{ V}} \times 100\%$$
$$= 5.22\%$$

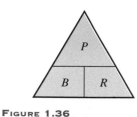

FIGURE 1.36

The triangle in Figure 1.36 can be used to help you remember the three percent formulas, as follows:

1. $P = BR$ To find the percentage, cover P; B and R are next to each other on the same line, as in multiplication.

2. $B = \dfrac{P}{R}$ To find the base, cover B; P is over R, as in division.

3. $R = \dfrac{P}{B}$ To find the rate, cover R; P is over B, as in division.

Exercises | 1.14

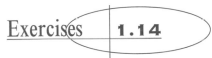

Identify the rate, R, the base, B, and the percentage, P, in each statement (do not solve the problem):

1. 60 is 25% of 240.

2. $33\frac{1}{3}$% of 300 is 100.

3. 40% of 270 is 108.

4. 72 is 15% of 480.

5. At plant A, 4% of the tires made are defective. Plant A made 28,000 tires. How many tires were defective?

6. On the last test, 25 of the 28 students got passing grades. What percent of students passed?

7. A girls' volleyball team won 60% of its games. The team won 21 games. How many games did it play?

8. A rancher loses 10% of his herd every winter due to weather. He has a herd of 15,000. How many does he expect to lose this winter?

9. An electronic firm finds that 6% of the resistors it makes are defective. There were 2050 defective resistors. How many resistors were made?

10. The interest on a $500 loan is $90. What is the rate of interest?

When finding the percent, round to the nearest tenth of a percent when necessary:

11. What percent of $2080 is $208?

12. The number 2040 is 7.5% of what number?

13. What percent of 5280 ft is 880 yd?

14. 0.35 mi is 4% of what amount?

15. $72 is 4.5% of what amount?

16. What percent of 7.15 is 3.5?

17. Find 235% of 48.

18. What percent of $\frac{1}{8}$ is $\frac{1}{15}$?

19. Find 28% of 32 volts (V).

20. Find 110% of 50.

21. The application rate of Aatrax 80W is $2\frac{3}{4}$ lb/acre. How many pounds are needed for 160 acres of corn? The "80W" means that it is 80% active ingredients by weight. How many pounds of active ingredients will be applied? How many pounds of inert ingredients will be applied?

22. U.S. soybeans average 39% protein. A bushel of soybeans weighs 60 lb. How many pounds of protein are in a bushel? A 120-acre field yields 45 bu/acre. How many pounds of protein does that field yield?

23. A dairy cow produced 7310 lb of milk in a year. A gallon of milk weighs 8.6 lb. How many gallons of milk did the cow produce? The milk tested at 4.2% butterfat. How many gallons of butterfat did the cow produce?

24. You need 15% of a 60-mg tablet. How many mg would you take?

25. Mary needs to give 40% of a 0.75-grain tablet. How many grains does she give?

26. You need 0.15% of 2000 mL. How many mL do you need?

27. What percent of 0.600 grain is 0.150 grain?

28. During a line voltage surge, the normal ac voltage increased from 115 V to 128 V. Find the percent increase.

29. During manufacturing, the pressure in a hydraulic line increases from 75 lb/in^2 to 115 lb/in^2. What is the percent increase in pressure?

30. The value of Caroline's house decreased from $72,500 to $58,500 when the area's major employer closed the local plant and moved to another state. Find the percent decrease in the value of her house.

31. Due to wage concessions, Bill's hourly wages dropped from $12.65 to $10.85. Find the percent decrease in his wages.

32. A building has 28,000 ft^2 of floor space. When an addition of 6500 ft^2 is built, what is the percent increase in floor space?

33. A business computer package that usually sells for $675 is on sale for $595. Find the percent decrease in its selling price.

34. A machinist is hired at $12.15 per hour. At the end of a six-month probationary period, the wage will increase by 32%. If the machinist successfully completes the apprenticeship, what will the pay be per hour?

35. An *invoice* is an itemized list of goods and services specifying the price and terms of sale. Illustration 1 shows an invoice for parts and labor for an addition to a home for the week indicated. Complete the invoice.

Jose's Plumbing Supply

120 East Main Street   Poughkeepsie, NY 12600

Satisfaction Guaranteed   Quality Since 1974

Date: _____ 6/25 — 6/29

Name: _____ Gary Jones

Address: _____ 2630 E. Elm St.

City: _____ Poughkeepsie, NY 12600

Quantity	Item	Cost/Unit	Total Cost
22 ea	3/4" fittings	$0.59	
14 ea	3/4" nozzles	$2.09	
12 ea	3/4" 90° ells	$0.77	
6 ea	3/4" faucets	$4.69	
6 ea	3/4" valves	$6.45	
6 ea	3/4" unions	$1.96	
5 ea	3/4" T-joints	$1.09	
4 ea	3/4" 45° ells	$1.29	
120 ft	3/4" type M copper pipe	$0.59/ft	
32 h	Labor	18.00/h	
		Total	
		Less 5% Cash Discount Net 30 Days	
		Net Total	

ILLUSTRATION 1

36. Illustration 2 shows an invoice for grain sold at a local elevator. Complete the invoice.

FARMERS TERMINAL GRAIN COMPANY

Beardstown, Illinois 62618   Since 1893

Customer name: ___Shaw Farms, Inc.___   Account No. ___3786___

Date	Gross wt–pounds	Weight of empty truck	Net wt–pounds	Type of grain	No. of bushels*	Price/bu	Amount
7/2	21560	9160	12400	Wheat	207	$3.17	$656.19
7/3	26720	9240		Wheat		3.23	
7/5	20240	7480		Wheat		3.20	
7/6	28340	9200		Wheat		3.22	
7/8	26760	9160		Wheat		3.25	
7/8	17880	7485		Wheat		3.25	
10/1	25620	9080		Soybeans		6.20	
10/1	21560	7640		Soybeans		6.20	
10/2	26510	9060		Soybeans		6.24	
10/2	22630	7635		Soybeans		6.24	
10/4	22920	9220		Soybeans		6.35	
10/5	20200	7660		Soybeans		6.28	
10/6	25880	9160		Soybeans		6.40	
10/7	21300	7675		Soybeans		6.41	
10/8	18200	7665		Soybeans		6.45	
10/12	26200	9150		Corn		2.31	
10/12	22600	7650		Corn		2.31	
10/13	27100	9080		Corn		2.37	
10/15	22550	7635		Corn		2.25	
10/15	23600	7680		Corn		2.25	
10/17	26780	9160		Corn		2.36	
10/18	28310	9200		Corn		2.43	
10/21	21560	7665		Corn		2.33	
10/22	25750	9160		Corn		2.31	
						TOTAL	

*Round to the nearest bushel. *Note:* Corn weighs 56 lb/bu; soybeans weigh 60 lb/bu; wheat weighs 60 lb/bu.

ILLUSTRATION 2

37. Complete the feed invoice shown in Illustration 3.

38. Complete the electronics parts invoice shown in Illustration 4.

SCFC — South County Feed Company

Highway 51 South   Janesville, Wisconsin 53545

CUSTOMER: Rolling Acres, Inc.   DATE: 8/5

QUANTITY	ITEM	PRICE	DISCOUNT	AMOUNT
600 lb	Pig starter	$24.00/100 lb	$0.25/100 lb	
500 lb	Beef grower	10.60/100 lb	0.11/100 lb	
14,000 lb	Bulk feed	14.75/100 lb	0.15/100 lb	
800 lb	Molasses	6.20/100 lb	—	
6	Salt blocks	0.85/50# block	—	
			Subtotal	
			2% discount (if paid in 30 days)	
			TOTAL	

ILLUSTRATION 3

APPLIANCE DISTRIBUTORS INCORPORATED
1400 West Elm Street St. Louis, Missouri 63100

Sold to: Ed's TV   Date: 9/26

Kampsville, IL 62053

Quantity	Description	Unit price	Discount	Net amount
3	67A76-1	$ 8.58	40%	
5	A8934-1	35.10	25%	
5	A8935-1	33.95	25%	
8	A8922-2	33.90	25%	
2	A8919-2X	24.60	20%	
5	700A256	18.80	15%	
Appliance Distributors Incorporated		SUBTOTAL		
		Less 5% if paid in 30 days		
		TOTAL		

ILLUSTRATION 4

39. Many lumberyards write invoices for their lumber by the piece. (See Illustration 5.) Complete the invoice, which is for the rough framing of the shell of a home.

KURT'S LUMBER 400 WEST OAK AKRON, OHIO 44300			
SOLD TO Robert Bennett 32 Park Pl E. Akron 44305			DATE 5/16

QUANTITY	DESCRIPTION	UNIT PRICE	TOTAL
66	2" x 4" x 16', fir, plate material	$ 4.89	
30	2" x 4" x 10', fir, plate studs	2.79	
14	2" x 4" x 8', fir, knee wall studs	1.99	
17	2" x 6" x 12', fir, kit. ceiling joists	5.29	
4	2" x 12" x 12', fir, kitchen girders	11.10	
9	2" x 6" x 10', fir, kitchen rafters	4.29	
7	2" x 4" x 12', fir, collar beams	3.59	
10	2" x 8" x 12', fir, 2nd floor joists	6.69	
6	2" x 8" x 16', fir, 2nd floor joists	8.58	
11	2" x 8" x 20', fir, 2nd floor joists	11.88	
15	4' x 8' x 3/4", T & G plywood	19.85	
27	2" x 8" x 18', fir, kitchen and living room rafters	11.35	
7	2" x 8" x 16', fir, kitchen and living room rafters	8.95	
1	2" x 10" x 22', fir, kitchen and living room ridge	18.90	
10	1" x 8" x 14', #2 white pine, sub facia	5.85	
27	2" x 8" x 22', fir, bedroom rafters	13.55	
7	2" x 8" x 16', fir, dormer rafters	8.95	
1	2" x 10" x 20', fir, bedroom ridge	15.85	
7	2" x 6" x 20', fir, bedroom ceiling joists	10.19	
8	2" x 6" x 8', fir, bedroom ceiling joists	3.29	
3	2" x 12" x 14', fir, stair stringers	12.89	
80	4' x 8' x 1/2", COX roof deck plywood	6.99	
7	rolls #15 felt building paper	9.95	
1	keg 50 lb #16 cement coated nails	34.50	
1	keg 50 lb simplex roofing nails	64.25	
1	keg 50 lb gold roofing nails	39.75	
250	precut fir studs	1.15	

KURT'S LUMBER

Subtotal _____

Less 2% cash discount _____

Subtotal _____

4 3/4% sales tax _____

NET TOTAL _____

ILLUSTRATION 5

Unit 1C   Review

1. Change $1\frac{5}{8}$ to a decimal.

Perform the indicated operations:

3. $4.206 + 0.023 + 5.9$

4. $120 - 3.065$

5. $12.1 - 6.25 + 0.004$

6. Find the missing dimension in the figure in Illustration 1.

7. Find the perimeter of the figure in Illustration 1.

8. Round 45.0649 to **a.** the nearest tenth and **b.** the nearest hundredth.

9. Round 45.0649 to **a.** three significant digits and **b.** four significant digits.

10. Multiply: 4.606×0.025

11. Divide: $45.24 \div 2.4$

12. A cable 18.5 in. long is to be cut into lengths of 2.75 in. each. How many cables of this length can be cut? How much of the cable is left?

2. Change 0.45 to a common fraction in lowest terms.

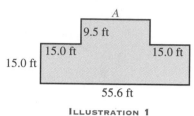

ILLUSTRATION 1

13. Change 25% to a decimal.

14. Change 0.724 to a percent.

15. Find 16.5% of 420.

16. 240 is 12% of what number?

17. What percent of 240 yd is 96 yd?

18. Jean makes $16.50/h. If she receives a raise of 6%, find her new wage.

Chapter 1   Review

1. Add: $435 + 2600 + 18 + 5184 + 6$

2. Subtract: $\begin{array}{r}60,000 \\ 4,803\end{array}$

3. Multiply: 7060×1300

4. Divide: $68,040 \div 300$

5. Evaluate: $12 - 3(5 - 2)$

6. Evaluate: $(6 + 4)8 \div 2 + 3$

7. Evaluate: $18 \div 2 \times 5 \div 3 - 6 + 4 \times 7$

8. Evaluate: $18/(5 - 3) + (6 - 2) \times 8 - 10$

9. Find the area of the figure in Illustration 1.

10. Find the volume of the figure in Illustration 2.

11. Given the formula $C = \frac{5}{9}(F - 32)$ and $F = 50$, find C.

12. Given the formula $P = \frac{Fs}{t}$, $F = 600$, $s = 50$, and $t = 10$, find P.

13. Is 460 divisible by 3?

14. Find the prime factorization of 54.

15. Find the prime factorization of 330.

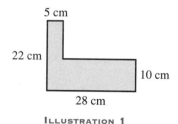

ILLUSTRATION 1

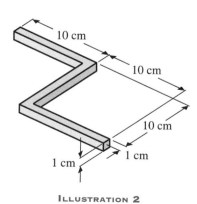

ILLUSTRATION 2

Simplify:

16. $\dfrac{36}{56}$

17. $\dfrac{180}{216}$

Change each to a mixed number in simplest form:

18. $\dfrac{25}{6}$

19. $3\dfrac{18}{5}$

Change each mixed number to an improper fraction:

20. $2\dfrac{5}{8}$

21. $3\dfrac{7}{16}$

Perform the indicated operations and simplify:

22. $\dfrac{3}{8} + \dfrac{7}{8} + \dfrac{6}{8}$

23. $\dfrac{1}{4} + \dfrac{5}{12} + \dfrac{5}{6}$

24. $\dfrac{29}{36} - \dfrac{7}{30}$

25. $5\dfrac{3}{16} + 9\dfrac{5}{12}$

26. $6\dfrac{3}{8} - 4\dfrac{7}{12}$

27. $18 - 6\dfrac{2}{5}$

28. $16\dfrac{2}{3} + 1\dfrac{1}{4} - 12\dfrac{11}{12}$

29. $\dfrac{5}{6} \times \dfrac{3}{10}$

30. $3\dfrac{6}{7} \times 4\dfrac{2}{3}$

31. $\dfrac{3}{8} \div 6$

32. $\dfrac{2}{3} \div 1\dfrac{7}{9}$

33. $1\dfrac{4}{5} \div 1\dfrac{9}{16} \times 11\dfrac{2}{3}$

34. Find dimensions A and B in the figure in Illustration 3.

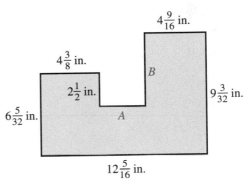

ILLUSTRATION 3

Fill in each blank:

35. 6 lb 9 oz = _____ oz

36. 168 ft = _____ in.

37. 72 ft = _____ yd

38. 36 mi = _____ yd

Write each common fraction as a decimal:

39. $\dfrac{9}{16}$

40. $\dfrac{5}{12}$

Change each decimal to a common fraction or a mixed number and simplify:

41. 0.45

42. 19.625

Perform the indicated operations:

43. 8.6 + 140 + 0.048 + 19.63

44. 25 + 16.3 − 18 + 0.05 − 6.1

45. 86.7 − 18.035

46. 34 − 0.28

47. 0.605 × 5300

48. 18.05 × 0.106

49. 74.73 ÷ 23.5

50. 9.27 ÷ 0.45

51. Round 248.1563 to **a.** the nearest hundred, **b.** the nearest tenth, and **c.** the nearest ten.

52. Round 5.64908 to **a.** the nearest tenth, **b.** the nearest hundredth, and **c.** the nearest ten-thousandth.

Change each percent to a decimal:

53. 15%

54. $8\frac{1}{4}\%$

Change each decimal to a percent:

55. 0.065

56. 1.2

57. What is $8\frac{3}{4}\%$ of $12,000?

58. In a small electronics business, the overhead is $16,000 and the gross income is $42,000. What percent is the overhead?

59. A new tire has a tread depth of $\frac{13}{32}$ in. At 16,000 miles, the tread depth is $\frac{11}{64}$ in. What percent of the tread is left?

60. A farmer bales 60 tons of hay, which are 20% moisture. How many tons of dry matter does he harvest?

Chapter 1 Test

1. Add:
47 + 4969 + 7 + 256

2. Subtract: 4000
        484

3. Multiply: 4070
         635

Evaluate each expression.

6. 8 + 2(5 × 6 + 8)

7. 15 − 9 ÷ 3 + 3 × 4

8. Find the area of the figure in Illustration 1.

9. Find the volume of the figure in Illustration 2.

10. Ohm's law states that current (in A) equals voltage (in V) divided by resistance (in Ω). Find the current in the circuit in Illustration 3.

4. Divide: 96,000 ÷ 60

5. Find the total farm income for the past year.

Corn sales	$22,482
Soybean sales	19,008
Wheat sales	5,610
Hog sales	57,659
Beef sales	2,563
Tax refunds	315

ILLUSTRATION 1   ILLUSTRATION 2

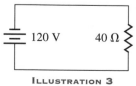

ILLUSTRATION 3

11. If $P = 2\ell + 2w$, $\ell = 20$, and $w = 15$, find P.

13. If $P = 2a + b$, $a = 36$, $b = 15$, find P.

12. If $t = \dfrac{d}{r}$, $d = 1050$, $r = 21$, find t.

Find the prime factorization of each number:

14. 90

15. 220

Simplify:

16. $\dfrac{30}{64}$

21. Subtract: $\dfrac{5}{16} - \dfrac{5}{32}$

17. $\dfrac{28}{42}$

22. Add: $3\dfrac{1}{8}$

$2\dfrac{1}{2}$

$4\dfrac{3}{4}$

18. Change $\dfrac{23}{6}$ to a mixed number.

19. Change $3\dfrac{1}{4}$ to an improper fraction.

23. Subtract: $10\dfrac{1}{8}$

$3\dfrac{5}{16}$

20. Add: $\dfrac{3}{8} + \dfrac{1}{4}$

Perform the indicated operations and simplify:

24. $3\dfrac{5}{8} + 2\dfrac{3}{16} - 1\dfrac{1}{4}$

30. Find the total current in the circuit in Illustration 4.

25. $\dfrac{3}{8} \times \dfrac{16}{27}$

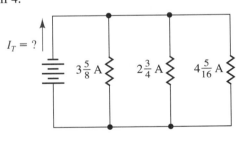

26. $\dfrac{3}{8} \div 3\dfrac{5}{16}$

27. $\dfrac{4}{3} \times \dfrac{1}{8} \times \dfrac{9}{20}$

ILLUSTRATION 4

28. $3\dfrac{5}{8} + 1\dfrac{3}{4} \times 6\dfrac{1}{5}$

29. Given the formula $P = 2\ell + 2w$, $\ell = 4\dfrac{3}{4}$, and $w = 2\dfrac{1}{2}$, find P.

Fill in each blank:

31. 120 ft = _____ yd

33. Express $\dfrac{5}{8}$ as a decimal.

32. 3 lb 5 oz = _____ oz

34. Express 2.12 as a mixed number and simplify.

35. Add: $2.147 + 2.04 + 60 + 0.007 + 0.83$

36. Subtract: $400 - 2.81$

37. Round 27.2847 to the nearest **a.** tenth and **b.** hundredth.

38. Multiply: $\begin{array}{r} 6.12 \\ \underline{1.32} \end{array}$

39. Divide: $6.3\overline{\smash{\big)}0.315}$

40. 58 is 41% of what number?

41. 88 is what percent of 284? (to the nearest tenth)

42. Rachel receives a 6.7% increase in salary. If her salary was $312 per week, what is her new salary?

Signed Numbers and Powers of 10

ALLIED HEALTH
A laboratory technician uses scientific notation in recording the results of medical tests.

MATHEMATICS AT WORK

Signed numbers are used in all areas of mathematics. Understanding the rules for signed numbers is essential to correctly evaluating the applied formulas used in industry. Machinists, draftspersons, engineering technicians, and many other professions use applied formulas involving signed numbers.

Scientific notation using powers of 10 is a way of expressing extremely large or small numerical values. Medical laboratories use these concepts when testing germ cultures and expressing the number of bacteria present in a culture. Environmental laboratories use special formulas involving scientific notation and powers of 10 when testing for toxic wastes and other pollutants in soil, water, and air—often working in parts per billion.

2.1 Addition of Signed Numbers

Technicians use negative numbers in many ways. In an experiment using low temperatures, for example, you would record 10° below zero as −10°. Or consider sea level as zero altitude. If a submarine dives 75 m, you could consider its depth as −75 m (75 m below sea level). See Figure 2.1.

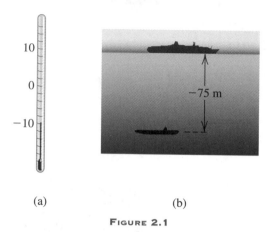

(a)   (b)

FIGURE 2.1

FIGURE 2.2

These measurements indicate a need for numbers other than positive integers, which are the only numbers that we have used up to now. First, draw a vertical number line and label the positive integers above a point representing zero, as shown in Figure 2.2. Using the same spacing, mark off the same number of points below zero. We get points that correspond to the negative integers. We represent negative integers by using the names of the positive integers preceded by a negative (−) sign. For example, −3 is read "negative 3," and −5 is read "negative 5." Each positive integer corresponds to a negative integer. For example, 3 and −3 are corresponding integers. Note that the distances from 3 to 0 and from 0 to −3 are equal.

The *absolute value* of a number is its distance from zero on the number line. Since distance is always considered positive, the absolute value of a number is never negative. We write the absolute value of a number x as $|x|$; it is read "the absolute value of x." Thus, $|x| \geq 0$. ("$\geq$" means "is greater than or equal to.") For example, $|+6| = 6$, $|4| = 4$, and $|0| = 0$. However, if a number is less than 0 (negative), its absolute value is the corresponding positive number. For example, $|-6| = 6$ and $|-7| = 7$.

Remember:

> The absolute value of a number is never negative.

• **EXAMPLE 1** Find the absolute value of each number: **a.** +3, **b.** −5, **c.** 0, **d.** −10, **e.** 15.

a. $|+3| = 3$ The distance between +3 and zero on the number line is 3 units.

b. $|-5| = 5$ The distance between −5 and zero on the number line is 5 units.

c. $|0| = 0$ The distance is 0 units.

d. $|-10| = 10$ The distance between −10 and zero on the number line is 10 units.

e. $|15| = |+15| = 15$ The distance between +15 and zero on the number line is 15 units.

One number is *larger* than another number if the first number is *above* the second on the number line in Figure 2.2. Thus, 5 is larger than 1, 0 is larger than −3, and 2 is larger than −4. Similarly, one number is *smaller* than another if the first number is *below* the second on the number line in Figure 2.2. Thus, 0 is smaller than 3, −1 is smaller than 4, and −5 is smaller than −2.

The use of *signed numbers* (positive and negative numbers) is one of the most important operations that we will study. Signed numbers are used in work with exponents and certain dials. Operations with signed numbers are also essential for success in the basic algebra that follows later.

ADDING TWO NUMBERS WITH LIKE SIGNS (THE SAME SIGNS)

1. To add two positive numbers, add their absolute values. The result is positive. A positive sign may or may not be used before the result. It is usually omitted.

2. To add two negative numbers, add their absolute values and place a negative sign before the result.

• **EXAMPLE 2**   Add:

a. $+2$
    $+3$
    $\overline{+5}$

b. $-\ 4$
    $-\ 6$
    $\overline{-10}$

c. $(+4) + (+5) = +9$

d. $(-8) + (-3) = -11$

ADDING TWO NUMBERS WITH DIFFERENT SIGNS

To add a negative number and a positive number, find the difference of their absolute values. The sign of the number having the larger absolute value is placed before the result.

• **EXAMPLE 3**   Add:

a. $+4$
    -7
    $\overline{-3}$

b. -3
    $+8$
    $\overline{+5}$

c. -9
    $+2$
    $\overline{-7}$

d. $+8$
    -5
    $\overline{+3}$

e. $(+6) + (-1) = +5$

f. $(-8) + (+6) = -2$

g. $(-2) + (+5) = +3$

h. $(+3) + (-11) = -8$

ADDING THREE OR MORE SIGNED NUMBERS

STEP 1 Add the positive numbers.

STEP 2 Add the negative numbers.

STEP 3 Add the sums from Steps 1 and 2 according to the rules for addition of signed numbers.

• EXAMPLE 4   Add $(-8) + (+12) + (-7) + (-10) + (+3)$.

STEP 1	STEP 2	STEP 3
$+12$	$-\ 8$	$+15$
$\underline{+\ 3}$	$-\ 7$	$\underline{-25}$
$+15$	$\underline{-10}$	-10
	-25	

Therefore, $(-8) + (+12) + (-7) + (-10) + (+3) = -10$.

• EXAMPLE 5   Add $(+4) + (-7) + (-2) + (+6) + (-3) + (-5)$.

STEP 1	STEP 2	STEP 3
$+\ 4$	$-\ 7$	$+10$
$\underline{+\ 6}$	$-\ 2$	$\underline{-17}$
$+10$	$-\ 3$	$-\ 7$
	$\underline{-\ 5}$	
	-17	

Therefore, $(+4) + (-7) + (-2) + (+6) + (-3) + (-5) = -7$.

Exercises | 2.1

Find the absolute value of each number:

1. 3   **3.** -6 **5.** $+4$ **7.** 17   **9.** -15
2. -4 **4.** 0   **6.** $+8$ **8.** -37 **10.** 49

Add:

11. $+4$ $\underline{+6}$	**15.** $+5$ $\underline{-7}$	**19.** 12 $\underline{-\ 6}$	**23.** 0 $\underline{4}$	**27.** $+15$ $\underline{-20}$
12. -5 $\underline{-9}$	**16.** -4 $\underline{+6}$	**20.** -12 $\underline{6}$	**24.** $-\ 5$ $\underline{-10}$	**28.** -11 $\underline{-\ 3}$
13. $+9$ $\underline{-2}$	**17.** -3 $\underline{-9}$	**21.** 3 $\underline{9}$	**25.** -7 $\underline{-7}$	**29.** 1 $\underline{-13}$
14. -10 $\underline{+\ 4}$	**18.** $+4$ $\underline{-9}$	**22.** -2 $\underline{0}$	**26.** 18 $\underline{-18}$	**30.** -22 $\underline{+14}$

31. $(-4) + (-5)$ **35.** $(-5) + (+2)$ **39.** $(-10) + (6)$ **43.** $(-2) + (0)$
32. $(+2) + (-11)$ **36.** $(-7) + (-6)$ **40.** $(+4) + (-11)$ **44.** $(0) + (+3)$
33. $(-3) + (+7)$ **37.** $(+7) + (-8)$ **41.** $(-8) + (2)$ **45.** $(9) + (-5)$
34. $(+4) + (+6)$ **38.** $(8) + (-3)$ **42.** $(+3) + (+7)$ **46.** $(+9) + (-9)$

47. $(16) + (-7)$ **53.** $(+1) + (+7) + (-1)$ **59.** $(-4) + (-7) + (-7) + (-2)$
48. $(-19) + (-12)$ **54.** $(-5) + (-9) + (-4)$ **60.** $(-3) + (-9) + (+5) + (+6)$
49. $(-6) + (+9)$ **55.** $(-9) + (+6) + (-4)$ **61.** $(-1) + (-2) + (+9) + (-8)$
50. $(+20) + (-30)$ **56.** $(+8) + (+7) + (-2)$ **62.** $(+6) + (+5) + (-7) + (-3)$
51. $(-1) + (-3) + (+8)$ **57.** $(+8) + (-8) + (+7) + (-2)$ **63.** $(-6) + (+2) + (+7) + (-3)$
52. $(+5) + (-3) + (+4)$ **58.** $(-6) + (+5) + (-8) + (+4)$ **64.** $(+8) + (-1) + (+9) + (+6)$

65. $(-5) + (+1) + (+3) + (-2) + (-2)$ **70.** $(-1) + (-2) + (-6) + (-3) + (-5)$
66. $(+5) + (+2) + (-3) + (-9) + (-9)$ **71.** $(-2) + 8 + (-4) + 6 + (-1)$
67. $(-5) + (+6) + (-9) + (-4) + (-7)$ **72.** $14 + (-5) + (-1) + 6 + (-3)$
68. $(-9) + (+7) + (-6) + (+5) + (-8)$ **73.** $5 + 6 + (-2) + 9 + (-7)$
69. $(+1) + (-4) + (-2) + (+2) + (-9)$ **74.** $(-5) + 4 + (-1) + 6 + (-7)$

75. $(-3) + 8 + (-4) + (-7) + 10$

76. $16 + (-7) + (-5) + 20 + (-5)$

77. $3 + (-6) + 7 + 4 + (-4)$

78. $(-8) + 6 + 9 + (-5) + (-4)$

79. $(-5) + 4 + (-7) + 2 + (-8)$

80. $7 + 9 + (-6) + (-4) + 9 + (-2)$

2.2 Subtraction of Signed Numbers

> ### SUBTRACTING TWO SIGNED NUMBERS
> To subtract signed numbers, change the sign of the number being subtracted and add according to the rules for addition of signed numbers.

• **EXAMPLE 1**   Subtract:

a. Subtract: $\begin{array}{r} +2 \\ +5 \\ \hline -3 \end{array}$ ⟷ Add: $\begin{array}{r} +2 \\ -5 \\ \hline -3 \end{array}$ To subtract, change the sign of the number being subtracted, + 5, and add.

b. Subtract: $\begin{array}{r} -7 \\ -6 \\ \hline -1 \end{array}$ ⟷ Add: $\begin{array}{r} -7 \\ +6 \\ \hline -1 \end{array}$ To subtract, change the sign of the number being subtracted, −6, and add.

c. Subtract: $\begin{array}{r} +\ 6 \\ -\ 4 \\ \hline +10 \end{array}$ ⟷ Add: $\begin{array}{r} +\ 6 \\ +\ 4 \\ \hline +10 \end{array}$

d. Subtract: $\begin{array}{r} -\ 8 \\ +\ 3 \\ \hline -11 \end{array}$ ⟷ Add: $\begin{array}{r} -\ 8 \\ -\ 3 \\ \hline -11 \end{array}$

e. $(+4) - (+6) = (+4) + (-6)$
$= -2$ To subtract, change the sign of the number being subtracted, +6, and add.

f. $(-8) - (-10) = (-8) + (+10) = +2$

g. $(+9) - (-6) = (+9) + (+6) = +15$

h. $(-4) - (+7) = (-4) + (-7) = -11$

> ### SUBTRACTING MORE THAN TWO SIGNED NUMBERS
> When more than two signed numbers are involved in subtraction, change the sign of *each* number being subtracted and add the resulting signed numbers.

• **EXAMPLE 2**   Subtract $(-4) - (-6) - (+2) - (-5) - (+7)$
$= (-4) + (+6) + (-2) + (+5) + (-7)$

STEP 1 $\begin{array}{r} +6 \\ +5 \\ \hline +11 \end{array}$ STEP 2 $\begin{array}{r} -4 \\ -2 \\ -7 \\ \hline -13 \end{array}$ STEP 3 $\begin{array}{r} +11 \\ -13 \\ \hline -2 \end{array}$

Therefore, $(-4) - (-6) - (+2) - (-5) - (+7) = -2$.

When combinations of additions and subtractions of signed numbers occur in the same problem, change *only* the sign of each number being subtracted. Then add the resulting signed numbers.

• **EXAMPLE 3**   Perform the indicated operations:

$$(+4) - (-5) + (-6) - (+8) - (-2) + (+5) - (+1)$$
$$= (+4) + (+5) + (-6) + (-8) + (+2) + (+5) + (-1)$$

STEP 1:   +4   STEP 2   −6   STEP 3   +16
          +5   −8   −15
          +2   −1   +1
          +5   −15
          +16

Therefore, $(+4) - (-5) + (-6) - (+8) - (-2) + (+5) - (+1) = +1$.

• **EXAMPLE 4**   Perform the indicated operations:

$$(-12) + (-3) - (-5) - (+6) - (-1) + (+4) - (+3) - (-8)$$
$$= (-12) + (-3) + (+5) + (-6) + (+1) + (+4) + (-3) + (+8)$$

STEP 1   +5   STEP 2   −12   STEP 3   +18
         +1   −3   −24
         +4   −6   −6
         +8   −3
         +18   −24

Therefore,

$$(-12) + (-3) - (-5) - (+6) - (-1) + (+4) - (+3) - (-8) = -6.$$

Exercises   2.2

Subtract:

1. +4	**5.** +5	**9.** 12	**13.** 0	**17.** +15
+6	−7	− 6	4	−20
2. −5	**6.** −4	**10.** −12	**14.** − 5	**18.** −11
−9	+6	6	−10	− 3
3. +9	**7.** −3	**11.** 3	**15.** −7	**19.** 1
−2	−9	9	−7	−13
4. −10	**8.** +4	**12.** −2	**16.** 18	**20.** −22
+ 4	−9	0	−18	+14

Perform the indicated operations:

21. $(-4) - (-5)$	**27.** $(+7) - (-8)$	**33.** $(-2) - (0)$	**39.** $(-6) - (+9)$
22. $(+2) - (-11)$	**28.** $(8) - (-3)$	**34.** $(0) - (+3)$	**40.** $(+20) - (-30)$
23. $(-3) - (+7)$	**29.** $(-10) - (6)$	**35.** $(9) - (-5)$	**41.** $(+6) - (-3) - (+1)$
24. $(+4) - (+6)$	**30.** $(+4) - (-11)$	**36.** $(+9) - (-9)$	**42.** $(+3) - (-7) - (+6)$
25. $(-5) - (+2)$	**31.** $(-8) - (+2)$	**37.** $(16) - (-7)$	**43.** $(-3) - (-7) - (+8)$
26. $(-7) - (-6)$	**32.** $(+3) - (+7)$	**38.** $(-19) - (-12)$	**44.** $(+3) - (4) - (-9)$

45. $(+3) - (-6) - (+9) - (-8)$

46. $(+10) - (-4) - (6) - (-9)$

47. $(+5) - (-5) + (-8)$

48. $(+1) + (-7) - (-7)$

49. $(-3) + (-5) - (0) - (+7)$

50. $(+4) - (-3) + (+6) - (8)$

51. $(+4) - (-11) - (+12) + (-6)$

52. $(8) - (-6) - (+18) - (4)$

53. $(-7) - (+6) + (-3) - (-2) - (+9)$

54. $(-3) + (-4) + (+7) - (-2) - (+6)$

55. $-9 + 8 - 5 + 6 - 4$

56. $-12 + 2 + 30 - 6$

57. $-8 + 12 - 7 - 4 + 6$

58. $7 + 4 - 8 - 9 + 3$

59. $16 - 18 + 4 - 7 - 2 + 9$

60. $3 - 7 + 5 - 6 - 7 + 2$

61. $8 + 10 - 20 + 4 - 5 - 6 + 1$

62. $5 - 6 - 7 + 2 - 8 + 10$

63. $9 - 7 + 4 + 3 - 8 - 6 - 6 + 1$

64. $-4 + 6 - 7 - 5 + 6 - 7 - 1$

2.3 Multiplication and Division of Signed Numbers

MULTIPLYING TWO SIGNED NUMBERS

1. If the two numbers have the same signs, multiply their absolute values. This product is always positive.
2. If the two numbers have different signs, multiply their absolute values and place a negative sign before the product.

• **EXAMPLE 1**   Multiply:

a. $\begin{array}{r} +2 \\ +3 \\ \hline +6 \end{array}$
b. $\begin{array}{r} -4 \\ -7 \\ \hline +28 \end{array}$
c. $\begin{array}{r} -2 \\ +4 \\ \hline -8 \end{array}$
d. $\begin{array}{r} -\ 6 \\ +\ 5 \\ \hline -30 \end{array}$

e. $(+3)(+4) = +12$ **g.** $(-5)(+7) = -35$

f. $(-6)(-9) = +54$ **h.** $(+4)(-9) = -36$

MULTIPLYING MORE THAN TWO SIGNED NUMBERS

1. If the number of negative factors is even (divisible by 2), multiply the absolute values of the numbers. This product is positive.
2. If the number of negative factors is odd, multiply the absolute values of the numbers and place a negative sign before the product.

• **EXAMPLE 2**   Multiply: $(-11)(+3)(-6) = +198$

The number of negative factors is 2, which is even; therefore, the product is positive.

• **EXAMPLE 3**   Multiply: $(-5)(-4)(+2)(-7) = -280$

The number of negative factors is 3, which is odd; therefore, the product is negative.

DIVIDING TWO SIGNED NUMBERS

1. If the two numbers have the same signs, divide their absolute values. This quotient is always positive.

2. If the two numbers have different signs, divide their absolute values and place a negative sign before the quotient.

Since multiplication and division are related operations, the same rules for signed numbers apply to both operations.

•   **EXAMPLE 4**   Divide:

a. $\dfrac{+12}{+2} = +6$ **b.** $\dfrac{-18}{-6} = +3$ **c.** $\dfrac{+20}{-4} = -5$ **d.** $\dfrac{-24}{+6} = -4$

e. $(+30) \div (+5) = (+6)$ **g.** $(+16) \div (-4) = -4$

f. $(-42) \div (-2) = +21$ **h.** $(-45) \div (+9) = -5$

Exercises | 2.3

Multiply:

1. +4
  +6

2. −5
  −9

3. +9
  −2

4. −10
  + 4

5. +5
  −7

6. −4
  +6

7. −3
  −9

8. +4
  −9

9. 12
  −6

10. −12
    6

11. 3
  9

12. −2
    0

13. 0
  4

14. − 5
  −10

15. −7
  −7

16.   18
  −18

17. +15
  −20

18. −11
  − 3

19.   1
  −13

20. −22
  +14

21. $(+3)(-2)$
22. $(+5)(+7)$
23. $(-6)(-8)$
24. $(-9)(+2)$
25. $(-7)(-3)$
26. $(-4)(+4)$

27. $(-8)(+2)$
28. $(+5)(-3)$
29. $(-6)(-9)$
30. $(+4)(+7)$
31. $(-6)(+1)$
32. $(+4)(-2)$

33. $(-8)(-3)$
34. $(-2)(+6)$
35. $(-3)(-9)$
36. $(8)(-4)$
37. $(-9)(7)$
38. $(-8)(0)$

39. $(9)(-1)$
40. $(-4)(-10)$
41. $(+3)(-2)(+1)$
42. $(-5)(-9)(+2)$
43. $(-3)(+3)(-4)$
44. $(-2)(-8)(-3)$

45. $(+5)(+2)(+3)$
46. $(-4)(+3)(0)(+3)$

47. $(-3)(-2)(4)(-7)$
48. $(-3)(-1)(+1)(+2)$

49. $(-9)(-2)(+3)(+1)(-3)$
50. $(-6)(-2)(-4)(-1)(-2)(+2)$

Divide:

51. $\dfrac{+10}{+2}$

52. $\dfrac{-8}{-4}$

53. $\dfrac{+27}{-3}$

54. $\dfrac{-48}{+6}$

55. $\dfrac{-32}{-4}$

56. $\dfrac{+39}{-13}$

57. $\dfrac{+14}{+7}$

58. $\dfrac{45}{+15}$

59. $\dfrac{-54}{-6}$

60. $\dfrac{+72}{-9}$

61. $\dfrac{-100}{+25}$

62. $\dfrac{+84}{+12}$

63. $\dfrac{-75}{-25}$

64. $\dfrac{+36}{-6}$

65. $\dfrac{+85}{+5}$

66. $\dfrac{-270}{+9}$

67. $\dfrac{+480}{+12}$

68. $\dfrac{-350}{+70}$

69. $\dfrac{-900}{-60}$

70. $\dfrac{+4800}{-240}$

71. $(-49) \div (-7)$
72. $(+9) \div (-3)$
79. $(-96) \div (-12)$

73. $(+80) \div (+20)$
74. $(-60) \div (+12)$

75. $(+45) \div (-15)$
76. $(+120) \div (-6)$
80. $(-800) \div (+25)$

77. $(-110) \div (-11)$
78. $(+84) \div (+6)$

2.4 Signed Fractions

The rules for operations with signed integers also apply to fractions.

• **EXAMPLE 1** $\left(-\dfrac{1}{4}\right) + \left(-\dfrac{3}{16}\right) = \left(-\dfrac{4}{16}\right) + \left(-\dfrac{3}{16}\right) = -\dfrac{7}{16}$

• **EXAMPLE 2** $\dfrac{3}{5} + \left(-\dfrac{2}{3}\right) = \dfrac{9}{15} + \left(-\dfrac{10}{15}\right) = \dfrac{9 - 10}{15} = -\dfrac{1}{15}$

• **EXAMPLE 3** $\left(-\dfrac{4}{9}\right) + \dfrac{2}{3} = \left(-\dfrac{4}{9}\right) + \dfrac{6}{9} = \dfrac{-4 + 6}{9} = \dfrac{2}{9}$

• **EXAMPLE 4** $\left(-2\dfrac{3}{4}\right) + \left(-1\dfrac{5}{6}\right) = \left(-2\dfrac{9}{12}\right) + \left(-1\dfrac{10}{12}\right)$

$$= -3\dfrac{19}{12}$$
$$= -\left(3 + \dfrac{12 + 7}{12}\right)$$
$$= -\left(3 + 1\dfrac{7}{12}\right)$$
$$= -4\dfrac{7}{12}$$

• **EXAMPLE 5** $\left(-\dfrac{5}{9}\right) - \left(-\dfrac{5}{12}\right) = \left(-\dfrac{20}{36}\right) + \left(+\dfrac{15}{36}\right) = \dfrac{-20 + 15}{36} = -\dfrac{5}{36}$

• **EXAMPLE 6** $\left(-1\dfrac{3}{8}\right) - \left(+2\dfrac{5}{6}\right) = \left(-1\dfrac{3}{8}\right) + \left(-2\dfrac{5}{6}\right)$

$$= \left(-1\dfrac{9}{24}\right) + \left(-2\dfrac{20}{24}\right)$$
$$= -3\dfrac{29}{24}$$
$$= -\left(3 + \dfrac{24 + 5}{24}\right)$$
$$= -\left(3 + 1\dfrac{5}{24}\right)$$
$$= -4\dfrac{5}{24}$$

• **EXAMPLE 7** $\left(-3\dfrac{5}{12}\right) - \left(-1\dfrac{2}{3}\right) = \left(-3\dfrac{5}{12}\right) + \left(+1\dfrac{2}{3}\right)$

$$= \left(-3\dfrac{5}{12}\right) + \left(1\dfrac{8}{12}\right)$$

$$= \left(-2\dfrac{17}{12}\right) + \left(1\dfrac{8}{12}\right)$$

$$= -1\dfrac{9}{12}$$

$$= -1\dfrac{3}{4}$$

• **EXAMPLE 8** $\left(-\dfrac{2}{5}\right)\left(-\dfrac{5}{8}\right) = \dfrac{1}{4}$

• **EXAMPLE 9** $\left(\dfrac{4}{15}\right)\left(-\dfrac{5}{2}\right) = -\dfrac{2}{3}$

• **EXAMPLE 10** $\left(-\dfrac{3}{7}\right) \div \left(-\dfrac{9}{14}\right) = \left(-\dfrac{3}{7}\right) \times \left(-\dfrac{14}{9}\right) = \dfrac{2}{3}$

• **EXAMPLE 11** $\left(-\dfrac{11}{15}\right) \div \dfrac{2}{3} = \left(-\dfrac{11}{15}\right) \times \dfrac{3}{2} = -\dfrac{11}{10}$ or $-1\dfrac{1}{10}$

One more rule about fractions will help you.

EQUIVALENT SIGNED FRACTIONS

$$\dfrac{a}{-b} = \dfrac{-a}{b} = -\dfrac{a}{b}$$

That is, a negative fraction may be written in three different but equivalent forms. However, the form $-\dfrac{a}{b}$ is the customary form.

For example, $\dfrac{3}{-4} = \dfrac{-3}{4} = -\dfrac{3}{4}$.

Note: $\dfrac{-a}{-b} = \dfrac{a}{b}$, using the rules for dividing signed numbers.

• **EXAMPLE 12** $\dfrac{3}{-4} + \dfrac{-2}{3} = \left(-\dfrac{3}{4}\right) + \left(-\dfrac{2}{3}\right) = \left(-\dfrac{9}{12}\right) + \left(-\dfrac{8}{12}\right) = \dfrac{-9 + -8}{12} = -\dfrac{17}{12}$

• **EXAMPLE 13** $\dfrac{3}{-4} + \dfrac{2}{3} = \left(-\dfrac{3}{4}\right) + \dfrac{2}{3} = \left(-\dfrac{9}{12}\right) + \left(\dfrac{8}{12}\right) = \dfrac{-9 + 8}{12} = -\dfrac{1}{12}$

• **EXAMPLE 14** $\dfrac{3}{-4} - \dfrac{2}{3} = \left(-\dfrac{3}{4}\right) + \left(-\dfrac{2}{3}\right) = \left(-\dfrac{9}{12}\right) + \left(-\dfrac{8}{12}\right) = \dfrac{-9 - 8}{12} = -\dfrac{17}{12}$

• **EXAMPLE 15** $\left(-\dfrac{1}{4}\right)\left(\dfrac{-3}{5}\right) = \left(-\dfrac{1}{4}\right)\left(-\dfrac{3}{5}\right) = \dfrac{3}{20}$

- **EXAMPLE 16** $\left(\dfrac{-1}{2}\right)\left(\dfrac{3}{4}\right) = \left(-\dfrac{1}{2}\right)\left(\dfrac{3}{4}\right) = -\dfrac{3}{8}$

- **EXAMPLE 17** $\left(\dfrac{-2}{3}\right) \div 3 = \left(-\dfrac{2}{3}\right) \div 3 = \left(-\dfrac{2}{3}\right)\left(\dfrac{1}{3}\right) = -\dfrac{2}{9}$

- **EXAMPLE 18** $\left(\dfrac{-3}{7}\right) \div \dfrac{-5}{6} = \left(-\dfrac{3}{7}\right) \div \left(-\dfrac{5}{6}\right) = \left(-\dfrac{3}{7}\right)\left(-\dfrac{6}{5}\right) = \dfrac{18}{35}$

Exercises | 2.4

Perform the indicated operations and simplify:

1. $\dfrac{1}{8} + \left(-\dfrac{5}{16}\right)$

2. $\left(-\dfrac{2}{3}\right) + \left(-\dfrac{2}{7}\right)$

3. $\dfrac{1}{2} + \left(-\dfrac{7}{16}\right)$

4. $\dfrac{2}{3} + \left(-\dfrac{7}{9}\right)$

5. $\left(-5\dfrac{3}{4}\right) + \left(-6\dfrac{2}{5}\right)$

6. $\left(-1\dfrac{3}{8}\right) + \left(5\dfrac{5}{12}\right)$

7. $\left(-3\dfrac{2}{3}\right) + \left(-\dfrac{4}{9}\right) + \left(4\dfrac{5}{6}\right)$

8. $\left(-\dfrac{3}{4}\right) + \left(-1\dfrac{1}{6}\right) + \left(-1\dfrac{1}{3}\right)$

9. $\left(-\dfrac{1}{4}\right) - \left(-\dfrac{1}{5}\right)$

10. $\left(-\dfrac{2}{9}\right) - \left(+\dfrac{1}{2}\right)$

11. $1\dfrac{3}{8} - \left(+\dfrac{5}{16}\right)$

12. $\left(-\dfrac{1}{3}\right) - \left(+3\dfrac{1}{2}\right)$

13. $\left(+1\dfrac{3}{4}\right) - (-4)$

14. $2\dfrac{3}{4} - \left(-3\dfrac{1}{4}\right)$

15. $\left(-\dfrac{2}{3}\right) + \left(-\dfrac{5}{6}\right) - \dfrac{1}{4}$

16. $\left(-\dfrac{3}{4}\right) - \left(-1\dfrac{2}{3}\right) + \left(-1\dfrac{5}{6}\right)$

17. $-\dfrac{1}{9} \times \dfrac{1}{7}$

18. $\left(-\dfrac{2}{3}\right)\left(-\dfrac{1}{2}\right)$

19. $\left(-3\dfrac{1}{3}\right)\left(-1\dfrac{4}{5}\right)$

20. $\dfrac{21}{8} \times 1\dfrac{7}{9}$

21. $\dfrac{4}{5} \div \left(-\dfrac{8}{9}\right)$

22. $\left(-1\dfrac{1}{4}\right) \div \dfrac{3}{5}$

23. $\left(-\dfrac{7}{9}\right) \div \left(-\dfrac{8}{3}\right)$

24. $2\dfrac{3}{4} \div \left(-3\dfrac{1}{6}\right)$

25. $\left(\dfrac{-1}{4}\right) + \left(\dfrac{1}{-5}\right)$

26. $\left(\dfrac{-4}{5}\right) + \left(-1\dfrac{1}{2}\right)$

27. $\dfrac{3}{4} + \left(\dfrac{-3}{8}\right)$

28. $\left(\dfrac{-3}{2}\right) + \left(\dfrac{-8}{3}\right)$

29. $\dfrac{5}{8} - \left(\dfrac{-5}{8}\right)$

30. $\left(\dfrac{-1}{4}\right) - \left(\dfrac{1}{-5}\right)$

31. $\left(\dfrac{-6}{8}\right) - (-4)$

32. $(-2) - \left(\dfrac{-1}{-4}\right)$

33. $\left(\dfrac{-1}{4}\right)\left(\dfrac{1}{-5}\right)$

34. $(-2)\left(\dfrac{-1}{-4}\right)$

35. $\left(\dfrac{-5}{-8}\right) \times \left(-5\dfrac{1}{3}\right)$

36. $\left(\dfrac{-3}{-4}\right)(-12)$

37. $32 \div \left(\dfrac{-2}{3}\right)$

38. $\left(\dfrac{-4}{9}\right) \div (-2)$

39. $\left(\dfrac{-2}{-3}\right) \div \left(\dfrac{2}{-3}\right)$

40. $\left(-1\dfrac{3}{5}\right) \div \left(-3\dfrac{1}{5}\right)$

41. $\left(\dfrac{-2}{3}\right) + \left(-\dfrac{5}{6}\right) + \dfrac{1}{4} + \dfrac{1}{8}$

42. $\left(\dfrac{-3}{4}\right) + \left(\dfrac{2}{-3}\right) - \left(\dfrac{-1}{-2}\right) - \left(\dfrac{-5}{6}\right)$

43. $\left(\dfrac{-2}{5}\right)\left(\dfrac{3}{-4}\right)\left(\dfrac{-15}{-18}\right)$

44. $\left(-2\dfrac{3}{4}\right) \div \left(1\dfrac{3}{5}\right) \times \left(\dfrac{-2}{5}\right)$

45. $\left(\dfrac{-2}{3}\right) + \left(-\dfrac{1}{2}\right)\left(\dfrac{5}{-6}\right)$

46. $\left(\dfrac{-4}{5}\right) \div \left(-1\dfrac{1}{2}\right) - \left(\dfrac{2}{-5}\right)$

2.5 Powers of 10

MULTIPLYING POWERS OF 10

To multiply two powers of 10, add the exponents as follows:

$$10^a \times 10^b = 10^{a+b}$$

Note: The rules for working with powers of 10 shown in this section also apply to other bases, as shown in Section 5.4.

• **EXAMPLE 1**   Multiply: $(10^2)(10^3)$

Method 1: $(10^2)(10^3) = (10 \cdot 10)(10 \cdot 10 \cdot 10) = 10^5$

Method 2: $(10^2)(10^3) = 10^{2+3} = 10^5$

• **EXAMPLE 2**   Multiply: $(10^3)(10^5)$

Method 1: $(10^3)(10^5) = (10 \cdot 10 \cdot 10) (10 \cdot 10 \cdot 10 \cdot 10 \cdot 10) = 10^8$

Method 2: $(10^3)(10^5) = 10^{3+5} = 10^8$

• **EXAMPLE 3**   Multiply each of the following powers of 10:

a. $(10^9)(10^{12}) = 10^{9+12} = 10^{21}$
b. $(10^{-12})(10^{-7}) = 10^{(-12)+(-7)} = 10^{-19}$
c. $(10^{-9})(10^6) = 10^{(-9)+6} = 10^{-3}$
d. $(10^{10})(10^{-6}) = 10^{10+(-6)} = 10^4$
e. $10^5 \cdot 10^{-8} \cdot 10^4 \cdot 10^{-3} = 10^{5+(-8)+4+(-3)} = 10^{-2}$

DIVIDING POWERS OF 10

To divide two powers of 10, subtract the exponents as follows:

$$10^a \div 10^b = 10^{a-b}$$

• **EXAMPLE 4**   Divide: $\dfrac{10^6}{10^2}$

Method 1: $\dfrac{10^6}{10^2} = \dfrac{\cancel{10} \cdot \cancel{10} \cdot 10 \cdot 10 \cdot 10 \cdot 10}{\cancel{10} \cdot \cancel{10}} = 10^4$

Method 2: $\dfrac{10^6}{10^2} = 10^{6-2} = 10^4$

• **EXAMPLE 5**   Divide: $\dfrac{10^5}{10^3}$

Method 1: $\dfrac{10^5}{10^3} = \dfrac{\cancel{10} \cdot \cancel{10} \cdot \cancel{10} \cdot 10 \cdot 10}{\cancel{10} \cdot \cancel{10} \cdot \cancel{10}} = 10^2$

Method 2: $\dfrac{10^5}{10^3} = 10^{5-3} = 10^2$

• **EXAMPLE 6**   Divide each of the following powers of 10:

   a. $\dfrac{10^{12}}{10^4} = 10^{12-4} = 10^8$

   b. $\dfrac{10^{-5}}{10^5} = 10^{(-5)-5} = 10^{-10}$

   c. $\dfrac{10^6}{10^{-9}} = 10^{6-(-9)} = 10^{15}$

   d. $10^{-8} \div 10^{-5} = 10^{-8-(-5)} = 10^{-3}$

   e. $10^5 \div 10^9 = 10^{5-9} = 10^{-4}$

RAISING A POWER OF 10 TO A POWER

To raise a power of 10 to a power, multiply the exponents as follows:

$$(10^a)^b = 10^{ab}$$

• **EXAMPLE 7**   Find the power $(10^2)^3$.

   Method 1: $(10^2)^3 = 10^2 \cdot 10^2 \cdot 10^2$
   $\qquad\qquad\quad = 10^{2+2+2}$ using the product of powers rule
   $\qquad\qquad\quad = 10^6$

   Method 2: $(10^2)^3 = 10^{(2)(3)} = 10^6$

• **EXAMPLE 8**   Find each power of 10:

   a. $(10^4)^3 = 10^{(4)(3)} = 10^{12}$

   b. $(10^{-5})^2 = 10^{(-5)(2)} = 10^{-10}$

   c. $(10^{-6})^{-3} = 10^{(-6)(-3)} = 10^{18}$

   d. $(10^4)^{-4} = 10^{(4)(-4)} = 10^{-16}$

   e. $(10^{10})^8 = 10^{(10)(8)} = 10^{80}$

In Section 1.10, we stated that $10^0 = 1$. Let's see why. To show this, we use the substitution principle, which states that if $a = b$ and $a = c$, then $b = c$.

$$a = b$$

$$\dfrac{10^n}{10^n} = 10^{n-n} \qquad \text{To divide powers, subtract the exponents.}$$

$$= 10^0$$

$$a = c$$

$$\dfrac{10^n}{10^n} = 1 \qquad \text{Any number other than zero divided by itself equals 1.}$$

Therefore, $b = c$; that is, $10^0 = 1$.

Zero Power of 10

$$10^0 = 1$$

We also have used the fact that $10^{-a} = \dfrac{1}{10^a}$. To show this, we start with $\dfrac{1}{10^a}$:

$$\frac{1}{10^a} = \frac{10^0}{10^a} \qquad (1 = 10^0)$$
$$= 10^{0-a} \qquad \text{To divide powers, subtract exponents.}$$
$$= 10^{-a}$$

Negative Power of 10

$$10^{-a} = \frac{1}{10^a}$$

For example, $10^{-3} = \dfrac{1}{10^3}$ and $10^{-8} = \dfrac{1}{10^8}$.

In a similar manner, we can also show that

$$\frac{1}{10^{-a}} = 10^a$$

For example, $\dfrac{1}{10^{-5}} = 10^5$ and $\dfrac{1}{10^{-2}} = 10^2$.

Combinations of multiplications and divisions of powers of 10 can also be done easily using the rules of exponents.

• **EXAMPLE 9**   Perform the indicated operations. Express the results using positive exponents.

a. $\dfrac{10^4 \cdot 10^0}{10^{-3}} = \dfrac{10^{4+0}}{10^{-3}}$

$$= \frac{10^4}{10^{-3}}$$
$$= 10^{4-(-3)}$$
$$= 10^7$$

b. $\dfrac{10^{-2} \cdot 10^5}{10^2 \cdot 10^{-5} \cdot 10^8} = \dfrac{10^{-2+5}}{10^{2+(-5)+8}}$

$$= \frac{10^3}{10^5}$$
$$= 10^{3-5}$$
$$= 10^{-2}$$
$$= \frac{1}{10^2}$$

c. $\dfrac{10^2 \cdot 10^{-3}}{10^4 \cdot 10^{-7}} = \dfrac{10^{2+(-3)}}{10^{4+(-7)}}$

$$= \frac{10^{-1}}{10^{-3}}$$
$$= 10^{(-1)-(-3)}$$
$$= 10^2$$

d. $\dfrac{10^{-5} \cdot 10^8 \cdot 10^{-6}}{10^3 \cdot 10^4 \cdot 10^{-1}} = \dfrac{10^{(-5)+8+(-6)}}{10^{3+4+(-1)}}$

$$= \frac{10^{-3}}{10^6}$$
$$= 10^{-3-6}$$
$$= 10^{-9}$$
$$= \frac{1}{10^9}$$

Exercises | 2.5

Perform the indicated operations using the laws of exponents. Express the results using positive exponents:

1. $10^4 \cdot 10^9$

2. $10^4 \div 10^{-6}$

3. $(10^4)^3$

4. $\dfrac{10^4}{10^8}$

5. $10^{-6} \cdot 10^{-4}$

6. $\dfrac{1}{10^{-5}}$

7. $10^{-2} \div 10^{-5}$

8. $(10^3)^{-2}$

9. $(10^{-3})^4$

10. $\dfrac{(10^0)^3}{10^{-2}}$

11. $10^{-15} \cdot 10^{10}$

12. $\dfrac{10^0 \cdot 10^{-3}}{10^{-6} \cdot 10^3}$

13. $\dfrac{10^3 \cdot 10^2 \cdot 10^{-7}}{10^5 \cdot 10^{-3}}$

14. $\dfrac{10^{-2} \cdot 10^{-3} \cdot 10^{-7}}{10^3 \cdot 10^4 \cdot 10^{-5}}$

15. $\dfrac{10^8 \cdot 10^{-6} \cdot 10^{10} \cdot 10^0}{10^4 \cdot 10^{-17} \cdot 10^8}$

16. $\dfrac{(10^{-4})^6}{10^4 \cdot 10^{-3}}$

17. $\dfrac{(10^{-9})^{-2}}{10^{16} \cdot 10^{-4}}$

18. $\left(\dfrac{10^4}{10^{-7}}\right)^3$

19. $\left(\dfrac{10^5 \cdot 10^{-2}}{10^{-4}}\right)^2$

20. $\left(\dfrac{10^{-7} \cdot 10^{-2}}{10^9}\right)^{-3}$

2.6   Scientific Notation

SCIENTIFIC NOTATION

Scientific notation is a method that is especially useful for writing very large or very small numbers. To write a number in scientific notation, write it as a product of a number between 1 and 10 and a power of 10.

• **EXAMPLE 1**   Write 226 in scientific notation.

$$226 = 2.26 \times 10^2$$

Remember that 10^2 is a short way of writing $10 \times 10 = 100$. Since multiplying 2.26 by 100 gives 226, you have simply moved the decimal point two places to the left.

• **EXAMPLE 2**   Write 52,800 in scientific notation.

$$52{,}800 = 5.28 \times 10{,}000 = 5.28 \times (10 \times 10 \times 10 \times 10)$$
$$= 5.28 \times 10^4$$

WRITING A DECIMAL NUMBER IN SCIENTIFIC NOTATION

1. Reading from left to right, place a decimal point after the first nonzero digit.
2. Place a caret ($\wedge$) at the position of the original decimal point.
3. If the decimal point is to the *left* of the caret, the exponent of 10 is the same as the number of places from the caret to the decimal point.

$$26{,}638 = 2.6638. \times 10^{④} = 2.6638 \times 10^4$$

4. If the decimal point is to the *right* of the caret, the exponent of 10 is the same as the negative of the number of places from the caret to the decimal point.

$$0.00986 = 0.009.86 \times 10^{-③} = 9.86 \times 10^{-3}$$

5. If the decimal point is already after the first nonzero digit, the exponent of 10 is zero.

$$2.15 = 2.15 \times 10^0$$

•   **EXAMPLE 3**   Write 2,738 in scientific notation.

$$2{,}738 = 2.738 \times 10^{③} = 2.738 \times 10^3$$

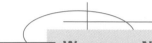

•   **EXAMPLE 4**   Write 0.0000003842 in scientific notation.

$$0.0000003842 = 0.0000003.842 \times 10^{-⑦} = 3.842 \times 10^{-7}$$

WRITING A NUMBER IN SCIENTIFIC NOTATION IN DECIMAL FORM

To change from scientific notation to a decimal, you must be able to multiply by a power of 10.

1. To multiply by a positive power of 10, move the decimal point *to the right* the same number of places as indicated by the exponent of 10. Write zeros when needed.
2. To multiply by a negative power of 10, move the decimal point *to the left* the same number of places as the absolute value of the exponent of 10. Write zeros when needed.

•   **EXAMPLE 5**   Write 2.67×10^2 as a decimal.

$$2.67 \times 10^2 = 267$$

Move the decimal point two places to the *right*, since the exponent of 10 is +2.

•   **EXAMPLE 6**   Write 8.76×10^4 as a decimal.

$$8.76 \times 10^4 = 87{,}600$$

Move the decimal point four places to the *right*, since the exponent of 10 is +4. It is necessary to write two zeros.

•   **EXAMPLE 7**   Write 5.13×10^{-4} as a decimal.

$$5.13 \times 10^{-4} = 0.000513$$

Move the decimal point four places to the *left*, since the exponent of 10 is −4. It is necessary to write three zeros.

You may find it useful to note that a number in scientific notation with

a. a positive exponent greater than 1 is *greater than* 10, and
b. a negative exponent is *less than* 1.

That is, a number in scientific notation with a positive exponent represents a relatively large number. A number in scientific notation with a negative exponent represents a relatively small number.

Scientific notation may be used to compare two positive numbers expressed as decimals. First, write both numbers in scientific notation. The number having the greater power of 10 is the larger. If the powers of 10 are equal, compare the parts of the numbers that are between 1 and 10.

• **EXAMPLE 8**   Which is greater, 0.000876 or 0.0004721?

$$0.000876 = 8.76 \times 10^{-4}$$
$$0.0004721 = 4.721 \times 10^{-4}$$

Since the exponents are the same, compare 8.76 and 4.721. Since 8.76 is greater than 4.721, 0.000876 is greater than 0.0004721.

• **EXAMPLE 9**   Which is greater, 0.0062 or 0.0382?

$$0.0062 = 6.2 \times 10^{-3}$$
$$0.0382 = 3.82 \times 10^{-2}$$

Since -2 is greater than -3, 0.0382 is greater than 0.0062.

Scientific notation is especially helpful for multiplying and dividing very large and very small numbers. To perform these operations, you must first know some rules for exponents. Many calculators perform multiplication, division, and powers of numbers entered in scientific notation and give the results, when very large or very small, in scientific notation.

MULTIPLYING NUMBERS IN SCIENTIFIC NOTATION

To multiply numbers in scientific notation, multiply the decimals between 1 and 10. Then add the exponents of the powers of 10.

• **EXAMPLE 10**   Multiply $(4.5 \times 10^8)(5.2 \times 10^{-14})$. Write the result in scientific notation.

$$(4.5 \times 10^8)(5.2 \times 10^{-14}) = (4.5)(5.2) \times (10^8)(10^{-14})$$
$$= 23.4 \times 10^{-6}$$
$$= (2.34 \times 10^1) \times 10^{-6}$$
$$= 2.34 \times 10^{-5}$$

Note that 23.4×10^{-6} is not in scientific notation, because 23.4 is not between 1 and 10.

To find this product using a calculator that accepts numbers in scientific notation, use the following procedure.

Flowchart	Buttons Pushed	Display
Enter 4.5×10^8	$4 \rightarrow . \rightarrow 5 \rightarrow \boxed{EXP}^* \rightarrow 8$	4.5 08
Push times	$\times$	4.5 08
Enter 5.2×10^{-14}	$5 \rightarrow . \rightarrow 2 \rightarrow \boxed{EXP} \rightarrow 1 \rightarrow 4 \rightarrow \boxed{+/-}^{**}$	5.2 −14
Push equals	$=$	2.34 −05

The product is 2.34×10^{-5}.

DIVIDING NUMBERS IN SCIENTIFIC NOTATION

To divide numbers in scientific notation, divide the decimals between 1 and 10. Then subtract the exponents of the powers of 10.

• **EXAMPLE 11**   Divide $\dfrac{4.8 \times 10^{-7}}{1.6 \times 10^{-11}}$. Write the result in scientific notation.

$$\frac{4.8 \times 10^{-7}}{1.6 \times 10^{-11}} = \frac{4.8}{1.6} \times \frac{10^{-7}}{10^{-11}}$$
$$= 3 \times 10^4$$

Using a calculator, we have:

Flowchart	Buttons Pushed	Display
Enter 4.8×10^{-7}	$4 \rightarrow . \rightarrow 8 \rightarrow \boxed{EXP} \rightarrow 7 \rightarrow \boxed{+/-}$	4.8 −07
Push divide	$\div$	4.8 −07
Enter 1.6×10^{-11}	$1 \rightarrow . \rightarrow 6 \rightarrow \boxed{EXP} \rightarrow 1 \rightarrow 1 \rightarrow \boxed{+/-}$	1.6 −11
Push equals	$=$	3. 04

or | 30000 |

The quotient is 3×10^4.

---

*Some calculators have this button marked $\boxed{EE}$.

**This button is used to change the sign of the *last number* entered.

• **EXAMPLE 12** Evaluate $\dfrac{(6 \times 10^{-6})(3 \times 10^9)}{(2 \times 10^{-10})(4 \times 10^{-5})}$. Write the result in scientific notation.

$$\frac{(6 \times 10^{-6})(3 \times 10^9)}{(2 \times 10^{-10})(4 \times 10^{-5})} = \frac{(6)(3)}{(2)(4)} \times \frac{(10^{-6})(10^9)}{(10^{-10})(10^{-5})} = \frac{18}{8} \times \frac{10^3}{10^{-15}}$$

$$= 2.25 \times 10^{18}$$

Again, use a calculator.

Flowchart	Buttons Pushed	Display
Enter 6×10^{-6}	$\boxed{6} \to \boxed{\text{EXP}} \to \boxed{6} \to \boxed{+/-}$	6.　　　　−06
Push times	$\boxed{\times}$	6.　　　　−06
Enter 3×10^9	$\boxed{3} \to \boxed{\text{EXP}} \to \boxed{9}$	3.　　　　09
Push divide	$\boxed{\div}$	1.8　　　　04
		or　18000
Enter 2×10^{-10}	$\boxed{2} \to \boxed{\text{EXP}} \to \boxed{1} \to \boxed{0} \to \boxed{+/-}$	2.　　　　−10
Push divide	$\boxed{\div}$	9.　　　　13
Enter 4×10^{-5}	$\boxed{4} \to \boxed{\text{EXP}} \to \boxed{5} \to \boxed{+/-}$	4.　　　　−05
Push equals	$\boxed{=}$	2.25　　　　18

The result is 2.25×10^{18}.

POWERS OF NUMBERS IN SCIENTIFIC NOTATION

To find the power of a number in scientific notation, find the power of the decimal between 1 and 10. Then multiply the exponent of the power of 10 by this same power.

• **EXAMPLE 13** Find the power $(4.5 \times 10^6)^2$. Write the result in scientific notation.

$$(4.5 \times 10^6)^2 = (4.5)^2 \times (10^6)^2$$
$$= 20.25 \times 10^{12} \qquad \text{Note that 20.25 is not between 1 and 10.}$$
$$= (2.025 \times 10^1) \times 10^{12}$$
$$= 2.025 \times 10^{13}$$

Flowchart	Buttons Pushed	Display

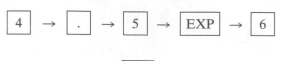

The result is 2.025×10^{13}.

• **EXAMPLE 14**   Find the power $(3 \times 10^{-8})^5$. Write the result in scientific notation.

$$(3 \times 10^{-8})^5 = 3^5 \times (10^{-8})^5$$
$$= 243 \times 10^{-40}$$
$$= (2.43 \times 10^2) \times 10^{-40}$$
$$= 2.43 \times 10^{-38}$$

The y^x button is used to raise a number to any power.

Flowchart	Buttons Pushed	Display

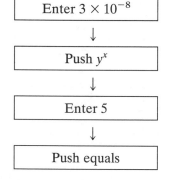

The result is 2.43×10^{-38}.

Exercises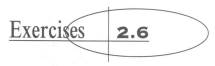

Write each number in scientific notation:

1. 356	**3.** 634.8	**5.** 0.00825	**7.** 7.4	**9.** 0.000072
2. 15,600	**4.** 24.85	**6.** 0.00063	**8.** 377,000	**10.** 0.00335

11. 710,000	**13.** 0.0000045	**15.** 0.000000034	**17.** 640,000
12. 1,200,000	**14.** 0.0000007	**16.** 4,500,000,000	**18.** 85,000

Write each number in decimal form:

19. 7.55×10^4	**22.** 5.14×10^5	**25.** 5.55×10^{-4}	**28.** 3.785×10^{-2}	**31.** 5.76×10^0
20. 8.76×10^2	**23.** 7.8×10^{-2}	**26.** 4.91×10^{-6}	**29.** 9.6×10^2	**32.** 6.8×10^{-5}
21. 5.31×10^3	**24.** 9.44×10^{-3}	**27.** 6.4×10^1	**30.** 7.3×10^3	**33.** 6.4×10^{-6}

34. 7×10^8	**37.** 6.2×10^{-7}	**40.** 1.5×10^{11}
35. 5×10^{10}	**38.** 2.1×10^{-9}	**41.** 3.3×10^{-11}
36. 5.05×10^0	**39.** 2.5×10^{12}	**42.** 7.23×10^{-8}

Find the larger number:

43. 0.0037; 0.0048

45. 0.000042; 0.00091

47. 0.00037; 0.000094

49. 0.0613; 0.00812

44. 0.029; 0.0083

46. 148,000; 96,988

48. 0.8216; 0.792

50. 0.0000613; 0.01200

Find the smaller number:

51. 0.008; 0.0009

53. 1.003; 1.0009

55. 0.00000000998; 0.01

57. 0.000314; 0.000271

52. 295,682; 295,681

54. 21.8; 30.2

56. 0.10108; 0.10102

58. 0.00812; 0.0318

Perform the indicated operations (write each result in scientific notation with the decimal part rounded to three significant digits when necessary):

59. $(4 \times 10^{-6})(6 \times 10^{-10})$

60. $(3 \times 10^{7})(3 \times 10^{-12})$

61. $\dfrac{4.5 \times 10^{16}}{1.5 \times 10^{-8}}$

62. $\dfrac{1.6 \times 10^{6}}{6.4 \times 10^{10}}$

63. $\dfrac{(4 \times 10^{-5})(6 \times 10^{-3})}{(3 \times 10^{-10})(8 \times 10^{8})}$

64. $\dfrac{(5 \times 10^{4})(3 \times 10^{-5})(4 \times 10^{6})}{(1.5 \times 10^{6})(2 \times 10^{-11})}$

65. $(1.2 \times 10^{6})^{3}$

66. $(2 \times 10^{-9})^{4}$

67. $(6.2 \times 10^{-5})(5.2 \times 10^{-6})(3.5 \times 10^{8})$

68. $\dfrac{(5 \times 10^{-6})^{2}}{4 \times 10^{6}}$

69. $\left(\dfrac{2.5 \times 10^{-4}}{7.5 \times 10^{8}}\right)^{2}$

70. $\left(\dfrac{2.5 \times 10^{-9}}{5 \times 10^{-7}}\right)^{4}$

71. $(18,000)(0.00005)$

72. $(4500)(69,000)(150,000)$

73. $\dfrac{2,400,000}{36,000}$

74. $\dfrac{(3500)(0.00164)}{2700}$

75. $\dfrac{84,000 \times 0.0004 \times 142,000}{}$

76. $\dfrac{(0.0025)^{2}}{3500}$

77. $\left(\dfrac{48,000 \times 0.0144}{0.0064}\right)^{2}$

78. $\left(\dfrac{0.0027 \times 0.16}{12,000}\right)^{3}$

79. $\left(\dfrac{1.3 \times 10^{4}}{(2.6 \times 10^{-3})(5.1 \times 10^{8})}\right)^{5}$

80. $\left(\dfrac{9.6 \times 10^{-3}}{(2.45 \times 10^{-4})(1.1 \times 10^{5})}\right)^{6}$

81. $\left(\dfrac{18.4 \times 2100}{0.036 \times 950}\right)^{8}$

82. $\left(\dfrac{0.259 \times 6300}{866 \times 0.013}\right)^{10}$

Chapter 2   Review

Find the absolute value of:

1. $+5$

2. -16

3. 13

Add:

4. $\begin{array}{r} -4 \\ +7 \\ \hline \end{array}$

5. $\begin{array}{r} -6 \\ -2 \\ \hline \end{array}$

6. $(+5) + (-8)$

7. $(-9) + (+2) + (-6)$

Subtract:

8. 3
   6
   ――

9. −7
   +4
   ――

10. $(+9) - (-10)$

11. $(-6) - (+4) - (-8)$

Perform the indicated operations:

12. $(-2) - (+7) + (+4) + (-5) - (-10)$

13. $5 - 6 + 4 - 9 + 4 + 3 - 12 - 8$

Multiply:

14. $(-6)(+4)$

15. $(+4)(+9)$

16. $(-9)(-8)$

17. $(-2)(-7)(+1)(+3)(-2)$

Divide:

18. $\dfrac{-18}{-3}$

19. $(+30) \div (-5)$

20. $\dfrac{+45}{+9}$

Perform the indicated operations and simplify:

21. $\left(-\dfrac{6}{7}\right) - \left(\dfrac{5}{-6}\right)$

22. $\dfrac{-3}{16} \div \left(-2\dfrac{1}{4}\right)$

23. $\dfrac{-5}{8} + \left(-\dfrac{5}{6}\right) - \left(+1\dfrac{2}{3}\right)$

24. $\left(-\dfrac{9}{16}\right) \times 2\dfrac{2}{3}$

Perform the indicated operations using the laws of exponents and express the results using positive exponents:

25. $10^9 \cdot 10^{-14}$

26. $10^6 \div 10^{-3}$

27. $(10^{-4})^3$

28. $\dfrac{(10^{-3} \cdot 10^5)^3}{10^6}$

Write each number in scientific notation:

29. 476,000

30. 0.0014

Write each number in decimal form:

31. 5.35×10^{-5}

32. 6.1×10^7

Find the larger number:

33. 0.00063; 0.00105

34. 0.056; 0.06

Find the smaller number:

35. 0.000075; 0.0006

36. 0.04; 0.00183

Perform the indicated operations and write each result in scientific notation:

37. $(9.5 \times 10^{10})(4.6 \times 10^{-13})$

39. $\dfrac{(50,000)(640,000,000)}{(0.0004)^2}$

41. $(2 \times 10^9)^4$

38. $\dfrac{8.4 \times 10^8}{3 \times 10^{-6}}$

40. $(4.5 \times 10^{-8})^2$

Chapter 2   Test

1. Add: −7
       +9
       ――

3. Multiply: $(-4)(-7)$

2. Subtract: − 6
       −10
       ――

4. Divide: $\dfrac{+20}{-4}$

Perform the indicated operations.

5. $(-2) + (-6) + (+5) + (-9) + (+10)$

6. $(-3)(-2)(+1)(+2)(-1)(+2)(+1)$

7. $(-5) + (-6) - (-7) - (+4) + (+3)$

8. $8 + (-1) + (-5) - (-3) + 10$

9. $-8 + 5 + 2 - 12 + 5 - 3$

Do as indicated and simplify.

10. $\left(-\dfrac{2}{3}\right)(-6) \div \left(-1\dfrac{1}{3}\right)$

11. $2\dfrac{1}{5} - \left(-1\dfrac{3}{10}\right) + 2\dfrac{3}{5}$

12. $\left(\dfrac{-5}{9}\right)\left(3\dfrac{2}{5}\right)$

13. Write 0.000182 in scientific notation.

14. Write 4.7×10^6 in decimal form.

Perform the indicated operations using the laws of exponents and express each result using positive exponents.

15. $(10^{-3})(10^6)$

16. $(10^3) \div 10^{-5}$

17. $(10^2)^4$

18. $\dfrac{10^8 \cdot 10^{-6}}{(10^{-3})^{-2}}$

Perform the indicated operations. Write each result in scientific notation.

19. $\dfrac{(7.6 \times 10^{13})(5.35 \times 10^{-6})}{4.64 \times 10^8}$

20. $\dfrac{150,000 \times 18,000 \times 0.036}{0.0056 \times 48,000}$

The Metric System

CHAPTER 3

AEROSPACE
In many kinds of high-tech manufacturing, metric dimensions are standard.

MATHEMATICS AT WORK

The United States is one of the few industrialized countries still using the English system of measurement. Many U.S. companies have adopted the metric system in order to compete in world trade.

Engineering/design drafting firms, auto and farm manufacturers, and cabinetmakers use metric dimensions. Dual dimensioning is becoming standard for blueprints, so that products can be marketed globally.

Medications are administered in milligrams or grams. Doctors, nurses, medical assistants, and pharmacists use metric measurements to compute dosages based on the patient's body weight in kilograms. Photographic film is measured in millimetres. Soft drinks are sold in one-litre and two-litre bottles. It is becoming increasingly common for products in the United States to be labeled with both metric and English measurements.

3.1 Introduction to the Metric System

In early recorded time, parts of the human body were used for standards of measurement. However, these standards were neither uniform nor acceptable to all. The next step was to define the various standards, such as the inch, the foot, the rod, and so on. But each country introduced or defined its own standards, which were often not related to those in other countries. Then, in 1670, as the need for a single worldwide measurement system became recognized, Gabriel Mouton, a Frenchman, proposed a uniform decimal measurement system. By the 1800s, metric standards were first adopted worldwide. In 1960, the International System of Units was adopted for the modern metric system. The abbreviation for the International System of Units is SI (from the French *Système International d'Unités*).

The U.S. Congress legalized the use of the metric system throughout the United States over 100 years ago on July 28, 1866. The United States has signed various metric treaties and has participated in many international conferences and conventions dealing with the metric system. In the early 1970s, the United States found that it was the only industrial nation not committed to converting to the metric system. The Metric Conversion Act of 1975 declares "a national policy of coordinating the increasing use of the metric system in the United States."

From a 1988 amendment to the Metric Conversion Act, the U.S. Congress found the following:

- World trade is increasingly geared toward the metric system of measurement.

- Industry in the United States is often at a competitive disadvantage when dealing in international markets because of its nonstandard measurement system, and is sometimes excluded when it is unable to deliver goods that are measured in metric terms.

- The inherent simplicity of the metric system of measurement and standardization of weights and measures has led to major cost savings in industries that have converted to that system.

- The federal government has a responsibility to develop procedures and techniques to assist industry (especially small business) as it voluntarily converts to the metric system of measurement.

- The metric system of measurement can provide substantial advantages to the federal government in its own operations.

The Executive Order regarding Metric Usage in Federal Government Programs (1991) declares that it is the policy of the United States:

1. to designate the metric system of measurement as the preferred system of weights and measures for United States trade and commerce;

2. to require that each federal agency use the metric system of measurement in its procurements, grants, and other business-related activities (by fiscal year 1992), except to the extent that such use is impractical or is likely to cause significant inefficiencies or loss of markets to United States firms, such as when foreign competitors are producing competing products in non-metric units;

3. to seek out ways to increase understanding of the metric system of measurement through educational information and guidance in government publications; and

4. to permit the continued use of traditional systems of weights and measures in nonbusiness activities.

At present, the United States is the only country that has not yet fully planned and implemented a move toward world uniformity in measurement. This huge imbalance makes it economically imperative that the United States convert as soon as possible so that it can compete more favorably for international trade. Most countries have restrictions on importing nonmetric products. Metric countries just naturally want metric products.

Many major U.S. industries, such as the automotive industry, have already effectively converted to the metric system completely. Most others are in the process of converting or are making major plans to do so.

The SI metric system has seven base units, as shown in Table 3.1.

TABLE 3.1		
Basic unit	SI abbreviation	For measuring
metre*	m	length
kilogram	kg	mass
second	s	time
ampere	A	electric current
kelvin	K	temperature
candela	cd	light intensity
mole	mol	molecular substance

Other commonly used metric units are shown in Table 3.2.

TABLE 3.2		
Unit	SI abbreviations	For measuring
litre*	L	volume
cubic metre	m^3	volume
square metre	m^2	area
newton	N	force
metre per second	m/s	speed
joule	J	energy
watt	W	power
radian	rad	plane angle
*At present, there is some difference of opinion on the spelling of *metre* and *litre*. We have chosen the "re" spelling because it is the internationally accepted spelling and because it distinguishes the unit from other meters, such as parking meters and electricity meters.		

The metric system, a decimal or base 10 system, is very similar to our decimal number system. It is an easy system to use, because calculations are based on the number 10 and its multiples. The SI system has special prefixes that name multiples and submultiples; these may be used with almost all SI units. Table 3.3 shows the prefixes and the corresponding symbols.

Multiple or submultiple* decimal form	Power of 10	Prefix	Prefix symbol	Pronun-ciation	Meaning
1,000,000,000,000	10^{12}	tera	T	tĕr′ă	one trillion times
1,000,000,000	10^{9}	giga	G	jĭg′ă	one billion times
1,000,000	10^{6}	mega	M	mĕg′ă	one million times
1,000	10^{3}	kilo**	k	kĭl′ō or kēl′ō	one thousand times
100	10^{2}	hecto	h	hĕk′tō	one hundred times
10	10^{1}	deka	da	dĕk′ă	ten times
0.1	10^{-1}	deci	d	dĕs′ĭ	one tenth of
0.01	10^{-2}	centi**	c	sĕnt′ĭ	one hundredth of
0.001	10^{-3}	milli**	m	mĭl′ĭ	one thousandth of
0.000001	10^{-6}	micro	μ	mī′krō	one millionth of
0.000000001	10^{-9}	nano	n	năn′ō	one billionth of
0.000000000001	10^{-12}	pico	p	pē′kō	one trillionth of

TABLE 3.3   PREFIXES FOR SI UNITS

* Factor by which the unit is multiplied.
** Most commonly used prefixes.

Since the same prefixes are used with most all SI metric units, it is not necessary to memorize long lists or many tables.

• **EXAMPLE 1**   Write the SI abbreviation for 45 kilometres.

The symbol for the prefix *kilo* is k.
The symbol for the unit *metre* is m.
The SI abbreviation for 45 kilometres is 45 km.

• **EXAMPLE 2**   Write the SI unit for the abbreviation 50 mg.

The prefix for m is *milli*.
The unit for g is *gram*.
The SI unit for 50 mg is 50 milligrams.

In summary, the English system is an ancient one based on standards initially determined by parts of the human body, which is why there is no consistent relationship between units. In the metric system, however, standard units are subdivided in multiples of 10, similar to our number system, and the names associated with each subdivision have prefixes that indicate a multiple of 10.

Exercises  3.1

Give the metric prefix for each value:

1. 1000   **3.** 0.01   **5.** 0.001   **7.** 1,000,000

2. 100   **4.** 0.1   **6.** 10   **8.** 0.000001

Give the SI symbol for each prefix:

9. hecto   **11.** deci   **13.** centi   **15.** micro

10. kilo   **12.** milli   **14.** deka   **16.** mega

Write the abbreviation for each quantity:

17. 65 milligrams   **19.** 82 centimetres   **21.** 36 microamps   **23.** 19 hectolitres

18. 125 kilolitres   **20.** 205 millilitres   **22.** 75 kilograms   **24.** 5 megawatts

Write the SI unit for each abbreviation:

25. 18 m   **27.** 36 kg   **29.** 24 ps   **31.** 135 mL   **33.** 45 mA

26. 15 L   **28.** 85 mm   **30.** 9 dam   **32.** 45 dL   **34.** 75 MW

35. The basic SI unit of length is _____.

36. The basic SI unit of mass is _____.

37. The basic SI unit of electric current is _____.

38. The basic SI unit of time is _____.

39. The common SI units of volume are _____ and _____.

40. The common SI unit of power is _____.

3.2   Length

The basic SI unit of length is the metre (m). The height of a door knob is about 1 m. (See Figure 3.1.) One metre is a little more than 1 yd. (See Figure 3.2.)

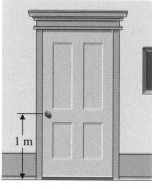

FIGURE 3.1

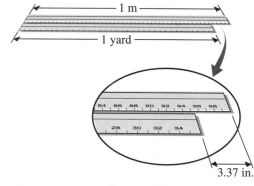

FIGURE 3.2

Long distances are measured in kilometres (km). (1 km = 1000 m.) The length of five city blocks is about 1 km. (See Figure 3.3.)

The centimetre (cm) is used to measure short distances, such as the width of this page (about 22 cm), or the width of a board. The width of your small fingernail is about 1 cm. (See Figure 3.4.)

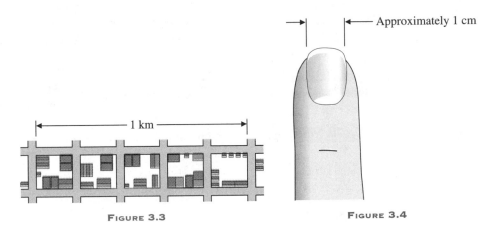

FIGURE 3.3   FIGURE 3.4

The millimetre (mm) is used to measure very small lengths, such as the thickness of a sheet of metal or the depth of a tire tread. The thickness of a dime is about 1 mm. (See Figure 3.5.)

FIGURE 3.5

Each of the large numbered divisions on the metric ruler in Figure 3.6 marks one centimetre (cm). The smaller lines indicate halves of centimetres, and the smallest divisions show tenths of centimetres, or millimetres.

FIGURE 3.6

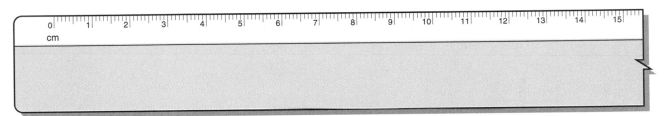

• **EXAMPLE 1**   Read *A*, *B*, *C*, and *D* on the metric ruler in Figure 3.7. Give each result in millimetres, centimetres, and metres.

FIGURE 3.7

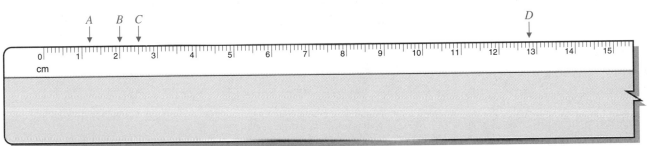

Answers: A = 12 mm, 1.2 cm, 0.012 m
B = 20 mm, 2.0 cm, 0.020 m
C = 25 mm, 2.5 cm, 0.025 m
D = 128 mm, 12.8 cm, 0.128 m

To convert from one metric unit to another, we could use the same conversion factor procedure that we used in the English system in Section 1.9.

> **CHOOSING CONVERSION FACTORS**
>
> The correct choice for a given conversion factor is the one in which the old units are in the numerator of the original expression and in the denominator of the conversion factor, or in the denominator of the original expression and in the numerator of the conversion factor. That is, set up the conversion factor so that the old units cancel each other.

• **EXAMPLE 2**   Change 3.6 km to metres.

Since *kilo* means 10^3 or 1000, 1 km = 1000 m. The two possible conversion factors are $\frac{1\,\text{km}}{1000\,\text{m}}$ and $\frac{1000\,\text{m}}{1\,\text{km}}$. Choose the one whose numerator is expressed in the new units (m) and whose denominator is expressed in the old units (km). This is $\frac{1000\,\text{m}}{1\,\text{km}}$.

$$3.6\ \cancel{\text{km}} \times \frac{1000\ \text{m}}{1\ \cancel{\text{km}}} = 3600\ \text{m}$$

• **EXAMPLE 3**   Change 4 m to centimetres.

First, *centi* means 10^{-2} or 0.01; and 1 cm = 10^{-2} m. Choose the conversion factor with metres in the denominator and centimetres in the numerator.

$$4\ \cancel{\text{m}} \times \frac{1\ \text{cm}}{10^{-2}\ \cancel{\text{m}}} = 400\ \text{cm}$$

Note: Conversions within the metric system only involve moving the decimal point.

Exercises 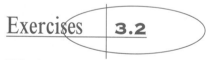 3.2

Which is longer?

1. 1 metre or 1 millimetre

2. 1 metre or 1 centimetre

3. 1 metre or 1 kilometre

4. 1 millimetre or 1 kilometre

5. 1 centimetre or 1 millimetre

6. 1 kilometre or 1 centimetre

Fill in each blank:

7. 1 m = _____ mm

8. 1 km = _____ m

9. 1 cm = _____ m

10. 1 m = _____ cm

11. 1 m = _____ km

12. 1 hm = _____ m

13. 1 cm = _____ mm

14. 1 dam = _____ dm

Which metric unit (km, m, cm, or mm) should you use to measure each item?

15. Diameter of an automobile tire

16. Thickness of sheet metal

17. Metric wrench sizes

18. Length of an auto race

19. Length of a discus throw in track and field

20. Length and width of a table top

21. Distance between Chicago and St. Louis

22. Thickness of plywood

23. Thread size of a pipe

24. Length and width of a house lot

Fill in each blank with the most reasonable unit (km, m, cm, or mm):

25. A standard film size for cameras is 35 _____.

26. The diameter of a wheel on a ten-speed bicycle is 56 _____.

27. A jet plane generally flies about 8–9 _____ high.

28. The width of a door in our house is 910 _____.

29. The length of the ridge on our roof is 24 _____.

30. Antonio's waist size is 95 _____.

31. The steering wheel on Brenda's car is 36 _____ in diameter.

32. Jan drives 12 _____ to school.

33. The standard metric size for plywood is 1200 _____ wide and 2400 _____ long.

34. The distance from home plate to the centerfield wall in a baseball park is 125 _____.

35. Read the measurements indicated by the letters on the metric ruler in Illustration 1 and give each result in millimetres, centimetres, and metres.

ILLUSTRATION 1

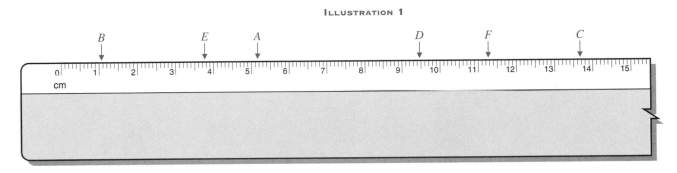

Use a metric ruler to measure each line segment. Give each result in millimetres, centimetres, and metres:

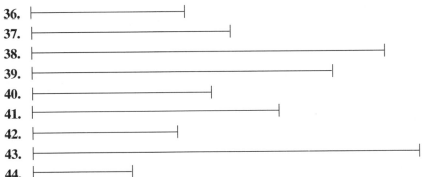

36.

37.

38.

39.

40.

41.

42.

43.

44.

45. Change 675 m to km.

46. Change 450 cm to m.

47. Change 1540 mm to m.

48. Change 3.2 km to m.

49. Change 65 cm to m.

50. Change 1.4 m to mm.

51. Change 7.3 m to cm.

52. Change 48 cm to mm.

53. Change 125 mm to cm.

54. Change 0.75 m to μm.

55. What is your height **a.** in metres? **b.** in centimetres? **c.** in millimetres?

3.3 Mass and Weight

The *mass* of an object is the quantity of material making up the object. One unit of mass in the SI system is the gram (g). The gram is defined as the mass contained in 1 cubic centimetre (cm^3) of water, at its maximum density. A common paper clip has a mass of about 1 g. Three aspirin have a mass of about 1 g.

Because the gram is so small, the kilogram is the basic unit of mass in the SI system. One kilogram is defined as the mass contained in 1 cubic decimetre (dm^3) of water, at its maximum density.

For very large quantities, such as a trainload of coal or grain or a shipload of ore, the metric ton (1000 kg) is used. The milligram (mg) is used to measure very, very small masses such as medicine dosages. One grain of salt has a mass of about 1 mg.

The *weight* of an object is a measure of the earth's gravitational force—or pull—acting on the object. The SI unit of weight is the newton (N). As you are no doubt aware, the terms *mass* and *weight* are commonly used interchangeably by the general public. We have presented them as technical terms, as they are used in the technical, engineering, and scientific professions. To further illustrate the difference, the mass of an astronaut remains relatively constant while his or her weight varies (the weight decreases as the distance from the earth increases). If the spaceship is in orbit or farther out in space, we say the crew is "weightless," because they seem to float freely in space. Their mass has not changed, although their weight is near zero. (See Figure 3.8.)

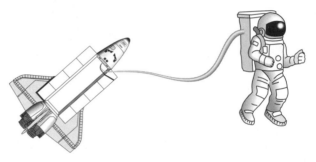

FIGURE 3.8
"Weightless" astronaut

EXAMPLE 1   Change 12 kg to grams.

First, *kilo* means 10^3 or 1000; and 1 kg = 1000 g. Choose the conversion factor with kilograms in the denominator and grams in the numerator.

$$12 \text{ kg} \times \frac{1000 \text{ g}}{1 \text{ kg}} = 12{,}000 \text{ g}$$

EXAMPLE 2   Change 250 mg to grams.

First, *milli* means 10^{-3} or 0.001; and 1 mg = 10^{-3} g. Choose the conversion factor with milligrams in the denominator and grams in the numerator.

$$250 \text{ mg} \times \frac{10^{-3} \text{ g}}{1 \text{ mg}} = 0.25 \text{ g}$$

Exercises 3.3

Which is larger?

1. 1 gram or 1 milligram

2. 1 gram or 1 kilogram

3. 1 milligram or 1 kilogram

4. 1 metric ton or 1 kilogram

5. 1 milligram or 1 microgram

6. 1 kilogram or 1 microgram

Fill in each blank:

7. 1 g = _____ mg

8. 1 kg = _____ g

9. 1 cg = _____ g

10. 1 mg = _____ g

11. 1 metric ton = _____ kg

12. 1 g = _____ cg

13. 1 mg = _____ μg

14. 1 μg = _____ mg

Which metric unit (kg, g, mg, or metric ton) should you use to measure the mass of each item?

15. A bar of handsoap

16. A vitamin capsule

17. A bag of flour

18. A four-wheel-drive tractor

19. A pencil

20. Your mass

21. A trainload of coal

22. A bag of potatoes

23. A contact lens

24. An apple

Fill in each blank with the most reasonable unit (kg, g, mg, or metric ton):

25. A slice of bread has a mass of about 25 _____.

26. Elevators in the college have a load limit of 2200 _____.

27. I take 1000 _____ of vitamin C every day.

28. My uncle's new truck can haul a load of 4 _____.

29. Postage rates for letters are based on the number of _____.

30. I take 1 _____ of vitamin C every day.

31. My best friend has a mass of 65 _____.

32. A jar of peanut butter contains 1200 _____.

33. The local grain elevator shipped 20,000 _____ of wheat last year.

34. One common size of aspirin tablets is 325 _____.

35. Change 875 g to kg.

36. Change 127 mg to g.

37. Change 85 g to mg.

38. Change 1.5 kg to g.

39. Change 3.6 kg to g.

40. Change 430 g to mg.

41. Change 270 mg to g.

42. Change 18 mg to μg.

43. Change 2.5 metric tons to kg.

44. Change 18,000 kg to metric tons.

45. What is your mass in kilograms?

3.4 Volume and Area

A common unit of volume in the metric system is the litre (L). One litre of milk is a little more than 1 quart. The litre is commonly used for liquid volume. (See Figure 3.9.)

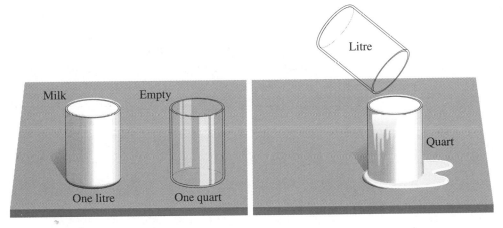

Milk Empty

One litre One quart

Litre

Quart

(a) (b)

FIGURE 3.9

The cubic metre (m^3) is used to measure large volumes. One cubic metre is the volume contained in a cube 1 m on an edge. The cubic centimetre (cm^3) is used to measure small volumes. It is the volume contained in a cube 1 cm on an edge.

Note: It is important to understand the relationship between the litre and the cubic centimetre. The litre is defined as the volume in 1 cubic decimetre (dm^3). That is, 1 L of liquid fills a cube 1 dm (10 cm) on an edge. (See Figure 3.10.)

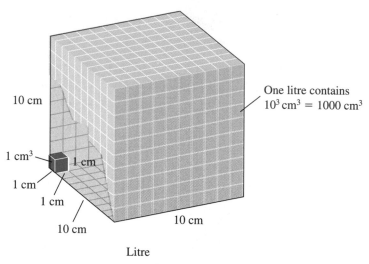

10 cm

One litre contains
$10^3 \text{ cm}^3 = 1000 \text{ cm}^3$

1 cm³

1 cm

1 cm

1 cm

10 cm

10 cm

10 cm

Litre

FIGURE 3.10

The volume of the cube in Figure 3.10 can also be found by the formula

$V = \ell w h$

$V = (10 \text{ cm})(10 \text{ cm})(10 \text{ cm})$

$\quad = 1000 \text{ cm}^3$

Thus, 1 L = 1000 cm^3. Dividing each side by 1000, we have

$$\frac{1}{1000} \text{ L} = 1 \text{ cm}^3$$

or $\boxed{1 \text{ mL} = 1 \text{ cm}^3}$ $1 \text{ mL} = \frac{1}{1000} \text{ L}$

Milk, soft drinks, and gasoline are sold by the litre. Liquid medicine and eye drops are sold by the millilitre. Large quantities of liquid are sold by the kilolitre (1000 L).

In Section 3.3, the kilogram was defined as the mass of 1 dm^3 of water. Since 1 dm^3 = 1 L, 1 litre of water has a mass of 1 kg.

•   **EXAMPLE 1**   Change 0.5 L to millilitres.

$$0.5 \cancel{\text{L}} \times \frac{1000 \text{ mL}}{1 \cancel{\text{L}}} = 500 \text{ mL}$$

•   **EXAMPLE 2**   Change 4.5 cm^3 to mm^3.

$$4.5 \text{ cm}^3 \times \left(\frac{10 \text{ mm}}{1 \text{ cm}}\right)^3 = 4500 \text{ mm}^3$$

Use the length conversion factor 1 cm = 10 mm, and first form the conversion factor with cm in the denominator and mm in the numerator. Then raise the conversion factor to the third power to obtain cubic units in both numerator and denominator. Since the numerator equals the denominator, both the length conversion factor $\frac{10 \text{ mm}}{1 \text{ cm}}$ and its third power $\left(\frac{10 \text{ mm}}{1 \text{ cm}}\right)^3$ equal 1.

Alternate Method:

$$4.5 \text{ cm}^3 \times \frac{1000 \text{ mm}^3}{1 \text{ cm}^3} = 4500 \text{ mm}^3$$

The alternate method conversion factor 1 cm^3 = 1000 mm^3 is taken directly from the metric volume conversion table on the reference card. The first method is preferred, because only the length conversion needs to be remembered or found in a table.

cm^2

mm^2

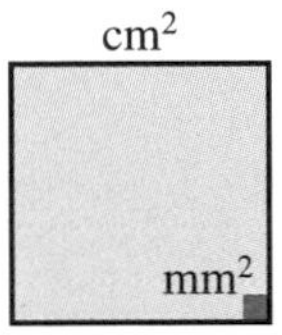

FIGURE 3.11

The basic unit of area in the metric system is the square metre (m^2), the area contained in a square whose sides are each 1 m long. The square centimetre (cm^2) and the square millimetre (mm^2) are smaller units of area. (See Figure 3.11.) The larger-area units are the square kilometre (km^2) and the hectare (ha).

• **EXAMPLE 3**   Change 2400 cm^2 to m^2.

$$2400 \text{ cm}^2 \times \left(\frac{1 \text{ m}}{100 \text{ cm}}\right)^2 = 0.24 \text{ m}^2$$

Use the length conversion factor 1 m = 100 cm, and first form the conversion factor with cm in the denominator and m in the numerator. Then raise the conversion factor to the second power to obtain square units in both numerator and denominator. Since the numerator equals the denominator, both the length conversion factor $\frac{1 \text{ m}}{100 \text{ cm}}$ and its second power $\left(\frac{1 \text{ m}}{100 \text{ cm}}\right)^2$ equal 1.

Alternate Method:

$$2400 \text{ cm}^2 \times \frac{1 \text{ m}^2}{10,000 \text{ cm}^2} = 0.24 \text{ m}^2$$

The alternate method conversion factor 1 m^2 = 10,000 cm^2 is taken directly from the metric area conversion table. The first method is again preferred, because only the length conversion needs to be remembered or found in a table.

• **EXAMPLE 4**   Change 1.2 km^2 to m^2.

$$1.2 \text{ km}^2 \times \left(\frac{1000 \text{ m}}{1 \text{ km}}\right)^2 = 1,200,000 \text{ m}^2$$

Alternate Method:

$$1.2 \text{ km}^2 \times \frac{10^6 \text{ m}^2}{1 \text{ km}^2} = 1,200,000 \text{ m}^2$$

The hectare (ha) is the basic metric unit of land area. The area of 1 hectare equals the area of a square 100 m on a side, whose area is 10,000 m^2 or 1 square hectometre (hm^2). (See Figure 3.12.)

1 hectare (ha) =
10,000 m² =   100 m
1 hm²

100 m

FIGURE 3.12

The hectare is used because it is more convenient to say and use than "square hectometre." The metric prefixes are *not* used with the hectare unit. That is, instead of saying "kilohectare," we say "1000 hectares."

• **EXAMPLE 5**   How many hectares are contained in a rectangular field 360 m by 850 m?

The area in m² is

$$(360 \text{ m})(850 \text{ m}) = 306{,}000 \text{ m}^2$$

$$306{,}000 \text{ m}^2 \times \frac{1 \text{ ha}}{10{,}000 \text{ m}^2} = 30.6 \text{ ha}$$

Exercises 3.4

Which is larger?

1. 1 litre or 1 millilitre

2. 1 millilitre or 1 kilolitre

3. 1 cubic millimetre or 1 cubic centimetre

4. 1 cubic metre or 1 litre

5. 1 square kilometre or 1 hectare

6. 1 square centimetre or 1 square millimetre

Fill in each blank.

7. 1 L = _____ mL

8. 1 mL = _____ L

9. 1 m³ = _____ cm³

10. 1 mm³ = _____ cm³

11. 1 cm² = _____ mm²

12. 1 km² = _____ ha

13. 1 m³ = _____ L

14. 1 cm³ = _____ mL

Which metric unit (m³, L, mL, m², cm², or ha) should you use to measure the following?

15. Oil in your car's crankcase

16. Cough syrup

17. Floor space in a warehouse

18. Size of a farm

19. Cross-sectional area of a piston

20. Piston displacement in an engine

21. Cargo space in a truck

22. Paint needed to paint a house

23. Eye drops

24. Page size of this book

25. Size of an industrial park

26. Gasoline in your car's gas tank

Fill in each blank with the most reasonable unit (m³, L, mL, m², cm², or ha):

27. Lateesha ordered 12 _____ of concrete for her new driveway.

28. I drink 250 _____ of orange juice each morning.

29. Juan, a farmer, owns a 2500- _____ storage tank for diesel fuel.

30. Dwight planted 75 _____ of wheat this year.

31. Our house has 195 _____ of floor space.

32. We must heat 520 _____ of living space in our house.

33. When I was a kid, I mowed 6 _____ of lawns each week.

34. Our community's water tower holds 650 _____ of water.

35. The cross section of a log is 2500 _____.

36. Darnell bought a 25- _____ tarpaulin for his truck.

37. I need copper tubing with a cross section of 4 _____.

38. We should each drink 2 _____ of water each day.

39. Change 1500 mL to L.

40. Change 0.60 L to mL.

41. Change 1.5 m^3 to cm^3.

42. Change 450 mm^3 to cm^3.

43. Change 85 cm^3 to mL.

44. Change 650 L to m^3.

45. Change 85,000 m^2 to km^2.

46. Change 18 m^2 to cm^2.

47. Change 85,000 m^2 to ha.

48. Change 250 ha to km^2.

49. What is the mass of 500 mL of water?

50. What is the mass of 1 m^3 of water?

51. How many hectares are contained in a rectangular field that measures 75 m by 90 m?

52. How many hectares are contained in a rectangular field that measures $\frac{1}{4}$ km by $\frac{1}{2}$ km?

3.5 Time, Current, and Other Units

The basic SI unit of time is the second (s), which is the same in all units of measurement. Time is also measured in minutes (min), hours (h), days, and years.

$$1 \text{ min} = 60 \text{ s}$$
$$1 \text{ h} = 60 \text{ min}$$
$$1 \text{ day} = 24 \text{ h}$$
$$1 \text{ year} = 365\frac{1}{4} \text{ days (approximately)}$$

• **EXAMPLE 1**   Change 4 h 15 min to seconds.

First, $4 \cancel{h} \times \dfrac{60 \text{ min}}{1 \cancel{h}} = 240 \text{ min}$

And,   4 h 15 min = 240 min + 15 min
         = 255 min

Then. $255 \cancel{\text{min}} \times \dfrac{60 \text{ s}}{1 \cancel{\text{min}}} = 15,300 \text{ s}$

Very short periods of time are commonly used in electronics. These are measured in parts of a second, given with the appropriate metric prefix.

• **EXAMPLE 2**   What is the meaning of each unit? **a.** 1 ms **b.** 1μs **c.** 1 ns **d.** 1 ps

a. 1 ms = 1 millisecond = 10^{-3} s   It means one-thousandth of a second.

b. 1 μs = 1 microsecond = 10^{-6} s   It means one-millionth of a second.

c. 1 ns = 1 nanosecond = 10^{-9} s   It means one-billionth of a second.

d. 1 ps = 1 picosecond = 10^{-12} s   It means one-trillionth of a second.

• **EXAMPLE 3**   Change 25 ms to seconds.

First, *milli* means 10^{-3}, and 1 ms = 10^{-3} s. Then

$$25 \cancel{\text{ms}} \times \frac{10^{-3} \text{ s}}{1 \cancel{\text{ms}}} = 0.025 \text{ s}$$

• **EXAMPLE 4**   Change 0.00000025 s to nanoseconds.

First, *nano* means 10^{-9}, and 1 ns = 10^{-9} s. Then

$$0.00000025 \text{ s} \times \frac{1 \text{ ns}}{10^{-9} \text{ s}} = 250 \text{ ns}$$

The basic SI unit of electric current is the ampere (A), sometimes called the amp. This same unit is used in the English system. The ampere is a fairly large amount of current, so smaller currents are measured in parts of an ampere and are given the appropriate SI prefix.

• **EXAMPLE 5**   What is the meaning of each unit?   **a.** 1 mA   **b.** 1 μA

a. 1 mA = 1 milliampere = 10^{-3} A   It means one-thousandth of an ampere.
b. 1 μA = 1 microampere = 10^{-6} A   It means one-millionth of an ampere.

• **EXAMPLE 6**   Change 275 μA to amperes.

First, *micro* means 10^{-6}, and 1 μA = 10^{-6} A. Then

$$275 \text{ μA} \times \frac{10^{-6} \text{ A}}{1 \text{ μA}} = 0.000275 \text{ A}$$

• **EXAMPLE 7**   Change 0.045 A to milliamps.

First, *milli* means 10^{-3}, and 1 mA = 10^{-3} A. Then

$$0.045 \text{ A} \times \frac{1 \text{ mA}}{10^{-3} \text{ A}} = 45 \text{ mA}$$

The common metric unit for both electrical and mechanical power is the watt (W).

• **EXAMPLE 8**   What is the meaning of each unit?   **a.** 1 mW   **b.** 1 kW   **c.** 1 MW

a. 1 mW = 1 milliwatt = 10^{-3} W   It means one-thousandth of a watt.
b. 1 kW = 1 kilowatt = 10^{3} W   It means one thousand watts.
c. 1 MW = 1 megawatt = 10^{6} W   It means one million watts.

• **EXAMPLE 9**   Change 0.025 W to milliwatts.

First, *milli* means 10^{-3}, and 1 mW = 10^{-3} W. Then

$$0.025 \text{ W} \times \frac{1 \text{ mW}}{10^{-3} \text{ W}} = 25 \text{ mW}$$

• **EXAMPLE 10**   Change 2.3 MW to watts.

First, *mega* means 10^{6}, and 1 MW = 10^{6} W. Then

$$2.3 \text{ MW} \times \frac{10^{6} \text{ W}}{1 \text{ MW}} = 2.3 \times 10^{6} \text{ W or } 2,300,000 \text{ W}$$

A few other units commonly used in electronics are listed in Table 3.4. The metric prefixes are used with each of these units in the same way as with the other metric units we have studied.

TABLE 3.4		
Unit	Symbol	Used to measure
volt	V	voltage
ohm	Ω	resistance
hertz	Hz	frequency
farad	F	capacitance
henry	H	inductance
coulomb	C	charge

Exercises 3.5

Which is larger?

1. 1 amp or 1 milliamp

2. 1 microsecond or 1 picosecond

3. 1 second or 1 nanosecond

4. 1 megawatt or 1 milliwatt

5. 1 kilovolt or 1 megavolt

6. 1 volt or 1 millivolt

Write the abbreviation for each unit:

7. 43 kilowatts

8. 7 millivolts

9. 17 picoseconds

10. 1.2 amperes

11. 3.2 megawatts

12. 55 microfarads

13. 450 ohms

14. 70 nanoseconds

Fill in each blank:

15. 1 kW = _____ W

16. 1 mA = _____ A

17. 1 ns = _____ s

18. 1 day = _____ s

19. 1 A = _____ μA

20. 1 F = _____ μF

21. 1 V = _____ MV

22. 1 Hz = _____ kHz

23. Change 0.35 A to mA.

24. Change 18 kW to W.

25. Change 350 ms to s.

26. Change 1 h 25 min 16 s to s.

27. Change 13,950 s to h, min, and s.

28. Change 15 MV to kV.

29. Change 175 μF to mF.

30. Change 145 ps to ns.

31. Change 1500 kHz to MHz.

32. Change 5×10^{12} W to MW.

3.6 Temperature

The basic SI unit for temperature is kelvin (K), which is used mostly in scientific and engineering work. Everyday temperatures are measured in degrees Celsius (°C). The United States also measures temperatures in degrees Fahrenheit (°F).

On the Celsius scale, water freezes at 0° and boils at 100°. Each degree Celsius is 1/100 of the difference between the boiling temperature and the freezing temperature of water. Figure 3.13 shows some approximate temperature readings in degrees Celsius and Fahrenheit and compares them with some related activity.

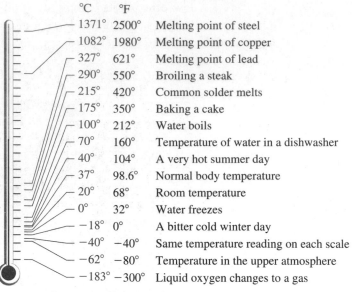

°C	°F	
1371°	2500°	Melting point of steel
1082°	1980°	Melting point of copper
327°	621°	Melting point of lead
290°	550°	Broiling a steak
215°	420°	Common solder melts
175°	350°	Baking a cake
100°	212°	Water boils
70°	160°	Temperature of water in a dishwasher
40°	104°	A very hot summer day
37°	98.6°	Normal body temperature
20°	68°	Room temperature
0°	32°	Water freezes
−18°	0°	A bitter cold winter day
−40°	−40°	Same temperature reading on each scale
−62°	−80°	Temperature in the upper atmosphere
−183°	−300°	Liquid oxygen changes to a gas

FIGURE 3.13
Related temperature readings
in degrees Celsius and degrees
Fahrenheit

The formulas for changing between degrees Celsius and degrees Fahrenheit
are:

$$C = \frac{5}{9}(F - 32°)$$

$$F = \frac{9}{5}C + 32°$$

• **EXAMPLE 1**   Change 68°F to degrees Celsius.

$$C = \frac{5}{9}(F - 32°)$$

$$C = \frac{5}{9}(68° - 32°)$$

$$= \frac{5}{9}(36°)$$

$$= 20°$$

Thus, 68°F = 20°C.

• **EXAMPLE 2**   Change 35°C to degrees Fahrenheit.

$$F = \frac{9}{5}C + 32°$$

$$F = \frac{9}{5}(35°) + 32°$$

$$= 63° + 32°$$

$$= 95°$$

That is, 35°C = 95°F.

• **EXAMPLE 3**   Change 10°F to degrees Celsius.

$$C = \frac{5}{9}(F - 32°)$$

$$C = \frac{5}{9}(10° - 32°)$$

$$= \frac{5}{9}(-22°)$$

$$= -12.2°$$

So 10°F = −12.2°C.

• **EXAMPLE 4**   Change −60°C to degrees Fahrenheit.

$$F = \frac{9}{5}C + 32°$$

$$F = \frac{9}{5}(-60°) + 32°$$

$$= -108° + 32°$$

$$= -76°$$

So −60°C = −76°F.

Exercises 3.6

Use Figure 3.13 to choose the most reasonable answer for each statement:

1. The boiling temperature of water is **a.** 212°C, **b.** 100°C, **c.** 0°C, or **d.** 50°C.

2. The freezing temperature of water is **a.** 32°C, **b.** 100°C, **c.** 0°C, or **d.** −32°C.

3. Normal body temperature is **a.** 100°C, **b.** 50°C, **c.** 37°C, or **d.** 98.6°C.

4. The body temperature of a person who has a fever is **a.** 102°C, **b.** 52°C, **c.** 39°C, or **d.** 37°C.

5. The temperature on a hot summer day in the California desert is **a.** 108°C, **b.** 43°C, **c.** 60°C, or **d.** 120°C.

6. The temperature on a cold winter day in Chicago is **a.** 20°C, **b.** 10°C, **c.** 30°C, or **d.** −10°C.

7. The thermostat in your home should be set at **a.** 70°C, **b.** 50°C, **c.** 19°C, or **d.** 30°C.

8. Solder melts at **a.** 215°C, **b.** 420°C, **c.** 175°C, or **d.** 350°C.

9. Freezing rain is most likely to occur at **a.** 32°C, **b.** 25°C, **c.** −18°C, or **d.** 0°C.

10. The weather forecast calls for a low temperature of 3°C. What should you plan to do? **a.** Sleep with the windows open. **b.** Protect your plants from frost. **c.** Sleep with the air conditioner on. **d.** Sleep with an extra blanket.

Fill in each blank:

11. 77°F − _____ °C

12. 45°C = _____ °F

13. 325°C = _____ °F

14. 140°F = _____ °C

15. −16°C = _____ °F

16. 5°F = _____ °C

17. −16°F = _____ °C

18. −40°C = _____ °F

19. −78°C = _____ °F

20. −10°F = _____ °C

3.7   Metric and English Conversion

In technical work, you must sometimes change from one system of measurement to another. The approximate conversions between metric units and English units are found in the Metric and English Conversion Table on your reference card. Most numbers are rounded to three or four significant digits. Due to this rounding and your choice of conversion factors, there may be a small difference in the last digit(s) of the answers involving conversion factors. This small difference is acceptable. In this section, round each result to three significant digits, when necessary. You may review significant digits in Section 1.11.

Figure 3.14 shows the relative sizes of each of four sets of common metric and English units of area.

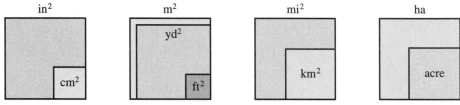

FIGURE 3.14
Relative sizes of some common metric and English units of area

•   **EXAMPLE 1**   Change 17 in. to centimetres.

$$17 \text{ in.} \times \frac{2.54 \text{ cm}}{1 \text{ in.}} = 43.2 \text{ cm}$$

•   **EXAMPLE 2**   Change 1950 g to pounds.

$$1950 \text{ g} \times \frac{1 \text{ lb}}{454 \text{ g}} = 4.30 \text{ lb}$$

Note: If you choose a different conversion factor, the result may vary slightly due to rounding. For example:

$$1950 \text{ g} \times \frac{0.00220 \text{ lb}}{1 \text{ g}} = 4.29 \text{ lb}$$

•   **EXAMPLE 3**   Change 0.85 qt to millilitres.

$$0.85 \text{ qt} \times \frac{0.946 \text{ L}}{1 \text{ qt}} \times \frac{10^3 \text{ mL}}{1 \text{ L}} = 804 \text{ mL}$$

•   **EXAMPLE 4**   Change 5 yd^2 to ft^2.

$$5 \text{ yd}^2 \times \left(\frac{3 \text{ ft}}{1 \text{ yd}}\right)^2 = 45 \text{ ft}^2$$

Use the length conversion factor 1 yd = 3 ft, and first form the conversion factor with yd in the denominator and ft in the numerator. Then raise the conversion factor to the second power to obtain square units in both numerator and denominator. Since the numerator equals the denominator, both the length conversion factor $\frac{3 \text{ ft}}{1 \text{ yd}}$ and its second power $\left(\frac{3 \text{ ft}}{1 \text{ yd}}\right)^2$ equal 1.

Alternate Method:

$$5 \text{ yd}^2 \times \frac{9 \text{ ft}^2}{1 \text{ yd}^2} = 45 \text{ ft}^2$$

The alternate method conversion factor 1 yd^2 = 9 ft^2 is taken directly from the English area conversion table on your reference card. The first method is again preferred, because only the length conversion needs to be remembered or found in a table.

• **EXAMPLE 5**   How many square inches are in a metal plate 14 cm^2 in area?

$$14 \text{ cm}^2 \times \left(\frac{1 \text{ in.}}{2.54 \text{ cm}} \right)^2 = 2.17 \text{ in}^2$$

Alternate Method:

$$14 \text{ cm}^2 \times \frac{0.155 \text{ in}^2}{1 \text{ cm}^2} = 2.17 \text{ in}^2$$

• **EXAMPLE 6**   Change 147 ft^3 to cubic yards.

$$147 \text{ ft}^3 \times \left(\frac{1 \text{ yd}}{3 \text{ ft}} \right)^3 = 5.44 \text{ yd}^3$$

Alternate Method:

$$147 \text{ ft}^3 \times \frac{1 \text{ yd}^3}{27 \text{ ft}^3} = 5.44 \text{ yd}^3$$

• **EXAMPLE 7**   How many cubic yards are in 12 m^3?

$$12 \text{ m}^3 \times \left(\frac{1.09 \text{ yd}}{1 \text{ m}} \right)^3 = 15.5 \text{ yd}^3$$

Alternate Method:

$$12 \text{ m}^3 \times \frac{1.31 \text{ yd}^3}{1 \text{ m}^3} = 15.7 \text{ yd}^3$$

In the English system, the acre is the basic unit of land area. Historically, the acre was the amount of ground that a yoke of oxen could plow in one day.

1 acre = 43,560 ft^2
1 mi^2 = 640 acres = 1 section

• **EXAMPLE 8**   How many acres are in a rectangular field that measures 1350 ft by 2750 ft?

The area in ft^2 is

$$(1350 \text{ ft})(2750 \text{ ft}) = 3{,}712{,}500 \text{ ft}^2$$

$$3{,}712{,}500 \text{ ft}^2 \times \frac{1 \text{ acre}}{43{,}560 \text{ ft}^2} = 85.2 \text{ acres}$$

Professional journals and publications in nearly all scientific areas, including agronomy and animal science, have been metric for several years, so that scientists around the world can better understand and benefit from U.S. research.

Land areas in the United States are typically measured in the English system. When converting between metric and English land-area units, use the following relationship:

1 hectare = 2.47 acres

A good approximation is

1 hectare = 2.5 acres

• **EXAMPLE 9**   How many acres are in 30.6 ha?

$$30.6 \ \cancel{ha} \times \frac{2.47 \text{ acres}}{1 \ \cancel{ha}} = 75.6 \text{ acres}$$

• **EXAMPLE 10**   How many hectares are in the rectangular field in Example 8?

$$85.2 \ \cancel{\text{acres}} \times \frac{1 \text{ ha}}{2.47 \ \cancel{\text{acres}}} = 34.5 \text{ ha}$$

Considerable patience and education will be necessary before the hectare becomes the common unit of land area in the United States. The mammoth task of changing all property documents is only one of many obstacles.

The following example shows how to use multiple conversion factors involving more complex units.

• **EXAMPLE 11**   Change 165 lb/in² to kg/cm².

This conversion requires a series of conversion factors, as follows:

a. from pounds to kilograms
b. from in² to cm²

$$165 \ \frac{\cancel{lb}}{\cancel{in^2}} \times \frac{1 \text{ kg}}{2.20 \ \cancel{lb}} \times \left(\frac{1 \ \cancel{in.}}{2.54 \text{ cm}}\right)^2 = 11.6 \text{ kg/cm}^2$$

Conversion factors for: **a.**   **b.**

Exercises | 3.7

Fill in each blank, rounding each result to three significant digits when necessary:

1. 8 lb = _____ kg
2. 16 ft = _____ m
3. 38 cm = _____ in.
4. 81 m = _____ ft
5. 4 yd = _____ cm
6. 17 qt = _____ L
7. 30 kg = _____ lb
8. 15 mi = _____ km
9. 3.2 in. = _____ mm
10. 2 lb 4 oz = _____ g

11. A road sign reads "75 km to Chicago." What is this distance in miles?
12. A camera uses 35-mm film. How many inches wide is each exposure?
13. The diameter of a bolt is 0.425 in. Express this diameter in mm.
14. Change $3\frac{13}{32}$ in. to cm.
15. A tank contains 8 gal of fuel. How many litres of fuel are in the tank?
16. How many pounds does a 150-kg satellite weigh?

17. An iron bar weighs 2 lb 6 oz. Express its weight in **a.** oz and **b.** kg.

18. A precision part is milled to 1.125 in. in width. What is the width in millimetres?

19. A football field is 100 yd long. What is its length **a.** in feet and **b.** in metres?

20. A micro wheel weighs 0.045 oz. What is its weight in mg?

21. A hole is to be drilled in a metal plate 5 in. in diameter. What is the diameter **a.** in cm and **b.** in mm?

22. A can contains 15 oz of tomato sauce. How many grams does the can contain?

23. Change 3 yd^2 to m^2.

24. Change 12 cm^2 to in^2.

25. How many ft^2 are in 140 yd^2?

26. How many m^2 are in 15 yd^2?

27. Change 18 in^2 to cm^2.

28. How many ft^2 are in a rectangle 12.6 yd long and 8.6 yd wide? ($A = \ell w$)

29. How many ft^2 are in a rectangle 12.6 m long and 8.6 m wide?

30. Find the area of the figure in Illustration 1 in in^2.

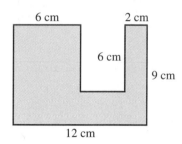

ILLUSTRATION 1

31. Change 15 yd^3 to m^3.

32. Change 5473 in^3 to cm^3.

33. How many mm^3 are in 17 in^3?

34. How many in^3 are in 25 cm^3?

35. Change 84 ft^3 to cm^3.

36. How many cm^3 are in 98 in^3?

37. A commercial lot 80 ft wide and 180 ft deep sold for $32,400. What was the price per square foot? What was the price per frontage foot?

38. A concrete sidewalk is to be built (as shown in Illustration 2) around the outside of a corner lot that measures 140 ft by 180 ft. The sidewalk is to be 5 ft wide. What is the surface area of the sidewalk? The sidewalk is to be 4 in. thick. How many yards (actually, cubic yards) of concrete are needed?

Concrete costs $58/yd^3 delivered. How much will the sidewalk cost?

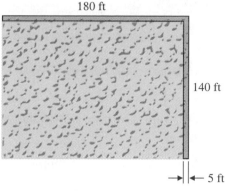

ILLUSTRATION 2

39. How many acres are in a rectangular field that measures 2400 ft by 625 ft?

40. How many acres are in a rectangular field that measures 820 yd by 440 yd?

41. How many hectares are in the field in problem 39?

42. How many hectares are in the field in problem 40?

43. A house lot measures 145 ft by 186 ft. What part of an acre is the lot?

44. How many acres are in $\frac{1}{4}$ mi^2?

45. How many acres are in $\frac{1}{8}$ section?

46. How many acres are in 520 square rods?

47. A corn yield of 10,550 kg/ha is equivalent to how many lb/acre? To how many bu/acre? (1 bu of corn weighs 56 lb.)

48. A soybean yield of 45 bu/acre is equivalent to how many kg/ha? To how many metric tons/hectare? (1 bu of soybeans weighs 60 lb.)

49. How many acres are in eight 30-in. rows 440 yards long?

50. a. How many rows 30 in. apart can be planted in a rectangular field 3300 ft long and 2600 ft wide? The rows run lengthwise.

b. Suppose seed corn is planted at 20 lb/acre. How many bushels of seed corn will be needed to plant the field?

c. Suppose 1 lb of seed corn contains 1200 kernels. How many bags, each containing 80,000 kernels, will be needed?

51. Change 25.6 kg/cm^2 to lb/in^2.

52. Change 1.5 g/cm^2 to mg/mm^2.

53. Change 65 mi/h to m/s.

54. Change 415 lb/ft^3 to g/cm^3.

Chapter 3   Review

Give the metric prefix for each value:

1. 0.001   **2.** 1000

Give the SI abbreviation for each prefix:

3. mega   **4.** micro

Write the SI abbreviation for each quantity:

5. 42 millilitres   **6.** 8.3 nanoseconds

Write the SI unit for each abbreviation:

7. 18 km   **8.** 350 mA   **9.** 50 μs

Which is larger?

10. 1 L or 1 mL   **11.** 1 kW or 1 MW   **12.** 1 km^2 or 1 ha   **13.** 1 m^3 or 1 L

Fill in each blank:

14. 650 m = _____ km   **21.** 25,000 m^2 = _____ ha   **28.** 180 lb = _____ kg
15. 750 mL = _____ L   **22.** 0.6 m^3 = _____ cm^3 **29.** 126 ft = _____ m
16. 6.1 kg = _____ g   **23.** 250 cm^3 = _____ mL   **30.** 360 cm = _____ in.
17. 4.2 A = _____ μA   **24.** 72°F = _____ °C   **31.** 275 in^2 = _____ cm^2
18. 18 MW = _____ W   **25.** −25°C = _____ °F   **32.** 18 yd^2 = _____ ft^2
19. 25 μs = _____ ns   **26.** Water freezes at _____ °C.   **33.** 5 m^3 = _____ ft^3
20. 250 cm^2 = _____ mm^2 **27.** Water boils at _____ °C.   **34.** 15.0 acres = _____ ha

Choose the most reasonable quantity:

35. Jorge and Maria drive **a.** 1600 cm, **b.** 470 m, **c.** 12 km, or **d.** 2400 mm to college each day.

36. Chuck's mass is **a.** 80 kg, **b.** 175 kg, **c.** 14 μg, or **d.** 160 Mg.

37. A car's gas tank holds **a.** 18 L, **b.** 15 kL, **c.** 240 mL, or **d.** 60 L of gasoline.

38. Jamilla, being of average height, is **a.** 5.5 m, **b.** 325 mm, **c.** 55 cm, or **d.** 165 cm tall.

39. A car's average gas consumption is **a.** 320 km/L, **b.** 15 km/L, **c.** 35 km/L, or **d.** 0.75 km/L.

40. On Illinois winter mornings, the temperature sometimes dips to **a.** −50°C, **b.** −30°C, **c.** 30°C, or **d.** −80°C.

41. Abdul drives **a.** 85 km/h, **b.** 50 km/h, **c.** 150 km/h, or **d.** 25 km/h on the interstate highway.

Chapter 3   Test

1. Give the metric prefix for 0.01.

2. Write the SI unit for the abbreviation 150 kg.

3. Which is larger, 1 g or 1 cg?

4. Convert 25°C to degrees Fahrenheit.

5. Convert 28°F to degrees Celsius.

6. What is the basic SI unit of time?

Fill in each blank:

7. 310 g = _____ cg

8. 4.250 km = _____ m

9. 1.52 dL = _____ L

10. 16.2 g = _____ mg

11. 7.236 metric tons = _____ kg

12. 1.7 L = _____ m^3

13. 2.7 m^3 = _____ cm^3

14. 400 ha = _____ km^2

15. 100 km = _____ mi

16. 200 cm = _____ in.

17. 1.8 ft^3 = _____ in^3

18. 1825 in^2 = _____ ft^2

19. 37.8 ha = _____ acres

20. 80.2 kg = _____ lb

Chapters 1–3   Cumulative Review

1. Evaluate: $16 \div 8 + 5 \times 2 - 3 + 7 \times 9$.

2. Find the area of the figure in Illustration 1.

3. Is 2306 divisible by 6?

4. Find the prime factorization of 630.

5. Change $\frac{32}{9}$ to a mixed number in simplest form.

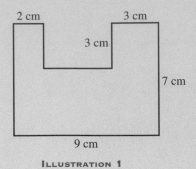

ILLUSTRATION 1

Evaluate and simplify:

6. $\frac{5}{16} + \frac{1}{16} + \frac{7}{16}$

7. $\frac{3}{8} + \frac{1}{4} + \frac{7}{16}$

8. $6\frac{1}{2} - 4\frac{5}{8}$

9. $\frac{2}{5} \times \frac{1}{8}$

10. 5 lb 3 oz = _____ oz

11. Round 615.2875 to the nearest **a.** hundred, **b.** tenth, **c.** ten, and **d.** thousandth.

Perform the indicated operations and simplify:

19. $-\frac{3}{8} + \left(-\frac{1}{4}\right) - 2\frac{5}{16}$

20. $\frac{5}{-8} \div 1\frac{3}{5}$

21. Write 318,180 in scientific notation.

22. Find the larger number: 0.000618 and 0.00215.

23. Simplify. Express the result using a positive expo-

Fill in each blank:

26. 250 m = _____ km

27. 20,050 cm^3 = _____ m^3

Choose the most reasonable answer.

30. In Minnesota, the average summer high temperature during the day is **a.** 60°C, **b.** 70°C, **c.** 32°C, or **d.** 10°C.

12. Change $7\frac{2}{5}\%$ to a decimal.

13. Find 28.5% of $14,000.

14. 212 is 32% of what number?

15. 58 is what percent of 615? (To the nearest hundredth.)

16. A used car is listed to sell at $4800. Joy bought it for $4500. What percent markdown did she receive?

17. Find the value of $-8 + (+9) + (-3) - (-12)$.

18. Multiply: $(-8)(-9)(+3)(-1)(+2)$.

nent:

$$\frac{(10^{-4} \times 10^3)^{-2}}{10^6}$$

24. Give the SI abbreviation for *milli*.

25. Write the SI unit for 250 μs.

28. 86°F = _____ °C

29. 163 in^2 = _____ cm^2

Measurement

MATHEMATICS AT WORK

In the highly paid work of manufacturing precision parts, a sound understanding of tolerances and measuring instruments is critical. The manufacture of automobile parts, hydraulic systems, electrical components, firearms and ammunition depends on the precise measurement of parts to very close tolerances. The degree of precision required varies from company to company but is usually to within thousandths of an inch or of a millimetre. The manufacture of airplane or aerospace parts usually requires precision to within ten-thousandths of an inch or of a millimetre, due to the vibration and stress that occur during takeoff, flight, and landing.

4.1 Approximate Numbers and Accuracy

Approximate Numbers (Measurements) versus Exact Numbers

A *tachometer* is used to measure the number of revolutions an object makes with respect to some unit of time. Since the unit of time is usually minutes, a tachometer usually measures revolutions per minute (rpm). This measurement is usually given in integral units, such as 10 rpm or 255 rpm. Tachometers are used in industry to test motors to see whether or not they turn at a specified rate. In the shop, both wood and metal lathes have specified rpm rates. Tachometers are also commonly used in sports cars to help drivers shift gears at the appropriate engine rpm. A tachometer normally measures the spindle speed or the rpm of a shaft, not the surface speed.

Consider the diagram of the tachometer in Figure 4.1. Each of the printed integral values on the dial indicates hundreds of rpm. That is, if the dial indicator is at 10, then the reading is 10 hundred rpm, or 1000 rpm. If the dial indicator is at 70, then the reading is 70 hundred rpm, or 7000 rpm. Each of the subdivisions between 0 and 10, 10 and 20, 20 and 30, and so forth, represents an additional one hundred rpm.

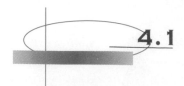

FIGURE 4.1
Tachometer

• **EXAMPLE 1**   Read the tachometer in Figure 4.2.

The indicator is at the sixth division past 30; so the reading is 36 thousand, or 36,000 rpm. ————————————————————•

Tachometer readings are only approximate. Tachometers are calibrated in tens, hundreds, or thousands of revolutions per minute, and it is impossible to read the exact number of rpm.

Next consider the measurement of the length of a metal block like the one in Figure 4.3.

FIGURE 4.2

1. First, measure the block with ruler A, graduated only in inches. This means measurements will be to the nearest inch. The measured length is 2 in.

2. Measure the same block with ruler B, graduated in half-inches. Measurements now are to the nearest half-inch. The measured length is $2\frac{1}{2}$ in. to the nearest half-inch.

3. Measure the block again with ruler C, graduated in fourths of an inch. To the nearest $\frac{1}{4}$ in., the measurement is $2\frac{1}{4}$ in.

4. Measure the block again with ruler D, graduated in eighths of an inch. To the nearest $\frac{1}{8}$ in., the measurement is $2\frac{3}{8}$ in.

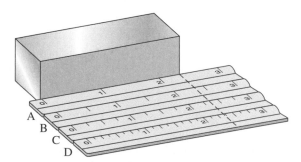

FIGURE 4.3
Measuring the length of a metal block with rulers of different precision

If you continue this process, by using finer and finer graduations on the ruler, will you ever find the "exact" length? No—since all measurements are only approximations, the "exact" length cannot be found. A measurement is only as good as the measuring instrument you use. It would be rather difficult for you to measure the diameter of a pinhead with a ruler.

Up to this time in your study of mathematics, all measurements have probably been treated as exact numbers. An *exact number* is a number that has been determined as a result of counting—such as 24 students enrolled in a class—or by some definition—such as 1 hour (h) = 60 minutes (min), or 1 in. = 2.54 cm. (These are conversion definitions agreed on by government bureaus of standards the world over.) Addition, subtraction, multiplication, and division of exact numbers usually make up the main content of elementary-school mathematics.

However, nearly all data of a technical nature involve *approximate numbers*; that is, numbers that have been determined by some measurement process. This process may be direct, as with a ruler, or indirect, as with a surveying transit. Before studying how to perform calculations with approximate numbers (measurements), we must determine the "correctness" of an approximate number. First, we must realize that no measurement can be found exactly. The length of the cover of this book can be found using many instruments. The better the measuring device used, the better the measurement.

In summary:

> **1.** Only counting numbers are exact.
> **2.** All measurements are approximations.

Accuracy and Significant Digits

The *accuracy* of a measurement means the number of digits, called *significant digits*, that it contains. These indicate the number of units we are reasonably sure of having counted when making a measurement. The greater the number of significant digits given in a measurement, the better the accuracy, and vice versa.

• **EXAMPLE 2**   The average distance between the moon and the earth is 239,000 mi. This measurement indicates measuring 239 thousands of miles. Its accuracy is indicated by 3 significant digits.

• **EXAMPLE 3**   A measurement of 10,900 m indicates measuring 109 hundreds of metres. Its accuracy is 3 significant digits.

• **EXAMPLE 4**   A measurement of 0.042 cm indicates measuring 42 thousandths of a centimetre. Its accuracy is 2 significant digits.

• **EXAMPLE 5**   A measurement of 12.000 m indicates measuring 12,000 thousandths of metres. Its accuracy is 5 significant digits.

Notice that sometimes a zero is significant and sometimes it is not. Apply the following rules to determine whether a digit is significant or not.

SIGNIFICANT DIGITS

1. All nonzero digits are significant.

   For example, the measurement 1765 kg has 4 significant digits. (This measurement indicates measuring 1765 units of kilograms.)

2. All zeros between significant digits are significant.

   For example, the measurement 30,060 m has 4 significant digits. (This measurement indicates measuring 3006 tens of metres.)

3. A zero in a whole-number measurement that is specially tagged, such as by a bar above it, is significant.

   For example, the measurement $3\overline{0},000$ ft has 2 significant digits. (This measurement indicates measuring $3\overline{0}$ thousands of feet.)

4. All zeros to the right of a significant digit *and* a decimal point are significant.

   For example, the measurement $6.100 \underline{L}$ has 4 significant digits. (This measurement indicates measuring $610\overline{0}$ thousandths of litres.)

5. Zeros to the right in a whole-number measurement that are not tagged are *not* significant.

   For example, the measurement 4600 V has 2 significant digits. (This measurement indicates measuring 46 hundreds of volts.)

6. Zeros to the left in a decimal measurement that is less than 1 are *not* significant.

   For example, the measurement 0.00960 s has 3 significant digits. (This measurement indicates $96\overline{0}$ hundred-thousandths of a second.)

• **EXAMPLE 6**   Determine the accuracy of each measurement.

Measurement	Accuracy (significant digits)
a. 109.006 m	6
b. 0.000589 kg	3
c. 75 V	2
d. 239,000 mi	3
e. $239,00\overline{0}$ mi	6
f. $239,0\overline{0}0$ mi	5
g. 0.03200 mg	4
h. 1.20 cm	3
i. 9.020 μA	4
j. 100.050 km	6

Exercises | 4.1

Determine the accuracy of each measurement; that is, give the number of significant digits for each measurement:

1. 115 V

2. 47,000 lb

3. 7009 ft

4. 420 m

5. 6972 m

6. 320,070 ft

7. $440\overline{0}$ ft

8. 4400 Ω

9. $44\overline{0}0$ m

10. 0.0040 g

11. 173.4 m

12. 2070 ft

13. $41,\overline{0}00$ mi

14. 0.025 A

15. 0.0350 in.

16. 6700 g

17. 173 m

18. 8060 ft

19. $240,\overline{0}00$ V

20. 2500 g

21. $72,0\overline{0}0$ mi

22. 137 V

23. 0.047000 A

24. 7.009 g

25. 0.20 mi	**28.** $61\overline{0}$ L	**31.** 100.020 in.	**34.** $15\overline{0}$ cm
26. 69.72 m	**29.** $15,0\overline{0}0$ mi	**32.** 250.0100 m	**35.** 16,000 W
27. 32.0070 g	**30.** 0.07050 mL	**33.** 900,200 ft	**36.** 0.001005 m

 Precision and Greatest Possible Error

PRECISION

The *precision* of a measurement means the smallest unit with which the measurement is made; that is, the position of the last significant digit.

• **EXAMPLE 1** The precision of the measurement 239,000 mi is 1000 mi. (The position of the last significant digit is in the thousands place.)

• **EXAMPLE 2** The precision of the measurement 10,900 m is 100 m. (The position of the last significant digit is in the hundreds place.)

• **EXAMPLE 3** The precision of the measurement 6.90 L is 0.01 L. (The position of the last significant digit is in the hundredths place.)

• **EXAMPLE 4** The precision of the measurement 0.0016 A is 0.0001 A. (The position of the last significant digit is in the ten-thousandths place.)

• **EXAMPLE 5** Determine the precision of each measurement (see Example 6 in Section 4.1).

Measurement	Accuracy (significant digits)	Precision
a. 109.006 m	6	0.001 m
b. 0.000589 kg	3	0.000001 kg
c. 75 V	2	1 V
d. 239,000 mi	3	1000 mi
e. $239,00\overline{0}$ mi	6	1 mi
f. $239,0\overline{0}0$ mi	5	10 mi
g. 0.03200 mg	4	0.00001 mg
h. 1.20 cm	3	0.01 cm
i. 9.020 μA	4	0.001 μA
j. 100.050 km	6	0.001 km

The precision of a measuring instrument is determined by the smallest unit or calibration on the instrument. The precision of the tachometer in Figure 4.4(a) is 100 rpm. The precision of the tachometer in Figure 4.4(b) is 1000 rpm.

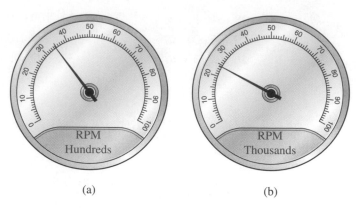

(a)   (b)

FIGURE 4.4

The precision of a ruler graduated in eighths of an inch is $\frac{1}{8}$ in. The precision of a ruler graduated in fourths of an inch is $\frac{1}{4}$ in. However, if a measurement is given as $4\frac{5}{8}$ in., you have no way of knowing what ruler was used. Therefore, you cannot tell whether the precision is $\frac{1}{8}$ in., $\frac{1}{16}$ in., $\frac{1}{32}$ in., or what. The measurement could have been $4\frac{5}{8}$ in., $4\frac{10}{16}$ in., $4\frac{20}{32}$ in., or a similar measurement of some other precision. *Unless it is stated otherwise, you should assume that the smallest unit used is the one that is recorded.* In this case, the precision is assumed to be $\frac{1}{8}$ in.

Now study closely the enlarged portions of the tachometer (tach) readings in Figure 4.5, given in hundreds of rpm. Note that in each case the measurement is 4100 rpm, although the locations of the pointer are slightly different. Any actual rpm between 4050 and 4150 is read 4100 rpm on the scale of the tachometer.

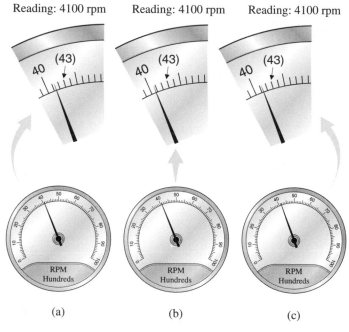

(a)   (b)   (c)

FIGURE 4.5

The *greatest possible error* is one-half of the smallest unit on the scale on which the measurement is read. We see this in the tach readings in Figure 4.5, where any reading within 50 rpm of 4100 rpm is read as 4100 rpm. Therefore, the greatest possible error is 50 rpm.

If you have a tach reading of 5300 rpm, the greatest possible error is $\frac{1}{2}$ of 100 rpm, or 50 rpm. This means that the actual rpm is between $5300 - 50$ and $5300 + 50$; that is, between 5250 rpm and 5350 rpm.

Next consider the measurements of the three metal rods shown in Figure 4.6. Note that in each case, the measurement of the length is $4\frac{5}{8}$ in., although it is obvious that the rods are of different lengths. Any rod that measures $4\frac{9}{16}$ in. to $4\frac{11}{16}$ in. will measure $4\frac{5}{8}$ in. on this scale. The greatest possible error is one-half of the smallest unit on the scale with which the measurement is made.

FIGURE 4.6

If the length of a metal rod is given as $3\frac{3}{16}$ in., the greatest possible error is $\frac{1}{2}$ of $\frac{1}{16}$ in., or $\frac{1}{32}$ in. This means that the actual length of the rod is between $3\frac{3}{16}$ in. $-\frac{1}{32}$ in. and $3\frac{3}{16}$ in. $+\frac{1}{32}$ in.; that is, between $3\frac{5}{32}$ in. and $3\frac{7}{32}$ in.

GREATEST POSSIBLE ERROR

The greatest possible error of a measurement is equal to one-half its precision.

• **EXAMPLE 6**   Find the precision and greatest possible error of the measurement 8.00 kg.

The position of the last significant digit is in the hundredths place; therefore, the precision is 0.01 kg.

The greatest possible error is one-half the precision.

$(0.5)(0.01 \text{ kg}) = 0.005 \text{ kg}$

• **EXAMPLE 7**   Find the precision and greatest possible error of the measurement 26,000 gal.

The position of the last significant digit is in the thousands place; therefore, the precision is 1000 gal.

The greatest possible error is one-half the precision.

$\frac{1}{2} \times 1000 \text{ gal} = 500 \text{ gal}$

• **EXAMPLE 8**   Find the precision and greatest possible error of the measurement 0.0460 mg.

Precision: 0.0001 mg

Greatest possible error: $(0.5)(0.0001 \text{ mg}) = 0.00005 \text{ mg}$

• **EXAMPLE 9**   Find the precision and the greatest possible error of the measurement $7\frac{5}{8}$ in.

The smallest unit is $\frac{1}{8}$ in., which is the precision.

The greatest possible error is $\frac{1}{2} \times \frac{1}{8}$ in. $= \frac{1}{16}$ in., which means that the actual length is within $\frac{1}{16}$ in. of $7\frac{5}{8}$ in.

Many calculations with measurements are performed by people who do not make the actual measurements. Therefore, it is often necessary to agree on a method of recording measurements to indicate the precision of the instrument used.

Exercises    4.2

Find **a.** *the precision and* **b.** *the greatest possible error of each measurement:*

1. 2.70 A	**7.** 17.50 mi	**13.** $14\overline{0}0\ \Omega$	**19.** 120 V	**25.** $9\frac{7}{32}$ in.
2. 13.0 ft	**8.** 6.100 m	**14.** 301,000 Hz	**20.** $30\overline{0}$ km	**26.** $4\frac{5}{8}$ mi
3. 14.00 cm	**9.** 0.040 A	**15.** $3\overline{0},000$ L	**21.** 67.500 m	**27.** $9\frac{5}{16}$ mi
4. 1.000 in.	**10.** 0.0001 in.	**16.** $7,0\overline{0}0,000$ g	**22.** $1\frac{7}{8}$ in.	**28.** $5\frac{13}{64}$ in.
5. 15 km	**11.** 0.0805 W	**17.** 428.0 cm	**23.** $3\frac{2}{3}$ yd	**29.** $9\frac{4}{9}$ in^2
6. 1.010 cm	**12.** $10,\overline{0}00$ W	**18.** 60.0 cm	**24.** $3\frac{3}{4}$ yd	**30.** $18\frac{4}{5}$ in^3

 4.3   The Vernier Caliper

In your use of English and metric rulers for making measurements, you have seen that very precise results are difficult to obtain. When more precise measurements are required, you must use a more precise instrument. One such instrument is the *vernier caliper*, which is used by technicians in machine shops, plant assembly lines, and many other work places. This instrument is a slide-type caliper used to take inside, outside, and depth measurements. It has two metric scales and two English scales. A vernier caliper is shown in Figure 4.7.

FIGURE 4.7
Vernier caliper

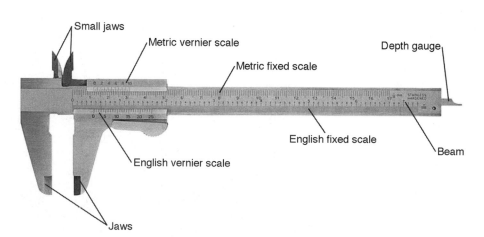

To make an outside measurement, the jaws are closed snugly around the outside of an object, as in Figure 4.8. For an inside measurement, the smaller jaws are placed inside the object to be measured, as in Figure 4.9.

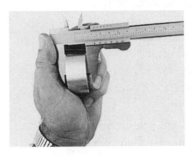

FIGURE 4.8
Outside measurement

FIGURE 4.9
Inside measurement

For a depth measurement, the depth gauge is inserted into the opening to be measured, as in Figure 4.10.

Here are some tips for using a vernier caliper (and the micrometer in Section 4.4):

1. Check that the instrument is held perpendicular (that is, at 90°) to the surface of the part being measured.

2. When measuring the diameter of a round piece, check that the *full* diameter is being measured.

3. On rounds, take two readings at approximately 90° to each other. Then average the two readings.

Let's first consider the metric scales (Figure 4.11). One of them is fixed and located on the upper part of the beam. This fixed scale is divided into centimetres and subdivided into millimetres, so record all readings in millimetres (mm). The other metric scale, called the vernier scale, is the upper scale on the slide. The vernier scale is divided into tenths of millimetres (0.1 mm) and subdivided into halves of tenths of millimetres $\left(\frac{1}{20} \text{ mm or } 0.05 \text{ mm}\right)$. The precision of the vernier scale is therefore 0.05 mm.

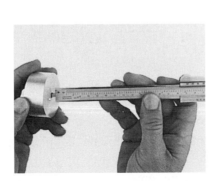

FIGURE 4.10
Depth measurement

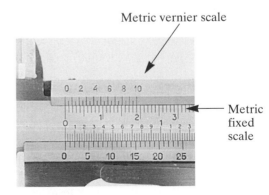

Metric vernier scale

Metric fixed scale

FIGURE 4.11

The figures in the examples and exercises have been computer generated for easier reading.

READING A VERNIER CALIPER IN MILLIMETRES

STEP 1 Determine the number of whole millimetres in a measurement by counting—on the fixed scale—the number of millimetre graduations that are to the left of the zero graduation on the vernier scale. (Remember that each numbered graduation on the fixed scale represents 10 mm.) The zero graduation on the vernier scale may be directly in line with a graduation on the fixed scale. If so, read the total measurement directly from the fixed scale. Write it in millimetres, followed by a decimal point and two zeros.

STEP 2 The zero graduation on the vernier scale may not be directly in line with a graduation on the fixed scale. In that case, find the graduation on the vernier scale that is most nearly in line with any graduation on the fixed scale.

    a. If the vernier graduation is a long graduation, it represents the number of tenths of millimetres between the last graduation on the fixed scale and the zero graduation on the vernier scale. Then insert a zero in the hundredths place.

    b. If the vernier graduation is a short graduation, add 0.05 mm to the vernier graduation that is on the immediate left of the short graduation.

STEP 3 Add the numbers from Steps 1 and 2 to determine the total measurement.

• **EXAMPLE 1**   Read the measurement in millimetres on the vernier caliper in Figure 4.12.

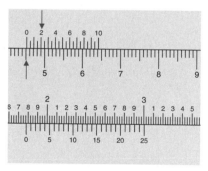

FIGURE 4.12

STEP 1 The first mark to the left of the zero mark is: 45.00 mm

STEP 2 The mark on the vernier scale that most nearly lines up with a mark on the fixed scale is: 0.20 mm

STEP 3 The total measurement is: 45.20 mm

• **EXAMPLE 2**   Read the measurement in millimetres on the vernier caliper in Figure 4.13.

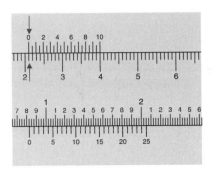

FIGURE 4.13

The total measurement is 21.00 mm, because the zero graduation on the vernier scale most nearly lines up with a mark on the fixed scale. ●

• **EXAMPLE 3** Read the measurement in millimetres on the vernier caliper in Figure 4.14.

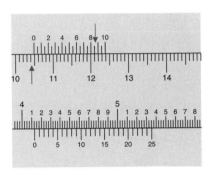

FIGURE 4.14

STEP 1 The first mark to the left of the zero mark is: 104.00 mm

STEP 2 The mark on the vernier scale that most nearly lines up with a mark on the fixed scale is: 0.85 mm

STEP 3 Total measurement: 104.85 mm

Exercises | 4.3A

Read the measurement shown on each vernier caliper in millimetres:

1.

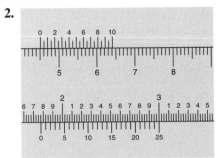

2.

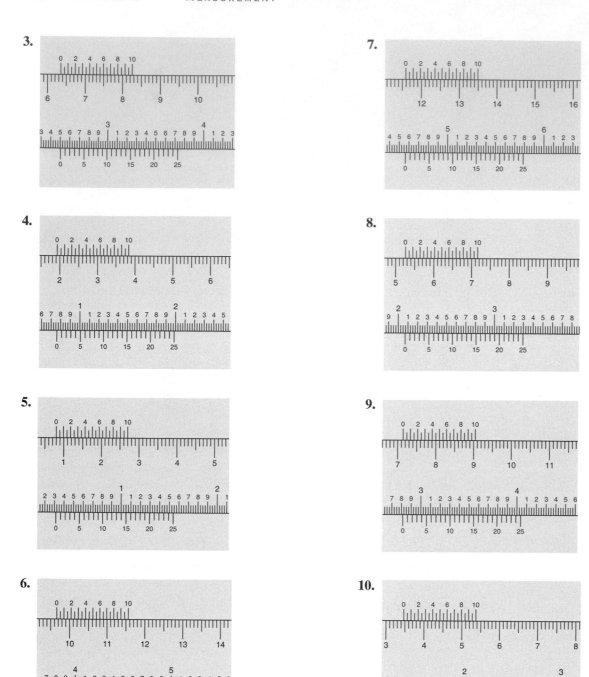

11–20. *Read in millimetres the measurement shown on each vernier caliper in Exercises 4.3B* (page 155).

Now consider the two English scales on the vernier caliper in Figure 4.15. One is fixed, the other movable. The fixed scale is located on the lower part of the beam, where each inch is divided into tenths (0.100 in., 0.200 in., 0.300 in., and so on). Each tenth is subdivided into four parts, each of which represents 0.025 in. The vernier scale is divided into 25 parts, each of which represents thousandths (0.001 in.). This means that the precision of this scale is 0.001 in.

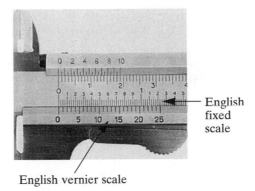

English fixed scale

English vernier scale

FIGURE 4.15

READING A VERNIER CALIPER IN THOUSANDTHS OF AN INCH

STEP 1 Determine the number of inches and tenths of inches by reading the first numbered division that is to the left of the zero graduation on the vernier scale.

STEP 2 Add 0.025 in. to the number from Step 1 for each graduation between the last numbered division on the fixed scale and the zero graduation on the vernier scale. (If this zero graduation is directly in line with a graduation on the fixed scale, read the total measurement directly from the fixed scale.)

STEP 3 If the zero graduation on the vernier scale is not directly in line with a graduation on the fixed scale, find the graduation on the vernier scale that is most nearly in line with any graduation on the fixed scale. This graduation determines the number of thousandths of inches in the measurement.

STEP 4 Add the numbers from Steps 1, 2, and 3 to determine the total measurement.

• **EXAMPLE 1** Read the measurement in inches shown on the vernier caliper in Figure 4.16.

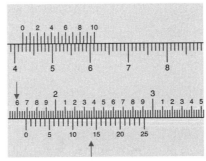

FIGURE 4.16

STEP 1 The first numbered mark to the left of the zero mark is: 1.600 in.

STEP 2 The number of 0.025-in. graduations is 3; 3×0.025 in. = 0.075 in.

STEP 3 The mark on the vernier scale that most nearly lines
        up with a mark on the fixed scale is: 0.014 in.
STEP 4 Total measurement: 1.689 in.

● EXAMPLE 2   Read the measurement in inches shown on the vernier caliper in Figure 4.17.

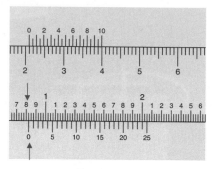

FIGURE 4.17

STEP 1 The first numbered mark to the left of the zero
        mark is: 0.800 in.
STEP 2 The number of 0.025-in. graduations is 1;
        1×0.025 in. =   0.025 in.
STEP 4 Total measurement: 0.825 in.

Note: The zero graduation is directly in line with a graduation on the fixed scale.

● EXAMPLE 3   Read the measurement in inches shown on the vernier caliper in Figure 4.18.

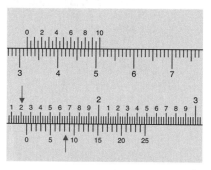

FIGURE 4.18

STEP 1 The first numbered mark to the left of the zero mark is: 1.200 in.
STEP 2 The number of 0.025-in. graduations is 2; 2×0.025 in. = 0.050 in.
STEP 3 The mark on the vernier scale that most nearly lines up
        with a mark on the fixed scale is: 0.008 in.
STEP 4 Total measurement: 1.258 in.

Exercises | 4.3B

Read the measurement shown on each vernier caliper in inches:

1.

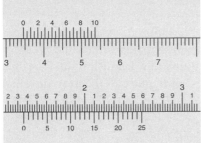

6.

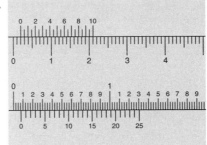

2.

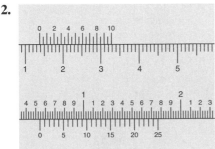

7.

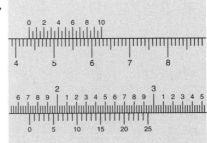

3.

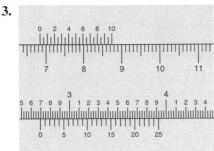

8.

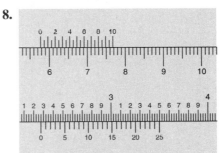

4.

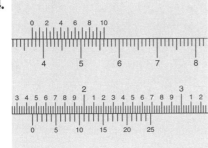

9.

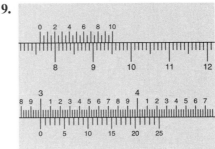

5.

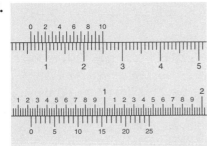

10.

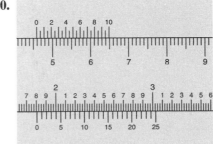

11–20. *Read in inches the measurement shown on each vernier caliper in Exercises 4.3A (page 151).*

4.4   The Micrometer Caliper

The *micrometer caliper* (micrometer or "mike") is a more precise measuring instrument than the vernier caliper and is used in technical fields where fine precision is required. Micrometers are available in metric units and English units. The metric "mike" is graduated and read in hundredths of a millimetre (0.01 mm); the English "mike" is graduated and read in thousandths of an inch (0.001 in.). The parts of a micrometer are labeled in Figure 4.19.

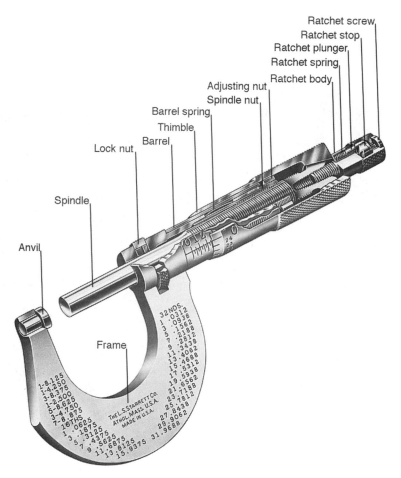

FIGURE 4.19
Basic parts of a micrometer

To use a micrometer properly, place the object to be measured between the anvil and spindle, and turn the thimble until the object fits snugly. *Do not force the turning of the thimble, since this may damage the very delicate threads on the spindle that are located inside the thimble.* Some calipers have a ratchet to protect the instrument; the ratchet prevents the thimble from being turned with too much force. A metric micrometer is shown in Figure 4.20, and the basic parts are labeled. The barrels of most metric micrometers are graduated in millimetres. The micrometer in

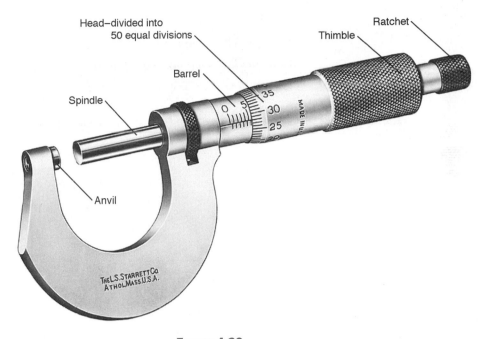

FIGURE 4.20
Metric micrometer

Figure 4.20 also has graduations of halves of millimetres, which are indicated by the lower set of graduations on the barrel. The threads on the spindle are made so that it takes two complete turns of the thimble for the spindle to move precisely one millimetre. The head is divided into 50 equal divisions—each division indicating 0.01 mm, which is the precision.

READING A METRIC MICROMETER IN MILLIMETRES

STEP 1 Find the *whole number* of mm in the measurement by counting the number of mm graduations on the barrel to the left of the head.

STEP 2 Find the *decimal part* of the measurement by reading the graduation on the head (see Figure 4.20) that is most nearly in line with the center line on the barrel. Then multiply this reading by 0.01. If the head is at, or immediately to the right of, the half-mm graduation, then add 0.50 mm to the reading on the head.

STEP 3 Add the numbers found in Step 1 and Step 2.

•   **EXAMPLE 1**   Read the measurement shown on the metric micrometer in Figure 4.21.

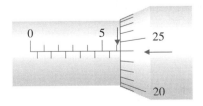

FIGURE 4.21

STEP 1 The barrel reading is   6.00 mm
STEP 2 The head reading is   0.24 mm
STEP 3 The total measurement is   6.24 mm

• **EXAMPLE 2**   Read the measurement shown on the metric micrometer in Figure 4.22.

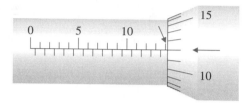

FIGURE 4.22

STEP 1 The barrel reading is   14.00 mm
STEP 2 The head reading is   0.12 mm
STEP 3 The total measurement is   14.12 mm

• **EXAMPLE 3**   Read the measurement shown on the metric micrometer in Figure 4.23.

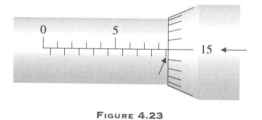

FIGURE 4.23

STEP 1 The barrel reading is   8.50 mm
        (Note that the head is past the half-mm mark.)
STEP 2 The head reading is   0.15 mm
STEP 3 The total measurement is   8.65 mm

<u>Exercises</u> | **4.4A**

Read the measurement shown on each metric micrometer:

1.   **2.**

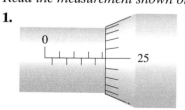   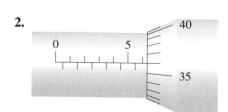

3.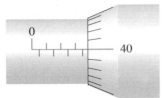

4.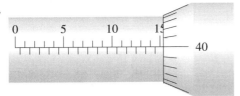

5.

6.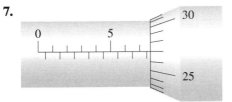

7.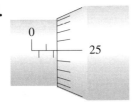

8.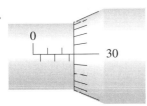

9.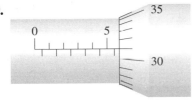

10.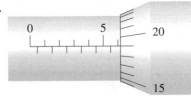

11.

12.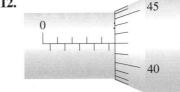

13.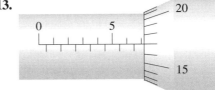

14.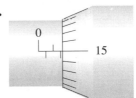

15.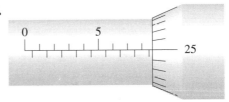

16.

17.

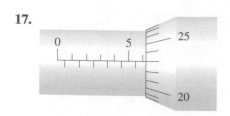

18.

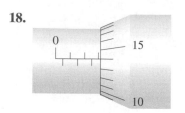

19.

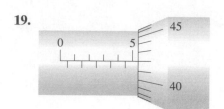

20.

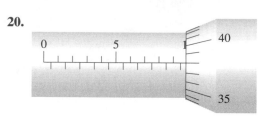

The barrel of the English micrometer shown in Figure 4.24 is divided into tenths of an inch. Each tenth is subdivided into four 0.025-in. parts. The threads on the spindle allow the spindle to move 0.025 in. in one complete turn of the thimble and 4×0.025 in., or 0.100 in., in four complete turns. The head is divided into 25 equal divisions—each division indicating 0.001 in., which is the precision.

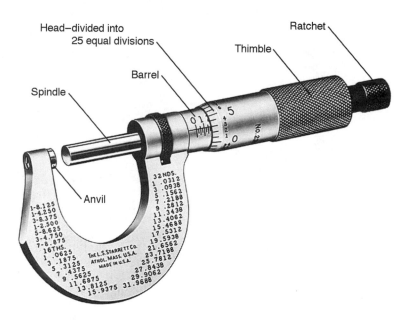

FIGURE 4.24
English micrometer

READING AN ENGLISH MICROMETER IN THOUSANDTHS OF AN INCH

STEP 1 Read the last numbered graduation showing on the barrel. Multiply this number by 0.100 in.

STEP 2 Find the number of smaller graduations between the last numbered graduation and the head. Multiply this number by 0.025 in.

> STEP 3 Find the graduation on the head that is most nearly in line with the center line on the barrel. Multiply the number represented by this graduation by 0.001 in.
>
> STEP 4 Add the numbers found in Steps 1, 2, and 3.

• **EXAMPLE 4**   Read the measurement shown on the English micrometer in Figure 4.25.

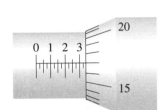

STEP 1 3 numbered divisions on the barrel
(3 × 0.100 in.):   0.300 in.

STEP 2 1 small division on the barrel
(1 × 0.025 in.):   0.025 in.

STEP 3 The head reading is 17
(17 × 0.001 in.):   0.017 in.

STEP 4 The total measurement is: 0.342 in.

FIGURE 4.25

• **EXAMPLE 5**   Read the measurement shown on the English micrometer in Figure 4.26.

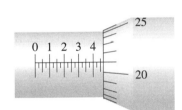

STEP 1 4 numbered divisions on the barrel
(4 × 0.100 in.):   0.400 in.

STEP 2 2 small divisions on the barrel
(2 × 0.025 in.):   0.050 in.

STEP 3 The head reading is 21
(21 × 0.001 in.):   0.021 in.

STEP 4 The total measurement is: 0.471 in.

FIGURE 4.26

Exercises | 4.4B

Read the measurement shown on each English micrometer:

1.

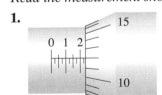

3.

2.

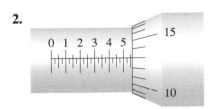

4.

5.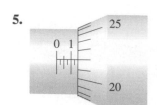

6.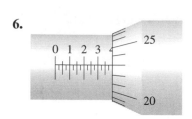

7.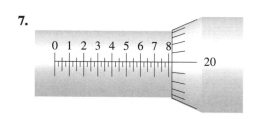

8.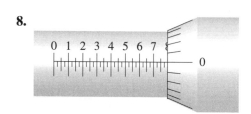

9.

10.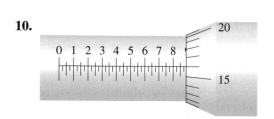

11.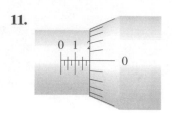

12.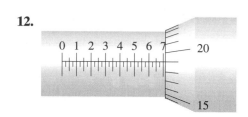

13.

14.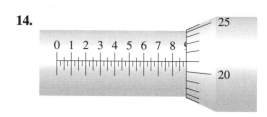

15.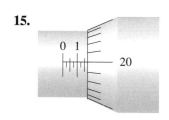

16.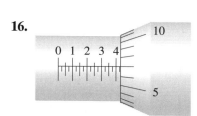

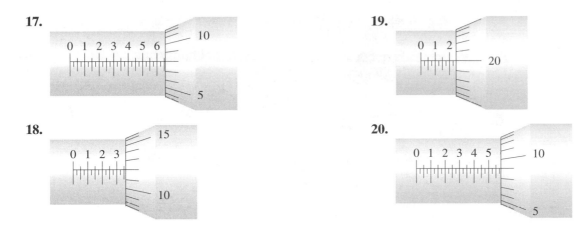

Other Micrometers

By adding a vernier scale on the barrel of a micrometer (as shown in Figure 4.27), we can increase the precision by one more decimal place. That is, the metric micrometer with vernier scale has a precision of 0.001 mm. The English micrometer with vernier scale has a precision of 0.0001 in. Of course, these micrometers cost more, because they require more precise threading than the ones previously discussed. Nevertheless, many jobs require this precision.

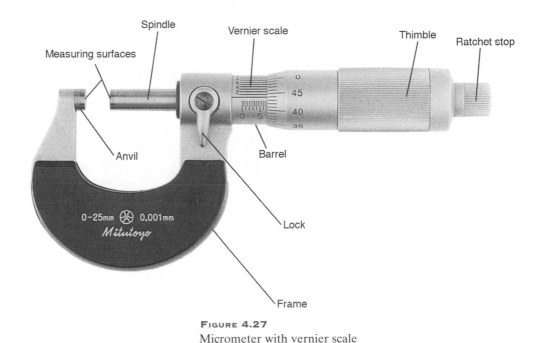

FIGURE 4.27
Micrometer with vernier scale

Micrometers are basic, useful, and important tools of the technician. Figure 4.28 shows just a few of their uses.

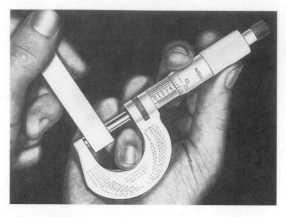

(a) Measuring a piece of die steel

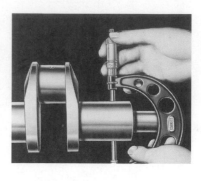

(b) Measuring the diameter of a crankshaft bearing

(c) Measuring tubing wall thickness with a round anvil micrometer

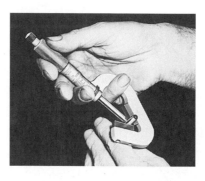

(d) Checking out-of-roundness on centerless grinding work

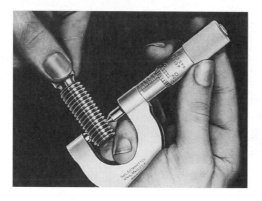

(e) Measuring the pitch diameter of a screw thread

(f) Measuring the depth of a shoulder with a micrometer depth gauge

FIGURE 4.28
Examples of how micrometers are used

4.5   Addition and Subtraction of Measurements

Precision versus Accuracy

Recall that the *precision* of a measurement is the smallest unit with which a measurement is made; that is, the *position* of the last significant digit. Recall also that the *accuracy* of a measurement is the *number* of digits, called significant digits, which indicate the number of units we are reasonably sure of having counted when making a measurement. Unfortunately, some people tend to use the terms *precision* and *accuracy* interchangeably. Each term expresses a different aspect of a given measurement.

• **EXAMPLE 1**   Compare the precision and the accuracy of the measurement 0.0004 mm.

Since the precision is 0.0001 mm, its precision is relatively good. However, since the accuracy is only one significant digit, its accuracy is relatively poor. ——————•

• **EXAMPLE 2**   Given the measurements 13.00 m, 0.140 m, 3400 m, and 0.006 m, find the measurement which is **a.** the least precise, **b.** the most precise, **c.** the least accurate, and **d.** the most accurate.

First, let's find the precision and the accuracy of each measurement.

Measurement	Precision	Accuracy (significant digits)
13.00 m	0.01 m	4
0.140 m	0.001 m	3
3400 m	100 m	2
0.006 m	0.001 m	1

From the table, we find:

a. The least precise measurement is 3400 m.

b. The most precise measurements are 0.140 m and 0.006 m.

c. The least accurate measurement is 0.006 m.

d. The most accurate measurement is 13.00 m. ——————•

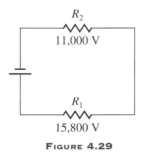

R_2

11,000 V

R_1

15,800 V

FIGURE 4.29

In a series circuit, the electromagnetic force (emf) of the source equals the sum of the separate voltage drops across each resistor in the circuit. Suppose that someone measures the voltage across the first resistor R_1 in Figure 4.29. He uses a voltmeter calibrated in hundreds of volts, and measures 15,800 V. Across the second resistor R_2, he uses a voltmeter in thousands of volts, and measures 11,000 V. Does the total emf equal 26,800 V? Note that the first voltmeter and its reading indicate a precision of 100 V and a greatest possible error of 50 V. This means that the actual reading lies between 15,750 V and 15,850 V. The second voltmeter and its reading indicate a precision of 1000 V and a greatest possible error of 500 V. The actual reading, therefore, lies between 10,500 V and 11,500 V. This means that we are not very certain of the digit in the hundreds place in the sum 26,800 V.

To be consistent when adding or subtracting measurements of different precision, the sum or difference can be no more *precise* than the least precise measurement.

ADDING OR SUBTRACTING MEASUREMENTS OF DIFFERENT PRECISION

1. Make certain that all measurements are expressed in the same units. If they are not, change them all to any common unit.
2. Add or subtract.
3. Then round the result to the same precision as the *least precise* measurement.

The total emf in the circuit shown in Figure 4.29 is therefore calculated as follows:

Measurement		*Rounded*

R_1:   15,800 V
R_2:   11,000 V
          26,800 V → 27,000 V

• **EXAMPLE 3**   Use the rules for addition of measurements to add 13,800 ft, 14,020 ft, 19,864 ft, 2490 ft and 14,700 ft.

Since all of the measurements are in the same unit (that is, ft), add them together:

    13,800 ft
    14,020 ft
    19,864 ft
     2,490 ft
    14,700 ft
    64,874 ft → 64,900 ft

Round this sum to the same precision as the least precise measurement. Since the precision of both 13,800 ft and 14,700 ft is 100 ft, round the sum to the nearest 100 ft. Thus, the sum is 64,900 ft.

• **EXAMPLE 4**   Use the rules for addition of measurements to add 735,000 V, 490,000 V, 86,000 V, 1,300,000 V, and $20\overline{0},000$ V.

Since all of the measurements are in the same unit, add:

      735,000 V
      490,000 V
       86,000 V
    1,300,000 V
      $20\overline{0},000$ V
    2,811,000 V → 2,800,000 V

The least precise measurement is 1,300,000 V, which has a precision of 100,000 V. Round the sum to the nearest hundred thousand volts: 2,800,000 V.

• **EXAMPLE 5**   Use the rules for addition of measurements to add 13.8 m, 140.2 cm, 1.853 m, and 29.95 cm.

First, change each measurement to a common unit (say, m) and add:

$$
\begin{aligned}
13.8 \ \ \text{m} &\rightarrow 13.8 \ \ \ \text{m} \\
140.2 \ \ \text{cm} &\rightarrow 1.402 \ \ \text{m} \\
1.853 \ \text{m} &\rightarrow 1.853 \ \ \text{m} \\
\underline{29.95 \ \ \text{cm}} &\rightarrow \underline{0.2995 \ \text{m}} \\
&\ \ \ 17.3545 \ \text{m} \rightarrow 17.4 \ \text{m}
\end{aligned}
$$

The least precise measurement is 13.8 m, which is precise to the nearest tenth of a metre. So round the sum to the nearest tenth of a metre: 17.4 m. ———————•

• **EXAMPLE 6**   Use the rules for subtraction of measurements to subtract 19.352 cm from 41.7 cm.

Since both measurements have the same units, subtract:

$$
\begin{aligned}
41.7 \ \ \ \text{cm} \\
\underline{19.352 \ \text{cm}} \\
22.348 \ \text{cm} \rightarrow 22.3 \ \text{cm}
\end{aligned}
$$

The least precise measurement is 41.7 cm, which is precise to the nearest tenth of a cm. Round the difference to the nearest tenth of a cm: 22.3 cm. ———————•

Exercises | **4.5**

*In each set of measurements, find the measurement that is **a.** the most accurate and **b.** the most precise:*

1. 14.7 in.; 0.017 in.; 0.09 in.

2. 459 ft; 600 ft; 190 ft

3. 0.737 mm; 0.94 mm; 16.01 mm

4. 4.5 cm; 9.3 cm; 7.1 cm

5. 0.0350 A; 0.025 A; 0.00050 A; 0.041 A

6. 134.00 g; 5.07 g; 9.000 g; 0.04 g

7. 145 cm; 73.2 cm; 2560 cm; 0.391 cm

8. 15.2 km; 631.3 km; 20.0 km; 37.7 km

9. 205,000 Ω; 45,000 Ω; 5$\overline{0}$0,000 Ω; 90,000 Ω

10. 1,500,000 V; 65,000 V; 30,$\overline{0}$00 V; 20,000 V

*In each set of measurements, find the measurement that is **a.** the least accurate and **b.** the least precise:*

11. 15.5 in.; 0.053 in.; 0.04 in.

12. 635 ft; 400 ft; 240 ft

13. 43.4 cm; 0.48 cm; 14.05 cm

14. 4.9 kg; 670 kg; 0.043 kg

15. 0.0730 A; 0.043 A; 0.00008 A; 0.91 A

16. 197.0 m; 5.43 m; 4.000 m; 0.07 m

17. 2.1 m; 31.3 m; 461.5 m; 0.6 m

18. 295 m; 91.3 m; 1920 m; 0.360 m

19. 405,000 Ω; 35,000 Ω; 8$\overline{0}$0,000 Ω; 500,000 Ω

20. 1,600,000 V; 36,000 V; 40,$\overline{0}$00 V; 60,000 V

Use the rules for addition of measurements to find the sum of each set of measurements:

21. 14.7 m; 3.4 m

22. 168 in.; 34.7 in.; 61 in.

23. 42.6 cm; 16.41 cm; 1.417 cm; 34.4 cm

24. 407 g; 1648.5 g; 32.74 g; 98.1 g

25. 26,000 W; 19,600 W; 8450 W; 42,500 W

26. 5420 km; 1926 km; 850 km; 2$\overline{0}$00 km

27. 140,000 V; 76,200 V; 4700 V; 254,000 V; 370,000 V

28. 19,200 m; 8930 m; 50,040 m; 137 m

29. 14 V; 1.005 V; 0.017 V; 3.6 V

30. 120.5 cm; 16.4 cm; 1.417 m

31. 10.555 cm; 9.55 mm; 13.75 cm; 206 mm

32. 1350 cm; 1476 mm; 2.876 m; 4.82 m

Use the rules for subtraction of measurements to subtract the second measurement from the first:

33. 140.2 cm
   $\underline{\ \ 13.8 \ \text{cm}}$

34. 14.02 mm
   $\underline{\ \ 13.8 \ \ \text{mm}}$

35. 9200 mi
   $\underline{\ \ 627 \ \text{mi}}$

36. 1,900,000 V
    645,000 V

37. 167 mm
    13.2 cm

38. 16.41 oz
    11.372 oz

39. 98.1 g
    32.743 g

40. 4.000 in.
    2.006 in.

41. 0.54361 in.
    0.214 in.

42. If you bolt four pieces of metal together, with thicknesses 0.136 in., 0.408 in., 1.023 in., and 0.88 in., what is the total thickness?

43. If you clamp five pieces of metal together, with thicknesses 2.38 mm, 10.5 mm, 3.50 mm, 1.455 mm, and 8.200 mm, what is the total thickness?

44. What is the current going through R_5 in the circuit in Illustration 1? (*Hint:* In a parallel circuit, the current is divided among its branches. That is, $I_T = I_1 + I_2 + I_3$.)

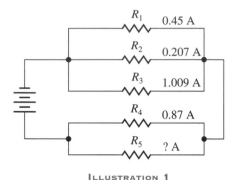

ILLUSTRATION 1

4.6   Multiplication and Division of Measurements

Suppose that you want to find the area of a plot of ground that measures 206 m by 84 m. The product, 17,304 m², shows five significant digits. The original measurements have three and two significant digits, respectively.

To be consistent when multiplying or dividing measurements, the product or quotient can be no more *accurate* than the least accurate measurement.

MULTIPLYING OR DIVIDING MEASUREMENTS

1. First, multiply and/or divide the measurements.
2. Then round the result to the same number of significant digits as the measurement that has the least number of significant digits. That is, round the result to the same accuracy as the *least accurate* measurement.

Using this procedure, the area of the plot of ground (206 m × 84 m = 17,304 m²) is rounded to 17,000 m².

•   **EXAMPLE 1**   Use the rules for multiplication of measurements: 20.41 g × 3.5 cm.

STEP 1   20.41 g × 3.5 cm = 71.435 g cm

STEP 2   Round this product to two significant digits, which is the accuracy of the least accurate measurement, 3.5 cm. That is,

20.41 g × 3.5 cm = 71 g cm

•   **EXAMPLE 2**   Use the rules for multiplication of measurements: 125 m × 345 m × 204 m.

STEP 1   125 m × 345 m × 204 m = 8,797,500 m³

STEP 2 Round this product to three significant digits, which is the accuracy of the least accurate measurement (which is the accuracy of each measurement in this example). That is,

$$125 \text{ m} \times 345 \text{ m} \times 204 \text{ m} = 8,8\overline{0}0,000 \text{ m}^3$$ ————●

• **EXAMPLE 3** Use the rules for division of measurements to divide 288,000 ft^3 by 216 ft.

STEP 1 $\dfrac{288,000 \text{ ft}^3}{216 \text{ ft}} = 1333.333 \ldots \text{ ft}^2$

STEP 2 Round this quotient to three significant digits, which is the accuracy of the least accurate measurement (which is the accuracy of each measurement in this example). That is,

$$\frac{288,000 \text{ ft}^3}{216 \text{ ft}} = 1330 \text{ ft}^2$$ ————●

• **EXAMPLE 4** Use the rules for multiplication and division of measurements to evaluate

$$\frac{4750 \text{ N} \times 4.82 \text{ m}}{1.6 \text{ s}}$$

STEP 1 $\dfrac{4750 \text{ N} \times 4.82 \text{ m}}{1.6 \text{ s}} = 14{,}309.375 \dfrac{\text{N m}}{\text{s}}$

STEP 2 Round this result to two significant digits, which is the accuracy of the least accurate measurement, 1.6 s. That is,

$$\frac{4750 \text{ N} \times 4.82 \text{ m}}{1.6 \text{ s}} = 14{,}000 \frac{\text{N m}}{\text{s}} \quad \text{or} \quad \text{or } 14{,}000 \text{ N m/s}$$ ————●

There are even more sophisticated methods for dealing with the calculations of measurements. The method one uses (and indeed, whether one should even follow any given procedure) depends on the number of measurements and the sophistication needed for a particular situation.

The procedures for addition, subtraction, multiplication, and division of measurements are based on methods followed and presented by the American Society for Testing and Materials.

Note: To multiply or divide measurements, the units do not need to be the same. (They must be the same in addition and subtraction of measurements.) Also note that the units are multiplied and/or divided in the same manner as the corresponding numbers.

Exercises │ 4.6

Use the rules for multiplication and/or division of measurements to evaluate:

1. 126 m $\times$ 35 m
2. 470 mi $\times$ 1200 mi

3. 1463 cm $\times$ 838 cm
4. 2.4 A $\times$ 3600 Ω

5. 18.7 m $\times$ 48.2 m
6. 560 cm $\times$ 28.0 cm

7. 4.7 Ω $\times$ 0.0281 A
8. 5.2 km $\times$ 6.71 km

9. 24.2 cm $\times$ 16.1 cm $\times$ 18.9 cm

10. 0.045 m $\times$ 0.0292 m $\times$ 0.0365 m

11. 2460 m $\times$ 960 m $\times$ 1970 m

12. 460 in. $\times$ 235 in. $\times$ 368 in.

13. $(0.480 \text{ A})^2 (150 \ \Omega)$

14. 360 ft^2 $\div$ 12 ft

15. 62,500 in^3 $\div$ 25 in.

16. 9180 yd^3 $\div$ 36 yd^2

17. $1520 \text{ m}^2 \div 40 \text{ m}$

18. $18.4 \text{ m}^3 \div 9.2 \text{ m}^2$

19. $4800 \text{ V} \div 14.2 \text{ A}$

20. $\dfrac{4800 \text{ V}}{6.72 \text{ } \Omega}$

21. $\dfrac{5.63 \text{ km}}{2.7 \text{ s}}$

22. $\dfrac{0.497 \text{ N}}{1.4 \text{ m} \times 8.0 \text{ m}}$

23. $\dfrac{(120 \text{ V})^2}{47.6 \text{ } \Omega}$

24. $\dfrac{19 \text{ kg} \times (3.0 \text{ m/s})^2}{2.46 \text{ m}}$

25. $\dfrac{140 \text{ g}}{3.2 \text{ cm} \times 1.7 \text{ cm} \times 6.4 \text{ cm}}$

26. Find the area of a rectangle measured as 6.5 cm by 28.3 cm. ($A = \ell w$)

27. $V = \ell w h$ is the formula for the volume of a rectangular solid, where ℓ = length, w = width, and h = height. Find the volume of a rectangular solid when $\ell = 16.4$ ft, $w = 8.6$ ft, and $h = 6.4$ ft.

28. Find the volume of a cube measuring 8.10 cm on each edge. ($V = e^3$, where e is the length of each edge.)

29. The formula $s = 4.90t^2$ gives the distance, s, in metres, that a body falls in a given time, t. Find the distance a ball falls in 2.4 seconds.

30. Given K.E. $= \frac{1}{2}mv^2$, $m = 2.87 \times 10^6$ kg, and $v = 13.4$ m/s. Find K.E.

31. A formula for finding the horsepower of an engine is $p = \frac{d^2 n}{2.50}$, where d is the diameter of each cylinder in inches and n is the number of cylinders. What is the horsepower of an eight-cylinder engine if each cylinder has a diameter of 3.00 in.? (*Note:* Eight is an exact number. Ignore the number of significant digits in an exact number when determining the number of significant digits in a product or quotient.)

32. Six pieces of metal, each 2.48 mm in thickness, are fitted together. What is the total thickness of the six pieces?

33. Find the volume of a cylinder having a radius of 6.2 m and a height of 8.5 m. The formula for the volume of a cylinder is $V = \pi r^2 h$.

34. In 1993, the United States harvested 6,336,470,000 bu of corn from 62,920,000 acres. In 1995, there were 7,373,880,000 bu harvested from 64,990,000 acres. What was the yield in bu/acre for each year? What was the increase in yield between 1993 and 1995?

35. A room 24 ft long and 14 ft wide, with a ceiling height of 8.0 ft, has its air changed six times per hour. What are its ventilation requirements in CFM (ft^3/min)?

4.7 Relative Error and Percent of Error

Technicians must determine the importance of measurement error, which may be expressed in terms of relative error. The *relative error* of a measurement is found by comparing the greatest possible error with the measurement itself.

$$\text{Relative error} = \frac{\text{Greatest possible error}}{\text{Measurement}}$$

• **EXAMPLE 1** Find the relative error of the measurement 0.08 cm.

The precision is 0.01 cm. The greatest possible error is one-half the precision, which is 0.005 cm.

$$\text{Relative error} = \frac{\text{Greatest possible error}}{\text{Measurement}} = \frac{0.005 \text{ cm}}{0.08 \text{ cm}} = 0.0625$$

Note that the units will always cancel, which means that the relative error is expressed as a unitless decimal. When this decimal is expressed as a percent, we have the percent of error.

> The *percent of error* is the relative error expressed as a percent.

Percent of error may be used to compare different measurements because, being a percent, it compares each error in terms of 100. (The percent of error in Example 1 is 6.25%.)

• **EXAMPLE 2**   Find the relative error and percent of error of the measurement 13.8 m.

The precision is 0.1 m and the greatest possible error is then 0.05 m. Therefore,

$$\text{Relative error} = \frac{\text{Greatest possible error}}{\text{Measurement}} = \frac{0.05 \text{ m}}{13.8 \text{ m}} = 0.00362$$

Percent of error = 0.362%

• **EXAMPLE 3**   Compare the measurements $3\frac{3}{4}$ in. and 16 mm. Which one is better? (Which one has the smaller percent of error?)

Measurement	$3\frac{3}{4}$ in.	16 mm
Precision	$\frac{1}{4}$ in.	1 mm
Greatest possible error	$\frac{1}{2} \times \frac{1}{4}$ in. $= \frac{1}{8}$ in.	$\frac{1}{2} \times 1$ mm $= 0.5$ mm
Relative error	$\dfrac{\frac{1}{8}\text{ in.}}{3\frac{3}{4}\text{ in.}} = \frac{1}{8} \div 3\frac{3}{4}$	$\dfrac{0.5\text{ mm}}{16\text{ mm}} = 0.03125$

$$= \frac{1}{8} \div \frac{15}{4}$$

$$= \frac{1}{8} \times \frac{4}{15}$$

$$= \frac{1}{30}$$

$$= 0.0333$$

Percent of error	3.33%	3.125%

Therefore, 16 mm is the better measurement, because its percent of error is smaller.

Tolerance

In industry, the *tolerance* of a part or component is the acceptable amount that the part or component may vary from a given size. For example, a steel rod may be specified as $14\frac{3}{8}$ in. $\pm \frac{1}{32}$ in. The symbol "$\pm$" is read "plus or minus." This means that the rod may be as long as $14\frac{3}{8}$ in. $+ \frac{1}{32}$ in.; that is, $14\frac{13}{32}$ in. This is called the *upper limit*. Or it may be as short as $14\frac{3}{8}$ in. $- \frac{1}{32}$ in.; that is, $14\frac{11}{32}$ in. This is called the *lower limit*. Therefore, the specification means that any rod between $14\frac{11}{32}$ in. and $14\frac{13}{32}$ in. would be acceptable. We say that the tolerance is $\pm \frac{1}{32}$ in. The *tolerance interval*—the difference between the upper limit and the lower limit—is $\frac{2}{32}$ in., or $\frac{1}{16}$ in.

A simple way to check the tolerance of the length of a metal rod would be to carefully mark off lengths that represent the lower limit and upper limit, as shown in Figure 4.30. To check the acceptability of a rod, place one end of the rod flush against the metal barrier on the left. If the other end is between the upper and lower limit marks, the part is acceptable. If the rod is longer than the upper limit, it can then be cut to the acceptable limits. If the rod is shorter than the lower limit, it must be rejected. (It can be melted down for another try.)

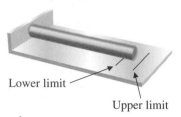

Lower limit

Upper limit

FIGURE 4.30
The tolerance interval is the difference between the upper limit and the lower limit.

EXAMPLE 4   The specifications for a stainless steel cylindrical piston are given as:

Diameter: 10.200 cm $\pm$ 0.001 cm
Height: 14.800 cm $\pm$ 0.005 cm

Find the upper limit, the lower limit, and the tolerance interval for each dimension.

	Given length	Tolerance	Upper limit	Lower limit	Tolerance interval
Diameter:	10.200 cm	$\pm$0.001 cm	10.201 cm	10.199 cm	0.002 cm
Height:	14.800 cm	$\pm$0.005 cm	14.805 cm	14.795 cm	0.010 cm

Tolerance may also be expressed as a percent. For example, resistors are color coded to indicate the tolerance of a given resistor. If the fourth band is silver, this indicates that the acceptable tolerance is $\pm$10% of the given resistance. If the fourth band is gold, this indicates that the acceptable tolerance is $\pm$5% of the given resistance.

Many times, bids may be accepted under certain conditions. They may be accepted, for example, when they are less than 10% over the architect's estimate.

EXAMPLE 5   If the architect's estimate for a given project is $356,200, and bids may be accepted if they are less than 10% over the estimate, what is the maximum acceptable bid?

10% of $356,200 = (0.10)($356,200) = $35,620

The upper limit or maximum acceptable bid is $356,200 + $35,620 = $391,820.

Exercises | 4.7

For each measurement, find the precision, the greatest possible error, the relative error, and the percent of error (to the nearest hundredth percent):

1. 1400 lb

2. 240,000 Ω

3. 875 rpm

4. 12,500 V

5. 0.085 g

6. 0.188 cm

7. 2 g

8. 2.2 g

9. 2.22 g

10. $18,0\overline{0}0$ W

11. 1.00 kg

12. 1.0 kg

13. 0.041 A

14. 0.08 ha

15. $11\frac{7}{8}$ in.

16. $1\frac{3}{4}$ in.

17. 12 ft 8 in.

18. 4 lb 13 oz

Compare each set of measurements by indicating which measurement is better or best:

19. 13.5 cm; $8\frac{3}{4}$ in.

20. 364 m; 36.4 cm

21. 16 mg; 19.7 g; $12\frac{3}{16}$ oz

22. 68,000 V; 3450 Ω; 3.2 A

Complete the table:

	Given measurement	Tolerance	Upper limit	Lower limit	Tolerance interval
23.	$3\frac{1}{2}$ in.	$\pm\frac{1}{8}$ in.	$3\frac{5}{8}$ in.	$3\frac{3}{8}$ in.	$\frac{1}{4}$ in.
24.	$5\frac{3}{4}$ in.	$\pm\frac{1}{16}$ in.			
25.	$6\frac{5}{8}$ in.	$\pm\frac{1}{32}$ in.			
26.	$7\frac{7}{16}$ in.	$\pm\frac{1}{32}$ in.			
27.	$3\frac{7}{16}$ in.	$\pm\frac{1}{64}$ in.			
28.	$\frac{9}{64}$ in.	$\pm\frac{1}{128}$ in.			
29.	$3\frac{3}{16}$ in.	$\pm\frac{1}{128}$ in.			
30.	$9\frac{3}{16}$ mi	$\pm\frac{1}{32}$ mi			
31.	1.19 cm	±0.05 cm			
32.	1.78 m	±0.05 m			
33.	0.0180 A	±0.0005 A			
34.	9.437 L	±0.001 L			
35.	24,000 V	±2000 V			
36.	375,000 W	$\pm10,000$ W			
37.	10.31 km	±0.05 km			
38.	21.30 kg	±0.01 kg			

Complete the table:

	Architect's estimate	Maximum rate above estimate	Maximum acceptable bid
39.	$48,250	10%	
40.	$259,675	7%	
41.	$1,450,945	8%	
42.	$8,275,625	5%	

4.8   Reading Scales

Circular Scales

In reading dials on meters, you must first determine which scale is being used and what the basic unit of that scale is. Consider, for example, the face of the water meter in Figure 4.31. Each dial has ten graduations on its scale. Each dial represents a different power of 10. You read the dial starting with the 100,000 scale and continue clockwise to the one-cubic-foot scale. If the indicator is between two digits, read the smaller. The reading is 475,603 ft^3.

FIGURE 4.31
Water meter

• EXAMPLE 1   Read the water meter shown in Figure 4.32.

The reading is 228,688 ft^3.

FIGURE 4.32

Consider the face of the electric meter in Figure 4.33, which measures electricity in kilowatt hours (kWh). The four dials on the meter are read from left to right; the first dial is in thousands, the next in hundreds, the third in tens, and the fourth in ones. Each dial is divided into ten equal parts. If the indicator is between two digits, always read the smaller one. The reading of the dial in Figure 4.33 is 2473 kWh.

FIGURE 4.33
Electric meter

• **EXAMPLE 2**   What is the reading on the electric meter shown in Figure 4.34?

Kilowatt Hours

FIGURE 4.34

The reading is 6250 kWh.   ————————————•

• **EXAMPLE 3**   The January water meter reading was 312,017 ft^3. The March water meter reading is 314,830 ft^3. The rate for water is:

$1.41 per 100 ft^3 for the first 1800 ft^3

$0.97 per 100 ft^3 for the next 4000 ft^3

$0.65 per 100 ft^3 for the next 14,000 ft^3

Find the amount of the water bill.

First, subtract the old reading from the new reading.

$$314,830 \text{ ft}^3$$
$$\underline{312,017 \text{ ft}^3}$$
$$2,813 \text{ ft}^3$$

Then round to the lowest hundred and divide by 100. The result is 28.

The cost of the first 1800 ft^3 is 18 × $1.41 = $25.38

The cost of the next 1000 ft^3 is 10 × $0.97 = $\underline{9.70}$

Subtotal $35.08

3% municipal tax $\underline{1.05}$

Net bill $36.13

A sample bill for this account is shown in Figure 4.35.   ————————————•

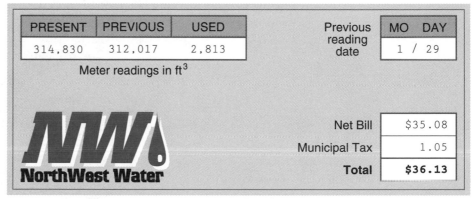

PRESENT	PREVIOUS	USED	Previous reading date	MO DAY
314,830	312,017	2,813		1 / 29

Meter readings in ft^3

Net Bill	$35.08
Municipal Tax	1.05
Total	**$36.13**

NW
NorthWest Water

FIGURE 4.35
Water bill

Dial gauges or dial indicators are useful for making very precise comparisons between a known measurement and some measurement that must be checked for precision. They are used for inspection operations, in toolrooms, and in machine shops in a wide variety of applications. Some of these uses are shown in Figure 4.36.

(a) To ensure positive vise alignment on milling machine, operator uses indicator against parallel clamped in vise jaws.

(b) Precise alignment of cutting bar is assured by using a dial indicator.

(c) Planer operator uses dial test indicator to check depth of cut on flat casting.

(d) Lathe operator uses dial indicator to check total indicator runout (TIR).

FIGURE 4.36
Uses of a dial indicator

Let's first study the metric dial indicator shown in Figure 4.37. Each graduation represents 0.01 mm. If the needle deflects six graduations to the right (+) of zero, the object being measured is 6×0.01 mm. This is 0.06 mm larger than the desired measurement. If the needle deflects 32 graduations to the left (−) of zero, the object being measured is 32×0.01 mm. This is 0.32 mm smaller than the desired measurement.

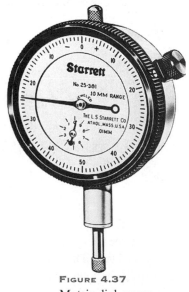

FIGURE 4.37
Metric dial gauge

Note the smaller dial on the lower left portion of the dial in Figure 4.37. This small needle records the number of complete revolutions that the large needle makes. Each complete revolution of the large needle corresponds to 1.00 mm.

• **EXAMPLE 4**   Read the metric dial in Figure 4.38.

FIGURE 4.38

The small needle reads +3 × 1.00 mm = +3.00 mm
The large needle reads +36 × 0.01 mm = +0.36 mm
The total reading is   +3.36 mm

This measurement is 3.36 mm more than the desired measurement.

Now look closely at the English dial indicator in Figure 4.39. Each graduation represents 0.001 in. If the needle deflects 7 graduations to the right (+) of zero, the object being measured is 7 × 0.001 in. This is 0.007 in. larger than the desired measurement. If the needle deflects to the left (−) 14 graduations, the object being measured is 14 × 0.001 in. This is 0.014 in. smaller than the desired measurement.

Note the smaller dial on the lower left portion of the dial in Figure 4.39. This small needle records the number of complete revolutions that the large needle makes. Each complete revolution of the large needle corresponds to 0.100 in. Other dials are read in a similar manner.

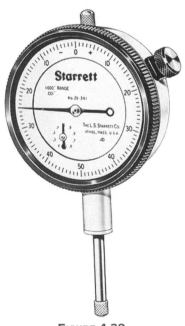

FIGURE 4.39
English dial gauge

FIGURE 4.40

• **EXAMPLE 5**   Read the English dial in Figure 4.40.

The small needle reads −2 × 0.100 in. = −0.200 in.
The large needle reads −23 × 0.001 in. = −0.023 in.
The total reading is   −0.223 in.

This measurement is 0.223 in. less than the desired measurement.

Uniform Scales

Figure 4.41 shows some of the various scales that may be found on a volt-ohm meter (VOM). This instrument is used to measure voltage (measured in volts, V) and resistance (measured in ohms, Ω) in electrical circuits. Note that the voltage scales are uniform, while the resistance scale is nonuniform. On a given voltage scale, the graduations are equally spaced, and each subdivision represents the same number of volts. On the resistance scale, the graduations are not equally spaced, and subdivisions on various intervals represent different numbers of ohms. To make things clear in the examples and exercises that follow, we show only one of the VOM scales at a time in a given figure.

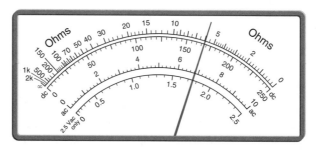

FIGURE 4.41
Volt-ohm meter (VOM) scales

The first uniform voltage scale that we study is shown in Figure 4.42. This scale has a range of 0–10 V. There are 10 large divisions, each representing 1 V. Each large division is divided into 5 equal subdivisions. Each subdivision is $\frac{1}{5}$ V, or 0.2 V.

• **EXAMPLE 6**   Read the scale shown in Figure 4.42.

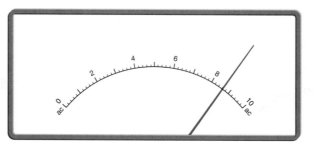

FIGURE 4.42

The needle is on the third graduation to the right of 8. Each subdivision is 0.2 V. Therefore, the reading is 8.6 V.   ————————•

Figure 4.43 shows a voltage scale that has a range of 0–2.5 V. There are 5 large divisions, each representing 0.5 V. Each division is divided into 5 subdivisions. Each subdivision is $\frac{1}{5} \times 0.5$ V = 0.1 V.

• **EXAMPLE 7**   Read the scale shown in Figure 4.43.

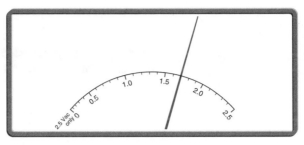

FIGURE 4.43

The needle is on the second graduation to the right of 1.5. Each subdivision is 0.1 V. Therefore, the reading is 1.7 V. ————————•

Figure 4.44 shows a voltage scale that has a range of 0–250 V. There are 10 large divisions, each representing 25 V. Each division is divided into 5 subdivisions. Each subdivision is $\frac{1}{5} \times 25$ V = 5 V.

• **EXAMPLE 8**   Read the scale shown in Figure 4.44.

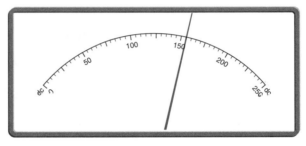

FIGURE 4.44

The needle is on the first graduation to the right of 150. Each subdivision is 5 V. Therefore, the reading is 155 V. ————————•

Nonuniform Scales

Figure 4.45 shows a nonuniform ohm scale usually found on a VOM. First, consider that part of the scale between 0 and 5. Each large division represents 1 ohm (Ω). Each large division is divided into 5 subdivisions. Therefore, each subdivision represents $\frac{1}{5} \times 1$ Ω, or 0.2 Ω.

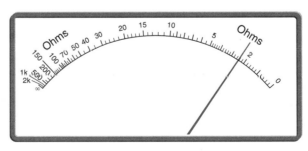

FIGURE 4.45

- Between 5 and 10, each subdivision is divided into 2 sub-subdivisions. Each sub-subdivision represents $\frac{1}{2} \times 1\ \Omega$, or 0.5 Ω.

- Between 10 and 20, each division represents 1 Ω.

- Between 20 and 100, each large division represents 10 Ω. Between 20 and 30, there are 5 subdivisions. Therefore, each subdivision represents $\frac{1}{5} \times 10\ \Omega$, or 2 Ω.

- Between 30 and 100, each large division has 2 subdivisions. Each subdivision represents $\frac{1}{2} \times 10\ \Omega$, or 5 Ω.

- Between 100 and 200, each large division represents 50 Ω.

- Between 100 and 150, there are 5 subdivisions. Each subdivision represents $\frac{1}{5} \times 50\ \Omega$, or 10 Ω.

- Between 200 and 500, there are 3 subdivisions. Each subdivision represents $\frac{1}{3} \times 300\ \Omega$, or 100 Ω.

The subdivisions on each part of the ohm scale may be summarized as follows:

Range	Each subdivision represents:
0–5 Ω	0.2 Ω
5–10 Ω	0.5 Ω
10–20 Ω	1 Ω
20–30 Ω	2 Ω
30–100 Ω	5 Ω
100–150 Ω	10 Ω
200–500 Ω	100 Ω

•   **EXAMPLE 9**   Read the scale shown in Figure 4.45.

The needle is on the second subdivision to the left of 2, where each subdivision represents 0.2 Ω. Therefore, the reading is 2.4 Ω. ────────────●

•   **EXAMPLE 10**   Read the scale shown in Figure 4.46.

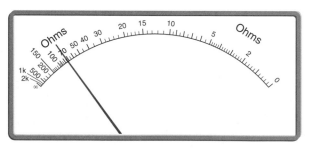

FIGURE 4.46

The needle is on the subdivision between 70 and 80. A subdivision represents 5 Ω on this part of the scale. Therefore, the reading is 75 Ω. ────────────●

Exercises | 4.8

Read each water meter:

1.

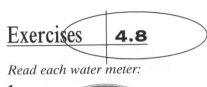

March 23

3.

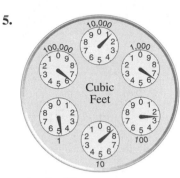

June 7

5.

October 29

2.

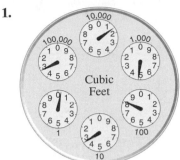

May 24

4.

August 10

6.

December 28

Using the water rates and the municipal tax rate given in Example 3, find the net water bill for the water used between the readings in:

7. Exercises 1 and 2   **8.** Exercises 3 and 4   **9.** Exercises 5 and 6

Read each electric meter:

10.

Kilowatt Hours

12.

Kilowatt Hours

11.

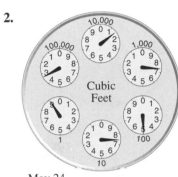

Kilowatt Hours

13.

Kilowatt Hours

14.

Kilowatt Hours

Read each metric dial (the arrow near the zero indicates the initial deflection of the needle):

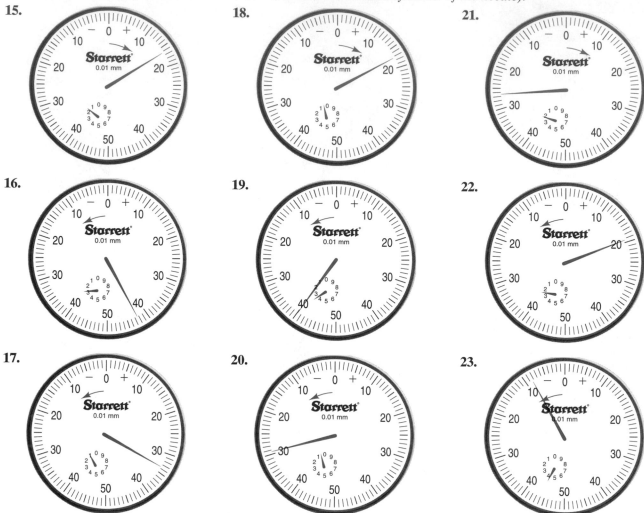

15. **18.** **21.**

16. **19.** **22.**

17. **20.** **23.**

Read each English dial (the arrow near the zero indicates the initial deflection of the needle):

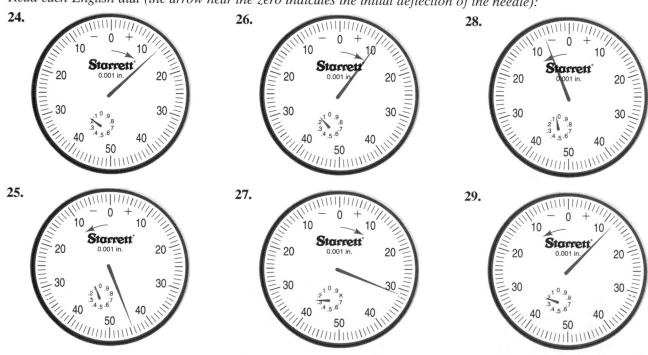

24. **26.** **28.**

25. **27.** **29.**

30.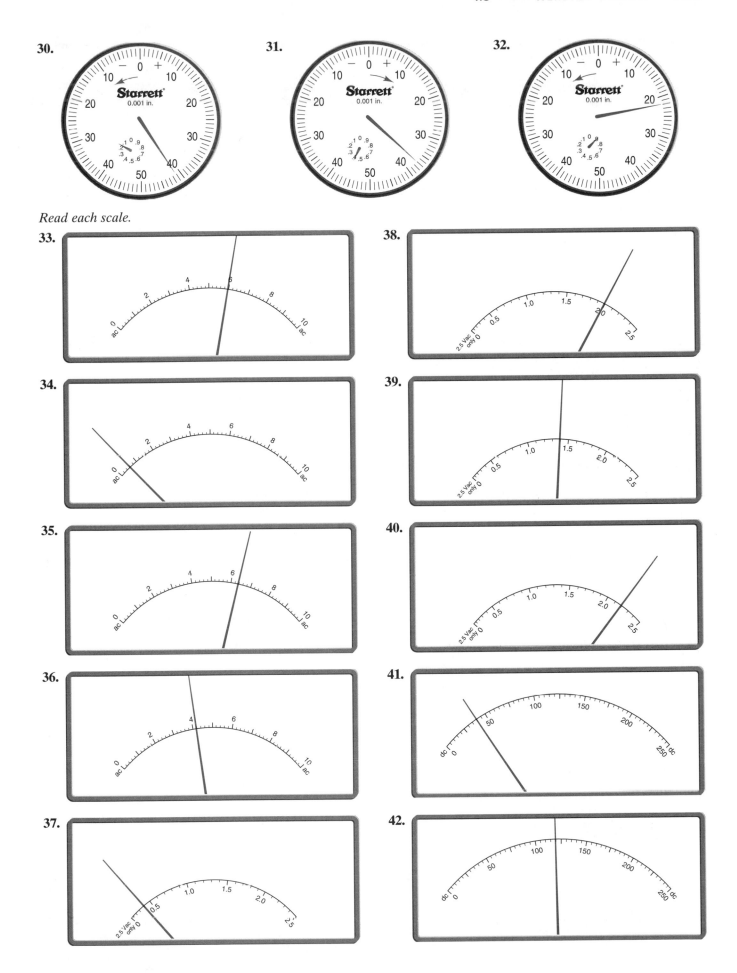

31.

32.

Read each scale.

33.

34.

35.

36.

37.

38.

39.

40.

41.

42.

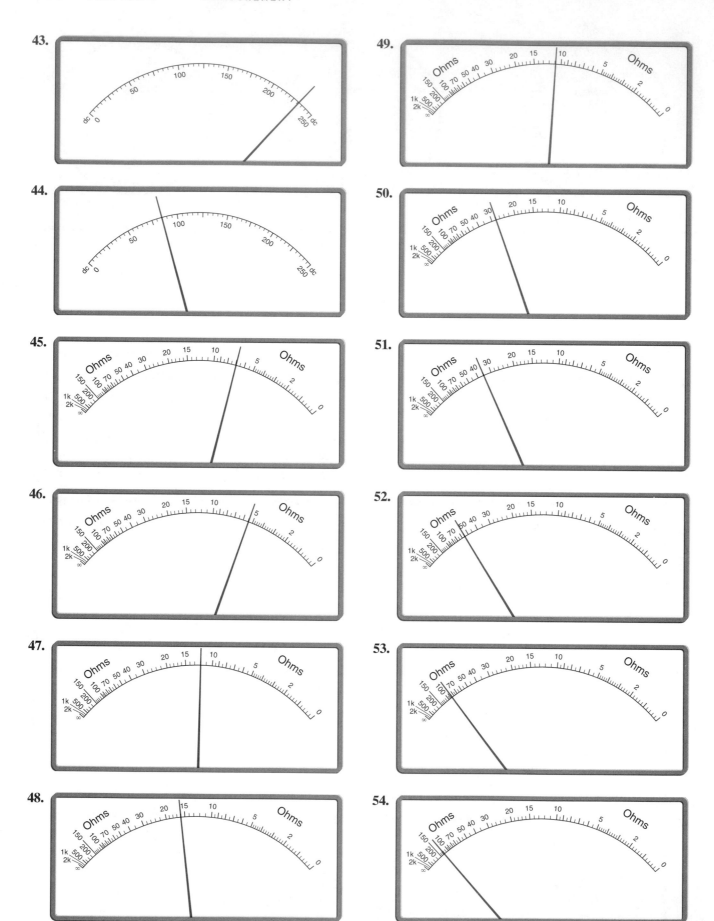

55.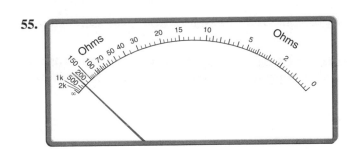

56.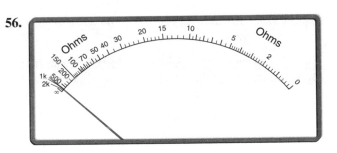

Chapter 4 Review

Give the number of significant digits (the accuracy) of each measurement:

1. 4.06 kg

2. 24,000 mi

3. $36\overline{0}0$ V

4. 5.60 cm

5. 0.0070 W

6. 0.0651 s

7. 20.00 m

8. 20.050 km

*Find **a.** the precision and **b.** the greatest possible error of each measurement:*

9. 6.05 m

10. 15.0 mi

11. 160,500 L

12. 2300 V

13. 17.00 cm

14. $13,0\overline{0}0,000$ V

15. $1\frac{5}{8}$ in.

16. $10\frac{3}{16}$ mi

Read the measurement shown on the vernier caliper in Illustration 1.

17. In metric units

18. In English units

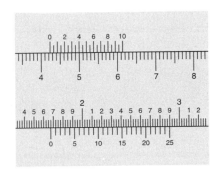

ILLUSTRATION 1

19. Read the measurement shown on the metric micrometer in Illustration 2.

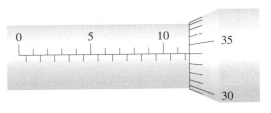

ILLUSTRATION 2

20. Read the measurement shown on the English micrometer in Illustration 3.

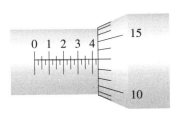

ILLUSTRATION 3

21. Find the measurement that is **a.** the most accurate and **b.** the most precise:

2500 V; 36,500 V; 60,000 V; 9.6 V; 120 V

22. Find the measurement that is **a.** the least accurate and **b.** the least precise:

0.0005 A; 0.0060 A; 0.425 A; 0.0105 A; 0.0055 A

Use the rules for addition of measurements to find the sum of each set of measurements:

23. 18,000 W; 260,000 W; 2300 W; 45,500 W; 398,000 W

24. 16.8 cm; 19.7 m; 0.14 km; 240 m

25. Use the rules for subtraction of measurements to subtract:

$$\begin{array}{r} 1{,}500{,}000 \text{ V} \\ \underline{1{,}125{,}000 \text{ V}} \end{array}$$

Use the rules for multiplication and/or division of measurements to evaluate:

26. 15.6 cm × 18.5 cm × 6.5 cm

27. $\dfrac{98.2 \text{ m}^3}{16.7 \text{ m}}$

28. $\dfrac{239 \text{ N}}{24.8 \text{ m} \times 6.7 \text{ m}}$

29. $\dfrac{(220 \text{ V})^2}{365 \ \Omega}$

*Find **a.** the relative error and **b.** the percent of error (to the nearest hundredth percent) for each measurement:*

30. $5\dfrac{7}{16}$ in.   **31.** 15.60 cm

32. Given a resistor of 2000 Ω with a tolerance of $\pm 10\%$, find the upper and lower limits.

Read each scale:

33.

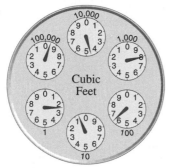

34.

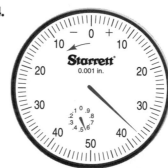

35.

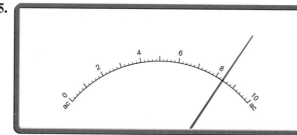

36.

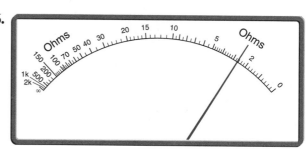

Chapter 4   Test

Give the number of significant digits in each measurement.

1. 1.806 g   **2.** 7.00 L   **3.** 0.00015 A

*Find **a.** the precision and **b.** the greatest possible error of each measurement:*

4. 6.13 mm   **5.** 2400 Ω **6.** $5\dfrac{3}{4}$ in.

Write the measurement shown on the vernier caliper in Illustration 1 in:

7. metric units

8. English units

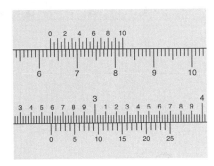

ILLUSTRATION 1

9. Write the measurement shown on the metric micrometer in Illustration 2.

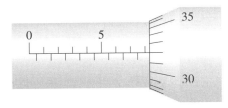

ILLUSTRATION 2

10. Write the measurement shown on the English micrometer in Illustration 3.

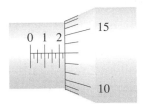

ILLUSTRATION 3

Read each scale.

19.

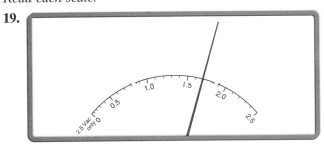

11. Find the measurement that is **a.** the most accurate, **b.** the most precise, **c.** the least accurate, and **d.** the least precise:

208 m; 17,060 m; 25.9 m; 0.167 m

12. Find the measurement that is **a.** the most accurate, **b.** the most precise, **c.** the least accurate, and **d.** the least precise:

360 V; 0.5 V; 125,000 V; 600,000 V

13. Use the rules of measurement to multiply:
(4.0 m)(12 m)(0.60 m)

14. Use the rules of measurement to add:
12.9 L + 341 L + 2104 L

15. Use the rules of measurement to subtract:
108.07 g − 56.1 g

16. Use the rules of measurement to divide:
$6.28 \text{ m}^2 \div 25 \text{ m}$

17. Use the rules of measurement to evaluate:
$$\frac{(56.3 \text{ m})(25 \text{ m})(112.5 \text{ m})}{(21.275 \text{ m})^2}$$

18. Find **a.** the relative error and **b.** the percent of error (to the nearest hundredth percent) for the measurement: 5.20 m

20.

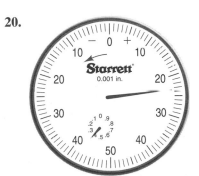

Polynomials: An Introduction to Algebra

CHAPTER 5

WELDING
A metalworker operates a laser welding machine.

MATHEMATICS AT WORK

The fundamentals of algebraic operations are all-important in arriving at the correct result when evaluating the applied formulas used in industry. These same rules and order of operations carry over to all areas of applied mathematics, physical sciences, and any other topics involving mathematics. Formulas are often given in a set form. The number of variables in the formula determines the number of different ways the formula can be rearranged. Understanding algebraic operations allows us to manipulate a given formula into the particular form that is most useful to solve for a missing variable. If there are four different variables in a formula, the formula can be rearranged in four different ways. Each of the arrangements allows us to find the numerical value of one of the variables in the original formula.

5.1 Fundamental Operations

In arithmetic, we perform mathematical operations with specific numbers. In algebra, we perform these same basic mathematical operations with numbers and *variables*—letters that represent unknown quantities. Algebra allows us to express and solve general as well as specific problems that cannot be solved using only arithmetic. As a result, employers in technical and scientific areas require a certain level of skill and knowledge of algebra. Your problem-solving skills will increase significantly as your algebra skills increase.

To begin our study of algebra, some basic mathematical principles that you will apply are listed below. Most of them you probably already know; the rest will be discussed.

> **BASIC MATHEMATICAL PRINCIPLES**
>
> **1.** $a + b = b + a$ (Commutative Property for Addition)
> **2.** $ab = ba$ (Commutative Property for Multiplication)
> **3.** $(a + b) + c = a + (b + c)$ (Associative Property for Addition)
> **4.** $(ab)c = a(bc)$ (Associative Property for Multiplication)
> **5.** $a(b + c) = ab + ac$, or $(b + c)a = ba + ca$ (Distributive Property)
> **6.** $a + 0 = a$
> **7.** $a \cdot 0 = 0$
> **8.** $a + (-a) = 0$
> **9.** $a \cdot 1 = a$
> **10.** $a \cdot \dfrac{1}{a} = 1$ $(a \neq 0)$

("$\neq$" means "is not equal to.")

In mathematics, letters are often used to represent numbers. Thus, it is necessary to know how to indicate arithmetic operations and carry them out using letters.

Addition: $x + y$ means add y to x.

Subtraction: $x - y$ means subtract y from x or add the negative of y to x; that is, $x + (-y)$.

Multiplication:
xy or $x \cdot y$ or $(x)(y)$ or $(x)y$ or $x(y)$ means multiply x by y.

Division: $x \div y$ or $\dfrac{x}{y}$ means divide x by y, or find a number z such that $zy = x$.

Exponents: $xxxx$ means use x as a factor 4 times, which is abbreviated by writing x^4. In the expression x^4, x is called the *base*, and 4 is called the *exponent*. For example, 2^4 means $2 \cdot 2 \cdot 2 \cdot 2 = 16$.

> **ORDER OF OPERATIONS**
>
> **1.** Perform all operations inside parentheses first. If the problem contains a fraction bar, treat the numerator and the denominator separately.

2. Evaluate all powers, if any. For example, $6 \cdot 2^3 = 6 \cdot 8 = 48$.

3. Perform any multiplications or divisions in order, from left to right.

4. Do any additions or subtractions in order, from left to right.

• **EXAMPLE 1**   Evaluate $4 - 9(6 + 3) \div (-3)$.

$$
\begin{aligned}
4 - 9(6 + 3) \div (-3) &= 4 - 9(9) \div (-3) \\
&= 4 - 81 \ \div (-3) \\
&= 4 - (-27) \\
&= 31
\end{aligned}
$$

• **EXAMPLE 2**   Evaluate $(-6) + 5(-2)^2(-9) - 7(3 - 5)^3$.

$$
\begin{aligned}
(-6) + 5(-2)^2(-9) - 7(3 - 5)^3 &= (-6) + 5(-2)^2(-9) - 7(-2)^3 \\
&= (-6) + 5(4)(-9) \ \ - 7(-8) \\
&= (-6) - 180 \ \ \ \ \ \ \ + 56 \\
&= -130
\end{aligned}
$$

To *evaluate an expression*, replace the letters with given numbers; then do the arithmetic in correct order. The result is the value of the expression.

• **EXAMPLE 3**   Evaluate $\dfrac{x^2 - y + 5}{2x - 2}$, if $x = 4$ and $y = 3$.

Replace x with 4 and y with 3 in the expression.

$$
\frac{x^2 - y + 5}{2x - 2} = \frac{4^2 - 3 + 5}{2(4) - 2} = \frac{16 - 3 + 5}{8 - 2} = \frac{18}{6} = 3
$$

Notice that in a fraction, the line between the numerator and denominator serves as parentheses for both. That is, do the operations in both numerator and denominator before the division.

• **EXAMPLE 4**   Evaluate $\dfrac{ab}{3c} + c$, if $a = 6$, $b = 10$, and $c = -5$.

Replace a with 6, b with 10, and c with -5 in the expression.

$$
\frac{ab}{3c} + c = \frac{6 \cdot 10}{3(-5)} + (-5) = \frac{60}{-15} + (-5) = -4 + (-5) = -9
$$

Exercises   5.1

Evaluate each expression:

1. $3(-5)^2 - 4(-2)$

2. $(-2)(-3)^2 + 3(-2) \div 6$

3. $4(-3) \div (-6) - (-18) \div 3$

4. $48 \div (-2)(-3) + (-2)^2$

5. $(-72) \div (-3) \div (-6) \div (-2) - (-4)(-2)(-5)$

6. $28 \div (-7)(2)^2 + 3(-4 - 2)^2 - (-3)^2$

7. $[(-2)(-3) + (-24) \div (-2)] \div [-10 + 7(-1)^2]$

8. $(-9)^2 \div 3^3(6) + [3(-2) - 5(-3)]$

9. $[(-2)(-8)^2 \div (-2)^3] - [-4 + (-2)^4]^2$

10. $[(-2)(3) + 5(-2)][5(-4) - 8(-3)]^2$

In Exercises 11–16, let x = 2 and y = 3, and evaluate each expression:

11. $2x - y$

13. $x^2 - y^2$

15. $\dfrac{3x + y}{3}$

12. $x - 2y$

14. $5y^2 - x^2$

16. $\dfrac{2(x + y) - 2x}{2(y - x)}$

In Exercises 17–26, let x = −1 and y = 5, and evaluate each expression:

17. $xy^2 - x$

19. $\dfrac{2y}{x} - \dfrac{2x}{y}$

21. $3 - 4(x + y)$

23. $\dfrac{1}{x} - \dfrac{1}{y} + \dfrac{2}{xy}$

18. $4x^3 - y^2$

20. $3 + 4(x + y)$

22. $1.7 - 5(2x - y)$

24. $(2.4 - x)(x - xy)$

25. $(y - 2x)(3x - 6xy)$

26. $\dfrac{(y - x)^2 - 4y}{4x^2}$

In Exercises 27–32, let x = −3, y = 4, and z = 6. Evaluate each expression:

27. $(2xy^2z)^2$

29. $(y^2 - 2x^2)z^2$

31. $(2x + 3y)(y + z)$

28. $(x^2 - y^2)z$

30. $\left(\dfrac{x + 3y}{z}\right)^2$

32. $(4 - x)^2(z - y)$

In Exercises 33–40, let x = −1, y = 2, and z = −3. Evaluate each expression:

33. $(2x + 6)(3y - 4)$

37. $(x - xy)^2(z - 2x)$

34. $z^2 - 5yx^2$

38. $3x^2(y - 3z)^2 - 6x$

35. $(3x + 5)(2y - 1)(5z + 2)$

39. $(x^2 + y^2)^2$

36. $(3x - 4z)(2x + 3z)$

40. $(3x^2 - z^2)^2$

5.2 ## Simplifying Algebraic Expressions

Parentheses are often used to clarify the order of operations when the order of operations is complicated or may be ambiguous. Sometimes it is easier to simplify such an expression by first removing the parentheses—before doing the indicated operations. Two rules for removing parentheses are as follows.

REMOVING PARENTHESES

1. Parentheses preceded by a "+" sign may be removed without changing the signs of the terms within. That is,

$$3w + (4x + y) = 3w + 4x + y$$

2. Parentheses preceded by a "−" sign may be removed if the signs of *all* the terms within the parentheses are changed; then the "−" sign that preceded the parentheses is dropped. That is,

$$3w - (4x - y) = 3w - 4x + y$$

(Notice that the sign of the term $4x$ inside the parentheses is not written. It is therefore understood to be "+".)

• **EXAMPLE 1**   Remove the parentheses from the expression $5x - (-3y + 2z)$.

$$5x - (-3y + 2z) = 5x + 3y - 2z$$

• **EXAMPLE 2**   Remove the parentheses from the expression $7x + (-y + 2z) - (w - 4)$.

$$7x + (-y + 2z) - (w - 4) = 7x - y + 2z - w + 4$$ ——————•

A *term* is a single number or a product of a number and one or more letters raised to powers. The following are examples of terms:

$$5x, \quad 8x^2, \quad -4y, \quad 15, \quad 3a^2b^3, \quad t$$

The *numerical coefficient* of the term $16x^2$ is 16. The numerical coefficient of the term $-6a^2b$ is -6. The numerical coefficient of y is 1.

Terms are parts of an algebraic expression separated by $+$ and $-$ signs. For example, $3xy + 2y + 8x^2$ is an expression consisting of three terms.

$$\boxed{3xy} \; + \; \boxed{2y} \; + \; \boxed{8x^2}$$

$$\begin{array}{ccc} \uparrow & \uparrow & \uparrow \\ \text{1st} & \text{2nd} & \text{3rd} \\ \text{term} & \text{term} & \text{term} \end{array}$$

Like Terms

Terms with exactly the same variables (including the same exponents) are called *like terms*. For example, $4x$ and $11x$ have the same variables and are like terms. The terms $-5x^2y^3$ and $8x^2y^3$ have the same variables with the same exponents and are like terms. The terms $8m$ and $5n$ have different variables, and the terms $7x^2$ and $4x^3$ have different exponents, so these are *unlike terms*.

• **EXAMPLE 3**   The following list gives examples of like terms and unlike terms.

Like Terms	**Unlike Terms**
a. $2x$ and $3x$	**e.** $2x^2$ and $3x$
b. $2ax$ and $5ax$	**f.** $2ax$ and $5bx$
c. $2x^3$ and $18x^3$	**g.** $2x^3$ and $18x^2$
d. $2a^2x^4$, a^2x^4, and $11a^2x^4$	**h.** $2a^2x^4$, $3ax^4$, and $11a^2x^3$ ——————•

Like terms that occur in a single expression can be combined into one term by combining coefficients (using the Distributive Property from Section 5.1). Thus, $ba + ca = (b + c)a$.

• **EXAMPLE 4**   Combine the like terms $2x + 3x$.

$$2x + 3x = (2 + 3)x = 5x$$ ——————•

• **EXAMPLE 5**   Combine the like terms $2ax + 3ax$.

$$2ax + 3ax = (2 + 3)ax = 5ax$$ ——————•

• **EXAMPLE 6**   Combine the like terms $2a^2x^4 + a^2x^4 + 11a^2x^4$.

$$2a^2x^4 + a^2x^4 + 11a^2x^4 = 2a^2x^4 + 1a^2x^4 + 11a^2x^4$$
$$= (2 + 1 + 11)a^2x^4 = 14a^2x^4$$ ——————•

• **EXAMPLE 7**   Combine the like terms $9a^3b^4 + 2a^2b^3 + 7a^3b^4$.

$$9a^3b^4 + 2a^2b^3 + 7a^3b^4 = (9 + 7)a^3b^4 + 2a^2b^3$$
$$= 16a^3b^4 + 2a^2b^3$$

Some expressions contain parentheses that must be removed before combining like terms.

• **EXAMPLE 8**   Simplify $4x - (x - 2)$.

$$4x - (x - 2) = 4x - x + 2 = 3x + 2$$

• **EXAMPLE 9**   Simplify $4x - (-2x - 3y) + 5y$.

$$4x - (-2x - 3y) + 5y = 4x + 2x + 3y + 5y = 6x + 8y$$

• **EXAMPLE 10**   Simplify $(7 - 2x) + (5x + 1)$.

$$(7 - 2x) + (5x + 1) = 7 - 2x + 5x + 1 = 3x + 8$$

From Section 5.1: $a(b + c) = ab + ac$. The Distributive Property is applied to remove parentheses when a number, a letter, or some product precedes the parentheses.

• **EXAMPLE 11**   Remove the parentheses from each expression.

a. $3(6x + 5) = (3)(6x) + (3)(5) = 18x + 15$
b. $-5(2a - 7) = (-5)(2a) - (-5)(7) = -10a + 35$
c. $\frac{1}{2}(10x^2 + 28x) = \left(\frac{1}{2}\right)(10x^2) + \left(\frac{1}{2}\right)(28x) = 5x^2 + 14x$

• **EXAMPLE 12**   Simplify $3x + 5(x - 3)$.

$$3x + 5(x - 3) = 3x + 5x - 15 = 8x - 15$$

• **EXAMPLE 13**   Simplify $4y - 6(-y + 2)$.

$$4y - 6(-y + 2) = 4y + 6y - 12 = 10y - 12$$

Exercises   5.2

Remove the parentheses from each expression:

1. $a + (b + c)$

2. $a - (b + c)$

3. $a - (-b - c)$

4. $a - (-b + c)$

5. $a + (-b - c)$

6. $x + (y + z + 3)$

7. $x - (-y + z - 3)$

8. $x - (-y - z + 3)$

9. $x - (y + z + 3)$

10. $x + (-y - z - 3)$

11. $(2x + 4) + (3y + 4r)$

12. $(2x + 4) - (3y + 4r)$

13. $(3x - 5y + 8) + (6z - 2w + 3)$

14. $(4x + 6y - 9) + (-2z + 5w + 3)$

15. $(-5x - 3y - 2) - (6z - 3w - 5)$

16. $(-9x + 6) - (3z + 3w - 1)$

17. $(2x + 3y - 5) + (-z - w + 2) - (-3r + 2s + 7)$
18. $(5x - 11y - 2) - (7z + 3) + (3r + 7) - (4s - 2)$
19. $-(2x - 3y) - (z + 4w) - (4r - s)$ **20.** $-(3x + y) - (2z + 7w) - (3r - 5s + 2)$

Combine the like terms:

21. $b + b$ **23.** $x^2 + 2x^2 + 3x + 7x$ **25.** $5m - 2m$ **27.** $3a + 5b - 2a + 7b$
22. $4h + 6h$ **24.** $9k + 3k$ **26.** $4x + 6x - 5x$ **28.** $11 + 2m - 6 + m$

29. $6a^2 + a + 1 - 2a$
30. $5x^2 + 3x^2 - 8x^2$
31. $2x^2 + 16x + x^2 - 13x$
32. $13x^2 + 14xy + 6y^2 - 3y^2 + x^2$
33. $1.3x + 5.6x - 13.2x + 4.5x$
34. $2.3x^2 - 4.7x + 0.92x^2 - 2.13x$

35. $\frac{5}{9}x + \frac{1}{4}y + \frac{1}{3}x - \frac{3}{8}y$

36. $\frac{1}{2}x - \frac{2}{3}y - \frac{3}{4}x + \frac{5}{6}y$
37. $4x^2y - 2xy - y^2 - 3x^2 - 2x^2y + 3y^2$
38. $3x^2 - 5x - 2 + 4x^2 + x - 4 + 5x^2 - x + 2$
39. $2x^3 + 4x^2y - 4y^3 + 3x^3 - x^2y + y - y^3$
40. $4x^2 - 5x - 7x^2 - 3x - y^2 + 2x^2 + 3xy - 2y^2$

Simplify by first removing the parentheses and then combining the like terms:

41. $y - (y - 1)$
42. $x + (2x + 1)$
43. $4x + (4 - x)$
44. $5x - (2 - 3x)$
45. $10 - (5 + x)$
46. $x - (-x - y) + 2y$
47. $2y - (7 - y)$
48. $-y - (y + 3)$
49. $(5y + 7) - (y + 2)$
50. $(2x + 4) - (x - 7)$
51. $(4 - 3x) + (3x + 1)$
52. $10 - (y + 6) + (3y - 2)$
53. $-5y + 9 - (-5y + 3)$
54. $0.5x + (x - 1) - (0.2x + 8)$
55. $0.2x - (0.2x - 28)$
56. $(0.3x - 0.5) - (-2.3x + 1.4)$
57. $\left(\frac{1}{2}x - \frac{2}{3}\right) - \left(2 - \frac{3}{4}x\right)$
58. $\left(\frac{3}{4}x - 1\right) + \left(-\frac{1}{2}x - \frac{2}{3}\right)$
59. $4(3x + 9y)$

60. $6(-2a + 8b)$
61. $-12(3x^2 - 4y^2)$
62. $-3(-a^2 - 4a)$
63. $(5x + 13) - 3(x - 2)$
64. $(7x + 8) - 5(x - 6)$
65. $-9y - 0.5(8 - y)$
66. $12(x + 1) - 3(4x - 2)$
67. $2y - 2(y + 21)$
68. $3x - 3(6 - x)$
69. $6n - (2n - 8)$
70. $14x - 8(2x - 8)$
71. $0.8x - (-x + 7)$
72. $-(x - 3) - 3(4 + x)$
73. $4(2 - 3n) - 2(5 - 3n)$
74. $(x + 4) - 2(2x - 7)$
75. $\frac{2}{3}(6x - 9) - \frac{3}{4}(12x - 16)$
76. $13\left(7x - 2\frac{1}{2}\right) - 9\left(8x + 9\frac{2}{3}\right)$
77. $0.45(x + 3) - 0.75(2x + 13)$
78. $0.6(0.5x^2 - 0.9x) + 0.4(x^2 + 0.4x)$

5.3   Addition and Subtraction of Polynomials

A *monomial*, or *term*, is any algebraic expression that contains only products of numbers and variables, which have nonnegative integer exponents. The following expressions are examples of monomials:

$$2x, \quad 5, \quad -3b, \quad \frac{3}{4}a^2bw, \quad \sqrt{315}mn$$

A *polynomial* is either a monomial or the sum of monomials. We consider two special types of polynomials. A *binomial* is a polynomial that is the sum of two unlike monomials. A *trinomial* is the sum of three unlike monomials.

The following expressions are examples of binomials:

$$a + b, \qquad 2a^2 + 3, \qquad 5mn^2 + 7wr^2$$

The following expressions are examples of trinomials:

$$3 + 5a + 7b, \qquad 2n + 4m + 6p, \qquad 2a^3b + 3a^2b^2 + 4ab^3$$

Expressions that contain variables in the denominator are *not* polynomials. For example,

$$\frac{3}{4x}, \quad \frac{8x}{3x-5}, \quad \text{and} \quad \frac{33}{4x^2} + \frac{8}{x-1}$$

are *not* polynomials.

> The *degree of a monomial in one variable* is the same as the exponent of the variable.

• **EXAMPLE 1**   Find the degree of each monomial: **a.** $-7m$, **b.** $6x^2$, **c.** $5y^3$, **d.** 5.

a. $-7m$ has degree 1.

b. $6x^2$ has degree 2.

c. $5y^3$ has degree 3.

d. 5 has degree 0 (since 5 may be written $5x^0$).

> The *degree of a polynomial in one variable* is the same as the highest-degree monomial contained in the polynomial.

• **EXAMPLE 2**   Find the degree of each polynomial: **a.** $5x^4 + x^2$ and **b.** $6y^3 + 4y^2 - y + 1$.

a. $5x^4 + x^2$ has degree 4.

b. $6y^3 + 4y^2 - y + 1$ has degree 3.

A polynomial is in *decreasing order* if each term is of some degree less than the preceding term. The following polynomial is written in decreasing order:

$$4x^5 - 3x^4 - 4x^2 - x + 5$$

exponents decrease

A polynomial is in *increasing order* if each term is of some degree larger than the preceding term. The following polynomial is written in increasing order:

$$5 - x - 4x^2 - 3x^4 + 4x^5$$

exponents increase

To add monomials and polynomials, write like terms under each other. Then add by columns, as in the examples below.

•   **EXAMPLE 3**   Add $3x + 5$ and $5x - 7$.

$$\begin{array}{r} 3x + 5 \\ 5x - 7 \\ \hline 8x - 2 \end{array}$$

•   **EXAMPLE 4**   Add $15x$ and $-12x + 3$.

$$\begin{array}{r} 15x \\ -12x + 3 \\ \hline 3x + 3 \end{array}$$

•   **EXAMPLE 5**   Add $(2x^2 - 5x) + (3x^2 + 2x - 4) + (-4x^2 + 5)$.

$$\begin{array}{r} 2x^2 - 5x \\ 3x^2 + 2x - 4 \\ -4x^2 \quad\ + 5 \\ \hline x^2 - 3x + 1 \end{array}$$

Note that the polynomials are usually written in decreasing order. Also note that the like terms are written in the same columns. If the terms are unlike, the addition must be left in the form of an indicated sum, as in the next example.

•   **EXAMPLE 6**   Add $2x^2 + 3x$.

$$\begin{array}{r} 2x^2 \\ + 3x \\ \hline 2x^2 + 3x \end{array}$$ is the sum because $2x^2$ and $3x$ are unlike terms.

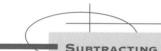

SUBTRACTING POLYNOMIALS

To subtract a second polynomial from a first,

1. Write the second polynomial under the first. Place like terms under each other where possible.
2. Change the sign of each term in the second polynomial.
3. Add the two resulting polynomials.

•   **EXAMPLE 7**   Find the difference: $(4a - 9b) - (2a - 4b)$.

Subtract: $\begin{array}{r} 4a - 9b \\ 2a - 4b \\ \hline \end{array}$ $\rightarrow$ Add: $\begin{array}{r} 4a - 9b \\ -2a + 4b \\ \hline 2a - 5b \end{array}$

The arrow indicates the change of the subtraction problem to an addition problem. Do this by changing the signs of each of the terms in the second polynomial.

•   **EXAMPLE 8**   Find the difference: $(5x^2 - 3x - 4) - (2x^2 - 5x + 5)$.

Subtract: $\begin{array}{r} 5x^2 - 3x - 4 \\ 2x^2 - 5x + 5 \\ \hline \end{array}$ $\rightarrow$ Add: $\begin{array}{r} 5x^2 - 3x - 4 \\ -2x^2 + 5x - 5 \\ \hline 3x^2 + 2x - 9 \end{array}$

To check subtraction, add the result to the polynomial in the second line of the original subtraction. If the sum is the same as the first polynomial, the result is correct.

• **EXAMPLE 9**   Check the result from Example 8.

$$\begin{array}{r} \text{Add:} \quad 2x^2 - 5x + 5 \\ 3x^2 + 2x - 9 \\ \hline 5x^2 - 3x - 4 \end{array}$$

This result is the same as the upper polynomial in Example 8. Therefore, the answer for Example 8 is correct. ————————●

A second method of subtracting polynomials uses the rules for removing parentheses given in Section 5.2.

• **EXAMPLE 10**   Subtract $x^2 - 2x$ from $3x^2 + 4x$.

Write the expression using parentheses:

$$(3x^2 + 4x) - (x^2 - 2x)$$

Remove the parentheses and combine the like terms:

$$3x^2 + 4x - x^2 + 2x = 2x^2 + 6x$$ ————————●

Exercises  5.3

Classify each expression as a monomial, a binomial, or a trinomial:

1. $3m + 27$
2. $4a^2bc^3$
3. $-5x - 7y$
4. $2x^2 + 7y + 3z^2$
5. $-5xy$
6. $a + b + c$
7. $2x + 3y - 5z$
8. $2a - 3b^3$

9. $-42x^3 - y^4$
10. $15x^{14} - 3x^2 + 5x$

Rearrange each polynomial in decreasing order and state its degree:

11. $1 - x + x^2$
12. $2x^3 - 3x^4 + 2x$
13. $4x + 7x^2 - 1$
14. $y^3 - 1 + y^2$
15. $-4x^2 + 5x^3 - 2$
16. $3x^3 + 6 - 2x + 4x^5$
17. $7 - 3y + 4y^3 - 6y^2$
18. $1 - x^5$

19. $x^3 - 4x^4 + 2x^2 - 7x^5 + 5x - 3$
20. $360x^2 - 720x - 120x^3 + 30x^4 + 1 - 6x^5 + x^6$

Add:

21. $\begin{array}{r} -5a \\ 7a \\ \hline \end{array}$

22. $\begin{array}{r} 3a - 5b + c \\ 4a + 6b - 2c \\ \hline \end{array}$

23. $\begin{array}{r} 4x^2 - 7x \\ -2x^2 + 5x \\ \hline \end{array}$

24. $\begin{array}{r} 5y^3 - 4y^2 - 2y \\ -7y^3 - 3y^2 - y \\ 2y^3 - y^2 + y \\ \hline \end{array}$

25. $\begin{array}{r} 4y^2 - 3y - 15 \\ 7y^2 - 6y + 8 \\ -3y^2 + 4y + 13 \\ \hline \end{array}$

26. $\begin{array}{r} 129a - 13b - 56c \\ -13a - 52b + 21c \\ 44a + 11c \\ \hline \end{array}$

27. $\begin{array}{r} 3a^3 + 2a^2 + 5 \\ a^3 - 7a - 2 \\ -5a^2 + 4a \\ - 2a - 3 \\ \hline \end{array}$

28.
$$\begin{array}{r} 4x^2 - 3xy \qquad + 5x - 6y \\ 9xy - 4y^2 \qquad + 6y - 4 \\ x^2 \qquad + y^2 - x \qquad + 3 \\ \hline -8x^2 + xy - 2y^2 + 3x \qquad - 10 \end{array}$$

29. $(5a^2 - 7a + 5) + (2a^2 - 3a - 4)$

30. $(2b - 5) + 3b + (-4b - 7)$

31. $(6x^2 - 7x + 5) + (3x^2 + 2x - 5)$

32. $(4x - 7y - z) + (2x - 5y - 3z) + (-3x - 6y - 4z)$

33. $(2a^3 - a) + (4a^2 + 7a) + (7a^3 - a - 5)$

34. $(5y - 7x + 4z) + (3z - 6y + 2x) + (13y + 7z - 6x)$

35. $(3x^2 + 4x - 5) + (-x^2 - 2x + 2) + (-2x^2 + 2x + 7)$

36. $(-x^2 + 6x - 8) + (10x^2 - 13x + 3) + (-12x^2 - 14x + 3)$

37. $(3x^2 + 7) + (6x - 7) + (2x^2 + 5x - 13) + (7x - 9)$

38. $(5x + 3y) + (-3x - 3y) + (-x - 6y) + (3x - 4y)$

39. $(5x^3 - 11x - 1) + (11x^2 + 3) + (3x + 7) + (2x^2 - 2)$

40. $(3x^4 - 5x^2 + 4) + (6x^4 - 6x^2 + 1) + (2x^4 - 7x^2)$

Find each difference and check your answer:

41.
$$\begin{array}{r} 3a - 4b \\ 2a - 5b \\ \hline \end{array}$$

42.
$$\begin{array}{r} 4x^2 - 3x - 5 \\ 2x^2 - 7x \\ \hline \end{array}$$

43.
$$\begin{array}{r} 57m - 16n + 32 \\ 41m + 42n - 13 \\ \hline \end{array}$$

44.
$$\begin{array}{r} 6a - 3b - 4c \\ -2a - b + 15c \\ \hline \end{array}$$

45.
$$\begin{array}{r} -12a^2 - 13a - 5 \\ 2a^2 - 2a - 2 \\ \hline \end{array}$$

46.
$$\begin{array}{r} -5a^2 - 3a \\ 2a^2 - 4a - 5 \\ \hline \end{array}$$

47.
$$\begin{array}{r} 4y^2 + 21y - 7 \\ -2y^2 - 27y + 8 \\ \hline \end{array}$$

48. $(5y^2 - 5y - 4) - (4y^2 - 2y + 4)$

49. $(3x^2 + 4x + 7) - (x^2 - 2x + 5)$

50. $(2x^2 + 5x - 9) - (3x^2 - 4x + 7)$

51. $(3x^2 - 5x + 4) - (6x^2 - 7x + 2)$

52. $(1 - 3x - 2x^2) - (-1 - 5x + x^2)$

53.
$$\begin{array}{r} 2x^3 + 4x - 1 \\ x^3 + x + 2 \\ \hline \end{array}$$

54.
$$\begin{array}{r} 7x^4 + 3x^3 + 5x \\ -2x^4 + x^3 - 6x + 6 \\ \hline \end{array}$$

55.
$$\begin{array}{r} 12x^5 - 13x^4 + 7x^2 \\ 4x^5 + 5x^4 + 2x^2 - 1 \\ \hline \end{array}$$

56.
$$\begin{array}{r} 8x^3 + 6x^2 - 15x + 7 \\ 14x^3 + 2x^2 + 9x - 1 \\ \hline \end{array}$$

Find each difference:

57. $(3a - 4b) - (2a - 7b)$

58. $(-13x^2 - 3y^2 - 4y) - (-5x - 4y + 5y^2)$

59. $(7a - 4b) - (3x - 4y)$

60. $(-16y^3 - 42y^2 - 3y - 5) - (12y^2 - 4y + 7)$

61. $(12x^2 - 3x - 2) - (11x^2 - 7)$

62. $(14z^3 - 6y^3) - (2y^2 + 4z^3)$

63. $(20w^2 - 17w - 6) - (13w^2 + 7w)$

64. $(y^2 - 2y + 1) - (2y^2 + 3y + 5)$

65. $(2x^2 - 5x - 2) - (x^2 - x + 8)$

66. $(3 - 5z + 3z^2) - (14 + z - 2z^2)$

67. Subtract $4x^2 + 2x - 7$ from $8x^2 - 2x + 5$.

68. Subtract $-6x^2 - 3x + 4$ from $2x^2 - 6x - 2$.

69. Subtract $9x^2 + 6$ from $3x^2 + 2x - 4$.

70. Subtract $-4x^2 - 6x + 2$ from $4x^2 + 6$.

5.4 Multiplication of Monomials

Earlier we used exponents to write products of repeated number factors as follows:

$$6^2 = 6 \cdot 6 = 36$$
$$5^3 = 5 \cdot 5 \cdot 5 = 125$$
$$2^4 = 2 \cdot 2 \cdot 2 \cdot 2 = 16$$

A power with a variable base may also be used as follows:

$$x^3 = x \cdot x \cdot x$$
$$m^4 = m \cdot m \cdot m \cdot m$$

In the expression 2^4, the number 2 is called the *base*, and 4 is called the *exponent*. The expression may also be called *the fourth power of 2*. In x^3, the letter x is called the base and 3 is called the exponent.

• **EXAMPLE 1**   Multiply:

a. $x^3 \cdot x^2 = (x \cdot x \cdot x)(x \cdot x) = x^5$

b. $a^4 \cdot a^2 = (a \cdot a \cdot a \cdot a)(a \cdot a) = a^6$

c. $m^5 \cdot m^3 = (m \cdot m \cdot m \cdot m \cdot m)(m \cdot m \cdot m) = m^8$ ⎯⎯⎯⎯●

RULE 1 FOR EXPONENTS: MULTIPLYING POWERS

$$x^a \cdot x^b = x^{a+b}$$

That is, to multiply powers with the same base, add the exponents.

To multiply two monomials, multiply their numerical coefficients and combine their variable factors according to the above rule for exponents.

• **EXAMPLE 2**   Multiply $(2x^3)(5x^4)$.

$$(2x^3)(5x^4) = 2 \cdot 5 \cdot x^3 \cdot x^4 = 10x^{3+4} = 10x^7$$ ⎯⎯⎯⎯●

• **EXAMPLE 3**   Multiply $(3a)(-15a^2)(4a^4b^2)$.

$$(3a)(-15a^2)(4a^4b^2) = (3)(-15)(4)(a)(a^2)(a^4)(b^2)$$
$$= -180a^7b^2$$ ⎯⎯⎯⎯●

A special note about the meaning of $-x^2$ is needed here. Note that x is squared, not $-x$. That is,

$$-x^2 = -(x \cdot x)$$

If $-x$ is squared, we have $(-x)^2 = (-x)(-x) = x^2$.

RULE 2 FOR EXPONENTS: RAISING A POWER TO A POWER

$$(x^a)^b = x^{ab}$$

That is, to raise a power to a power, multiply the exponents.

• **EXAMPLE 4**   Find $(x^3)^5$.

By Rule 1:

$$(x^3)^5 = x^3 \cdot x^3 \cdot x^3 \cdot x^3 \cdot x^3 = x^{15}$$

By Rule 2:

$$(x^3)^5 = x^{3 \cdot 5} = x^{15}$$ ⎯⎯⎯⎯●

• **EXAMPLE 5**   Find $(x^5)^9$.

$$(x^5)^9 = x^{45}$$

RULE 3 FOR EXPONENTS: RAISING A PRODUCT TO A POWER

$$(xy)^a = x^a y^a$$

That is, to raise a product to a power, raise each factor to that same power.

• **EXAMPLE 6**   Find $(xy)^3$.

$$(xy)^3 = x^3 y^3$$

• **EXAMPLE 7**   Find $(2x^3)^2$.

$$(2x^3)^2 = 2^2(x^3)^2 = 4x^6$$

• **EXAMPLE 8**   Find $(-3x^4)^5$.

$$(-3x^4)^5 = (-3)^5(x^4)^5 = -243x^{20}$$

• **EXAMPLE 9**   Find $(ab^2c^3)^4$.

$$(ab^2c^3)^4 = a^4(b^2)^4(c^3)^4 = a^4 b^8 c^{12}$$

• **EXAMPLE 10**   Find $(2a^2bc^3)^2$.

$$(2a^2bc^3)^2 = 2^2(a^2)^2(b)^2(c^3)^2 = 4a^4 b^2 c^6$$

• **EXAMPLE 11**   Evaluate $(2a^2)(-3ab^2)$ when $a = 2$ and $b = 3$.

$$(2a^2)(-3ab^2) = 2(-3)(a^2)(a)(b^2)$$
$$= -6a^3 b^2$$
$$= -6(2)^3(3)^2$$
$$= -6(8)(9)$$
$$= -432$$

Exercises | 5.4

Find each product:

1. $(3a)(-5)$

2. $(7x)(2x)$

3. $(4a^2)(7a)$

4. $(4x)(6x^2)$

5. $(-9m^2)(-6m^2)$

6. $(5x^2)(-8x^3)$

7. $(8a^6)(4a^2)$

8. $(-4y^4)(-9y^3)$

9. $(13p)(-2pq)$

10. $(4ab)(10a)$

11. $(6n)(5n^2m)$

12. $(-9ab^2)(6a^2b^3)$

13. $(-42a)\left(-\dfrac{1}{2}a^3b\right)$

14. $(28m^3)\left(\dfrac{1}{4}m^2\right)$

15. $\left(\dfrac{2}{3}x^2y^2\right)\left(\dfrac{9}{16}xy^2\right)$

16. $\left(-\dfrac{5}{6}a^2b^6\right)\left(\dfrac{9}{20}a^5b^4\right)$

17. $(8a^2bc)(3ab^3c^2)$

18. $(-4xy^2z^3)(4x^5z^3)$

19. $\left(\dfrac{2}{3}x^2y\right)\left(\dfrac{9}{32}xy^4z^3\right)$

20. $\left(\dfrac{3}{5}m^4n^7\right)\left(\dfrac{20}{9}m^2nq^3\right)$

21. $(32.6mnp^2)(-11.4m^2n)$

22. $(5.6a^2b^3c)(6.5a^4b^5)$ **23.** $(5a)(-17a^2)(3a^3b)$ **24.** $(-4a^2b)(-5ab^3)(-2a^4)$

Use the rules for exponents to simplify:

25. $(x^3)^2$ **27.** $(x^4)^6$ **29.** $(-3x^4)^2$ **31.** $(-x^3)^3$ **33.** $(x^2 \cdot x^3)^2$ **35.** $(x^5)^6$

26. $(xy)^4$ **28.** $(2x^2)^5$ **30.** $(5x^2)^3$ **32.** $(-x^2)^4$ **34.** $(3x)^4$ **36.** $(-3xy)^3$

37. $(-5x^3y^2)^2$ **40.** $(-7w^2)^3$ **43.** $(3x^2 \cdot x^4)^2$ **46.** $(-4a^2b^3c^4)^4$

38. $(-x^2y^4)^5$ **41.** $(25n^4)^3$ **44.** $(16x^4 \cdot x^5)^3$ **47.** $(-2h^3k^6m^2)^5$

39. $(15m^2)^2$ **42.** $(36a^5)^3$ **45.** $(2x^3y^4z)^3$ **48.** $(4p^5q^7r)^3$

Evaluate each expression when $a = 2$ and $b = -3$:

49. $(4a)(17b)$ **53.** $(41a^3)(-2b^3)$ **57.** $(4b)^3$ **61.** $(5ab)(a^2b^2)$ **65.** $-a^2b^4$

50. $(3a)(-5b^2)$ **54.** $(ab)^2$ **58.** $(-2ab^2)^2$ **62.** $(-3ab)^3$ **66.** $(-ab^2)^2$

51. $(9a^2)(-2a)$ **55.** $(a^2)^2$ **59.** $(5a^2b^3)^2$ **63.** $(9a)(ab^2)$ **67.** $(-a)^4$

52. $(a^2)(ab)$ **56.** $(3a)^2$ **60.** $(-7ab)^2$ **64.** $(-2a^2)(6ab)$ **68.** $-b^4$

5.5   Multiplication of Polynomials

To multiply a polynomial by a monomial, multiply each term of the polynomial by the monomial, and then add the products as shown in the following examples.

• **EXAMPLE 1**   Multiply $3a(a^2 - 2a + 1)$.

$$3a(a^2 - 2a + 1) = 3a(a^2) + 3a(-2a) + 3a(1)$$
$$= 3a^3 - 6a^2 + 3a$$

• **EXAMPLE 2**   Multiply $(-5a^3b)(3a^2 - 4ab + 5b^3)$.

$$(-5a^3b)(3a^2 - 4ab + 5b^3)$$
$$= (-5a^3b)(3a^2) + (-5a^3b)(-4ab) + (-5a^3b)(5b^3)$$
$$= -15a^5b + 20a^4b^2 - 25a^3b^4$$

To multiply a polynomial by another polynomial, multiply each term of the first polynomial by each term of the second polynomial. Then add the products. Arrange the work as shown in the example below.

• **EXAMPLE 3**   Multiply $(5x - 3)(2x + 4)$.

STEP 1   Write each polynomial in decreasing order, one under the other.

$$5x - 3$$
$$2x + 4$$

STEP 2   Multiply each term of the upper polynomial by the first term in the lower one.

$$10x^2 - 6x$$

STEP 3   Multiply each term of the upper polynomial by the second term in the lower one. Place like terms in the same columns.

$$20x - 12$$

STEP 4   Add the like terms.

$$10x^2 + 14x - 12$$

To multiply two binomials, such as $(x + 3)(2x + 5)$, you may think of finding the area of a rectangle with sides $(x + 3)$ and $(2x + 5)$ shown in Figure 5.1.

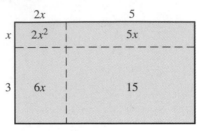

FIGURE 5.1

Note that the total area is $2x^2 + 6x + 5x + 15$ or $2x^2 + 11x + 15$. Using the four-step process as above, we have

STEP 1 $2x + 5$
$\underline{\qquad x + 3}$

STEP 2 $2x^2 + 5x$

STEP 3 $\underline{\qquad\quad 6x + 15}$

STEP 4 $2x^2 + 11x + 15$

• **EXAMPLE 4**   Multiply $(x + 3)(x^2 + 2x - 4)$.

STEP 1 $x^2 + 2x \ - 4$
$\underline{\qquad\quad x \ + 3\qquad}$

STEP 2 $x^3 + 2x^2 - 4x$

STEP 3 $\underline{\qquad\quad 3x^2 + 6x - 12}$

STEP 4 $x^3 + 5x^2 + 2x - 12$

• **EXAMPLE 5**   Multiply $(3a + b)(c + 2d)$.

STEP 1 $3a + b$
$\underline{\qquad c + 2d}$

STEP 2 $3ac + bc$

STEP 3 $\underline{\qquad\qquad\qquad 6ad + 2bd}$

STEP 4 $3ac + bc + 6ad + 2bd$

Note that there are no like terms in Steps 2 and 3.

Exercises | 5.5

Find each product.

1. $4(a + 6)$

2. $3(a^2 - 5)$

3. $-6(3x^2 + 2y)$

4. $-5(8x - 4y^2)$

5. $a(4x^2 - 6y + 1)$

6. $c(2a + b + 3c)$

7. $x(3x^2 - 2x + 5)$

8. $y(3x + 2y^2 + 4y)$

9. $2a(3a^2 + 6a - 10)$

10. $5x(8x^2 - x + 5)$

11. $-3x(4x^2 - 7x - 2)$

12. $-6x(8x^2 + 5x - 9)$

13. $4x(-7x^2 - 3y + 2xy)$

14. $7a(2a + 3b - 4ab)$

15. $3xy(x^2y - xy^2 + 4xy)$

16. $-2ab(3a^2 + 4ab - 2b^2)$

17. $-6x^3(1 - 6x^2 + 9x^4)$

18. $5x^4(2x^3 + 8x^2 - 1)$

19. $5ab^2(a^3 - b^3 - ab)$

20. $7w^2y(w^2 - 4y^2 + 6w^2y^3)$

21. $\frac{2}{3}m(14n - 12m)$

22. $\frac{1}{2}a^2b(8ab^2 - 2a^2b)$

23. $\frac{4}{7}yz^3\left(28y - \frac{2}{5}z\right)$

24. $-\frac{1}{8}rs(3s - 16t)$

25. $-4a(1.3a^5 + 2.5a^2 + 1)$

26. $1.28m(2.3m^2 + 4.7n^2)$

27. $417a(3.2a^2 + 4a)$

28. $1.2m^2n^3(9.7m + 6.5mn - 13n^2)$

29. $4x^2y(6x^2 - 4xy + 5y^2)$

30. $x^2y^3z(x^4 - 3x^2y - 3yz + 4z^2)$

31. $\frac{2}{3}ab^3\left(\frac{3}{4}a^2 - \frac{1}{2}ab^2 + \frac{5}{6}b^3\right)$

32. $-\frac{5}{9}a^2b^4\left(\frac{3}{7}a^3b^2 - \frac{3}{5}ab - \frac{15}{16}b^4\right)$

33. $3x(x - 4) + 2x(1 - 5x) - 6x(2x - 3)$

34. $x(x - 2) - 3x(x + 8) - 2(x^2 + 3x - 5)$

35. $xy(3x + 2xy - y^2) - 2xy^2(2x - xy + 3y)$

36. $ab^2(2a - 3a^2b + b) - a^2b(1 + 2ab^2 - 4b)$

37. $(x + 1)(x + 6)$

38. $(x + 10)(x - 3)$

39. $(x + 7)(x - 2)$

40. $(x - 3)(x - 7)$

41. $(x - 5)(x - 8)$

42. $(x + 9)(x + 4)$

43. $(3a - 5)(a - 4)$

44. $(5x - 2)(3x - 4)$

45. $(6a + 4)(2a - 3)$

46. $(3x + 5)(6x - 7)$

47. $(4a + 8)(6a + 9)$

48. $(5x - 4)(5x - 4)$

49. $(3x - 2y)(5x + 2y)$

50. $(4x - 6y)(6x + 9y)$

51. $(2x - 3)(2x - 3)$

52. $(5m - 9)(5m + 9)$

53. $(2c - 5d)(2c + 5d)$

54. $(3a + 2b)(2a - 3b)$

55. $(-7m - 3)(-13m + 1)$

56. $(w - r)(w - s)$

57. $(x^5 - x^2)(x^3 - 1)$

58. $(7w^4 - 6r^2)(7w^4 + 5r^2)$

59. $(2y^2 - 4y - 8)(5y - 2)$

60. $(m^2 + 2m + 4)(m - 2)$

61. $(4x - 2y - 13)(6x + 3y)$

62. $(4y - 3z)(2y^2 - 5yz + 6z^2)$

63. $(g + h - 6)(g - h + 3)$

64. $(2x - 3y + 4)(4x - 5y - 2)$

65. $(8x - x^3 + 2x^4 - 1)(x^2 + 2 + 5x^3)$

66. $(y^5 - y^4 + y^3 - y^2 + y - 1)(y + 1)$

5.6 Division by a Monomial

To divide a monomial by a monomial, first write the quotient in fraction form. Then factor both numerator and denominator into prime factors. Reduce to lowest terms by dividing both the numerator and the denominator by their common factors. The remaining factors in the numerator and the denominator give the quotient.

• **EXAMPLE 1**   Divide $6a^2 \div 2a$.

$$6a^2 \div 2a = \frac{6a^2}{2a}$$

$$= \frac{\overset{1}{\cancel{2}} \cdot 3 \cdot \overset{1}{\cancel{a}} \cdot a}{\underset{1}{\cancel{2}} \cdot \underset{1}{\cancel{a}}} = \frac{3a}{1} = 3a$$

• **EXAMPLE 2**   Divide $4a \div 28a^3$.

$$4a \div 28a^3 = \frac{4a}{28a^3}$$

$$= \frac{\overset{1}{\cancel{2}} \cdot \overset{1}{\cancel{2}} \cdot \overset{1}{\cancel{a}}}{\underset{1}{\cancel{2}} \cdot \underset{1}{\cancel{2}} \cdot 7 \cdot \underset{1}{\cancel{a}} \cdot a \cdot a} = \frac{1}{7a^2}$$

• **EXAMPLE 3**   Divide $\dfrac{6a^2bc^3}{10ab^2c}$.

$$\frac{6a^2bc^3}{10ab^2c} = \frac{\cancel{2} \cdot 3 \cdot \cancel{a} \cdot a \cdot \cancel{b} \cdot \cancel{c} \cdot c \cdot c}{\cancel{2} \cdot 5 \cdot \cancel{a} \cdot \cancel{b} \cdot b \cdot \cancel{c}} = \frac{3ac^2}{5b}$$

Note: Since division by zero is undefined, we will assume that there are no zero denominators here and for the remainder of this chapter.

To divide a polynomial by a monomial, divide each term of the polynomial by the monomial.

• **EXAMPLE 4**   Divide $\dfrac{15x^3 - 3x^2 + 21x}{3x}$.

$$\frac{15x^3 - 3x^2 + 21x}{3x} = \frac{15x^3}{3x} - \frac{3x^2}{3x} + \frac{21x}{3x}$$

$$= \frac{\cancel{3} \cdot 5 \cdot \cancel{x} \cdot x \cdot x}{\cancel{3} \cdot \cancel{x}} - \frac{\cancel{3} \cdot \cancel{x} \cdot x}{\cancel{3} \cdot \cancel{x}} + \frac{\cancel{3} \cdot 7 \cdot \cancel{x}}{\cancel{3} \cdot \cancel{x}}$$

$$= 5x^2 - x + 7$$

• **EXAMPLE 5**   Divide $(30d^4y - 28d^2y^2 + 12dy^2) \div (-6dy^2)$.

$$\frac{30d^4y - 28d^2y^2 + 12dy^2}{-6dy^2}$$

$$= \frac{30d^4y}{-6dy^2} + \frac{-28d^2y^2}{-6dy^2} + \frac{12dy^2}{-6dy^2}$$

$$= \frac{\cancel{6} \cdot 5 \cdot \cancel{d} \cdot d \cdot d \cdot d \cdot \cancel{y}}{\cancel{/} \; \cancel{/} \; \cancel{/}} + \frac{-\cancel{2} \cdot 14 \cdot \cancel{d} \cdot d \cdot \cancel{y} \cdot \cancel{y}}{\cancel{/} \; \cancel{/} \; \cancel{/} \; \cancel{/}} + \frac{\cancel{6} \cdot 2 \cdot \cancel{d} \cdot \cancel{y} \cdot \cancel{y}}{-\cancel{6} \cdot \cancel{d} \cdot \cancel{y} \cdot \cancel{y}}$$

$$= \frac{5d^3}{-y} + \frac{14d}{3} + \frac{2}{-1} = -\frac{5d^3}{y} + \frac{14d}{3} - 2$$

Exercises   5.6

Divide:

1. $\dfrac{9x^5}{3x^3}$ **3.** $\dfrac{18x^{12}}{12x^4}$ **5.** $\dfrac{18x^3}{3x^5}$ **7.** $\dfrac{8x^2}{12x}$ **9.** $\dfrac{x^2y}{xy}$

2. $\dfrac{15x^6}{5x^4}$ **4.** $\dfrac{20x^7}{4x^5}$ **6.** $\dfrac{4x^2}{12x^6}$ **8.** $\dfrac{-6x^3}{2x}$ **10.** $\dfrac{xy^2}{x^2y}$

11. $(15x) \div (6x)$ **17.** $0 \div (113w^2r^3)$ **23.** $\dfrac{252}{7r^2}$

12. $(14x^2) \div (2x^3)$ **18.** $(-148wr^3) \div (148wr^3)$ **24.** $\dfrac{118a^3}{-2a^4}$

13. $(15a^3b) \div (3ab^2)$ **19.** $(207p^3) \div (9p)$ **25.** $\dfrac{92x^3y}{-28xy^3}$

14. $(-13a^5) \div (7a)$ **20.** $(42x^2y^3) \div (-14x^2y^4)$ **26.** $\dfrac{45x^6}{-72x^3y^2}$

15. $(16m^2n) \div (2m^3n^2)$ **21.** $\dfrac{92mn}{-46mn}$ **27.** $\dfrac{-16a^5b^2}{-14a^8b^4}$

16. $(-108m^4n) \div (27m^3n)$ **22.** $\dfrac{-132rs^3}{-33r^2s^2}$ **28.** $\dfrac{35a^3b^4c^6}{63a^3b^2c^8}$

29. $(-72x^3yz^4) \div (-162xy^2)$

30. $(-144x^2z^3) \div (216x^5y^2z)$

31. $(4x^2 - 8x + 6) \div 2$

32. $(18y^3 + 12y^2 + 6y) \div 6$

33. $(x^4 + x^3 + x^2) \div x^2$

34. $(20r^2 - 16r - 12) \div (-4)$

35. $(ax - ay - az) \div a$

36. $(14c^3 - 28c^2 - 2c) \div (-2c)$

37. $\dfrac{24a^4 - 16a^2 - 8a}{8}$

38. $\dfrac{88x^5 - 110x^4 + 11x^3}{11x^3}$

39. $\dfrac{b^{12} - b^9 - b^6}{b^3}$

40. $\dfrac{27a^3 - 18a^2 + 36a}{-9a}$

41. $\dfrac{bx^4 - bx^3 + bx^2 - 4bx}{-bx}$

42. $\dfrac{4a^5 - 32a^4 + 8a^3 - 12a^2}{-4a^2}$

43. $\dfrac{24x^2y^3 + 12x^3y^4 - 6xy^3}{2xy^3}$

44. $\dfrac{3.5ax^2 - 0.42a^2x + 14a^2x^2}{0.07ax}$

45. $\dfrac{224x^4y^2z^3 - 168x^3y^3z^4 - 112xy^4z^2}{28xy^2z^2}$

46. $\dfrac{55w^2 - 11w - 33}{11w}$

47. $\dfrac{24y^5 - 18y^3 - 12y}{6y^2}$

48. $\dfrac{3a^2b + 4a^2b^2 - 6ab^2}{2ab^2}$

49. $\dfrac{1 - 6x^2 - 4x^4}{2x^2}$

50. $\dfrac{18w^4r^4 + 27w^3r^3 - 36w^2r^2}{9w^3r^3}$

5.7   Division by a Polynomial

Dividing a polynomial by a polynomial of more than one term is very similar to long division in arithmetic. We use the same names, as shown below.

$$
\begin{array}{r}
\phantom{25\overline{)5}}23 \quad \leftarrow \text{quotient} \\
\text{divisor} \rightarrow \; 25\overline{)593} \quad \leftarrow \text{dividend} \\
\underline{50} \\
93 \\
\underline{75} \\
18 \quad \leftarrow \text{remainder}
\end{array}
$$

$$
\begin{array}{r}
x + 3 \quad \leftarrow \text{quotient} \\
\text{divisor} \rightarrow \; x + 1\overline{)x^2 + 4x + 4} \quad \leftarrow \text{dividend} \\
\underline{x^2 + \; x} \\
3x + 4 \\
\underline{3x + 3} \\
1 \quad \leftarrow \text{remainder}
\end{array}
$$

As a check, you may use the relationship *dividend = divisor × quotient + remainder*. A similar procedure is followed in both cases. Compare the solutions of the two problems in the example that follows.

• **EXAMPLE 1**   Divide $337 \div 16$ (arithmetic) and $(2x^2 + x - 14) \div (x + 3)$ (algebra).

Arithmetic

$$
\begin{array}{r}
21 \text{ r } 1 \\
16\overline{)337} \\
\underline{32} \\
17 \\
\underline{16} \\
1
\end{array}
$$

1.
2.
3.
4.
5.
6.

Algebra

$$
\begin{array}{r}
2x - \;\; 5 \text{ r } 1 \\
x + 3\overline{)2x^2 + \;\; x - 14} \\
\underline{2x^2 + 6x} \\
-5x - 14 \\
\underline{-5x - 15} \\
1
\end{array}
$$

1. Divide 16 into 33. It will go at most 2 times, so write the 2 above the line over the dividend (337).

2. Multiply the divisor (16) by 2. Write the result (32) under the first two digits of 337.

3. Subtract 32 from 33, leaving 1. Bring down the 7, giving 17.

4. Divide 16 into 17. It will go at most 1 time, so write the 1 to the right of the 2 above the dividend.

5. Multiply the divisor (16) by 1. Write the result (16) under 17.

6. Subtract 16 from 17, leaving 1. The remainder (1) is less than 16, so the problem is finished. The quotient is 21 with remainder 1.

1. Divide x into $2x^2$, the first term in the dividend. It will go exactly $2x$ times, so write the $2x$ on the line above the dividend ($2x^2 + x - 14$).

2. Multiply the divisor ($x + 3$) by $2x$. Write the result ($2x^2 + 6x$) under the first two terms of the dividend.

3. Subtract this result ($2x^2 + 6x$) from the first two terms of the dividend, leaving $-5x$. Bring down the last term of the dividend (-14).

4. Divide x into $-5x$ (the first term in line 4). It goes exactly -5 times, so write -5 to the right of $2x$ above the dividend.

5. Multiply the divisor ($x + 3$) by the -5. Write the result ($-5x - 15$) on line 5.

6. Subtract line 5 from line 4, leaving 1. The remainder (1) is of a lower degree than the divisor ($x + 3$), so the problem is finished. The quotient is $2x - 5$ with remainder 1. ————————•

• **EXAMPLE 2**   Divide $8x^3 - 22x^2 + 27x - 18$ by $2x - 3$.

$$
\begin{array}{r}
4x^2 - 5x + 6 \\
\hline
2x - 3 \overline{\smash{\big)}\, 8x^3 - 22x^3 + 27x - 18} \\
8x^3 - 12x^2 \\
\hline
-10x^2 + 27x \\
-10x^2 + 15x \\
\hline
12x - 18 \\
12x - 18 \\
\hline
0
\end{array}
$$

1.
2.
3.
4.
5.
6.
7.
8.

STEP 1 Divide the first term in the dividend, $8x^3$, by the first term of the divisor, $2x$. Write $4x^2$ above the dividend in line 1.

STEP 2 Multiply the divisor, $2x - 3$, by $4x^2$ and write the result in line 3, as shown.

STEP 3 Subtract in line 4 and bring down the next term.

STEP 4 Divide $-10x^2$ by $2x$ and write $-5x$ above the dividend in line 1.

STEP 5 Multiply $2x - 3$ by $-5x$ and write the result in line 5.

STEP 6 Subtract in line 6 and bring down the next term.

STEP 7 Divide $12x$ by $2x$ and write 6 above the dividend in line 1.

STEP 8 Multiply $2x - 3$ by 6 and write the result in line 7.

STEP 9 Subtract in line 8; the 0 indicates that there is no remainder.

CHECK:

$$
\begin{array}{l}
4x^2 - 5x + 6 \qquad \text{quotient} \\
2x - 3 \qquad\qquad \text{divisor} \\
\hline
8x^3 - 10x^2 + 12x \\
-12x^2 + 15x - 18 \\
\hline
8x^3 - 22x^2 + 27x - 18 \qquad \text{dividend}
\end{array}
$$

————————•

The dividend should always be arranged in decreasing order of degree. Any missing powers of x should be filled in by using zeros as coefficients. For example, $x^3 - 1 = x^3 + 0x^2 + 0x - 1$.

• **EXAMPLE 3**   Divide $(x^3 - 1) \div (x - 1)$.

$$
\begin{array}{r}
x^2 + x + 1 \\
x - 1 \overline{\smash{\big)}\, x^3 + 0x^2 + 0x - 1} \\
\underline{x^3 - x^2} \\
x^2 + 0x \\
\underline{x^2 - x} \\
x - 1 \\
\underline{x - 1} \\
0
\end{array}
$$

The remainder is 0, so the quotient is $x^2 + x + 1$.

CHECK:

$$
\begin{array}{l}
x^2 + x + 1 \qquad \text{quotient} \\
\underline{ x - 1} \qquad \text{divisor} \\
x^3 + x^2 + x \\
\underline{ - x^2 - x - 1} \\
x^3 - 1 \qquad \text{dividend}
\end{array}
$$

When you subtract, be especially careful to change all of the signs of each term in the expression being subtracted. Then follow the rules for addition.

Exercises | 5.7

Find each quotient and check:

1. $(x^2 + 3x + 2) \div (x + 1)$

2. $(y^2 - 5y + 6) \div (y - 2)$

3. $(6a^2 - 3a + 2) \div (2a - 3)$

4. $(21y^2 + 2y - 10) \div (3y + 2)$

5. $(12x^2 - x - 9) \div (3x + 2)$

6. $(20x^2 + 57x + 30) \div (4x + 9)$

7. $\dfrac{2y^2 + 3y - 5}{2y - 1}$

8. $\dfrac{3x^2 - 5x - 10}{3x - 8}$

9. $\dfrac{6b^2 + 13b - 28}{2b + 7}$

10. $\dfrac{8x^2 + 13x - 27}{x + 3}$

11. $(6x^3 + 13x^2 + x - 2) \div (x + 2)$

12. $(8x^3 - 18x^2 + 7x + 3) \div (x - 1)$

13. $(8x^3 - 14x^2 - 79x + 110) \div (2x - 7)$

14. $(3x^3 - 17x^2 + 18x + 10) \div (3x + 1)$

15. $\dfrac{2x^3 - 14x - 12}{x + 1}$

16. $\dfrac{x^3 + 7x^2 - 36}{x + 3}$

17. $\dfrac{4x^3 - 24x^2 + 128}{2x + 4}$

18. $\dfrac{72x^3 + 22x + 4}{6x - 1}$

19. $\dfrac{3x^3 + 4x^2 - 6}{x - 2}$

20. $\dfrac{2x^3 + 3x^2 - 9x + 5}{x + 3}$

21. $\dfrac{4x^3 + 2x^2 + 30x + 20}{2x - 5}$

22. $\dfrac{18x^3 + 6x^2 + 4x}{6x - 2}$

23. $\dfrac{8x^4 - 10x^3 + 16x^2 + 4x - 30}{4x - 5}$

24. $\dfrac{9x^4 + 12x^3 - 6x^2 + 10x + 24}{3x + 4}$

25. $\dfrac{8x^3 - 1}{2x + 1}$

26. $\dfrac{x^3 + 1}{x + 1}$

27. $\dfrac{x^4 - 16}{x + 2}$

28. $\dfrac{16x^4 + 1}{2x - 1}$

29. $\dfrac{3x^4 + 5x^3 - 17x^2 + 11x - 2}{x^2 + 3x - 2}$

30. $\dfrac{6x^4 + 5x^3 - 11x^2 + 9x - 5}{2x^2 - x + 1}$

Chapter 5 Review

1. For any number a, $a \cdot 1 = ?$

2. For any number a, $a \cdot 0 = ?$

3. For any number a except 0, $a \cdot \dfrac{1}{a} = ?$

Evaluate:

4. $10 - 4(3)$

5. $2 + 3 \cdot 4^2$

6. $(4)(12) \div 6 - 2^3 + 18 \div 3^2$

In Exercises 7–12, let $x = 3$ and $y = -2$. Evaluate each expression:

7. $x + y$

8. $x - 3y$

9. $5xy$

10. $\dfrac{x^2}{y}$

11. $y^3 - y^2$

12. $\dfrac{2x^3 - 3y}{xy^2}$

In Exercises 13–16, simplify by removing the parentheses and combining like terms:

13. $(5y - 3) - (2 - y)$

14. $(7 - 3x) - (5x + 1)$

15. $11(2x + 1) - 4(3x - 4)$

16. $(x^3 + 2x^2y) - (3y^3 - 2x^3 + x^2y + y)$

17. Is $1 - 8x^2$ a monomial, a binomial, or a trinomial?

18. What is the degree of the polynomial $x^4 + 2x^3 - 6$?

Perform the indicated operations:

19. $(3a^2 + 7a - 2) + (5a^2 - 2a + 4)$

20. $(6x^3 + 3x^2 + 1) - (-3x^3 - x^2 - x - 1)$

21. $(3x^2 + 5x + 2) + (9x^2 - 6x - 2) - (2x^2 + 6x - 4)$

22. $(6x^2)(4x^3)$

23. $(-7x^2y)(8x^3y^2)$

24. $(3x^2)^3$

25. $5a(3a + 4b)$

26. $-4x^2(8 - 2x + 3x^2)$

27. $(5x + 3)(3x - 4)$

28. $(3x^2 - 6x + 1)(2x - 4)$

29. $(49x^2) \div (7x^3)$

30. $\dfrac{15x^3y}{3xy}$

31. $\dfrac{36a^3 - 27a^2 + 9a}{9a}$

32. $\dfrac{6x^2 + x - 12}{2x + 3}$

33. $\dfrac{3x^3 + 2x^2 - 6x + 4}{x + 2}$

Chapter 5 Test

Evaluate:

1. $3 \cdot 5 - 2 \cdot 4^2$

2. $12 \div 2 \cdot 3 \div 2 + 3^3 - 16 \div 2^2$

3. Evaluate $\dfrac{3x^2y - 4x}{2y}$ when $x = 4$ and $y = -1$.

Perform the indicated operations and simplify:

4. $3a^2 - 17a + 6a^2 + 4a$

6. Add: $3x^2 + 6x - 8$
$\qquad\quad\ -9x^2 + 6x$
$\qquad\quad\ -3x^2 \qquad + 15$
$\qquad\qquad\quad\ \ - 7x + 4$
$\qquad\overline{\qquad\qquad\qquad\qquad}$

5. $5(2 + x) - 2(x + 4)$

7. $(5a - 5b + 7) - (2a - 5b - 3)$

8. $(7a^2)(-4a^4)(a)$

9. $(6x^4y^2)^3$

10. $-5x(2x - 3)$

11. $\dfrac{85x^4y^2}{17x^2y^5}$

12. $(4a + 6)(a - 5)$

13. $(x + y - 5)(x - y)$

14. $\dfrac{36a^4 - 20a^3 - 16a^2}{4a}$

16. $(5x^2y^3)(-7x^3y)$

15. $\dfrac{9x^4y^3 - 12x^2y + 18y^2}{3x^2y^3}$

17. $\dfrac{3x^2 - 13x - 10}{x - 5}$

18. $\dfrac{4x^5y^3}{-2x^3y^5}$

19. $\dfrac{6x^2 - 7x - 6}{3x - 5}$

20. $\dfrac{x^3 + 2x^2 + x + 12}{x + 3}$

Equations
and Formulas

CHAPTER 6

ELECTRICAL WORK
*An electrician installs an air
conditioning unit on the roof of a
commercial building.*

MATHEMATICS AT WORK

Knowing the rules for solving equations is
essential to evaluating formulas on the job.
Each business or trade uses special formulas
that apply directly to specific tasks. The formulas are usually given in one form,
which can be rearranged in certain ways to solve for a particular variable.

Electricians and electronic technicians need to be very proficient with Ohm's
law, $V = IR$, where V = volts, I = amps, and R = resistance. By changing the
values of I and R, the electrician can determine different voltages; by rearranging
the formula, he or she can determine what size of resistor is needed or how many
amps are flowing through the electrical system.

Maintenance workers in automated manufacturing need to be proficient with
formulas from many different areas. For example, pneumatics and hydraulics formulas can help determine what diameter a valve must have in order to increase or
decrease pressure in a hydraulic line. Technicians in any trade who are comfortable with solving and evaluating formulas are a valuable asset to a company; they
are often promoted to supervisors be or lead persons.

6.1 Equations

In technical work, the ability to use equations and formulas is essential. A *variable* is a symbol (usually a letter of the alphabet) to represent an unknown number. An *algebraic expression* is a combination of numbers, variables, symbols for operations (plus, minus, times, etc.), and symbols for grouping (parentheses or a fraction bar). Examples of algebraic expressions are:

$$4x - 9, \quad 3x^2 + 6x + 9, \quad 5x(6x + 4), \quad \frac{2x + 5}{-3x}$$

An *equation* is a statement that two quantities are equal. The symbol "=" is read "equals" and separates an equation into two parts, the left member and the right member. For example, in the equation

$$2x + 3 = 11$$

the left member is $2x + 3$ and the right member is 11. Other examples of equations are:

$$x - 5 = 6, \quad 3x = 12, \quad 4m + 9 = 3m - 2, \quad \text{and} \quad \frac{x - 2}{4} = 3(x + 1)$$

To solve an equation means to find what number or numbers can replace the variable to make the equation a true statement. In the equation $2x + 3 = 11$, the solution is 4. That is, when x is replaced by 4, the resulting equation is a true statement.

$$2x + 3 = 11$$
$$\text{Let } x = 4: \quad 2(4) + 3 = 11 \quad ?$$
$$8 + 3 = 11 \quad \text{True}$$

A replacement number (or numbers) that produces a true statement is called a *solution* or a *root* of the equation.

Note that replacing x by any other number, such as 5, results in a false statement.

$$2x + 3 = 11$$
$$\text{Let } x = 5: \quad 2(5) + 3 = 11 \quad ?$$
$$10 + 3 = 11 \quad \text{False}$$

One method of solving equations involves changing the given equation to an *equivalent equation* by performing the same arithmetic operation on both sides of the equation. The basic arithmetic operations used are addition, subtraction, multiplication, and division.

Equivalent equations are equations that have the same roots. For example, $3x = 6$ and $x = 2$ are equivalent equations, since 2 is the root of each. In solving an equation by this method, continue to change the given equation to an equivalent equation until you find an equation whose root is obvious.

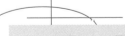

FOUR BASIC RULES USED TO SOLVE EQUATIONS

1. If the same quantity is added to both sides of an equation, the resulting equation is equivalent to the original equation.

Example 1: Solve $x - 2 = 8$.
$$x - 2 + 2 = 8 + 2$$
$$x = 10$$

2. If the same quantity is subtracted from both sides of an equation, the resulting equation is equivalent to the original equation.

Example 2: Solve $x + 5 = 2$.
$$x + 5 - 5 = 2 - 5$$
$$x = -3$$

3. If both sides of an equation are multiplied by the same (nonzero) quantity, the resulting equation is equivalent to the original equation.

Example 3: Solve $\dfrac{x}{9} = 4$.
$$9\left(\frac{x}{9}\right) = (4)9$$
$$x = 36$$

4. If both sides of an equation are divided by the same (nonzero) quantity, the resulting equation is equivalent to the original equation.

Example 4: Solve $4x = 20$.
$$\frac{4x}{4} = \frac{20}{4}$$
$$x = 5$$

Basically, to solve an equation, use one of the rules and use a number that will *undo* what has been done to the variable.

• **EXAMPLE 5**   Solve $x + 3 = 8$.

Since 3 has been added to the variable, use Rule 2 and subtract 3 from both sides of the equation.

$$x + 3 = 8$$
$$x + 3 - 3 = 8 - 3$$
$$x = 5$$

A check is recommended, since an error could have been made. To check, replace the variable in the original equation by 5, the apparent root, to make sure that the resulting statement is true.

CHECK: $x + 3 = 8$
$$5 + 3 = 8 \quad \text{True}$$

Thus, the root is 5.

• **EXAMPLE 6**   Solve $x - 4 = 7$.

Since 4 has been subtracted from the variable, use Rule 1 and add 4 to both sides of the equation.

$$x - 4 = 7$$
$$x - 4 + 4 = 7 + 4$$
$$x = 11$$

The apparent root is 11.

CHECK:
$$x - 4 = 7$$
$$11 - 4 = 7 \quad \text{True}$$

Thus, the root is 11.

• **EXAMPLE 7** Solve $2x = 9$.

Since the variable has been multiplied by 2, use Rule 4 and divide both sides of the equation by 2.

$$2x = 9$$
$$\frac{2x}{2} = \frac{9}{2}$$
$$x = \frac{9}{2}$$

Note: Each solution should be checked by substituting it into the original equation. When a check is not provided in this text, the check is left for you to do.

• **EXAMPLE 8** Solve $\frac{x}{3} = 9$.

Since the variable has been divided by 3, use Rule 3 and multiply both sides of the equation by 3.

$$\frac{x}{3} = 9$$
$$3\left(\frac{x}{3}\right) = (9)3$$
$$x = 27$$

• **EXAMPLE 9** Solve $-4 = -6x$.

Since the variable has been multiplied by -6, use Rule 4 and divide both sides of the equation by -6.

$$-4 = -6x$$
$$\frac{-4}{-6} = \frac{-6x}{-6}$$
$$\frac{2}{3} = x$$

The apparent root is $\frac{2}{3}$.

CHECK:
$$-4 = -6x$$
$$4 = 6\left(\frac{2}{3}\right) \quad ?$$
$$-4 = -4 \quad \text{True}$$

Thus, the root is $\frac{2}{3}$.

Some equations have more than one operation indicated on the variable. For example, the equation $2x + 5 = 6$ has both addition of 5 and multiplication by 2 indicated. Use the following procedure to solve equations like this.

> When more than one operation is indicated on the variable, undo the additions and subtractions first, then undo the multiplications and divisions.

• **EXAMPLE 10**   Solve $2x + 5 = 6$.

$$2x + 5 - 5 = 6 - 5 \qquad \text{Subtract 5 from both sides.}$$
$$2x = 1$$
$$\frac{2x}{2} = \frac{1}{2} \qquad \text{Divide both sides by 2.}$$
$$x = \frac{1}{2}$$

The apparent root is $\frac{1}{2}$.

CHECK:
$$2x + 5 = 6$$
$$2\left(\frac{1}{2}\right) + 5 = 6 \qquad ?$$
$$1 + 5 = 6 \qquad \text{True}$$

Thus, the root is $\frac{1}{2}$.

• **EXAMPLE 11**   Solve $\frac{x}{3} - 6 = 9$.

$$\frac{x}{3} - 6 + 6 = 9 + 6 \qquad \text{Add 6 to both sides.}$$
$$\frac{x}{3} = 15$$
$$3\left(\frac{x}{3}\right) = (15)3 \qquad \text{Multiply both sides by 3.}$$
$$x = 45$$

• **EXAMPLE 12**   Solve $118 - 22m = 30$.

Think of $118 - 22m$ as $118 + (-22m)$.

$$118 - 22m - 118 = 30 - 118 \qquad \text{Subtract 118 from both sides.}$$
$$-22m = -88$$
$$\frac{-22m}{-22} = \frac{-88}{-22} \qquad \text{Divide both sides by } -22.$$
$$m = 4$$

The apparent root is 4.

CHECK:
$$118 - 22m = 30$$
$$118 - 22(4) = 30 \qquad ?$$
$$118 - 88 = 30 \qquad ?$$
$$30 = 30 \qquad \text{True}$$

Thus, the root is 4.

Here is another approach to solving this equation.

$$118 - 22m = 30$$
$$118 - 22m + 22m = 30 + 22m \qquad \text{Add } 22m \text{ to both sides.}$$
$$118 = 30 + 22m$$
$$118 - 30 = 30 + 22m - 30 \qquad \text{Subtract 30 from both sides.}$$
$$88 = 22m$$
$$\frac{88}{22} = \frac{22m}{22} \qquad \text{Divide both sides by 22.}$$
$$4 = m$$

Exercises | 6.1

Solve each equation and check:

1. $x + 2 = 8$

2. $3a = 7$

3. $y - 5 = 12$

4. $\frac{2}{3}n = 6$

5. $w - 7\frac{1}{2} = 3$

6. $2m = 28.4$

7. $\frac{x}{13} = 1.5$

8. $n + 12 = -5$

9. $3b = 15.6$

10. $y - 17 = 25$

11. $17x = 5117$

12. $28 + m = 3$

13. $2 = x - 5$

14. $-29 = -4y$

15. $17 = -3 + w$

16. $49 = 32 + w$

17. $14b = 57$

18. $y + 28 = 13$

19. $5m = 0$

20. $28 + m = 28$

21. $x + 5 = 5$

22. $y + 7 = -7$

23. $4x = 64$

24. $5x - 125 = 0$

25. $\frac{x}{7} = 56$

26. $\frac{y}{5} = 35$

27. $-48 = 12y$

28. $13x = -78$

29. $-x = 2$

30. $-y = 7$

31. $5y + 3 = 13$

32. $4x - 2 = 18$

33. $10 - 3x = 16$

34. $8 - 2y = 4$

35. $\frac{x}{4} - 5 = 3$

36. $\frac{x}{5} + 4 = 9$

37. $2 - x = 6$

38. $8 - y = 3$

39. $\frac{2}{3}y - 4 = 8$

40. $5 - \frac{1}{4}x = 7$

41. $3x - 5 = 12$

42. $5y + 7 = 28$

43. $\frac{m}{3} - 6 = 8$

44. $\frac{w}{5} + 7 = 13$

45. $\frac{2x}{3} = 7$

46. $\frac{4b}{5} = 15$

47. $-3y - 7 = -6$

48. $28 = -7 - 3r$

49. $5 - x = 6$

50. $17 - 5w = -68$

51. $54y - 13 = 17.8$

52. $37a - 7 = 67$

53. $28w - 56 = -8$

54. $52 - 4x = -8$

55. $29r - 13 = 57$

56. $15x - 32 = 18$

57. $31 - 3y = 41$

58. $62 = 13y - 3$

59. $-83 = 17 - 4x$

60. $58 = 5m + 52$

6.2 Equations with Variables in Both Members

To solve equations with variables in both members (both sides), such as

$$3x + 4 = 5x - 12$$

do the following:

First, add or subtract either variable term from both sides of the equation.

$$3x + 4 = 5x - 12$$
$$3x + 4 - 3x = 5x - 12 - 3x \qquad \text{Subtract } 3x \text{ from both sides.}$$
$$4 = 2x - 12$$

Then take the constant term (which now appears on the same side of the equation with the variable term) and add it to, or subtract it from, both sides. Solve the resulting equation.

$$4 + 12 = 2x - 12 + 12 \qquad \text{Add 12 to both sides.}$$
$$16 = 2x$$
$$\frac{16}{2} = \frac{2x}{2} \qquad \text{Divide both sides by 2.}$$
$$8 = x$$

This equation could also have been solved as follows:

$$3x + 4 = 5x - 12$$
$$3x + 4 - 5x = 5x - 12 - 5x \qquad \text{Subtract } 5x \text{ from both sides.}$$
$$-2x + 4 = -12$$
$$-2x + 4 - 4 = -12 - 4 \qquad \text{Subtract 4 from both sides.}$$
$$-2x = -16$$
$$\frac{-2x}{-2} = \frac{-16}{-2} \qquad \text{Divide both sides by } -2.$$
$$x = 8$$

• **EXAMPLE 1**   Solve $5x - 4 = 8x - 13$.

$$5x - 4 - 8x = 8x - 13 - 8x \qquad \text{Subtract } 8x \text{ from both sides.}$$
$$-3x - 4 = -13$$
$$-3x - 4 + 4 = -13 + 4 \qquad \text{Add 4 to both sides.}$$
$$-3x = -9$$
$$\frac{-3x}{-3} = \frac{-9}{-3} \qquad \text{Divide both sides by } -3.$$
$$x = 3$$

CHECK:
$$5x - 4 = 8x - 13$$
$$5(3) - 4 = 8(3) - 13 \qquad ?$$
$$15 - 4 = 24 - 13 \qquad ?$$
$$11 = 11 \qquad \text{True}$$

Therefore, 3 is a root.

• **EXAMPLE 2**   Solve $-2x + 5 = 6x - 11$.

$$-2x + 5 + 2x = 6x - 11 + 2x \qquad \text{Add } 2x \text{ to both sides.}$$
$$5 = 8x - 11$$
$$5 + 11 = 8x - 11 + 11 \qquad \text{Add 11 to both sides.}$$
$$16 = 8x$$
$$\frac{16}{8} = \frac{8x}{8} \qquad \text{Divide both sides by 8.}$$
$$2 = x$$

CHECK:
$$-2x + 5 = 6x - 11$$
$$-2(2) + 5 = 6(2) - 11 \quad ?$$
$$-4 + 5 = 12 - 11 \quad ?$$
$$1 = 1 \qquad \text{True}$$

Thus, 2 is a root.

• **EXAMPLE 3**   Solve $5x + 7 = 2x - 14$.

$$5x + 7 - 2x = 2x - 14 - 2x \qquad \text{Subtract } 2x \text{ from both sides.}$$
$$3x + 7 = -14$$
$$3x + 7 - 7 = -14 - 7 \qquad \text{Subtract 7 from both sides.}$$
$$3x = -21$$
$$\frac{3x}{3} = \frac{-21}{3} \qquad \text{Divide both sides by 3.}$$
$$x = -7$$

CHECK:
$$5x + 7 = 2x - 14$$
$$5(-7) + 7 = 2(-7) - 14 \quad ?$$
$$-35 + 7 = -14 - 14 \quad ?$$
$$-28 = -28 \qquad \text{True}$$

So -7 is a root.

• **EXAMPLE 4**   Solve $4 - 5x = 28 + x$.

$$4 - 5x + 5x = 28 + x + 5x \qquad \text{Add } 5x \text{ to both sides.}$$
$$4 = 28 + 6x$$
$$4 - 28 = 28 + 6x - 28 \qquad \text{Subtract 28 from both sides.}$$
$$-24 = 6x$$
$$\frac{-24}{6} = \frac{6x}{6} \qquad \text{Divide both sides by 6.}$$
$$-4 = x$$

Exercises 6.2

Solve each equation and check:

1. $4y + 9 = 7y - 15$

2. $2y - 45 = -y$

3. $5x + 3 = 7x - 5$

4. $2x - 3 = 3x - 13$

5. $-2x + 7 = 5x - 21$

6. $5x + 3 = 2x - 15$

7. $3y + 5 = 5y - 1$

8. $3x - 4 = 7x - 32$

9. $-3x + 17 = 6x - 37$

10. $3x + 13 = 2x - 12$

11. $7x + 9 = 9x - 3$

12. $-5y + 12 = 12y - 5$

13. $3x - 2 = 5x + 8$

14. $13y + 2 = 20y - 5$

15. $-4x + 25 = 6x - 45$

16. $5x - 7 = 6x - 5$

17. $5x + 4 = 10x - 7$

18. $3x - 2 = 5x - 20$

19. $27 + 5x = 9 + 3x$

20. $2y + 8 = 5y - 1$

21. $-7x + 18 = 11x - 36$

22. $4x + 5 = 2x - 7$

23. $4y + 11 = 7y - 28$

24. $4x = 2x - 12$

25. $-4x + 2 = 8x - 7$

26. $6x - 1 = 9x - 9$

27. $13x + 6 = 6x - 1$

28. $6y + 7 = 18y - 1$

29. $3x + 1 = 17 - x$

30. $17 - 4y = 14 - y$

6.3   Equations with Parentheses

To solve an equation having parentheses in one or both members, always remove the parentheses first. Then combine like terms. Then use the previously explained methods to solve the resulting equation.

•   **EXAMPLE 1**   Solve $5 - (2x - 3) = 7$.

$$5 - 2x + 3 = 7 \qquad \text{Remove parentheses.}$$
$$8 - 2x = 7 \qquad \text{Combine like terms.}$$
$$8 - 2x - 8 = 7 - 8 \qquad \text{Subtract 8 from both sides.}$$
$$-2x = -1$$
$$\frac{-2x}{-2} = \frac{-1}{-2} \qquad \text{Divide both sides by } -2.$$
$$x = \frac{1}{2}$$

CHECK:
$$5 - (2x - 3) = 7$$
$$5 - \left[2\left(\frac{1}{2}\right) - 3\right] = 7 \qquad ?$$
$$5 - (1 - 3) = 7 \qquad ?$$
$$5 - (-2) = 7 \qquad \text{True}$$

Therefore, $\frac{1}{2}$ is the root.

•   **EXAMPLE 2**   Solve $7x - 6(5 - x) = 9$.

$$7x - 30 + 6x = 9 \qquad \text{Remove parentheses.}$$
$$13x - 30 = 9 \qquad \text{Combine like terms.}$$
$$13x - 30 + 30 = 9 + 30 \qquad \text{Add 30 to both sides.}$$
$$13x = 39$$
$$\frac{13x}{13} = \frac{39}{13} \qquad \text{Divide both sides by 13.}$$
$$x = 3$$

In the following examples we have parentheses as well as the variable in both members.

•   **EXAMPLE 3**   Solve $3(x - 5) = 2(4 - x)$.

$$3x - 15 = 8 - 2x \qquad \text{Remove parentheses.}$$
$$3x - 15 + 2x = 8 - 2x + 2x \qquad \text{Add } 2x \text{ to both sides.}$$
$$5x - 15 = 8 \qquad \text{Combine like terms.}$$
$$5x - 15 + 15 = 8 + 15 \qquad \text{Add 15 to both sides.}$$
$$5x = 23$$
$$\frac{5x}{5} = \frac{23}{5} \qquad \text{Divide both sides by 5.}$$
$$x = \frac{23}{5}$$

CHECK:

$$3(x - 5) = 2(4 - x)$$

$$3\left(\frac{23}{5} - 5\right) = 2\left(4 - \frac{23}{5}\right) \quad ?$$

$$3\left(-\frac{2}{5}\right) = 2\left(-\frac{3}{5}\right) \quad ?$$

$$-\frac{6}{5} = -\frac{6}{5} \qquad \text{True}$$

Therefore, $\dfrac{23}{5}$ is the root.

• **EXAMPLE 4**   Solve $8x - 4(x + 2) = 12(x + 1) - 14$.

$$8x - 4x - 8 = 12x + 12 - 14 \qquad \text{Remove parentheses.}$$

$$4x - 8 = 12x - 2 \qquad \text{Combine like terms.}$$

$$4x - 8 - 12x = 12x - 2 - 12x \qquad \text{Subtract } 12x \text{ from both sides.}$$

$$-8x - 8 = -2$$

$$-8x - 8 + 8 = -2 + 8 \qquad \text{Add 8 to both sides.}$$

$$-8x = 6$$

$$\frac{-8x}{-8} = \frac{6}{-8} \qquad \text{Divide both sides by } -8.$$

$$x = -\frac{3}{4}$$

CHECK:

$$8x - 4(x + 2) = 12(x + 1) - 14$$

$$8\left(-\frac{3}{4}\right) - 4\left(-\frac{3}{4} + 2\right) = 12\left(-\frac{3}{4} + 1\right) - 14 \quad ?$$

$$-6 - 4\left(\frac{5}{4}\right) = 12\left(\frac{1}{4}\right) - 14 \quad ?$$

$$-6 - 5 = 3 - 14 \quad ?$$

$$-11 = -11 \qquad \text{True}$$

Therefore, $-\dfrac{3}{4}$ is the root.

Exercises 6.3

Solve each equation and check:

1. $2(x + 3) - 6 = 10$
2. $-3x + 5(x - 6) = 32$
3. $3n + (2n + 4) = 6$
4. $5m - (2m - 7) = -5$
5. $16 = -3(x - 4)$
6. $5y + 6(y - 3) = 15$
7. $5a - (3a + 4) = 8$
8. $2(b + 4) - 3 = 15$
9. $5a - 4(a - 3) = 7$
10. $29 = 4 + (2m + 1)$

11. $5(x - 3) = 21$
12. $27 - 8(2 - y) = -13$
13. $2a - (5a - 7) = 22$
14. $2(5m - 6) - 13 = -1$
15. $2(w - 3) + 6 = 0$
16. $6r - (2r - 3) + 5 = 0$
17. $3x - 7 + 17(1 - x) = -6$
18. $4y - 6(2 - y) = 8$
19. $6b = 27 + 3b$
20. $2a + 4 = a - 3$

21. $4(25 - x) = 3x + 2$
22. $4x - 2 = 3(25 - x)$
23. $x + 3 = 4(57 - x)$
24. $2(y + 1) = y - 7$
25. $6x + 2 = 2(17 - x)$
26. $6(17 - x) = 2 - 4x$
27. $5(x - 8) - 3x - 4 = 0$
28. $5(28 - 2x) - 7x = 4$
29. $3(x + 4) + 3x = 6$
30. $8x - 4(x + 2) + 11 = 0$

31. $y - 4 = 2(y - 7)$

32. $7(w - 4) = w + 2$

33. $9m - 3(m - 5) = 7m - 3$

34. $4(x + 18) = 2(4x + 18)$

35. $3(2x + 7) = 13 + 2(4x + 2)$

36. $5y - 3(y - 2) = 6(y + 1)$

37. $8(x - 5) = 13x$

38. $4(x + 2) = 30 - (x - 3)$

39. $5 + 3(x + 7) = 26 - 6(5x + 11)$

40. $2(y - 3) = 4 + (y - 14)$

41. $5(2y - 3) = 3(7y - 6) + 19(y + 1) + 14$

42. $3(7y - 6) - 11(y + 1) = 58 - 7(9y + 4)$

43. $16(x + 3) = 7(x - 5) - 9(x + 4) - 7$

44. $31 - 2(x - 5) = -3(x + 4)$

45. $4(y + 2) = 8(y - 4) + 7$

46. $12x - 13(x + 4) = 4x - 6$

47. $4(5y - 2) + 3(2y + 6) = 25(3y + 2) - 19y$

48. $6(3x + 1) = 5x - (2x + 2)$

49. $12 + 8(2y + 3) = (y + 7) - 16$

50. $-2x + 6(2 - x) - 4 = 3(x + 1) - 6$

51. $5x - 10(3x - 6) = 3(24 - 9x)$

52. $4y + 7 - 3(2y + 3) = 4(3y - 4) - 7y + 7$

53. $6(y - 4) - 4(5y + 1) = 3(y - 2) - 4(2y + 1)$

54. $2(5y + 1) + 16 = 4 + 3(y - 7)$

55. $-6(x - 5) + 3x = 6x - 10(-3 + x)$

56. $14x + 14(3 - 2x) + 7 = 4 - x + 5(2 - 3x)$

57. $2.3x - 4.7 + 0.6(3x + 5) = 0.7(3 - x)$

58. $5.2(x + 3) + 3.7(2 - x) = 3$

59. $0.089x - 0.32 + 0.001(5 - x) = 0.231$

60. $5x - 2.5(7 - 4x) = x - 7(4 + x)$

6.4   Equations with Fractions

TO SOLVE AN EQUATION WITH FRACTIONS

1. Find the least common denominator (LCD) of all the fractional terms on both sides of the equation.

2. Multiply both sides of the equation by the LCD. (If this step has been done correctly, no fractions should now appear in the resulting equation.)

3. Solve the resulting equation from Step 2 using the methods introduced earlier in this chapter.

•   **EXAMPLE 1**   Solve $\dfrac{3x}{4} = \dfrac{45}{20}$.

The LCD of 4 and 20 is 20; therefore, multiply both sides of the equation by 20.

$$\frac{3x}{4} = \frac{45}{20}$$

$$20\left(\frac{3x}{4}\right) = \left(\frac{45}{20}\right)20$$

$$15x = 45$$

$$\frac{15x}{15} = \frac{45}{15} \qquad \text{Divide both sides by 15.}$$

$$x = 3$$

•   **EXAMPLE 2**   Solve $\dfrac{3}{4} + \dfrac{x}{6} = \dfrac{13}{12}$.

The LCD of 4, 6, and 12 is 12; multiply both sides of the equation by 12.

$$\frac{3}{4} + \frac{x}{6} = \frac{13}{12}$$

$$12\left(\frac{3}{4} + \frac{x}{6}\right) = \left(\frac{13}{12}\right)12$$

$$12\left(\frac{3}{4}\right) + 12\left(\frac{x}{6}\right) = \left(\frac{13}{12}\right)12$$

$$9 + 2x = 13$$

$$9 + 2x - 9 = 13 - 9 \qquad \text{Subtract 9 from both sides.}$$

$$2x = 4$$

$$\frac{2x}{2} = \frac{4}{2} \qquad \text{Divide both sides by 2.}$$

$$x = 2$$

CHECK:

$$\frac{3}{4} + \frac{x}{6} = \frac{13}{12}$$

$$\frac{3}{4} + \frac{2}{6} = \frac{13}{12} \qquad ?$$

$$\frac{9}{12} + \frac{4}{12} = \frac{13}{12} \qquad \text{True}$$

• **EXAMPLE 3**   Solve $\frac{2x}{9} - 4 = \frac{x}{6}$.

The LCD of 9 and 6 is 18; multiply both sides of the equation by 18.

$$\frac{2x}{9} - 4 = \frac{x}{6}$$

$$18\left(\frac{2x}{9} - 4\right) = \left(\frac{x}{6}\right)18$$

$$18\left(\frac{2x}{9}\right) - 18(4) = \left(\frac{x}{6}\right)18$$

$$4x - 72 = 3x$$

$$4x - 72 - 4x = 3x - 4x \qquad \text{Subtract } 4x \text{ from both sides.}$$

$$-72 = -x$$

$$\frac{-72}{-1} = \frac{-x}{-1} \qquad \text{Divide both sides by } -1.$$

$$72 = x$$

CHECK:

$$\frac{2x}{9} - 4 = \frac{x}{6}$$

$$\frac{2(72)}{9} - 4 = \frac{72}{6} \qquad ?$$

$$16 - 4 = 12 \qquad \text{True}$$

• **EXAMPLE 4**   Solve $\frac{2}{3}x + \frac{3}{4}(36 - 2x) = 32$.

The LCD of 3 and 4 is 12; multiply both sides of the equation by 12.

$$\frac{2}{3}x + \frac{3}{4}(36 - 2x) = 32$$

$$12\left[\frac{2}{3}x + \frac{3}{4}(36 - 2x)\right] = (32)12$$

$$12\left(\frac{2}{3}x\right) + 12\left(\frac{3}{4}\right)(36 - 2x) = (32)12 \qquad \text{Remove brackets.}$$

$$8x + 9(36 - 2x) = 384$$

$$8x + 324 - 18x = 384 \qquad \text{Remove parentheses.}$$

$$-10x + 324 = 384 \qquad \text{Combine like terms.}$$

$$-10x + 324 - 324 = 384 - 324 \qquad \text{Subtract 324 from both sides.}$$

$$-10x = 60$$

$$\frac{-10x}{-10} = \frac{60}{-10} \qquad \text{Divide both sides by } -10.$$

$$x = -6$$

CHECK:

$$\frac{2}{3}x + \frac{3}{4}(36 - 2x) = 32$$

$$\frac{2}{3}(-6) + \frac{3}{4}[36 - 2(-6)] = 32 \qquad ?$$

$$-4 + \frac{3}{4}(36 + 12) = 32 \qquad ?$$

$$-4 + \frac{3}{4}(48) = 32 \qquad ?$$

$$-4 + 36 = 32 \qquad \text{True}$$

• **EXAMPLE 5**   Solve $\dfrac{2x + 1}{3} - \dfrac{x - 6}{4} = \dfrac{2x + 4}{8} + 2$.

The LCD of 3, 4, and 8 is 24; multiply both sides of the equation by 24.

$$\frac{2x + 1}{3} - \frac{x - 6}{4} = \frac{2x + 4}{8} + 2$$

$$24\left(\frac{2x + 1}{3} - \frac{x - 6}{4}\right) = \left(\frac{2x + 4}{8} + 2\right)24$$

$$24\left(\frac{2x + 1}{3}\right) - 24\left(\frac{x - 6}{4}\right) = \left(\frac{2x + 4}{8}\right)24 + 2(24)$$

$$8(2x + 1) - 6(x - 6) = 3(2x + 4) + 48$$

$$16x + 8 - 6x + 36 = 6x + 12 + 48 \qquad \text{Remove parentheses.}$$

$$10x + 44 = 6x + 60 \qquad \text{Combine like terms.}$$

$$10x + 44 - 6x = 6x + 60 - 6x \qquad \text{Subtract } 6x \text{ from both sides.}$$

$$4x + 44 = 60$$

$$4x + 44 - 44 = 60 - 44 \qquad \text{Subtract 44 from both sides.}$$

$$4x = 16$$

$$\frac{4x}{4} = \frac{16}{4} \qquad \text{Divide both sides by 4.}$$

$$x = 4$$

CHECK:
$$\frac{2x + 1}{3} - \frac{x - 6}{4} = \frac{2x + 4}{8} + 2$$

$$\frac{2(4) + 1}{3} - \frac{4 - 6}{4} = \frac{2(4) + 4}{8} + 2 \qquad ?$$

$$\frac{9}{3} - \frac{-2}{4} = \frac{12}{8} + 2 \qquad ?$$

$$3 + \frac{1}{2} = \frac{3}{2} + 2 \qquad ?$$

$$\frac{7}{2} = \frac{7}{2} \qquad \text{True}$$

When the variable appears in the denominator of a fraction in an equation, multiply both members by the LCD. Be careful that the replacement for the variable does not make the denominator zero.

• EXAMPLE 6   Solve: $\dfrac{3}{x} = 2$

$$x\left(\frac{3}{x}\right) = (2)x \qquad \text{Multiply both sides by the LCD, } x.$$

$$3 = 2x$$

$$\frac{3}{2} = \frac{2x}{2} \qquad \text{Divide both sides by 2.}$$

$$\frac{3}{2} = x$$

CHECK:
$$\frac{3}{x} = 2$$

$$\frac{3}{\frac{3}{2}} = 2 \qquad ?$$

$$3 \div \frac{3}{2} = 2 \qquad ?$$

$$3 \cdot \frac{2}{3} = 2 \qquad ?$$

$$2 = 2 \qquad \text{True}$$

Thus, the root is $\dfrac{3}{2}$.

• EXAMPLE 7   Solve: $\dfrac{5}{x} - 2 = 3$

$$\frac{5}{x} - 2 + 2 = 3 + 2 \qquad \text{Add 2 to both sides.}$$

$$\frac{5}{x} = 5$$

$$x\left(\frac{5}{x}\right) = (5)x \qquad \text{Multiply both sides by } x.$$

$$5 = 5x$$

$$\frac{5}{5} = \frac{5x}{5} \qquad \text{Divide both sides by 5.}$$

$$1 = x$$

Exercises | 6.4

Solve each equation and check:

1. $\dfrac{2x}{3} = \dfrac{32}{6}$ **4.** $\dfrac{5}{3}x = -13\dfrac{1}{3}$ **7.** $\dfrac{2}{3} + \dfrac{x}{4} = \dfrac{28}{6}$ **10.** $1\dfrac{1}{4} + \dfrac{x}{3} = \dfrac{7}{12}$ **13.** $\dfrac{y}{3} - 1 = \dfrac{y}{6}$

2. $\dfrac{5x}{7} = \dfrac{20}{14}$ **5.** $2\dfrac{1}{2}x = 7\dfrac{1}{2}$ **8.** $\dfrac{x}{7} - \dfrac{1}{14} = \dfrac{70}{28}$ **11.** $\dfrac{3}{5}x - 25 = \dfrac{x}{10}$ **14.** $\dfrac{1}{2}x - 3 = \dfrac{x}{5}$

3. $\dfrac{3}{8}y = 1\dfrac{14}{16}$ **6.** $\dfrac{1}{2} + \dfrac{x}{3} = \dfrac{5}{2}$ **9.** $\dfrac{3}{4} - \dfrac{x}{3} = \dfrac{5}{12}$ **12.** $\dfrac{2x}{3} - 7 = -\dfrac{x}{2}$ **15.** $\dfrac{3x}{4} - \dfrac{7}{20} = \dfrac{2}{5}x$

16. $\dfrac{5x}{6} + \dfrac{1}{3}(6 + x) = 37$ **22.** $\dfrac{4x + 3}{15} - \dfrac{2x - 3}{9} = \dfrac{6x + 4}{6} - x$

17. $\dfrac{x}{2} + \dfrac{2}{3}(2x + 3) = 46$ **23.** $5x + \dfrac{6x - 8}{14} + \dfrac{10x + 6}{6} = 43$

18. $\dfrac{1}{6}x - \dfrac{1}{9}(2x - 3) = 1$ **24.** $\dfrac{3x}{5} - \dfrac{9 - 3x}{10} = \dfrac{6}{10} - \dfrac{3x + 6}{10}$

19. $0.96 = 0.06(12 + x)$ **25.** $\dfrac{4x}{6} - \dfrac{x + 5}{2} = \dfrac{6x - 6}{8}$

20. $\dfrac{1}{2}(x + 2) + \dfrac{3}{8}(28 - x) = 11$ **26.** $\dfrac{x}{3} + \dfrac{2x + 4}{4} = \dfrac{x - 1}{6} - \dfrac{3 - 2x}{2}$

21. $\dfrac{3x - 24}{16} - \dfrac{3x - 12}{12} = 3$

27. $\dfrac{4}{x} = 6$ **31.** $\dfrac{5}{y} - 1 = 4$ **35.** $7 - \dfrac{6}{x} = 5$ **39.** $1 - \dfrac{2}{x} = \dfrac{14}{3x} - \dfrac{1}{3}$

28. $\dfrac{2}{x} - 8 = -7$ **32.** $\dfrac{17}{x} = 8$ **36.** $9 + \dfrac{3}{x} = 10\dfrac{1}{2}$ **40.** $\dfrac{3}{x} + 2 = \dfrac{5}{x} - 4$

29. $5 - \dfrac{1}{x} = 7$ **33.** $\dfrac{3}{x} - 8 = 7$ **37.** $\dfrac{6}{x} + 5 = 14$ **41.** $\dfrac{7}{2x} + 14\dfrac{1}{2} = \dfrac{7}{x} - 10$

30. $\dfrac{3}{x} - 6 = 8$ **34.** $\dfrac{5}{2x} + 8 = 17$ **38.** $\dfrac{3}{x} - 3 = \dfrac{5}{2x} - 2$ **42.** $\dfrac{8}{x} + \dfrac{1}{4} = \dfrac{5}{x} + \dfrac{1}{3}$

6.5   Translating Words into Algebraic Symbols

The ability to translate English words into algebra is very important for solving "applied" problems. To help you, we provide the following table of common English words for the common mathematical symbols:

+	−	×	÷	=
plus	minus	times	divide	equal or equals
increased by	decreased by	product	quotient	is or are
added to	subtract	multiply by	divided by	is equal to
more than	less than	double or twice	half of	result is
sum of	difference	triple or thrice	one-third of	
	subtract from			

• **EXAMPLE 1**   Translate into algebra: One number is four times another, and their sum is twenty.

$$\text{Let } x = \text{first number}$$
$$4x = \text{four times the number}$$
$$x + 4x = \text{their sum}$$

Sentence in algebra: $x + 4x = 20$ ———————•

• **EXAMPLE 2**   Translate into algebra: The sum of a number and the number decreased by six is five.

$$\text{Let } x = \text{the number}$$
$$x - 6 = \text{number decreased by six}$$
$$x + (x - 6) = \text{sum}$$

Sentence in algebra: $x + (x - 6) = 5$ ———————•

• **EXAMPLE 3**   Translate into algebra: Fifteen more than twice a number is twenty-four.

$$\text{Let } x = \text{the number}$$
$$2x = \text{twice the number}$$
$$2x + 15 = \text{fifteen more than twice the number}$$

Sentence in algebra: $2x + 15 = 24$ ———————•

• **EXAMPLE 4**   Translate into algebra: Twice the sum of a number and five is eighty.

$$\text{Let } x = \text{the number}$$
$$x + 5 = \text{sum of a number and five}$$
$$2(x + 5) = \text{twice the sum of a number and five}$$

Sentence in algebra: $2(x + 5) = 80$ ———————•

Exercises | 6.5

Translate each phrase or sentence into algebraic symbols:

1. A number decreased by twenty
2. A number increased by five
3. A number divided by six
4. A number times eighteen
5. The sum of a number and sixteen
6. Subtract twenty-six from a number
7. Subtract a number from twenty-six
8. One-half a number
9. Twice a number
10. The difference between four and a number
11. The sum of six times a number and twenty-eight is forty.
12. The difference between twice a number and thirty is fifty.

13. The quotient of a number and six is five.
14. If seven is added to a number, the sum is 32.
15. If a number is increased by 28 and then multiplied by five, the result is 150.
16. The sum of a number and the number decreased by five is 25.
17. The quotient of a number and six, decreased by seven is two.
18. The product of five and five more than a number is 50.
19. The difference between thirty and twice a number is four.
20. Double the difference between a number and six is thirty.

21. The product of a number decreased by seven and the same number increased by five is thirteen.

22. Seven times a number decreased by eleven is 32.

23. The product of a number and six decreased by seventeen is seven.

24. If twelve is added to the product of a number and twelve, the sum is 72.

25. Seventeen less than four times a number is 63.

6.6 Applications Involving Equations*

An applied problem can often be expressed mathematically as a simple equation. The problem can then be solved by solving the equation. To solve such an application problem, we suggest the following steps.

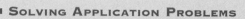

SOLVING APPLICATION PROBLEMS

STEP 1 Read the problem carefully at least twice.

STEP 2 If possible, draw a diagram. This will often help you visualize the mathematical relationship needed to write the equation.

STEP 3 Choose a symbol to represent the unknown quantity in the problem, and write what it represents.

STEP 4 Write an equation that expresses the information given in the problem and that involves the unknown.

STEP 5 Solve the equation from Step 4.

STEP 6 Check your solution both in the equation from Step 4 and in the original problem itself.

•   **EXAMPLE 1**   You need to tile the floor of a rectangular room with a wooden outer border of 6 in. The floor of the room is 10 ft by 8 ft 2 in. How many rows of 4 in. by 4 in. tiles are needed to fit across the length of the room?

The sketch shown in Figure 6.1 is helpful in solving the problem.

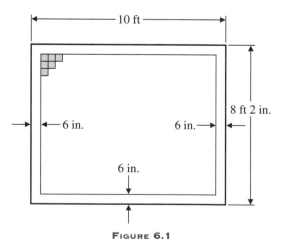

FIGURE 6.1

Note: In this chapter, do not use the rules for calculating with measurements.

Let x = the number of tiles across the length of the room

$4x$ = the number of inches in x tiles

10 ft = 120 in.

$4x + 6 + 6 = 120$

$4x + 12 = 120$

$4x = 108$ Subtract 12 from both sides.

$x = 27$ Divide both sides by 4.

So there are 27 rows of tiles.

EXAMPLE 2   An interior wall measures 30 ft 4 in. long. It is to be divided by 10 evenly spaced posts; each post is 4 in. by 4 in. (Posts are to be located in the corners.) What is the distance between posts? Note that there are 9 spaces between posts. (See Figure 6.2).

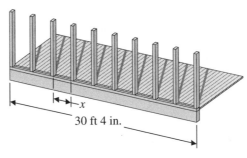

30 ft 4 in.

FIGURE 6.2

Let x = the distance between posts

$9x$ = the distance of 9 spaces

$(10)(4 \text{ in.})$ = the distance used up by ten 4-in. posts

$9x + (10)(4 \text{ in.}) = 364 \text{ in.}$ 30 ft 4 in. = 364 in.

$9x + 40 = 364$

$9x = 324$ Subtract 40 from both sides.

$x = 36 \text{ in.}$ Divide both sides by 9.

CHECK: $9x + 40 = 364$

$9(36) + 40 = 364$?

$324 + 40 = 364$?

$364 = 364$ True

EXAMPLE 3   Two different automotive batteries cost a total of $117. One costs $12 more than twice the other. Find the cost of each battery.

Let x = the cost of one battery

$2x + 12$ = the cost of the other battery

$x + 2x + 12 = 117$

$3x + 12 = 117$ Combine like terms.

$3x = 105$ Subtract 12 from both sides.

$x = \$35$, the cost of the first battery

$2x + 12 = 2(35) + 12 = \$82$, the cost of the other battery

• **EXAMPLE 4**   One side of a triangle is twice another. The third side is 5 more than the shortest side. The perimeter is 33. Find the length of each side.

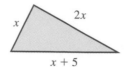

FIGURE 6.3

First, draw and label a triangle as in Figure 6.3.

$$\text{Let } x = \text{the length of the first side}$$
$$2x = \text{the length of the second side}$$
$$x + 5 = \text{the length of the third side}$$
$$x + 2x + x + 5 = 33$$
$$4x + 5 = 33$$
$$4x = 28$$
$$x = 7, \text{ the length of the first side}$$
$$2x = 2(7) = 14, \text{ the length of the second side}$$
$$x + 5 = (7) + 5 = 12, \text{ the length of the third side} \quad \underline{\hspace{2cm}}\bullet$$

• **EXAMPLE 5**   Forty acres of land were sold for $27,000. Some was sold at $800 per acre, and the rest at $600 per acre. How much was sold at each price?

$$\text{Let } x = \text{the amount of land sold at \$800/acre}$$
$$40 - x = \text{the amount of land sold at \$600/acre}$$

Then

$$800x = \text{the value of the land sold at \$800/acre}$$
$$600(40 - x) = \text{the value of the land sold at \$600/acre}$$
$$27{,}000 = \text{the total value of the land}$$

Therefore, the equation is

$$800x + 600(40 - x) = 27{,}000$$
$$800x + 24{,}000 - 600x = 27{,}000$$
$$200x + 24{,}000 = 27{,}000$$
$$200x = 3000$$
$$x = 15$$
$$40 - x = 25$$

Thus, 15 acres were sold at $800/acre, and 25 acres were sold at $600/acre.

$$\underline{\hspace{2cm}}\bullet$$

Note: When you know the total of two parts, one possible equation solving strategy is to

$$\text{let } x = \text{one part} \quad \text{and}$$
$$\text{total} - x = \text{the other part}$$

• **EXAMPLE 6**   How much pure alcohol must be added to 200 cm^3 of a solution that is 15% alcohol to make a solution that is 40% alcohol?

$$\text{Let } x = \text{the amount of pure alcohol (100\%) added}$$

You may find Figure 6.4 helpful.

Amount of alcohol to start	Amount of pure alcohol added	Total amount of alcohol
15% of 200 cm^3	100% of x	40% of (200 + x)

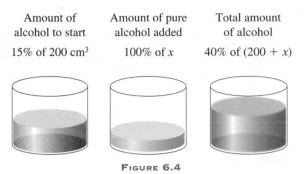

FIGURE 6.4

$$0.15(200) + 1.00x = 0.40(200 + x)$$
$$30 + x = 80 + 0.4x$$
$$x = 50 + 0.4x \qquad \text{Subtract 30 from both sides.}$$
$$0.6x = 50 \qquad \text{Subtract } 0.4x \text{ from both sides.}$$
$$x = 83.3 \qquad \text{Divide both sides by 0.6.}$$

Thus, 83.3 cm^3 of pure alcohol must be added.

Exercises 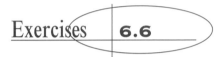 6.6

1. A set of eight built-in bookshelves is to be constructed in a room. The floor-to-ceiling clearance is 8 ft 2 in. Each shelf is 1 in. thick. An equal space is to be left between shelves. What space should there be between each shelf and the next? (There is no shelf against the ceiling and no shelf on the floor.)

2. Saw a board 8 ft 4 in. long into nine equal pieces. If the loss per cut is $\frac{1}{8}$ in., how long will each piece be?

3. Separate an order of 256 light fixtures so that the number of fluorescent light fixtures will be 20 fewer than twice the number of incandescent light fixtures.

4. Distribute $1000 into three parts so that one part will be three times as large as the second, and the third part will be as large as the sum of the other two.

5. Distribute $4950 among John, Maria, and Betsy so that Maria receives twice as much as John, and Betsy receives three times as much as John.

6. Distribute $4950 among John, Maria, and Betsy so that Maria receives twice as much as John, and Betsy receives three times as much as Maria.

7. A rectangle is twice as long as it is wide. Its perimeter (the sum of the lengths of its sides) is 60 cm. Find its length and width.

8. The length of a rectangle is 4 cm less than twice its width. Its perimeter is 40 cm. Find its length and width.

9. One side of a rectangular yard is bounded by the side of a house. The other three sides are to be fenced with 345 ft of fencing. The length of fence opposite the house is 15 ft less than either of the other two sides. Find the length and width of the yard.

10. A given type of concrete contains twice as much sand as cement and 1.5 times as much gravel as sand. How many yd^3 of each must be used to make 9 yd^3 of concrete? Assume no loss of volume in mixing.

11. The perimeter of a triangle is 122 ft. The lengths of two sides are the same. The length of the third side is 4 ft shorter than either of the other two sides. Find the lengths of the three sides.

12. Cut a board 20 ft long into three pieces so that the longest piece will be three times as long as each of the other two of equal lengths. Find the length of each piece.

13. Cut a 12-ft beam into two pieces so that one piece is 18 in. longer than the other. Find the length of the two pieces.

14. The total cost of three automobile batteries is $210. The most expensive one is three times the cost of

the least expensive. The third is $15 more than the least expensive. Find the cost of each battery.

15. The total cost of 20 boards is $166. One size costs $6.50, and the second size costs $9.50. How many boards were purchased at each price?

16. Amy and Kim earn a total of $308 by working a total of 30 hours. If Amy earns $8/h and Kim earns $12/h, how many hours did each work?

17. Joyce invests $7500 in two savings accounts. One account earns interest at 4% per year, while the other earns 6% per year. The total interest earned from both accounts after one year is $390. How much was originally deposited in each account?

18. Chuck receives loans totaling $12,000 from two banks. One bank charges 7.5% annual interest, and the second bank charges 9% annual interest. He paid $960 in total interest in one year. How much was loaned at each bank?

19. Regular milk has 4% butterfat. How many litres of regular milk must be mixed with 40 L of milk with 1% butterfat to get milk with 2% butterfat?

20. How much pure alcohol must be added to 750 mL of a solution that is 40% alcohol to make a solution that is 60% alcohol?

21. Mix a solution that is 30% alcohol with a solution that is 80% alcohol to make 800 mL of a solution that is 60% alcohol. How much of each solution should you use?

22. Mix a solution that is 50% acid with a solution that is 100% water to make 4 L of a solution that is 10% acid. How much of each solution should you use?

23. A 12-quart cooling system is checked and found to be filled with a solution that is 40% antifreeze. The desired strength of the solution is 60% antifreeze. How many quarts of solution need to be drained and replaced with pure antifreeze to reach the desired strength?

24. In testing an engine, various mixtures of gasoline and methanol are being tried. How much of a 90% gasoline mixture and a 75% gasoline mixture are needed for 1200 L of an 85% gasoline mixture?

6.7   Formulas

A *formula* is a general rule written as an equation, usually expressed in letters, which shows the relationship between two or more quantities. For example, the formula

$$d = rt$$

states that the distance, d, that a body travels equals the product of its rate, r, of travel and the time, t, of travel. The formula

$$p = \frac{F}{A}$$

states that the pressure, p, equals the quotient of the force, F, and the area, A, over which the force is applied.

Sometimes the letters in a formula do not match the first letter of the name of the quantity. For example, Ohm's law is often written

$$E = IR$$

where E is the voltage, I is the current, and R is the resistance.

Sometimes subscripts are used to distinguish between different readings of the same quantity. For example, the final velocity of an object equals the sum of the initial velocity and the product of its acceleration and the time of the acceleration. This is written

$$v_f = v_i + at$$

where v_f is the final velocity, v_i is the initial velocity, a is the acceleration, and t is the time.

Sometimes Greek letters are used. For example, the resistance of a wire is given by the formula

$$R = \frac{\rho L}{A}$$

where R is the resistance of the wire, ρ is the resistivity constant of the wire, L is the length of the wire, and A is the cross-sectional area of the wire.

Sometimes the formula is written with the letters and symbols used by the person who discovered the relationship. The letters may have no obvious relationship with the quantity.

To solve a formula for a given letter means to isolate the given letter on one side of the equation and express it in terms of all the remaining letters. This means that the given letter appears on one side of the equation by itself; all the other letters appear on the opposite side of the equation. We solve a formula using the same methods we use in solving an equation.

•   **EXAMPLE 1**   Solve **a.** $d = rt$ for t and **b.** $15 = 5t$ for t.

a. $d = rt$

$\dfrac{d}{r} = \dfrac{rt}{r}$ Divide both sides by r.

$\dfrac{d}{r} = t$

b. $15 = 5t$

$\dfrac{15}{5} = \dfrac{5t}{5}$ Divide both sides by 5.

$3 = t$

Note that we use the same techniques for solving a formula as we learned earlier for solving equations.

•   **EXAMPLE 2**   Solve $p = \dfrac{F}{A}$ for F and then for A.

First, solve for F.

$$p = \frac{F}{A}$$

$$pA = \left(\frac{F}{A}\right) A \qquad \text{Multiply both sides by } A.$$

$$pA = F$$

Now solve for A.

$$p = \frac{F}{A}$$

$$pA = F \qquad \text{Multiply both sides by } A.$$

$$\frac{pA}{p} = \frac{F}{p} \qquad \text{Divide both sides by } p.$$

$$A = \frac{F}{p}$$

• **EXAMPLE 3**   Solve $V = E - Ir$ for I.

One way:

$$V = E - Ir$$

$$V - E = E - Ir - E \qquad \text{Subtract } E \text{ from both sides.}$$

$$V - E = -Ir$$

$$\frac{V - E}{-r} = \frac{-Ir}{-r} \qquad \text{Divide both sides by } -r.$$

$$\frac{V - E}{-r} = I$$

Alternate way:

$$V = E - Ir$$

$$V + Ir = E - Ir + Ir \qquad \text{Add } Ir \text{ to both sides.}$$

$$V + Ir = E$$

$$V + Ir - V = E - V \qquad \text{Subtract } V \text{ from both sides.}$$

$$Ir = E - V$$

$$\frac{Ir}{r} = \frac{E - V}{r} \qquad \text{Divide both sides by } r.$$

$$I = \frac{E - V}{r}$$

Note that the two results are equivalent. Take the first result,

$$\frac{V - E}{-r}$$

and multiply numerator and denominator by -1:

$$\left(\frac{V - E}{-r}\right)\left(\frac{-1}{-1}\right) = \frac{-V + E}{r} = \frac{E - V}{r}$$

• **EXAMPLE 4**   Solve $S = \dfrac{n(a + l)}{2}$ for n and then for l.

First, solve for n.

$$S = \frac{n(a + l)}{2}$$

$$2S = n(a + l) \qquad \text{Multiply both sides by 2.}$$

$$\frac{2S}{a + l} = n \qquad \text{Divide both sides by } (a + l).$$

Now solve for l.

$$S = \frac{n(a + l)}{2}$$

$$2S = n(a + l) \qquad \text{Multiply both sides by 2.}$$

$$2S = na + nl \qquad \text{Remove parentheses.}$$

$$2S - na = nl \qquad \text{Subtract } na \text{ from both sides.}$$

$$\frac{2S - na}{n} = l \qquad \text{Divide both sides by } n.$$

Or $$\frac{2S}{n} - a = l$$

• **EXAMPLE 5**   Solve $(\triangle L) = kL(T - T_0)$ for T.

$$(\triangle L) = kL(T - T_0).$$ Note: Treat $(\triangle L)$ as one variable.

$$(\triangle L) = kLT - kLT_0$$ Remove parentheses.

$$(\triangle L) + kLT_0 = kLT$$ Add kLT_0 to both sides.

$$\frac{(\triangle L) + kLT_0}{kL} = T$$ Divide both sides by kL.

Can you show that $T = \dfrac{(\triangle L)}{kL} + T_0$ is an equivalent solution?

Exercises 6.7

Solve each formula for the given letter:

1. $E = Ir$ for r

2. $A = bh$ for b

3. $F = ma$ for a

4. $w = mg$ for m

5. $C = \pi d$ for d

6. $V = IR$ for R

7. $V = lwh$ for w

8. $X_L = 2\pi fL$ for f

9. $A = 2\pi rh$ for h

10. $C = 2\pi r$ for r

11. $v^2 = 2gh$ for h

12. $V = \pi r^2 h$ for h

13. $I = \dfrac{Q}{t}$ for t

14. $I = \dfrac{Q}{t}$ for Q

15. $v = \dfrac{s}{t}$ for s

16. $I = \dfrac{E}{Z}$ for Z

17. $I = \dfrac{V}{R}$ for R

18. $P = \dfrac{w}{t}$ for w

19. $E = \dfrac{I}{4\pi r^2}$ for I

20. $R = \dfrac{\pi}{2P}$ for P

21. $X_C = \dfrac{1}{2\pi fC}$ for f

22. $R = \dfrac{\rho L}{A}$ for L

23. $A = \dfrac{1}{2}bh$ for b

24. $V = \dfrac{1}{3}\pi r^2 h$ for h

25. $Q = \dfrac{I^2 Rt}{J}$ for R

26. $R = \dfrac{kl}{D^2}$ for l

27. $F = \dfrac{9}{5}C + 32$ for C

28. $C = \dfrac{5}{9}(F - 32)$ for F

29. $C_T = C_1 + C_2 + C_3 + C_4$ for C_2

30. $R_T = R_1 + R_2 + R_3 + R_4$ for R_4

31. $Ax + By + C = 0$ for x

32. $A = P + Prt$ for r

33. $Q_1 = P(Q_2 - Q_1)$ for Q_2

34. $v_f = v_i + at$ for v_i

35. $A = \left(\dfrac{a + b}{2}\right)h$ for h

36. $A = \left(\dfrac{a + b}{2}\right)h$ for b

37. $l = a + (n - 1)d$ for d

38. $A = ab + \dfrac{d}{2}(a + c)$ for d

39. $Ft = m(V_2 - V_1)$ for m

40. $l = a + (n - 1)d$ for n

41. $Q = wc(T_1 - T_2)$ for c

42. $Ft = m(V_2 - V_1)$ for V_1

43. $V = \dfrac{2\pi(3960 + h)}{P}$ for h

44. $Q = wc(T_1 - T_2)$ for T_2

6.8   ## Substituting Data into Formulas

Problem solving skills are essential in all technical fields. Working with formulas is one of the most important tools that you can gain from this course.

PROBLEM SOLVING

Necessary parts of problem solving include:

1. Analyzing the given data.
2. Finding an equation or formula that relates the given quantities with the unknown quantity.
3. Solving the formula for the unknown quantity.
4. Substituting the given data into this solved formula.

Actually, you may solve the formula for the unknown quantity, and then substitute the data. Or you may substitute the data into the formula first, and then solve for the unknown quantity. If you use a calculator, the first method is more helpful.

•   **EXAMPLE 1**   Given the formula $V = IR$, $V = 120$, and $R = 2\overline{0}0$. Find I.

First, solve for I.

$$V = IR$$

$$\frac{V}{R} = \frac{IR}{R} \qquad \text{Divide both sides by } R.$$

$$\frac{V}{R} = I$$

Then substitute the data.

$$I = \frac{V}{R} = \frac{120}{2\overline{0}0} = 0.60$$

───────●

•   **EXAMPLE 2**   Given $v = v_0 + at$, $v = 60.0$, $v_0 = 20.0$, and $t = 5.00$. Find a.

First, solve for a.

$$v = v_0 + at$$

$$v - v_0 = at \qquad\qquad \text{Subtract } v_0 \text{ from both sides.}$$

$$\frac{v - v_0}{t} = \frac{at}{t} \qquad\qquad \text{Divide both sides by } t.$$

$$\frac{v - v_0}{t} = a$$

Then substitute the data.

$$a = \frac{v - v_0}{t} = \frac{60.0 - 20.0}{5.00} = \frac{40.0}{5.00} = 8.00$$

───────●

•   **EXAMPLE 3**   Given the formula $S = \dfrac{MC}{l}$, $S = 47.5$, $M = 190$, and $C = 8.0$. Find l.

First, solve for l.

$$S = \frac{MC}{l}$$

$$Sl = MC \qquad \text{Multiply both sides by } l.$$

$$l = \frac{MC}{S} \qquad \text{Divide both sides by } S.$$

Then substitute the data.

$$l = \frac{MC}{S} = \frac{190(8.0)}{47.5} = 32$$

• **EXAMPLE 4** Given the formula $Q = WC(T_1 - T_2)$, $Q = 15$, $W = 3.0$, $T_1 = 11\overline{0}$, and $T_2 = 6\overline{0}$. Find C.

First, solve for C.

$$Q = WC(T_1 - T_2)$$

$$\frac{Q}{W(T_1 - T_2)} = C \qquad \text{Divide both sides by } W(T_1 - T_2).$$

Then substitute the data.

$$\frac{15}{3.0(11\overline{0} - 6\overline{0})} = C$$

$$C = \frac{15}{150} = 0.10$$

• **EXAMPLE 5** Given the formula $V = \frac{1}{2}lw(D + d)$, $V = 156.8$, $D = 2.00$, $l = 8.37$, and $w = 7.19$. Find d.

First, solve for d.

$$V = \frac{1}{2}lw(D + d)$$

$$2V = lw(D + d) \qquad \text{Multiply both sides by 2.}$$

$$2V = lwD + lwd \qquad \text{Remove parentheses.}$$

$$2V - lwD = lwd \qquad \text{Subtract } lwD \text{ from both sides.}$$

$$\frac{2V - lwD}{lw} = d \qquad \text{Divide both sides by } lw.$$

$$\frac{2V}{lw} - D = d \qquad \text{Simplify the left side.}$$

Then substitute the data.

$$d = \frac{2V}{lw} - D = \frac{2(156.8)}{(8.37)(7.19)} - 2.00 = 3.21$$

Exercises  6.8

a. *Solve for the indicated letter.* **b.** *Then substitute the given values to find the value of the indicated letter (use the rules for working with measurements):*

	Formula	Given	Find
1.	$A = lw$	$A = 414$, $w = 18.0$	l
2.	$V = IR$	$I = 9.20$, $V = 5.52$	R
3.	$V = \dfrac{\pi r^2 h^*}{3}$	$V = 753.6$, $r = 6.00$	h

Note: Use the π button on your calculator.

4.	$I = \dfrac{V}{R}$	$R = 44, I = 2.5$	V
5.	$E = \dfrac{mv^2}{2}$	$E = 484{,}000; v = 22.0$	m
6.	$v_f = v_i + at$	$v_f = 88, v_i = 10.0, t = 12$	a
7.	$v_f = v_i + at$	$v_f = 193.1, v_i = 14.9, a = 18.0$	t
8.	$y = mx + b$	$x = 3, y = 2, b = 9$	m
9.	$v_f^2 = v_i^2 + 2gh$	$v_f = 192, v_i = 0, g = 32.0$	h
10.	$A = P + Prt$	$r = 0.07, P = 1500, A = \$2025$	t
11.	$L = \pi(r_1 + r_2) + 2d$	$L = 37.68, d = 6.28, r_2 = 5.00$	r_1
12.	$C = \dfrac{5}{9}(F - 32)$	$C = -5$	F
13.	$Fgr = Wv^2$	$F = 12{,}000; W = 24{,}000;$ $v = 176; g = 32$	r
14.	$Q = WC(T_1 - T_2)$	$Q = 18.9, W = 3.0, C = 0.18,$ $T_2 = 59$	T_1
15.	$A = \dfrac{1}{2}h(a + b)$	$A = 1160, h = 22.0, a = 56.5$	b
16.	$A = \dfrac{1}{2}h(a + b)$	$A = 5502, h = 28.0, b = 183$	a
17.	$V = \dfrac{1}{2}lw(D + d)$	$V = 226.8, l = 9.00, w = 6.30,$ $D = 5.00$	d
18.	$S = \dfrac{n}{2}(a + l)$	$S = 575, n = 25, l = 15$	a
19.	$S = \dfrac{n}{2}(a + l)$	$S = 147.9, n = 14.5, l = 3.80$	a
20.	$S = \dfrac{n}{2}(a + l)$	$S = 96\dfrac{7}{8}, n = 15, a = 8\dfrac{2}{3}$	l

21. A drill draws a current, I, of 4.50 A. The resistance, R, is 16.0 Ω. Find its power, P, in watts. $P = I^2R$.

22. A flashlight bulb is connected to a 1.50 V source. Its current, I, is 0.250 A. What is its resistance, R, in ohms? $V = IR$.

23. The area of a rectangle is 84.0 ft^2. Its length is 12.5 ft. Find its width. $A = lw$.

24. The volume of a box is given by $V = lwh$. Find the width if the volume is 3780 ft^3, its length is 21.0 ft, and its height is 15.0 ft.

25. The volume of a cylinder is given by the formula $V = \pi r^2 h$, where r is the radius and h is the height. Find the height in m if the volume is 8550 m^3 and the radius is 15.0 m.

26. The pressure at the bottom of a lake is found by the formula $P = hD$, where h is the depth of the water and D is the density of the water. Find the pressure in lb/in^2 at 175 ft below the surface. $D = 62.4$ lb/ft^3.

27. The equivalent resistance R of two resistances connected in parallel is given by $\dfrac{1}{R} = \dfrac{1}{R_1} + \dfrac{1}{R_2}$. Find the equivalent resistance for two resistances of 20.0 Ω and 60.0 Ω connected in parallel.

28. The R value of insulation is given by the formula $R = \dfrac{L}{K}$, where L is the thickness of the insulating material and K is the thermal conductivity of the material. Find the R value of 8.0 in. of mineral wool insulation. $K = 0.026$. *Note:* L must be in feet.

29. A steel railroad rail expands and contracts according to the formula $\triangle l = \alpha l \triangle T$, where $\triangle l$ is the change in length, α is a constant called the *coefficient of linear expansion*, l is the original length, and $\triangle T$ is the change in temperature. If a 50.0 ft steel rail is installed at 0°F, how many inches will it expand when the temperature rises to 110°F? $\alpha = 6.5 \times 10^{-6}/°F$.

30. The inductive reactance, X_L, of a coil is given by $X_L = 2\pi fL$, where f is the frequency and L is the inductance. If the inductive reactance is 245 Ω and the frequency is 60.0 cycles/s, find the inductance L in henrys, H.

6.9 Reciprocal Formulas Using a Calculator

Note: Most scientific calculators are programmed so that several keys will perform more than one function. These calculators have what is called a *second function key*. This key is probably denoted in one of the following ways:

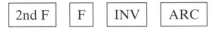

Any function that is above the key on the face of the calculator is the "second function" of that key. To access this function, press the second function key first.

The reciprocal of a number may be found by using the $1/x$ or x^{-1} button. This function, as well as other functions, may require you to use the second function key on your calculator.

• **EXAMPLE 1**   Find the reciprocal of 12 rounded to three significant digits.

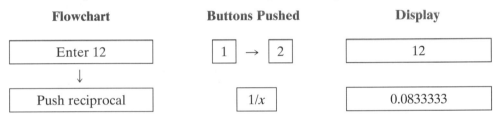

Thus, $\dfrac{1}{12} = 0.0833$ rounded to three significant digits.

• **EXAMPLE 2**   Find $\dfrac{1}{41.2}$ rounded to three significant digits.

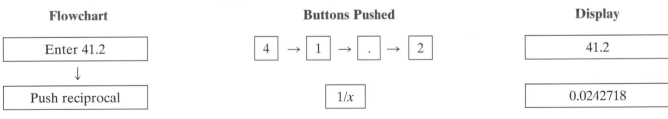

The reciprocal of 41.2 is 0.0243 rounded to three significant digits.

Formulas involving reciprocals are often used in electronics and physics. We next consider an alternative method for substituting data into such formulas and solving for a specified letter using a calculator.

First, solve for the reciprocal of the specified letter. Then substitute the given data and follow the calculator flowchart as shown in the following examples.

•   **EXAMPLE 3**   Given the formula

$$\frac{1}{R} = \frac{1}{R_1} + \frac{1}{R_2}$$

where $R_1 = 6.00\ \Omega$ and $R_2 = 12.0\ \Omega$, find R.

Since the formula is already solved for the reciprocal of R, substitute the data.

$$\frac{1}{R} = \frac{1}{R_1} + \frac{1}{R_2}$$

$$\frac{1}{R} = \frac{1}{6.00\ \Omega} + \frac{1}{12.0\ \Omega}$$

Then use your calculator as follows.

Flowchart	Buttons Pushed	Display
Enter 6	6	6
Push reciprocal	1/x	0.166667
Push plus	+	0.166667
Enter 12	1 → 2	12
Push reciprocal	1/x	0.083333
Push equals	=	0.25
Push reciprocal	1/x	4

So $R = 4.00\ \Omega$.

Note: This formula relates the electrical resistances in a parallel circuit.

———————•

•   **EXAMPLE 4**   Given the formula $\dfrac{1}{f} = \dfrac{1}{s_0} + \dfrac{1}{s_i}$, where $f = 8.00$ cm and $s_0 = 12.0$ cm, find s_i.

First, solve the formula for the reciprocal of s_i.

$$\frac{1}{f} = \frac{1}{s_0} + \frac{1}{s_i}$$

$$\frac{1}{s_i} = \frac{1}{f} - \frac{1}{s_0} \qquad \text{Subtract } \frac{1}{s_0} \text{ from both sides.}$$

Next, substitute the data.

$$\frac{1}{s_i} = \frac{1}{8.00 \text{ cm}} - \frac{1}{12.0 \text{ cm}}$$

Then use your calculator as follows.

Flowchart	Buttons Pushed	Display
Enter 8	8	8
Push reciprocal	1/x	0.125
Push minus	−	0.125
Enter 12	1 → 2	12
Push reciprocal	1/x	0.083333
Push equals	=	0.041667
Push reciprocal	1/x	24

So $s_i = 24.0$ cm.

• **EXAMPLE 5**   Given the formula

$$\frac{1}{C} = \frac{1}{C_1} + \frac{1}{C_2} + \frac{1}{C_3}$$

where $C = 2.00$ μF, $C_1 = 3.00$ μF, and $C_3 = 18.0$ μF, find C_2.

First, solve the formula for the reciprocal of C_2.

$$\frac{1}{C} = \frac{1}{C_1} + \frac{1}{C_2} + \frac{1}{C_3}$$

$$\frac{1}{C_2} = \frac{1}{C} - \frac{1}{C_1} - \frac{1}{C_3} \qquad \text{Subtract } \frac{1}{C_1} \text{ and } \frac{1}{C_3} \text{ from both sides.}$$

Next, substitute the data.

$$\frac{1}{C_2} = \frac{1}{2.00 \text{ μF}} - \frac{1}{3.00 \text{ μF}} - \frac{1}{18.0 \text{ μF}}$$

Then use your calculator as follows.

Flowchart	Buttons Pushed	Display
Enter 2	2	2
↓		
Push reciprocal	1/x	0.5
↓		
Push minus	−	0.5
↓		
Enter 3	3	3
↓		
Push reciprocal	1/x	0.333333
↓		
Push minus	−	0.166667
↓		
Enter 18	1 → 8	18
↓		
Push reciprocal	1/x	0.055556
↓		
Push equals	=	0.111111
↓		
Push reciprocal	1/x	9

Therefore, $C_2 = 9.00\ \mu F$.

Note: This formula relates the electrical capacitances of capacitors in a series circuit.

Exercises   6.9

Use the formula $\dfrac{1}{R} = \dfrac{1}{R_1} + \dfrac{1}{R_2}$ *for Exercises 1–6:*

1. Given $R_1 = 8.00\ \Omega$ and $R_2 = 12.0\ \Omega$, find R.

2. Given $R = 5.76\ \Omega$ and $R_1 = 9.00\ \Omega$, find R_2.

3. Given $R = 12.0\ \Omega$ and $R_2 = 36.0\ \Omega$, find R_1.

4. Given $R_1 = 24.0\ \Omega$ and $R_2 = 18.0\ \Omega$, find R.

5. Given $R = 15.0\ \Omega$ and $R_2 = 24.0\ \Omega$, find R_1.

6. Given $R = 90.0\ \Omega$ and $R_1 = 125\ \Omega$, find R_2.

Use the formula $\dfrac{1}{f} = \dfrac{1}{s_0} + \dfrac{1}{s_i}$ *for Exercises 7–10:*

7. Given $s_0 = 3.00$ cm and $s_i = 15.0$ cm, find f.

8. Given $f = 15.0$ cm and $s_i = 25.0$ cm, find s_0.

9. Given $f = 14.5$ cm and $s_0 = 21.5$ cm, find s_i.

10. Given $s_0 = 16.5$ cm and $s_i = 30.5$ cm, find f.

Use the formula $\dfrac{1}{R} = \dfrac{1}{R_1} + \dfrac{1}{R_2} + \dfrac{1}{R_3}$ *for Exercises 11–16:*

11. Given $R_1 = 30.0\ \Omega$, $R_2 = 18.0\ \Omega$, and $R_3 = 45.0\ \Omega$, find R.

12. Given $R_1 = 75.0\ \Omega$, $R_2 = 50.0\ \Omega$, and $R_3 = 75.0\ \Omega$, find R.

13. Given $R = 80.0\ \Omega$, $R_1 = 175\ \Omega$, and $R_2 = 275\ \Omega$, find R_3.

14. Given $R = 145\ \Omega$, $R_2 = 875\ \Omega$, and $R_3 = 645\ \Omega$, find R_1.

15. Given $R = 1250\ \Omega$, $R_1 = 3750\ \Omega$, and $R_3 = 4450\ \Omega$, find R_2.

16. Given $R = 1830\ \Omega$, $R_1 = 4560\ \Omega$, and $R_2 = 9150\ \Omega$, find R_3.

Use the formula $\dfrac{1}{C} = \dfrac{1}{C_1} + \dfrac{1}{C_2} + \dfrac{1}{C_3}$ *for Exercises 17–22:*

17. Given $C_1 = 12.0\ \mu F$, $C_2 = 24.0\ \mu F$, and $C_3 = 24.0\ \mu F$, find C.

18. Given $C = 45.0\ \mu F$, $C_1 = 85.0\ \mu F$, and $C_3 = 115\ \mu F$, find C_2.

19. Given $C = 1.25 \times 10^{-6}\ F$, $C_1 = 8.75 \times 10^{-6}\ F$, and $C_2 = 6.15 \times 10^{-6}\ F$, find C_3.

20. Given $C = 1.75 \times 10^{-12}\ F$, $C_2 = 7.25 \times 10^{-12}\ F$, and $C_3 = 5.75 \times 10^{-12}\ F$, find C_1.

21. Given $C_1 = 6.56 \times 10^{-7}\ F$, $C_2 = 5.05 \times 10^{-6}\ F$, and $C_3 = 1.79 \times 10^{-8}\ F$, find C.

22. Given $C = 4.45 \times 10^{-9}\ F$, $C_1 = 5.08 \times 10^{-8}\ F$, and $C_3 = 7.79 \times 10^{-9}\ F$, find C_2.

Use the formula $\dfrac{1}{R} = \dfrac{1}{R_1} + \dfrac{1}{R_2} + \dfrac{1}{R_3} + \dfrac{1}{R_4}$ *for Exercises 23–24:*

23. Given $R_1 = 655\ \Omega$, $R_2 = 775\ \Omega$, $R_3 = 1050\ \Omega$, and $R_4 = 1250\ \Omega$, find R.

24. Given $R = 155\ \Omega$, $R_1 = 625\ \Omega$, $R_3 = 775\ \Omega$, and $R_4 = 1150\ \Omega$, find R_2.

Chapter 6   Review

Solve each equation and check:

1. $2x + 4 = 7$

2. $11 - 3x = 23$

3. $\dfrac{x}{3} - 7 = 12$

4. $5 - \dfrac{x}{6} = 1$

5. $78 - 16y = 190$

6. $25 = 3x - 2$

7. $2x + 9 = 5x - 15$

8. $-6x + 5 = 2x - 19$

9. $3 - 2x = 9 - 3x$

10. $4x + 1 = 4 - x$

11. $7 - (x - 5) = 11$

12. $4x + 2(x + 3) = 42$

13. $3y - 5(2 - y) = 22$

14. $6(x + 7) - 5(x + 8) = 0$

15. $3x - 4(x - 3) = 3(x - 4)$

16. $4(x + 3) - 9(x - 2) = x + 27$

17. $\dfrac{2x}{3} = \dfrac{16}{9}$

18. $\dfrac{x}{3} - 2 = \dfrac{3x}{5}$

19. $\dfrac{3x}{4} - \dfrac{x - 1}{5} = \dfrac{3 + x}{2}$

20. $\dfrac{7}{x} - 3 = \dfrac{1}{x}$

21. $5 - \dfrac{7}{x} = 3\dfrac{3}{5}$

22. The length of a rectangle is 6 more than twice its width. Its perimeter is 48 in. Find its length and width.

23. Mix a solution that is 60% acid with a solution that is 100% acid to make 12 L of a solution that is 75% acid. How much of each solution should you use?

Solve each formula for the given letter:

24. $F = Wg$ for g

25. $P = \dfrac{W}{A}$ for A

26. $L = A + B + \dfrac{1}{2}t$ for t

27. $k = \dfrac{1}{2}mv^2$ for m

28. $P_2 = \dfrac{P_1 T_2}{T_1}$ for T_1

29. $v = \dfrac{v_t + v_0}{2}$ for v_0

30. $K = \dfrac{5}{9}(F - 32) + 273$; find F if $K = 175$.

31. $P = 2(l + w)$; find w if $P = 112.8$ and $l = 36.9$.

32. $k = \dfrac{1}{2}mv^2$; find m if $k = 460$ and $v = 5.0$.

33. Given $\dfrac{1}{R} = \dfrac{1}{R_1} + \dfrac{1}{R_2}$, $R_1 = 50.0\ \Omega$, and $R_2 = 75.0\ \Omega$, find R.

34. Given $\dfrac{1}{C} = \dfrac{1}{C_1} + \dfrac{1}{C_2} + \dfrac{1}{C_3}$, $C = 25.0\ \mu F$, $C_1 = 75.0\ \mu F$, and $C_3 = 80.0\ \mu F$, find C_2.

Chapter 6   Test

Solve each equation:

1. $x - 8 = -6$

2. $4x = 60$

3. $10 - 2x = 42$

4. $3x + 14 = 29$

5. $7x - 20 = 5x + 4$

6. $-2(x + 10) = 3(5 - 2x) + 5$

7. $\dfrac{1}{2}(3x - 6) = 3(x - 2)$

8. $\dfrac{8x}{9} = \dfrac{5}{6}$

9. $\dfrac{3x}{5} - 2 = \dfrac{x}{5} - \dfrac{x}{10}$

10. $\dfrac{8}{x} + 6 = 2$

11. $\dfrac{x}{2} - \dfrac{2}{5} = \dfrac{2x}{5} - \dfrac{3}{4}$

12. Distribute $2700 among Jose, Maria, and George so that Maria receives $200 more than Jose and George receives half of what Jose receives.

13. How much pure antifreeze must be added to 20 L of a solution that is 60% antifreeze to make a solution that is 80% antifreeze?

14. Solve $P = 2(l + w)$ for l.

15. Solve $C_T = C_1 + C_2 + C_3$ for C_2.

16. Solve $V = lwh$ for w.

17. Given $P = I^2R$, $P = 480$, and $I = 5.0$, find R.

18. Given $A = \dfrac{h}{2}(a + b)$, $A = 260$, $h = 13$, and $a = 15$, find b.

19. Given $\dfrac{1}{C} = \dfrac{1}{C_1} + \dfrac{1}{C_2}$, $C = 20.0\ \mu F$ and $C_2 = 30.0\ \mu F$, find C_1.

20. Given $\dfrac{1}{R} = \dfrac{1}{R_1} + \dfrac{1}{R_2} + \dfrac{1}{R_3}$, $R_1 = 225\ \Omega$, $R_2 = 475\ \Omega$, and $R_3 = 925\ \Omega$, find R.

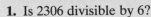

Chapters 1–6   Cumulative Review

1. Is 2306 divisible by 6?

2. Find the prime factorization of 696.

3. Find the absolute value of **a.** +6 and **b.** −28.

4. Solve: $\dfrac{2}{3} \times 3\dfrac{1}{2} \div 4\dfrac{1}{3}$.

5. 2650 yd = ___?___ mi

6. Write 3.015×10^{-4} in decimal form.

7. Change 0.081 to a percent.

8. Maria is paid $13.20 per hour. She receives a 5.5% raise. Find her new hourly pay rounded to the nearest cent.

9. 5 ha = ___?___ m^2

Give the number of significant digits (accuracy) of each measurement:

10. 110 cm

11. 6000 mi

12. 24.005 s

13. Find **a.** the precision and **b.** the greatest possible error of 3.81 in.

14. Read the measurement shown on the vernier caliper in Illustration 1 in metric units.

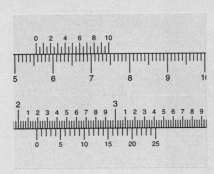

ILLUSTRATION 1

15. Read the measurement shown on the English micrometer in Illustration 2.

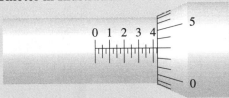

ILLUSTRATION 2

16. Find **a.** the most accurate and **b.** the most precise measurement from the following: 2600 s; 14,200 s; 20,000 s; 1.2 s; 780 s.

17. Use the rules for addition of measurements to find the sum of 25,000 W; 17,900 W; 13,962 W; 8752 W; and 428,000 W.

Simplify by removing parentheses and combining like terms:

18. $(2x - 5y) + (3y - 4x) - 2(3x - 5y)$

19. $(4y^3 + 3y - 5) - (2y^3 - 4y^2 - 2y + 6)$

20. $(3y^3)^3$

21. $-2x(x^2 - 3x + 4)$

22. $(6y^3 - 5y^2 - y + 2)(2y - 1)$

23. $(4x - 3y)(5x + 2y)$

24. $\dfrac{x^3 + 2x^3 - 11x - 20}{x + 5}$

Solve:

25. $4x - 2 = 12$

26. $\dfrac{x}{4} - 5 = 9$

27. $4x - 3 = 7x + 15$

28. $5 - (x - 3) = (2 + x) - 5$

29. $C = \dfrac{1}{2}(a + b + c)$ for a

30. $A = lw$; find w if $l = 8.20$ and $A = 91.3$.

Ratio and Proportion

CHAPTER 7

MACHINE WORK
*Ratio and proportion are all-
important in determining the
dimensions of parts to be
machined.*

MATHEMATICS AT WORK

Ratio and proportion are used by carpenters, bricklayers, culinary workers, machinists, and other tradespeople. For example, a bricklayer precalculates the number of bricks and the amount of mortar needed to complete an 8-ft section of wall. By using methods of direct proportion, he or she can easily compute the number of bricks and the amount of mortar for any given size of wall. Carpenters can predetermine the length of each stud on a rake wall or shed roof. In the culinary trade, chefs use direct proportion to calculate amounts of ingredients in a recipe that need to be increased or decreased.

Direct proportion is used when an increase in one thing causes an increase in another thing, or a decrease causes a decrease. On the other hand, inverse proportion is used when a decrease in something causes an increase in something else, or when an increase causes a decrease.

Auto mechanics, machinists, and workers in automated maintenance use inverse proportion when a change in rpm is needed in a pulley or gear system. If two pulleys are belted together or two gears are meshed together, what happens to the rpm as a result of increasing or decreasing the diameter of a pulley or changing the number of teeth on a gear?

7.1 Ratio*

The comparison of two numbers is a very important concept, and one of the most important of all comparisons is the ratio. The *ratio* of two numbers, *a* and *b*, is the first number divided by the second number. Ratios may be written in several different ways. For example, the ratio of 3 to 4 may be written as $\frac{3}{4}$, 3/4, 3 : 4, or 3 ÷ 4. Each of these forms is read "the ratio of 3 to 4."

If the quantities to be compared include units, the units should be the same whenever possible. To find the ratio of 1 ft to 15 in., first express both quantities in inches and then find the ratio:

$$\frac{1 \text{ ft}}{15 \text{ in.}} = \frac{12 \text{ in.}}{15 \text{ in.}} = \frac{12}{15} = \frac{4}{5}$$

Ratios are usually given in lowest terms.

• **EXAMPLE 1** Express the ratio of 18 : 45 in lowest terms.

$$18 : 45 = \frac{18}{45} = \frac{\cancel{9} \cdot 2}{\cancel{9} \cdot 5} = \frac{2}{5}$$

• **EXAMPLE 2** Express the ratio of 3 ft to 18 in. in lowest terms.

$$\frac{3 \text{ ft}}{18 \text{ in.}} = \frac{36 \text{ in.}}{18 \text{ in.}} = \frac{\cancel{18} \times 2 \cancel{\text{ in.}}}{\cancel{18} \times 1 \cancel{\text{ in.}}} = \frac{2}{1} \text{ or } 2$$

Note: $\frac{2}{1}$ and 2 indicate the same ratio, "the ratio of 2 to 1."

• **EXAMPLE 3** Express the ratio of 50 cm to 2 m in lowest terms.

First express the measurements in the same units.
1 m = 100 cm, so 2 m = 200 cm.

$$\frac{50 \text{ cm}}{2 \text{ m}} = \frac{50 \text{ cm}}{200 \text{ cm}} = \frac{1}{4}$$

To find the ratio of two fractions, use the technique for dividing fractions.

• **EXAMPLE 4** Express the ratio of $\frac{2}{3} : \frac{8}{9}$ in lowest terms.

$$\frac{2}{3} : \frac{8}{9} = \frac{2}{3} \div \frac{8}{9} = \frac{2}{3} \times \frac{9}{8} = \frac{18}{24} = \frac{\cancel{6} \cdot 3}{\cancel{6} \cdot 4} = \frac{3}{4}$$

• **EXAMPLE 5** Express the ratio of $2\frac{1}{2}$ to 10 in lowest terms.

$$2\frac{1}{2} \text{ to } 10 = \frac{5}{2} \div 10 = \frac{5}{2} \times \frac{1}{10} = \frac{5}{20} = \frac{\cancel{5} \cdot 1}{\cancel{5} \cdot 4} = \frac{1}{4}$$

• **EXAMPLE 6** Steel can be worked in a lathe at a cutting speed of 25 ft/min. Stainless steel can be worked in a lathe at a cutting speed of 15 ft/min. What is the ratio of the cutting speed of steel to the cutting speed of stainless steel?

$$\frac{\text{cutting speed of steel}}{\text{cutting speed of stainless steel}} = \frac{25 \text{ ft/min}}{15 \text{ ft/min}} = \frac{5}{3}$$

---

*Note: In this chapter, do not use rules for calculating with measurements.

• **EXAMPLE 7**   A construction crew uses 4 buckets of cement and 12 buckets of sand to mix a supply of concrete. What is the ratio of cement to sand?

$$\frac{\text{amount of cement}}{\text{amount of sand}} = \frac{4 \text{ buckets}}{12 \text{ buckets}} = \frac{1}{3}$$

In a ratio, we compare like or related quantities; for example,

$$\frac{18 \text{ ft}}{12 \text{ ft}} = \frac{3}{2} \quad \text{or} \quad \frac{50 \text{ cm}}{2 \text{ m}} = \frac{1}{4}$$

A ratio simplified into its lowest terms is a pair of unitless numbers.

Suppose you drive 75 miles and use 3 gallons of gasoline. Your mileage would be found as follows:

$$\frac{75 \text{ mi}}{3 \text{ gal}} = \frac{25 \text{ mi}}{1 \text{ gal}}$$

We say that your mileage is 25 miles per gallon. Note that each of these two fractions compares unlike quantities: miles and gallons. A *rate* is the comparison of two unlike quantities whose units do not cancel.

• **EXAMPLE 8**   Express the rate of $\dfrac{250 \text{ gal}}{50 \text{ acres}}$ in lowest terms.

$$\frac{250 \text{ gal}}{50 \text{ acres}} = \frac{5 \text{ gal}}{1 \text{ acre}} \quad \text{or} \quad 5 \text{ gal/acre}$$

The symbol "/" is read "per." The rate is read "5 gallons per acre."

A common medical practice is to give nourishment and/or medication to a patient by IV (intravenous). The number of drops per minute is related to the equipment that one uses. The number of drops per mL is called the *drop factor*. Common drop factors are 10 drops/mL, 12 drops/mL, and 15 drops/mL.

• **EXAMPLE 9**   A doctor orders 500 mL of glucose to be given to an adult patient by IV in 6 h. The drop factor of the equipment is 15. Determine the number of drops per minute in order to set up the IV.

First, change 6 h to minutes.

$$6 \text{ h} \times \frac{60 \text{ min}}{1 \text{ h}} = 360 \text{ min (time for IV)}$$

Thus, 500 mL of glucose is to be given during a 360 min time period that gives us a rate of $\frac{500 \text{ mL}}{360 \text{ min}}$. Since the equipment has a drop factor of 15 drops/mL, the flow rate is

$$\frac{500 \text{ mL}}{360 \text{ min}} \times \frac{15 \text{ drops}}{\text{mL}} = 21 \text{ drops/min (rounded to the nearest whole number)}$$

Sometimes the doctor orders an IV as a rate of flow, and the nurse must find the time needed to administer the IV.

• **EXAMPLE 10**   Give 1500 mL of saline solution IV at a rate of 50 drops/min to an adult patient with a drop factor of 10. Find how long the IV should be administered.

First, determine the total number of drops to be administered.

$$1500 \, \cancel{mL} \times \frac{10 \text{ drops}}{\cancel{mL}} = 15{,}000 \text{ drops}$$

Then, divide the total number of drops by the flow rate to find the time.

$$\frac{15{,}000 \text{ drops}}{50 \text{ drops/min}} = 300 \text{ min} \qquad \frac{\text{drops}}{\frac{\text{drops}}{\text{min}}} = \text{drops} \div \frac{\text{drops}}{\text{min}}$$

$$= \cancel{\text{drops}} \times \frac{\text{min}}{\cancel{\text{drops}}} = \text{min}$$

Exercises | 7.1

Express each ratio in lowest terms:

1. 3 to 15

2. 6 : 12

3. 7 : 21

4. $\dfrac{4}{22}$

5. $\dfrac{80}{48}$

6. 28 to 20

7. 3 in. to 15 in.

8. 3 ft to 15 in.

9. 3 cm to 15 mm

10. 1 in. to 8 ft

11. 9 in² : 2 ft²

12. 4 m : 30 cm

13. $\dfrac{3}{4}$ to $\dfrac{7}{6}$

14. $\dfrac{2}{3} : \dfrac{22}{9}$

15. $2\dfrac{3}{4} : 4$

16. $\dfrac{18\frac{1}{2}}{2\frac{1}{4}}$

17. $\dfrac{5\frac{1}{3}}{2\frac{2}{3}}$

18. 6 to $4\dfrac{2}{3}$

19. 10 to $2\dfrac{1}{2}$

20. $\dfrac{7}{8} : \dfrac{9}{16}$

21. $3\dfrac{1}{2}$ to $2\dfrac{1}{2}$

22. $2\dfrac{2}{3} : 3\dfrac{3}{4}$

23. $1\dfrac{3}{4}$ to 7

24. $4\dfrac{4}{5} : 12$

Express each rate in lowest terms:

25. $\dfrac{240 \text{ mi}}{8 \text{ gal}}$

26. $\dfrac{360 \text{ gal}}{18 \text{ acres}}$

27. $\dfrac{276 \text{ gal}}{6 \text{ h}}$

28. $\dfrac{\$36}{3 \text{ h}}$

29. $\dfrac{625 \text{ mi}}{12\frac{1}{2} \text{ h}}$

30. $\dfrac{150 \text{ mi}}{3\frac{3}{4} \text{ gal}}$

31. $\dfrac{2\frac{1}{4} \text{ lb}}{6 \text{ gal}}$

32. $\dfrac{\$64{,}800}{1800 \text{ ft}^2}$

33. A bearing bronze mix includes 96 lb of copper and 15 lb of lead. Find the ratio of copper to lead.

34. What is the alternator-to-engine-drive ratio if the alternator turns at 1125 rpm when the engine is idling at 500 rpm?

35. Suppose 165 gal of oil flow through a feeder pipe in 5 min. Find the flow rate in gallons per minute.

36. A flywheel has 72 teeth, and a starter drive-gear has 15 teeth. Find the ratio of flywheel teeth to drive-gear teeth.

37. A transformer has a voltage of 18 V in the primary circuit and 4950 V in the secondary circuit. Find the ratio of the primary voltage to the secondary voltage.

38. The ratio of the voltage drops across two resistors wired in series equals the ratio of their resistances. Find the ratio of a 720-Ω resistor to a 400-Ω resistor.

39. A transformer has 45 turns in the primary coil and 540 turns in the secondary coil. Find the ratio of secondary turns to primary turns.

40. The resistance in ohms of a resistor is the ratio of the voltage drop across the resistor, in volts, to the current through the resistor, in amperes. A resistor has a voltage drop across it of 117 V and a current through it of 2.6 A. What is the resistance in ohms of the resistor?

41. A 150-bu wagon holds 2.7 tons of grain. Express the weight of grain in pounds per bushel.

42. The total yield from a 55-acre field is 7425 bu. Express the yield in bushels per acre.

43. A 350-gal spray tank covers 14 acres. Find the rate of application in gallons per acre.

44. Suppose 12 gal of herbicide concentrate are used for 28 acres. Find the ratio of gallons of concentrate to acres.

45. Suppose 16 ft of copper tubing costs $13.60. Find its cost per foot.

46. A structure has 3290 ft² of wall area (excluding windows) and 1880 ft² of window area. Find the ratio of wall area to window area.

47. A 1750-ft² home sells for $73,500. Find the ratio of cost to area (price per ft²).

48. You need 15 ft³ of cement to make 80 ft³ of concrete. Find the ratio of volume of concrete to volume of cement.

49. Suppose 2.8 cm³ of medication are drawn from a vial of hydrocortisone that contains 140 mg of medication. How many milligrams of medication per cubic centimetre are in the vial?

50. A 250-cm³ bottle contains 4000 mg of aminophylline. Find the ratio of milligrams of aminophylline to each cubic centimetre.

51. A 45-cm³ vial contains 180 mg of Demerol. Find the ratio of milligrams of Demerol to each cubic centimetre.

52. Over a period of 5 h, 1200 cm³ of a solution will be administered intravenously. How many cubic centimetres per minute is this?

53. *Current ratio* is defined as the ratio of a firm's current assets to its current liabilities. A hardware store has current assets of $129,500 and current liabilities of $92,500. Compute the current ratio of the store.

54. A realty firm sold 81 commercial properties and 378 residential properties. Find the ratio of residential to commercial transactions.

55. From gross sales of $1650, profits of $900 were taken. Find the ratio of sales to profits.

56. The ratio of net income to net worth (*stockholders' equity*) indicates the rate of return on invested capital. Compute the rate of return on equity for a firm with $4000 of net income and $32,000 of stockholders' equity.

Find the flow rate for each given IV (assume a drop factor of 15 drops/mL):

57. 1200 mL in 6 h   **58.** 900 mL in 3 h   **59.** 1 L in 5.5 h   **60.** 2 L in 5 h

Find the length of time each IV should be administered (assume a drop factor of 10 drops/mL):

61. 1000 mL at a rate of 50 drops/min   **63.** 2 L at a rate of 40 drops/min

62. 1600 mL at a rate of 40 drops/min   **64.** 1.4 L at a rate of 35 drops/min

7.2 Proportion

A *proportion* states that two ratios or two rates are equal. Thus,

$$\frac{3}{4} = \frac{9}{12}, \quad 2:3 = 4:6, \quad \text{and} \quad \frac{a}{b} = \frac{c}{d}$$

are proportions. A proportion has four terms. In the proportion $\frac{2}{5} = \frac{4}{10}$, the first term is 2, the second term is 5, the third term is 4, and the fourth term is 10.

The first and fourth terms of a proportion are called the *extremes*, and the second and third terms are called the *means* of the proportion. This is more easily seen when the proportion $\frac{a}{b} = \frac{c}{d}$ is written in the form

means
$$a : b = c : d$$
extremes

• **EXAMPLE 1** Given the proportion $\frac{2}{3} = \frac{4}{6}$.

a. The first term is 2.   **e.** The means are 3 and 4.

b. The second term is 3.   **f.** The extremes are 2 and 6.

c. The third term is 4.   **g.** The product of the means $= 3 \cdot 4 = 12$.

d. The fourth term is 6.   **h.** The product of the extremes $= 2 \cdot 6 = 12$.

We see in **g** and **h** that the product of the means (that is, 12) equals the product of the extremes (also 12). Let us look at another proportion and see if this is true again.

• **EXAMPLE 2** Given the proportion $\frac{5}{13} = \frac{10}{26}$, find the product of the means and the product of the extremes.

The extremes are 5 and 26, and the means are 13 and 10. The product of the extremes is 130, and the product of the means is 130. Here again, the product of the means equals the product of the extremes. As a matter of fact, this will always be the case.

In any proportion, the *product of the means* equals the *product of the extremes*.

To determine whether two ratios are equal, put the two ratios in the form of a proportion. If the product of the means equals the product of the extremes, the ratios are equal.

• **EXAMPLE 3** Determine whether or not the ratios $\frac{13}{36}$ and $\frac{29}{84}$ are equal.

If $36 \times 29 = 13 \times 84$, then $\frac{13}{36} = \frac{29}{84}$.

However, $36 \times 29 = 1044$ and $13 \times 84 = 1092$. Therefore, $\frac{13}{36} \neq \frac{29}{84}$.

To *solve* a proportion means to find the missing term. To do this, form an equation by setting the product of the means equal to the product of the extremes. Then solve the resulting equation.

•   **EXAMPLE 4**   Solve the proportion $\dfrac{x}{3} = \dfrac{8}{12}$.

$$\frac{x}{3} = \frac{8}{12}$$
$$12x = 24 \qquad \text{The product of the means equals the}$$
$$x = 2 \qquad\qquad \text{product of the extremes.}$$

────────

•   **EXAMPLE 5**   Solve the proportion $\dfrac{5}{x} = \dfrac{10}{3}$.

$$\frac{5}{x} = \frac{10}{3}$$
$$10x = 15$$
$$x = \frac{3}{2} \quad \text{or} \quad 1.5$$

A calculator is helpful in solving a proportion with decimal fractions.

────────

•   **EXAMPLE 6**   Solve $\dfrac{32.3}{x} = \dfrac{17.9}{25.1}$.

$$17.9x = (32.3)(25.1)$$
$$x = \frac{(32.3)(25.1)}{17.9} = 45.3, \text{ rounded to three significant digits}$$

────────

•   **EXAMPLE 7**   If 125 bolts cost $7.50, how much do 75 bolts cost?

First, let's find the rate of dollars/bolts in each case.

$$\frac{\$7.50}{125 \text{ bolts}} \quad \text{and} \quad \frac{x}{75 \text{ bolts}} \qquad \text{where } x = \text{the cost of 75 bolts}$$

Since these two rates are equal, we have the proportion:

$$\frac{7.5}{125} = \frac{x}{75}$$
$$125x = (7.5)(75) \qquad \text{The product of the means equals the}$$
$$\qquad\qquad\qquad\qquad \text{product of the extremes.}$$
$$x = \frac{(7.5)(75)}{125} \qquad \text{Divide both sides by 125.}$$
$$x = 4.5$$

That is, the cost of 75 bolts is $4.50.

────────

Note: A key to solving proportions like the one in Example 7 is to set up the proportion with the same units in each ratio—in this case:

$$\frac{\$}{\text{bolts}} = \frac{\$}{\text{bolts}}$$

•   **EXAMPLE 8**   The pitch of a roof is the ratio of the rise to the run of a rafter. (See Figure 7.1.) The pitch of the roof shown in the figure is 2 : 7. Find the rise if the run is 21 ft.

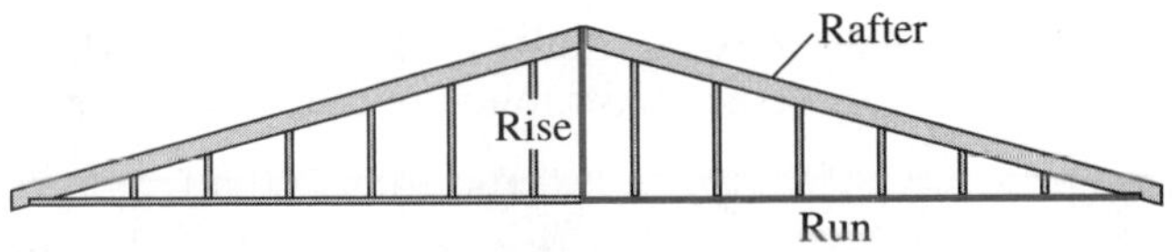

FIGURE 7.1

$$\text{pitch} = \frac{\text{rise}}{\text{run}}$$

$$\frac{2}{7} = \frac{x}{21 \text{ ft}}$$

$$7x = (2)(21 \text{ ft})$$

$$x = \frac{(2)(21 \text{ ft})}{7}$$

$x = 6$ ft, which is the rise ———————●

In Section 1.14, you studied percent, using the formula $P = BR$, where R is the rate written as a decimal. Knowing this formula and knowing the fact that the percent means per hundred, we can write the proportion

$$\frac{P}{B} = \frac{R}{100}$$

where R is the rate written as a percent. We can use this proportion to solve percent problems.

Note: You may find it helpful to review the meanings of P (percentage), B (base), and R (rate) in Section 1.14.

• **EXAMPLE 9**   A student answered 27 out of 30 questions correctly. What percent of the answers were correct?

$$P \text{ (percentage)} = 27$$
$$B \text{ (base)} = 30$$
$$\frac{27}{30} = \frac{R}{100}$$
$$30R = 2700$$
$$R = 90$$

Therefore, the student answered 90% of the questions correctly. ———————●

• **EXAMPLE 10**   A factory produces bearings used in automobiles. After inspecting 4500 bearings, the inspectors find that 127 are defective. What percent are defective?

$$P \text{ (percentage)} = 127$$
$$B \text{ (base)} = 4500$$
$$\frac{127}{4500} = \frac{R}{100}$$
$$4500R = 12,700$$
$$R = 2.8$$

Therefore, 2.8% of the bearings are defective. ———————●

• **EXAMPLE 11**   A nurse must prepare 300 mL of 10% glucose solution from pure crystalline glucose. How much pure crystalline glucose is needed?

$$B \text{ (base)} = 300 \text{ mL}$$
$$R \text{ (rate)} = 10\%$$
$$P \text{ (percentage)} = x$$
$$\frac{x}{300} = \frac{10}{100}$$
$$100x = 3000$$
$$x = 30 \text{ mL}$$

• **EXAMPLE 12**   Prepare 2000 mL of a Lysol solution containing 1 part Lysol and 20 parts water from pure Lysol. How much pure Lysol is needed?

$$R = 1 : 20 = 1/20 = 0.05 = 5\%$$
$$B = 2000 \text{ mL}$$
$$P = x$$

$$\frac{x}{2000} = \frac{5}{100}$$
$$100x = 10{,}000$$
$$x = 100 \text{ mL}$$

Exercises | 7.2

In each proportion, find **a.** *the means,* **b.** *the extremes,* **c.** *the product of the means, and* **d.** *the product of the extremes:*

1. $\dfrac{1}{2} = \dfrac{3}{6}$ **2.** $\dfrac{3}{4} = \dfrac{6}{8}$ **3.** $\dfrac{7}{9} = \dfrac{28}{36}$ **4.** $\dfrac{x}{3} = \dfrac{6}{9}$ **5.** $\dfrac{x}{7} = \dfrac{w}{z}$ **6.** $\dfrac{a}{b} = \dfrac{4}{5}$

Determine whether or not each pair of ratios is equal:

7. $\dfrac{2}{3}, \dfrac{10}{15}$ **8.** $\dfrac{2}{3}, \dfrac{9}{6}$ **9.** $\dfrac{3}{5}, \dfrac{18}{20}$ **10.** $\dfrac{3}{7}, \dfrac{9}{21}$ **11.** $\dfrac{1}{3}, \dfrac{4}{12}$ **12.** $\dfrac{125}{45}, \dfrac{25}{9}$

Solve each proportion (round each result to three significant digits when necessary):

13. $\dfrac{x}{4} = \dfrac{9}{12}$ **19.** $\dfrac{5}{7} = \dfrac{3x}{14}$ **25.** $\dfrac{1}{0.0004} = \dfrac{700}{x}$ **31.** $\dfrac{12}{y} = \dfrac{84}{144}$ **37.** $\dfrac{30.1}{442} = \dfrac{55.7}{x}$

14. $\dfrac{1}{a} = \dfrac{4}{16}$ **20.** $\dfrac{x}{18} = \dfrac{7}{9}$ **26.** $\dfrac{x}{9} = \dfrac{2}{0.6}$ **32.** $\dfrac{13}{169} = \dfrac{27}{x}$ **38.** $\dfrac{9.4}{291} = \dfrac{44.1}{x}$

15. $\dfrac{5}{7} = \dfrac{4}{y}$ **21.** $\dfrac{5}{7} = \dfrac{25}{y}$ **27.** $\dfrac{3x}{27} = \dfrac{0.5}{9}$ **33.** $\dfrac{x}{48} = \dfrac{56}{72}$ **39.** $\dfrac{36.9}{104} = \dfrac{3210}{x}$

16. $\dfrac{12}{5} = \dfrac{x}{10}$ **22.** $\dfrac{1.1}{6} = \dfrac{x}{12}$ **28.** $\dfrac{0.25}{2x} = \dfrac{8}{48}$ **34.** $\dfrac{124}{67} = \dfrac{149}{x}$ **40.** $\dfrac{0.0417}{0.355} = \dfrac{26.9}{x}$

17. $\dfrac{2}{x} = \dfrac{4}{28}$ **23.** $\dfrac{-5}{x} = \dfrac{2}{3}$ **29.** $\dfrac{17}{28} = \dfrac{153}{2x}$ **35.** $\dfrac{472}{x} = \dfrac{793}{64.2}$ **41.** $\dfrac{x}{4.2} = \dfrac{19.6}{3.87}$

18. $\dfrac{10}{15} = \dfrac{y}{75}$ **24.** $\dfrac{4x}{9} = \dfrac{12}{7}$ **30.** $\dfrac{3x}{10} = \dfrac{7}{50}$ **36.** $\dfrac{94.7}{6.72} = \dfrac{x}{19.3}$ **42.** $\dfrac{0.120}{3x} = \dfrac{0.575}{277}$

43. You need $2\frac{3}{4}$ ft^3 of sand to make 8 ft^3 of concrete. How much sand would you need to make 128 ft^3 of concrete?

44. The pitch of a roof is $\frac{1}{3}$. If the run is 15 ft, find the rise. (See Example 8.)

45. A builder sells a 1500-ft^2 home for $70,000.

What would be the price of a 2100-ft^2 home of similar quality? Assume the price per square foot remains constant.

46. Suppose 826 bricks are used in constructing a wall 14 ft long. How many bricks will be needed for a similar wall 35 ft long?

47. A buyer purchases 75 yd of material for $120. Then an additional 90 yd are ordered. What is the additional cost?

48. A salesperson is paid a commission of $75 for selling $300 worth of goods. What is the commission on $760 of sales at the same rate of commission?

49. A business realizes a profit of $780 on $6500 of sales. What volume of sales is needed to earn $1020 in profits? Assume the same profit-to-sales ratio.

50. A building assessed at $32,000 is billed $638.40 for property taxes. A similar building next door is assessed at $56,000. How much should this building be billed for property taxes?

51. Suppose 20 gal of water and 3 lb of pesticide are applied per acre. How much pesticide should you put in a 350-gal spray tank? Assume the pesticide dissolves in the water and has no volume.

52. A farmer uses 150 lb of Aatrex 80W on a 40-acre field. How many pounds will he need for a 220-acre field? Assume the same rate of application.

53. Suppose a yield of 100 bu of corn per acre removes 90 lb of nitrogen, potassium, and potash (N, P, and K). How many pounds of N, P, and K would be removed by a yield of 120 bu per acre?

54. A farmer has a total yield of 42,000 bu of corn from a 350-acre farm. What total yield should he expect from a similar 560-acre farm?

55. A copper wire 750 ft long has a resistance of 1.563 Ω. How long is a copper wire of the same area whose resistance is 2.605 Ω? (The resistance of these wires is proportional to their length.)

56. The voltage drop across a 28-Ω resistor is 52 V. What is the voltage drop across a 63-Ω resistor that is in series with the first one? (Resistors in series have voltage drops proportional to their resistances.)

57. The ratio of secondary turns to primary turns in a transformer is 35 to 4. How many secondary turns are there if the primary coil has 68 turns?

58. A piece of cable 180 ft long costs $63. What will 450 ft cost at the same rate?

59. An engine with displacement of 380 in^3 develops 212 hp. How many horsepower would be developed by a 318 in^3 engine of the same design?

60. An 8-V automotive coil has 250 turns of wire in the primary circuit. The secondary voltage is 15,000 V. How many secondary turns are there in the coil? (The ratio of secondary voltage to primary voltage equals the ratio of secondary turns to primary turns.)

61. The clutch linkage on a vehicle has an overall advantage of 24 : 1. The pressure plate applies a force of 504 lb. How much force must the driver apply to release the clutch?

62. A fuel pump delivers 35 mL of fuel in 420 strokes. How many strokes are needed to get 50 mL of fuel?

63. A label reads: "4 cm^3 of solution contains gr X [10 gr] of potassium chloride." How many cubic centimetres are needed to give gr XXXV [35 gr]?

64. A multiple-dose vial has been mixed and labeled "200,000 units in 1 cm^3." How many cubic centimetres are needed to give 900,000 units?

65. You are to administer 150 mg of aminophylline from a bottle marked 250 mg/10 cm^3. How many cubic centimetres should you draw?

66. A label reads: "Gantrisin, 1.5 g in 20 cm^3." How many cubic centimetres are needed to give 10.5 g of Gantrisin?

When finding the percent, round to the nearest tenth of a percent when necessary:

67. Carla bought a used car for $15,000. She paid $3500 down. What percent of the price was her down payment?

68. A baseball team last year won 18 games and lost 12. What percent did they win? What percent did they lose?

69. A car is listed to sell at $20,400, and the salesman offers to sell it to you for $19,200. What percent of the list price is the reduction?

70. A live hog weighs 254 lb, and its carcass weighs 198 lb. What percent of the live hog is carcass? What percent is waste?

71. In a 100-g sample of beef, there are 18 g of fat.
  a. What is the percent of fat in the beef?
  b. How many pounds of fat would there be in a 650-lb beef carcass? Assume the same percent of fat.

72. At the beginning of a trip, a tire on a car has a pressure of 32 psi (lb/in²). At the end of the trip, the pressure is 38 psi. What is the percent increase in pressure?

73. A gasoline tank contains 5.7 hectolitres (hL) when it is 30% full. What is the capacity of the tank?

74. Jamal had a pay raise from $1840 to $2208 per month. Find the percent increase.

75. A concrete mix is composed of 1 part cement, 2.5 parts sand, and 4 parts gravel by volume. What is the percent by volume in the dry mix of **a.** cement, **b.** sand, and **c.** gravel?

76. You are to put 4 qt of pure antifreeze in a tractor radiator and then fill the radiator with 5 gal of water. What percent of antifreeze will be in the radiator?

Find the amount of pure ingredient needed to prepare each solution as indicated:

77. 150 mL of 3% cresol solution from pure cresol

78. 1000 mL of 5% Lysol solution from pure Lysol

79. 500 mL of 1% sodium bicarbonate solution from pure powdered sodium bicarbonate

80. 600 mL of 10% glucose solution from pure crystalline glucose

81. 1.5 L of 1 : 1000 epinephrine solution from pure epinephrine

82. 20 mL of 1 : 200 silver nitrate solution from pure silver nitrate

83. 300 mL of 1 : 10 glucose solution from pure glucose

84. 400 mL of 1 : 50 sodium bicarbonate solution from pure sodium bicarbonate

7.3 Direct Variation

When two quantities, x and y, change so that their ratios are constant—that is,

$$\frac{y_1}{x_1} = \frac{y_2}{x_2}$$

they are said to *vary directly*. This relationship between the two quantities is called *direct variation*. If one quantity increases, the other increases by the same factor. Likewise, if one decreases, the other decreases by the same factor.

Consider the following data:

y	6	24	15	18	9	30
x	2	8	5	6	3	10

Note that y varies directly with x because the ratio $\frac{y}{x}$ is always 3, a constant. This relationship may also be written $y = 3x$.

Direct Variation

$$\frac{y_1}{x_1} = \frac{y_2}{x_2}$$

Examples of direct variation are scale drawings such as maps and blueprints where

$$\frac{\text{scale measurement 1}}{\text{scale measurement 2}} = \frac{\text{actual measurement 1}}{\text{actual measurement 2}}$$

or

$$\frac{\text{scale measurement 1}}{\text{actual measurement 1}} = \frac{\text{scale measurement 2}}{\text{actual measurement 2}}$$

A *scale drawing* of an object has the same shape as the actual object, but the scale drawing may be smaller than, equal to, or larger than the actual object. The scale used in a drawing indicates what the ratio is between the size of the scale drawing and the size of the object drawn.

A portion of a map of the state of Illinois is shown in Figure 7.2. The scale is 1 in. = 32 mi.

• **EXAMPLE 1**   Find the approximate distance between Champaign and Kankakee using the map in Figure 7.2.

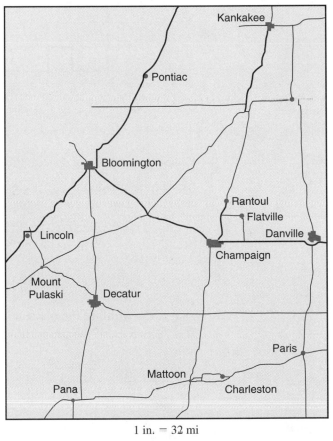

1 in. = 32 mi

FIGURE 7.2

The distance on the map measures $2\frac{3}{8}$ in. Set up a proportion that has as its first ratio the scale drawing ratio and as its second ratio the length measured on the map to the actual distance.

$$\frac{1}{32} = \frac{2\frac{3}{8}}{d}$$

$$1d = 32\left(2\frac{3}{8}\right)$$

$$d = 32\left(\frac{19}{8}\right) = 76 \text{ mi}$$

Square-ruled paper may also be used to represent scale drawings. Each square represents a unit of length according to some scale.

• **EXAMPLE 2**   The scale drawing in Figure 7.3 represents a metal plate a machinist is to make.

a. How long is the plate?

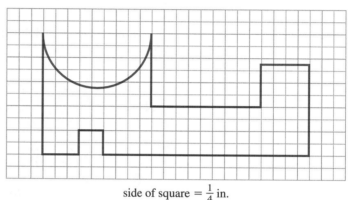

side of square $= \frac{1}{4}$ in.

FIGURE 7.3

Count the number of spaces, then set up the proportion

$$\frac{\text{scale measurement 1}}{\text{actual measurement 1}} = \frac{\text{scale measurement 2}}{\text{actual measurement 2}}$$

$$\frac{1 \text{ space}}{\frac{1}{4} \text{ in.}} = \frac{22 \text{ spaces}}{x \text{ in.}}$$

$$x = 22\left(\frac{1}{4}\right) = 5\frac{1}{2} \text{ in.}$$

b. What is the width of the plate at its right end?
It is $7\frac{1}{2}$ spaces, so the proportion is

$$\frac{1 \text{ space}}{\frac{1}{4} \text{ in.}} = \frac{7\frac{1}{2} \text{ spaces}}{x \text{ in.}}$$

$$x = \left(7\frac{1}{2}\right)\left(\frac{1}{4}\right) = 1\frac{7}{8} \text{ in.}$$

c. What is the diameter of the semicircle?

$$\frac{1 \text{ space}}{\frac{1}{4} \text{ in.}} = \frac{9 \text{ spaces}}{x \text{ in.}}$$

$$x = 9\left(\frac{1}{4}\right) = 2\frac{1}{4} \text{ in.}$$

Another example of direct variation is the hydraulic press or hydraulic pump, which allows one to exert a small force to move or raise a large object, such as a car. Other uses of hydraulics include compressing junk cars, stamping metal sheets to form car parts, and lifting truck beds.

A hydraulic press is shown in Figure 7.4. When someone presses a force of 50 lb on the small piston, a force of 5000 lb is exerted by the large piston. The *mechanical advantage* (MA) of the hydraulic press is the ratio of the force from the large piston (F_l) to the force on the small piston (F_s). The formula is

$$MA = \frac{F_l}{F_s}$$

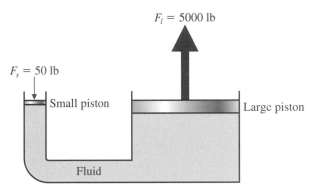

$F_l = 5000$ lb

$F_s = 50$ lb

Small piston

Large piston

Fluid

FIGURE 7.4

• **EXAMPLE 3** Find the mechanical advantage of the press shown in Figure 7.4.

$$MA = \frac{F_l}{F_s}$$
$$= \frac{5000 \text{ lb}}{50 \text{ lb}} = \frac{100}{1}$$

Thus, for every pound exerted on the small piston, 100 lb is exerted by the large piston. ────────•

The mechanical advantage can also be calculated when the radii of the pistons are known.

$$MA = \frac{r_l^2}{r_s^2}$$

• **EXAMPLE 4** The radius of the large piston of a hydraulic press is 12 in. The radius of the small piston is 2 in. Find the MA.

$$MA = \frac{r_l^2}{r_s^2}$$
$$= \frac{(12 \text{ in.})^2}{(2 \text{ in.})^2}$$
$$= \frac{144 \text{ in}^2}{4 \text{ in}^2}$$
$$= \frac{36}{1}$$

That is, for every pound exerted on the small piston, 36 lb is exerted by the large piston. ────────•

You now have two ways of finding mechanical advantage: when F_l and F_s are known and when r_l and r_s are known. From this knowledge, you can find a relationship among F_l, F_s, r_l, and r_s.

Since MA $= \dfrac{F_l}{F_s}$ and MA $= \dfrac{r_l^2}{r_s^2}$ then $\dfrac{F_l}{F_s} = \dfrac{r_l^2}{r_s^2}$

• **EXAMPLE 5**   Given $F_s = 240$ lb, $r_l = 16$ in., and $r_s = 2$ in., find F_l.

$$\frac{F_l}{F_s} = \frac{r_l^2}{r_s^2}$$

$$\frac{F_l}{240 \text{ lb}} = \frac{(16 \text{ in.})^2}{(2 \text{ in.})^2}$$

$$F_l = \frac{(240 \text{ lb})(16 \text{ in.})^2}{(2 \text{ in.})^2}$$

$$= \frac{(240 \text{ lb})(256 \text{ in}^2)}{4 \text{ in}^2}$$

$$= 15{,}360 \text{ lb}$$

Exercises | 7.3

Use the map in Figure 7.2 (page 255) to find the approximate distance between each pair of cities (find straight-line (air) distances only):

1. Champaign and Bloomington
2. Bloomington and Decatur
3. Rantoul and Kankakee
4. Rantoul and Bloomington
5. Pana and Rantoul

6. Champaign and Mattoon
7. Charleston and Pontiac
8. Paris and Watseka
9. Lincoln and Danville
10. Flatville and Mt. Pulaski

Use the map in Figure 7.5 to find the approximate air distance between each pair of cities:

11. St. Louis and Kansas City
12. Memphis and St. Louis
13. Memphis and Little Rock

14. Sedalia, MO and Tulsa, OK
15. Fort Smith, AR and Springfield, MO
16. Pine Bluff, AR and Jefferson City, MO

Use the scale drawing of a metal plate cover in Figure 7.6 in Exercises 17–26:

17. What is the length of the plate cover?
18. What is the width of the plate cover?
19. What is the area of the plate cover?
20. What is the diameter of the circular holes?
21. What are the dimensions of the square hole?
22. What are the dimensions of the rectangular holes?
23. What is the distance between the rectangular holes, center to center?

24. What is the distance between the centers of the upper pair of circular holes?
25. What is the distance between the centers of the right pair of circular holes?
26. Answer each of the questions in Exercises 17–25 if the scale were changed so that the side of the square $= 2\frac{1}{16}$ in.

With pencil and ruler make line drawings on square-ruled paper to fit each description in Exercises 27–30:

27. A rectangle 8 ft by 6 ft. Use the scale: side of a square = 1 ft.

28. A square 16 cm on a side. Use the scale: side of a square = 2 cm.

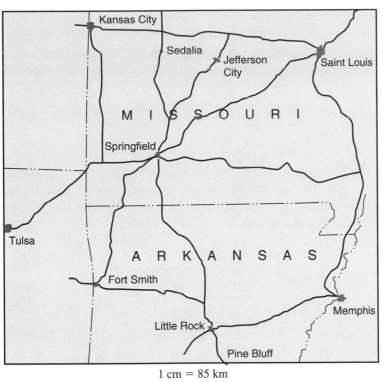

FIGURE 7.5

1 cm = 85 km

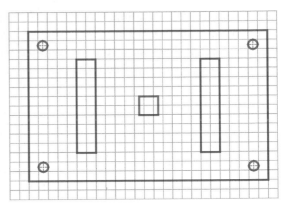

Side of square = 0.5 cm

FIGURE 7.6

29. A circle 36 mm in diameter. Use the scale: side of a square = 3 mm.

30. A rectangle 12 in. by 8 in. with a circle in its center 3 in. in diameter. Use the scale: side of a square = 2 in.

31. Can the actual circle in Exercise 29 be placed within the actual rectangle in Exercise 27? Can the scale drawing of the circle be placed within the scale drawing of the rectangle?

32. Can the actual circle in Exercise 29 be placed within the actual square in Exercise 28? Can the scale drawing of the circle be placed within the scale drawing of the square?

Use the formulas for the hydraulic press to find each value in Exercises 33–50:

33. $F_l = 4000$ lb and $F_s = 200$ lb. Find MA.

34. When a force of 160 lb is applied to the small piston of a hydraulic press, a force of 4800 lb is exerted by the large piston. Find its mechanical advantage.

35. A 400-lb force applied to the small piston of a hydraulic press produces a 3600-lb force by the large piston. Find its mechanical advantage.

36. $F_l = 2400$ lb and MA $= \frac{50}{1}$. Find F_s.

37. $F_l = 5100$ lb and MA $= \frac{75}{1}$. Find F_s.

38. A hydraulic press has a mechanical advantage of 36 : 1. If a force of 2750 lb is applied to the small piston, what force is produced by the large piston?

39. A hydraulic press with an MA of 90 : 1 has a force of 2650 lb applied to its small piston. What force is produced by its large piston?

40. A hydraulic system with an MA of 125 : 1 has a force of 2450 lb exerted by its large piston. What force is applied to its small piston?

41. $r_l = 27$ in. and $r_s = 3$ in. Find MA.

42. $r_l = 36$ in. and $r_s = 4$ in. Find MA.

43. The radii of the pistons of a hydraulic press are 3 in. and 15 in. Find its mechanical advantage.

44. The radius of the small piston of a hydraulic press is 2 in., and the radius of its large piston is 18 in. What is its mechanical advantage?

45. $F_s = 25$ lb, $r_l = 8$ in., and $r_s = 2$ in. Find F_l.

46. $F_s = 81$ lb, $r_l = 9$ in., and $r_s = 1$ in. Find F_l.

47. $F_l = 6400$ lb, $r_l = 16$ in., and $r_s = 4$ in. Find F_s.

48. $F_l = 7500$ lb, $r_l = 15$ in., and $r_s = 3$ in. Find F_s.

49. A force of 40 lb is applied to a piston of radius 7 in. of a hydraulic press. The large piston has a radius of 28 in. What force is exerted by the large piston?

50. A force of 8100 lb is exerted by a piston of radius 30 in. of a hydraulic press. What force was applied to its piston of radius 3 in.?

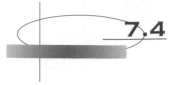

7.4 Inverse Variation

If two quantities, y and x, change so that their product is constant (that is, if $y_1x_1 = y_2x_2$), they are said to vary inversely. This relationship between the two quantities is called *inverse variation*. This means that if one quantity increases, the other decreases so that their product is always the same. Compare this with direct variation, where the ratio of the two quantities is always the same.

Consider the following data:

y	8	24	12	3	6	48
x	6	2	4	16	8	1

Note that y varies inversely with x because the product is always 48, a constant. This relationship may also be written $xy = 48$ or $y = \frac{48}{x}$.

> **Inverse Variation**
>
> $$x_1y_1 = x_2y_2$$

An inverse variation relationship may also be written in the form

$$\frac{y_1}{y_2} = \frac{x_2}{x_1}$$

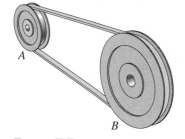

FIGURE 7.7
Pulley system

One example of inverse variation is the relationship between two rotating pulleys connected by a belt (see Figure 7.7). This relationship is given by the following formula.

> (diameter of A)(rpm of A) = (diameter of B)(rpm of B)

• **EXAMPLE 1**   A small pulley is 11 in. in diameter, and a larger one is 20 in. in diameter. How many rpm does the smaller pulley make if the larger one revolves at 44 rpm?

$$\text{(diameter of } A)(\text{rpm of } A) = (\text{diameter of } B)(\text{rpm of } B)$$
$$11 \quad \cdot \quad x \quad = \quad 20 \quad \cdot \quad 44$$
$$x = \frac{(20)(44)}{11} = 80 \text{ rpm}$$

Another example of inverse variation is the relationship between the number of teeth and the number of rpm of two rotating gears, as shown in Figure 7.8.

$$(\text{no. of teeth in } A)(\text{rpm of } A) = (\text{no. of teeth in } B)(\text{rpm of } B)$$

FIGURE 7.8
Gears

• **EXAMPLE 2**   A large gear with 14 teeth revolves at 40 rpm. It turns a small gear with 8 teeth. How fast does the small gear rotate?

(no. of teeth in A)(rpm of A) = (no. of teeth in B)(rpm of B)

(14   )(40 rpm) = (8   )(x)

$$\frac{(14)(40 \text{ rpm})}{8} = x$$

$$70 \text{ rpm} = x$$

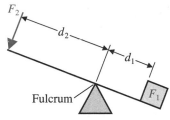

FIGURE 7.9
Lever

Figure 7.9 shows a lever, which is a rigid bar, pivoted to turn on a point (or edge) called a *fulcrum*. The parts of the lever on either side of the fulcrum are called *lever arms*.

To lift the box requires a force F_1. This is produced by pushing down on the other end of the lever with a force F_2. The distance from F_2 to the fulcrum is d_2. The distance from F_1 to the fulcrum is d_1. The principle of the lever is another example of inverse variation. It can be expressed by the following formula.

$$F_1 d_1 = F_2 d_2$$

When the equation is true, the lever is balanced.

• **EXAMPLE 3**   A man places one end of a lever under a large rock, as in Figure 7.10. He places a second rock under the lever, 2 ft from the first rock, to act as a fulcrum. He exerts a force of 180 pounds at a distance of 6 ft from the fulcrum. Find F_1, the maximum weight of rock that could be lifted.

$$F_1 d_1 = F_2 d_2$$

$$(F_1)(2 \text{ ft}) = (180 \text{ lb})(6 \text{ ft})$$

$$F_1 = \frac{(180 \text{ lb})(\overset{3}{\cancel{6} \text{ ft}})}{\cancel{2} \text{ ft}}$$

$$F_1 = 540 \text{ lb}$$

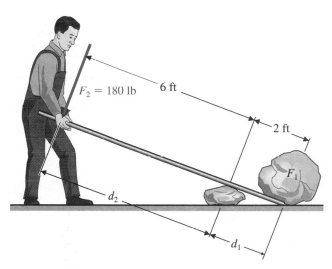

FIGURE 7.10

Exercises 7.4

Fill in the blanks:

	Pulley A Diameter	rpm	Pulley B Diameter	rpm
1.	25 cm	72	50 cm	
2.	18 cm		12 cm	96
3.	10 cm	120	15 cm	
4.		84	8 in.	48
5.	34 cm	440		680
6.	25 cm	600	48 cm	
7.		225	15 in.	465
8.	98 cm	240		360

9. A small pulley is 13 in. in diameter, and a larger one is 18 in. in diameter. How many rpm does the larger pulley make if the smaller one revolves at 720 rpm?

10. A 21-in. pulley, revolving at 65 rpm, turns a smaller pulley at 210 rpm. What is the diameter of the smaller pulley?

11. A large pulley turns at 48 rpm. A smaller pulley 8 in. in diameter turns at 300 rpm. What is the diameter of the larger pulley?

12. A pulley 32 in. in diameter turns at 825 rpm. At how many rpm will a pulley 25 in. in diameter turn?

Fill in the blanks:

	Gear A Number of teeth	rpm	Gear B Number of teeth	rpm
17.	50	400		125
18.	220		45	440
19.	42	600	25	
20.	50	64		80
21.		$6\frac{1}{4}$	120	30
22.	80	$1\frac{1}{4}$		$3\frac{1}{3}$

23. A small gear with 25 teeth turns a large gear with 75 teeth at 32 rpm. How many rpm does the small gear make?

13. A motor turning at 1870 rpm has a 4.0-in. pulley driving a fan that must turn at 680 rpm. What diameter pulley must be put on the fan?

14. A hydraulic pump is driven with an electric motor (see Figure 7.11). The pump must rotate at 1200 rpm. The pump is equipped with a 6.0-in.-diameter belt pulley. The motor runs at 1800 rpm. What diameter pulley is required on the motor?

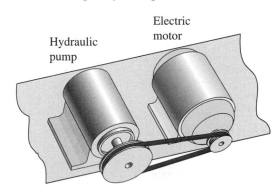

FIGURE 7.11

15. One pulley is 7 cm larger in diameter than a second pulley. The larger pulley turns at 80 rpm, and the smaller pulley turns at 136 rpm. What is the diameter of each pulley?

16. One pulley is twice as large in diameter as a second pulley. If the larger pulley turns at 256 rpm, what is the rpm of the smaller?

24. A large gear with 180 teeth running at 600 rpm turns a small gear 900 rpm. How many teeth does the small gear have?

25. A large gear with 60 teeth turning at 72 rpm turns a small gear with 30 teeth. At how many rpm does the small gear turn?

26. A large gear with 80 teeth turning at 150 rpm turns a small gear with 12 teeth. At how many rpm does the small gear turn?

27. A large gear with 120 teeth turning at 30 rpm turns a small gear at 90 rpm. How many teeth does the small gear have?

28. A large gear with 200 teeth turning at 17 rpm turns a small gear at 100 rpm. How many teeth does the small gear have?

Complete the table:

	F_1	d_1	F_2	d_2
29.	18 lb	5 in.	9 lb	
30.	30 lb		70 lb	8 in.
31.	40 lb	9 in.		3 in.
32.		6.3 ft	458.2 lb	8.7 ft

In Exercises 33–37, draw a sketch for each and solve:

33. An object is 6 ft from the fulcrum and balances a second object 8 ft from the fulcrum. The first object weighs 180 lb. How much does the second object weigh?

34. A block of steel weighing 1800 lb is to be raised by a lever extending under the block 9 in. from the fulcrum. How far from the fulcrum must a 150-lb man apply his weight to the bar to balance the steel?

35. A rocker arm raises oil from an oil well. On each stroke, it lifts a weight of 1 ton on a weight arm 4 ft long. What force is needed to lift the oil on a force arm 8 ft long?

36. A carpenter needs to raise one side of a building with a lever 3.65 m in length. The lever, with one end under the building, is placed on a fulcrum 0.45 m from the building. A 90-kg force pulls down on the other end. What weight is being lifted when the building begins to rise?

37. A 1200-g weight is placed 72 cm from the fulcrum of a lever. How far from the fulcrum is a 1350-g weight that balances it?

38. A lever is in balance when a force of 2000 g is placed 28 cm from a fulcrum. An unknown force is placed 20 cm from the fulcrum on the other side. What is the unknown force?

39. A 210-lb object is placed on a lever. It balances a 190-lb weight that is 28 in. from the fulcrum. How far from the fulcrum should the 210-lb weight be placed?

40. A piece of machinery weighs 3 tons. It is to be balanced by two men whose combined weight is 330 lb. The piece of machinery is placed 11 in. from the fulcrum. How far from the fulcrum must the two men exert their weight in order to balance it?

Chapter 7   Review

Write each ratio in lowest terms:

1. 7 to 28

2. $60 : 40$

3. 1 g to 500 mg

4. $\dfrac{5 \text{ ft } 6 \text{ in.}}{9 \text{ ft}}$

Determine whether or not each pair of ratios is equal:

5. $\dfrac{7}{2}, \dfrac{35}{10}$

6. $\dfrac{5}{18}, \dfrac{30}{115}$

Solve each proportion (round each to three significant digits when necessary):

7. $\dfrac{x}{4} = \dfrac{5}{20}$

8. $\dfrac{10}{25} = \dfrac{x}{75}$

9. $\dfrac{3}{x} = \dfrac{8}{64}$

10. $\dfrac{72}{96} = \dfrac{30}{x}$

11. $\dfrac{73.4}{x} = \dfrac{25.9}{37.4}$

12. $\dfrac{x}{19.7} = \dfrac{144}{68.7}$

13. $\dfrac{61.1}{81.3} = \dfrac{592}{x}$

14. $\dfrac{243}{58.3} = \dfrac{x}{127}$

15. A piece of cable 180 ft long costs $67.50. How much will 500 ft cost at the same unit price?

16. A copper wire 750 ft long has a resistance of 1.89 Ω. How long is a copper wire of the same area whose resistance is 3.15 Ω?

17. A crew of electricians can wire 6 houses in 144 h. How many hours will it take them to wire 9 houses?

18. An automobile braking system has a 12 to 1 lever advantage on the master cylinder. A 25-lb force is applied to the pedal. What force is applied to the master cylinder?

19. Jones invests $6380 and Hernandez invests $4620 in a partnership business. What percent of the total investment does each have?

20. One gallon of a pesticide mixture weighs 7 lb 13 oz. It contains 11 oz of pesticide. What percent of the mixture is pesticide?

21. Indicate what kind of variation is shown by the equation

$$\frac{y_1}{x_1} = \frac{y_2}{x_2}$$

22. What kind of variation is indicated when one quantity increases while the other increases?

23. What kind of variation is shown by the equation $y_1x_1 = y_2x_2$?

24. Suppose $\frac{1}{4}$ in. on a map represents 25 mi. What distance is represented by $3\frac{5}{8}$ in.?

25. The scale on a map is 1 in. = 200 ft. Two places are known to be 2 mi apart. What distance will show between them on the map?

26. Two pulleys are connected by a belt. The numbers of rpm of the two pulleys vary inversely as their diameters. A pulley having a diameter of 25 cm is turning at 900 rpm. What is the number of rpm of the second pulley, which has a diameter of 40 cm?

27. A large gear with 42 teeth revolves at 25 rpm. It turns a small gear with 14 teeth. How fast does the small gear rotate?

28. In hydraulics, the formula relating the forces and the radii of the pistons is

$$\frac{F_l}{F_s} = \frac{r_l^2}{r_s^2}$$

Given $F_l = 6050$ kg, $r_l = 22$ cm, and $r_s = 2$ cm, find F_s.

29. An object 9 ft from the fulcrum of a lever balances a second object 12 ft from the fulcrum. The first object weighs 240 lb. How much does the second object weigh?

30. The current I varies directly as the voltage E. Suppose $I = 0.6$ A when $E = 30$ V. Find the value of I when $E = 100$ V.

31. The number of workers needed to complete a particular job is inversely proportional to the number of hours that they work. If 12 electricians can complete a job in 72 h, how long will it take 8 electricians to complete the same job? Assume that each person works at the same rate, no matter how many persons are assigned to the job.

Chapter 7   Test

Write each ratio in lowest terms:

1. 16 m to 64 m

2. 3 ft to 6 in.

3. 400 mL to 5 L

Solve each proportion:

4. $\dfrac{x}{8} = \dfrac{18}{48}$

5. $\dfrac{8}{x} = \dfrac{24}{5}$

6. $\dfrac{7200}{84} = \dfrac{x}{252}$

7. If 60 ft of fencing costs $45, how much does 48 ft cost?

8. Five quarts of pure antifreeze are added to 10 quarts of water to fill a radiator. What percent of antifreeze is in the mixture?

9. The scale on a map is 1 cm = 10 km. If two cities are 4.8 cm apart on the map, what is the actual distance between them?

10. A used car sells for $3250. The down payment is $390. What percent of the selling price is the down payment?

11. A small gear with 36 teeth turns a large gear with 48 teeth at 150 rpm. What is the speed (in rpm) of the small gear?

12. A pulley is 20 cm in diameter and rotating at 150 rpm. Find the diameter of a smaller pulley that must rotate at 200 rpm.

13. Given the lever formula, $F_1d_1 = F_2d_2$, and $F_1 = 800$ kg, $d_1 = 8$ m, $d_2 = 3.6$ m, find F_2.

14. A man who weighs 200 lb is to be raised by a lever extending under the man 15 in. from the fulcrum. How much force must be applied at a distance of 24 in. from the fulcrum in order to lift him?

Graphing Linear Equations

ENVIRONMENTAL ENGINEERING
The toxicity of some hazardous wastes can be expressed in terms of a linear equation.

MATHEMATICS AT WORK

A graph of a linear equation with two variables is a pictorial representation of two outcomes where one variable is dependent on the other. For example, a hardware store can show the cost of nails, screws, bolts, etc. on a linear graph where the *x*-axis represents number of pounds and the

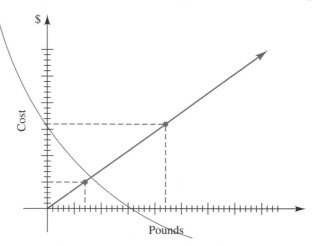

y-axis represents the cost per pound. Any customer can look at the graph and be able to see the cost for any given number of pounds.

Businesses can use linear graphs to plot marketing efforts against resulting sales. Such graphs can also be used to show revenue, material cost, production cost, and profit. Combining this information, a business can link the linear equation into a system of equations (discussed in the next chapter).

 8.1 Linear Equations with Two Variables

In Chapter 6, we studied linear equations with one variable, such as $2x + 4 = 10$ and $3x - 7 = 5$. We found that most linear equations in one variable have only one root. In this chapter, we study equations with *two* variables, such as

$$3x + 4y = 12 \quad \text{or} \quad x + y = 7$$

How many solutions does the equation $x + y = 7$ have? Any two numbers whose sum is 7 is a solution—for example, 1 for *x* and 6 for *y*, 2 for *x* and 5 for *y*, -2 for *x* and 9 for *y*, $5\frac{1}{2}$ for *x* and $1\frac{1}{2}$ for *y*, and so on. Most linear equations with *two variables* have *many possible solutions*.

Since it is very time-consuming to write pairs of replacements in this manner, we use *ordered pairs* in the form (x, y) to write solutions of equations with two variables. Therefore, instead of writing the solutions of the equation $x + y = 7$ as above, we write them as $(1, 6), (2, 5), (-2, 9), \left(5\frac{1}{2}, 1\frac{1}{2}\right)$, and so on.

> **LINEAR EQUATION WITH TWO VARIABLES**
>
> A **linear equation** with two variables can be written in the form
>
> $$ax + by = c$$
>
> where the numbers *a*, *b*, and *c* are such that *a* and *b* are not both 0.

• **EXAMPLE 1** Determine whether the given ordered pair is a solution of the given equation.

a. $(5, 2)$; $3x + 4y = 23$

To determine whether the ordered pair $(5, 2)$ is a solution to $3x + 4y = 23$, substitute 5 for *x* and 2 for *y* as follows:

$$3x + 4y = 23$$
$$3(5) + 4(2) = 23 \qquad \text{Substitute } x = 5 \text{ and } y = 2.$$
$$15 + 8 = 23$$
$$23 = 23 \qquad \text{True}$$

The result is true, so $(5, 2)$ is a solution of $3x + 4y = 23$.

b. $(-5, 6)$; $2x - 4y = -32$

$$2(-5) - 4(6) = -32 \qquad \text{Substitute } x = -5 \text{ and } y = 6$$
$$-10 - 24 = -32$$
$$-34 = -32 \qquad \text{False}$$

The result is false, so $(-5, 6)$ is not a solution of $2x - 4y = -32$.

To find solutions of a linear equation with two variables, replace one variable with a number you have chosen and then solve the resulting linear equation for the remaining variable.

• EXAMPLE 2   Complete the three ordered-pair solutions of $2x + y = 5$.

a. $(4, \quad)$

Replace x with 4. Any number could be used, but for this example we will use 4. The resulting equation is

$$2(4) + y = 5$$
$$8 + y = 5$$
$$8 + y - 8 = 5 - 8 \qquad \text{Subtract 8 from both sides.}$$
$$y = -3$$

CHECK   Replace x with 4 and y with -3.

$$2(4) + (-3) = 5 \qquad ?$$
$$8 - 3 = 5 \qquad \text{True}$$

Therefore, $(4, -3)$ is a solution.

b. $(-2, \quad)$

Replace x with -2. The resulting equation is

$$2(-2) + y = 5$$
$$-4 + y = 5$$
$$-4 + y + 4 = 5 + 4 \qquad \text{Add 4 to both sides.}$$
$$y = 9$$

CHECK   Replace x with -2 and y with 9.

$$2(-2) + 9 = 5 \qquad ?$$
$$-4 + 9 = 5 \qquad \text{True}$$

Thus, $(-2, 9)$ is a solution.

c. $(0, \quad)$

Replace x with 0. The resulting equation is

$$2(0) + y = 5$$
$$0 + y = 5$$
$$y = 5$$

CHECK   Replace x with 0 and y with 5.

$$2(0) + (5) = 5 \qquad ?$$
$$0 + 5 = 5 \qquad \text{True}$$

Therefore, $(0, 5)$ is a solution.   ————————•

You may find it easier first to solve the equation for y and then make each replacement for x.

• EXAMPLE 3   Complete the three ordered-pair solutions of $3x - y = 4$ by first solving the equation for y.

a. $(5, \quad)$ **b.** $(-2, \quad)$ **c.** $(0, \quad)$

$$3x - y = 4$$
$$3x - y - 3x = 4 - 3x \qquad \text{Subtract } 3x \text{ from both sides.}$$
$$-y = 4 - 3x$$
$$y = -4 + 3x \qquad \text{Divide both sides by } -1.$$

You may make a table to keep your work in order.

	x	$3x - 4$	y
a.	5	$3(5) - 4 = 15 - 4$	11
b.	-2	$3(-2) - 4 = -6 - 4$	-10
c.	0	$3(0) - 4 = 0 - 4$	-4

Therefore, the three solutions are $(5, 11)$, $(-2, -10)$, and $(0, -4)$. ———————•

EXAMPLE 4 Complete the three ordered pair solutions of $5x + 3y = 7$.

a. $(2, \quad)$ **b.** $(0, \quad)$ **c.** $(-1, \quad)$

We will first solve for y.

$$5x + 3y = 7$$
$$5x + 3y - 5x = 7 - 5x \qquad \text{Subtract } 5x \text{ from both sides.}$$
$$3y = 7 - 5x$$
$$\frac{3y}{3} = \frac{7 - 5x}{3} \qquad \text{Divide both sides by 3.}$$
$$y = \frac{7 - 5x}{3}$$

	x	$\dfrac{7 - 5x}{3}$	y
a.	2	$\dfrac{7 - 5(2)}{3} = \dfrac{-3}{3}$	-1
b.	0	$\dfrac{7 - 5(0)}{3} = \dfrac{7}{3}$	$\dfrac{7}{3}$
c.	-1	$\dfrac{7 - 5(-1)}{3} = \dfrac{12}{3}$	4

The three solutions are $(2, -1)$, $\left(0, \dfrac{7}{3}\right)$, and $(-1, 4)$. ———————•

Solutions to linear equations with two variables may be shown visually by graphing them in a number plane. To construct a number plane, draw a horizontal number line, which is called the *x axis*, as in Figure 8.1. Then draw a second number line intersecting the first line at right angles so that both number lines have the same zero point, called the *origin*. The vertical number line is called the *y axis*.

Each number line, or axis, has a scale. The numbers on the *x* axis are *positive to the right* of the origin and *negative to the left* of the origin. Similarly, the numbers on the *y* axis are *positive above* the origin and *negative below* the origin.

All the points in the plane determined by these two intersecting axes make up the *number plane*. The axes divide the number plane into four regions, called *quadrants*. The quadrants are numbered as shown in Figure 8.1.

Points in the number plane are usually indicated by an *ordered pair* of numbers written in the form (x, y), where x is the first number in the ordered pair and y is the

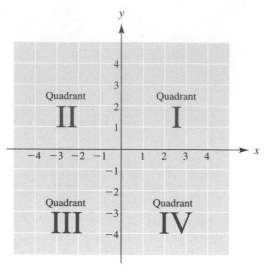

FIGURE 8.1
Rectangular coordinate system

second number in the ordered pair. The numbers *x* and *y* are also called the *coordinates* of a point in the number plane. Figure 8.1 is often called the *rectangular coordinate system.*

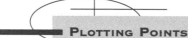

PLOTTING POINTS

To locate the point in the number plane which corresponds to an ordered pair:

STEP 1 Count right or left, from 0 (the origin) along the *x* axis, the number of spaces corresponding to the first number of the ordered pair (right if positive, left if negative).

STEP 2 Count up or down, from the point reached on the *x* axis in Step 1, the number of spaces corresponding to the second number of the ordered pair (up if positive, down if negative).

STEP 3 Mark the last point reached with a dot.

•   **EXAMPLE 5**   Plot the point corresponding to the ordered pair (3, −4) in Figure 8.2.

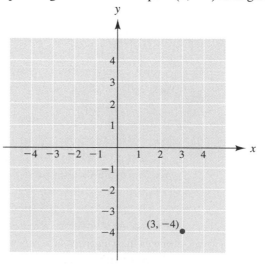

FIGURE 8.2

First, count three spaces to the right along the *x* axis. Then count down four spaces from that point. Mark the final point with a dot. ————————•

• **EXAMPLE 6** Plot the points corresponding to the ordered pairs in the number plane in Figure 8.3: A(1, 2), B(3, −2), C(−4, 7), D(5, 0), E(−2, −3), F(−5, −1), G(−2, 4).

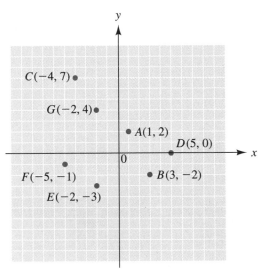

FIGURE 8.3 ————————•

Exercises | 8.1

Complete the three ordered pair solutions of each equation:

Equation	Ordered Pairs		
1. $x + y = 5$	(3,)	(8,)	(−2,)
2. $-2x + y = 8$	(2,)	(7,)	(−4,)
3. $6x + 2y = 10$	(2,)	(0,)	(−2,)
4. $6x - y = 0$	(3,)	(5,)	(−2,)
5. $3x - 4y = 8$	(0,)	(2,)	(−4,)
6. $5x - 3y = 8$	(1,)	(0,)	(−2,)
7. $-2x + 5y = 10$	(5,)	(0,)	(−3,)
8. $-4x - 7y = -3$	(−1,)	(0,)	(−8,)
9. $9x - 2y = 10$	(2,)	(0,)	(−4,)
10. $2x + 3y = 6$	(3,)	(0,)	(−6,)
11. $y = 3x + 4$	(2,)	(0,)	(−3,)
12. $y = 4x - 8$	(3,)	(0,)	(−4,)

Equation	Ordered Pairs		
13. $5x + y = 7$	(2,)	(0,)	(−4,)
14. $4x - y = 8$	(1,)	(0,)	(−3,)
15. $2x = y - 4$	(3,)	(0,)	(−1,)
16. $3y - x = 5$	(1,)	(0,)	(−4,)
17. $5x - 2y = -8$	(4,)	(0,)	(−2,)
18. $2x - 3y = 1$	(2,)	(0,)	(−4,)
19. $9x - 2y = 5$	(1,)	(0,)	(−3,)
20. $2x + 7y = -12$	(1,)	(0,)	(−8,)
21. $y = 3$	(2,)	(0,)	(−4,)
(Think: $0x + 1y = 3$)			
22. $y + 4 = 0$	(3,)	(0,)	(−7,)
23. $x = 5$	(, 4)	(, 0)	(, −2)
(Think: $1x + 0y = 5$)			
24. $x + 7 = 0$	(, 5)	(, 0)	(, −6)

Solve for y in terms of x:

25. $2x + 3y = 6$	**28.** $2x + 2y = 5$	**31.** $2x - 3y = 9$	**34.** $-3x + 5y = 25$
26. $4x + 5y = 10$	**29.** $x - 2y = 6$	**32.** $4x - 5y = 10$	**35.** $-2x - 3y = -15$
27. $x + 2y = 7$	**30.** $x - 3y = 9$	**33.** $-2x + 3y = 6$	**36.** $-3x - 4y = -8$

Write the ordered pair corresponding to each point in Illustration 1:

37. A

38. B

39. C

40. D

41. E

42. F

43. G

44. H

45. I

46. J

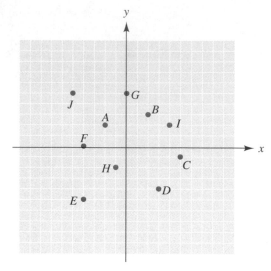

ILLUSTRATION 1

Plot each point in the number plane. Label each point by writing its ordered pair and letter:

47. A $(1, 3)$

48. B $(4, 0)$

49. C $(-6, -2)$

50. D $(-2, 4)$

51. E $(5, -4)$

52. F $(-4, -4)$

53. G $(0, 9)$

54. H $(3, 7)$

55. I $(5, -5)$

56. J $(-5, 5)$

57. K $(-6, -3)$

58. L $(3, -7)$

59. M $(-4, 5)$

60. N $(-2, -6)$

61. O $(1, -3)$

62. P $(5, 2)$

63. Q $\left(3, \dfrac{1}{2}\right)$

64. R $\left(4\dfrac{1}{2}, -3\dfrac{1}{2}\right)$

65. S $\left(-6, 2\dfrac{1}{2}\right)$

66. T $\left(-4\dfrac{1}{2}, 6\dfrac{1}{2}\right)$

8.2   Graphing Linear Equations

In Section 8.1, you learned that a linear equation with two variables has many solutions. In Example 2, you found that three of the solutions of $2x + y = 5$ were $(4, -3)$, $(-2, 9)$, and $(0, 5)$. Now plot the points corresponding to these ordered pairs and connect the points, as shown in Figure 8.4.

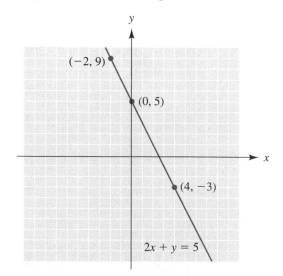

FIGURE 8.4

You can see from the figure that the three points lie on the same straight line. If you find another solution of $2x + y = 5$—say, $(1, 3)$—the point corresponding to this ordered pair also lies on the same straight line. The solutions of a linear equation with two variables always correspond to points lying on a straight line. Therefore, *the graph of the solutions of a linear equation with two variables is always a straight line.* Only part of this line can be shown on the graph; the line actually extends without limit in both directions.

GRAPHING LINEAR EQUATIONS

To draw the graph of a linear equation with two variables:

STEP 1 Find any three solutions of the equation.
(Note: Two solutions would be enough, since two points determine a straight line. However, a third solution gives a third point as a check. If the three points do not lie on the same straight line, you have made an error.)

STEP 2 Plot the points corresponding to the three ordered pairs that you found in Step 1.

STEP 3 Draw a line through the three points. If it is not a straight line, check your solutions.

• **EXAMPLE 1** Draw the graph of $3x + 4y = 12$.

STEP 1 Find any three solutions of $3x + 4y = 12$. First, solve for y:

$$3x + 4y - 3x = 12 - 3x \qquad \text{Subtract } 3x \text{ from both sides.}$$
$$4y = 12 - 3x$$
$$\frac{4y}{4} = \frac{12 - 3x}{4} \qquad \text{Divide both sides by 4.}$$
$$y = \frac{12 - 3x}{4}$$

Choose any three values of x and solve for y. Here we have chosen $x = 4$, $x = 0$, and $x = -2$.

x	$\dfrac{12 - 3x}{4}$	y
4	$\dfrac{12 - 3(4)}{4} = \dfrac{0}{4}$	0
0	$\dfrac{12 - 3(0)}{4} = \dfrac{12}{4}$	3
−2	$\dfrac{12 - 3(-2)}{4} = \dfrac{18}{4}$	$\dfrac{9}{2}$

Three solutions are $(4, 0)$, $(0, 3)$, and $\left(-2, \dfrac{9}{2}\right)$.

STEP 2 Plot the points corresponding to $(4, 0)$, $(0, 3)$, and $\left(-2, \dfrac{9}{2}\right)$.

STEP 3 Draw a straight line through these three points. (See Figure 8.5.)

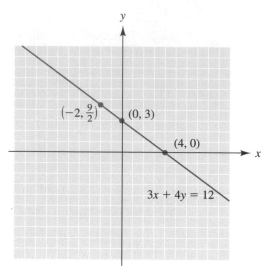

FIGURE 8.5

An alternative method is shown in Example 2.

•   **EXAMPLE 2**   Draw the graph of $2x - 3y = 6$.

STEP 1 Set up a table and write the values you choose for x—say, 3, 0, and −3.

	a.	**b.**	**c.**
x	3	0	−3
y			

STEP 2 Substitute the chosen values of x in the given equation and solve for y.

a.	**b.**	**c.**
$2x - 3y = 6$	$2x - 3y = 6$	$2x - 3y = 6$
$2(3) - 3y = 6$	$2(0) - 3y = 6$	$2(-3) - 3y = 6$
$6 - 3y = 6$	$0 - 3y = 6$	$-6 - 3y = 6$
$-3y = 0$	$-3y = 6$	$-3y = 12$
$y = 0$	$y = -2$	$y = -4$

STEP 3 Write the values for y that correspond to the chosen values for x in the table, thus:

	a.	**b.**	**c.**
x	3	0	−3
y	0	−2	−4

That is, three solutions of $2x - 3y = 6$ are the ordered pairs (3, 0), (0, −2), and (−3, −4).

STEP 4 Plot the points from Step 3 and draw a straight line through them, as in Figure 8.6.

The line in Figure 8.6 crosses the x axis at the point (3, 0). The number 3 is called the *x* intercept—the *x* coordinate of the point where the line crosses the *x* axis.

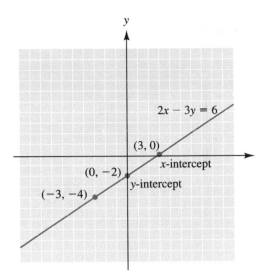

FIGURE 8.6

The line in Figure 8.6 crosses the *y* axis at the point (0, −2). The number −2 is called the *y intercept*—the *y* coordinate of the point where the line crosses the *y* axis.

The graphs of two special cases of linear equations are often helpful. The equation *y* = 5 is a linear equation with an *x* coefficient of 0. (This equation may also be written as 0*x* + 1*y* = 5.) Similarly, *x* = −7 is a linear equation with a *y* coefficient of 0. (This equation may also be written as 1*x* + 0*y* = −7.) These equations have graphs that are horizontal or vertical straight lines, as shown in the next two examples.

• **EXAMPLE 3**   Draw the graph of *y* = 5.

Set up a table and write the values you choose for *x*—say, 3, 0, and −4. As the equation states, *y* is always 5 for any value of *x* that you choose.

x	3	0	−4
y	5	5	5

Plot the points from the table: (3, 5), (0, 5), and (−4, 5). Then draw a straight line through them, as in Figure 8.7.

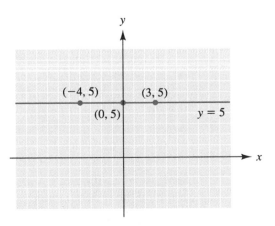

FIGURE 8.7

HORIZONTAL LINE

The graph of the linear equation $y = k$, where k is a constant, is the horizontal line through the point $(0, k)$. That is, $y = k$ is a horizontal line with a y intercept of k.

•   **EXAMPLE 4**   Draw the graph of $x = -7$.

All ordered pairs that are solutions of $x = -7$ have an x value of -7. You can choose any number for y. Three ordered pairs that satisfy $x = -7$ are $(-7, 3)$, $(-7, 1)$, and $(-7, -4)$. Plot these three points and draw a straight line through them, as in Figure 8.8.

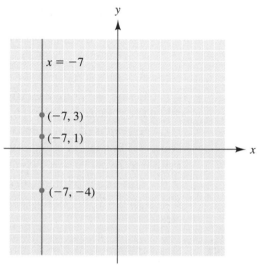

FIGURE 8.8

VERTICAL LINE

The graph of the linear equation $x = k$, where k is a constant, is the vertical line through the point $(k, 0)$. That is, $x = k$ is a vertical line with an x intercept of k.

Solve the equation $2x - y = 8$ for y. The solution is $y = 2x - 8$. To graph this equation, assign values for x and find the corresponding y values. We call x the *independent variable*, because we may choose any value for x that we wish. Since the value of y depends on the value of x, we call y the *dependent variable*.

In many technical classes, variables other than x and y are often used. These other variables are usually related to formulas. Recall that a formula can be solved for one variable in terms of another. For example, Ohm's law can be expressed as $V = IR$, or as $V = 10I$ when $R = 10$. For the equation $V = 10I$, I is called the *independent variable* and V is called the *dependent variable*. The dependent variable is the variable for which the formula is solved.

When graphing an equation, the horizontal axis corresponds to the independent variable; the vertical axis corresponds to the dependent variable. Think of graphing the ordered pairs

(independent variable, dependent variable)

• **EXAMPLE 5**   Draw the graph of $V = 10I$.

Since I is the independent variable, graph ordered pairs in the form (I, V). Set up a table and write the values you choose for I—say, 0, 5, and 8. For this example, limit your values of I to the positive numbers.

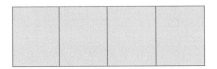

Next, choose a suitable scale for the vertical axis and graph the ordered pairs $(0, 0)$, $(5, 50)$, and $(8, 80)$, as shown in Figure 8.9.

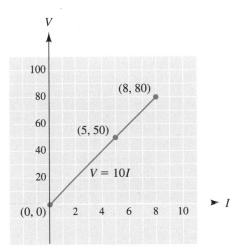

FIGURE 8.9

Exercises   8.2

Draw the graph of each equation:

1. $x + y = 7$
2. $x + 3y = 9$
3. $y = 2x + 3$
4. $y = 4x - 5$
5. $4y = x$
6. $2x + y = 6$

7. $6x - 2y = 10$
8. $2x + 3y = 9$
9. $3x - 4y = 12$
10. $3x - 5y = 15$
11. $5x + 4y = 20$
12. $2x - 3y = 18$

13. $2x + 7y = 14$
14. $2x - 5y = 20$
15. $y = 2x$
16. $y = -3x$
17. $3x + 5y = 11$
18. $4x - 7y = 15$

19. $y = -\dfrac{1}{2}x + 4$
20. $y = \dfrac{2}{3}x - 6$
21. $y = 3$
22. $y = -2$
23. $x = -4$
24. $x = 5$

25. $y - 6 = 0$
26. $y + 10 = 0$
27. $x + 3\dfrac{1}{2} = 0$
28. $x - 4 = 0$
29. $y = 0$
30. $x = 0$

Identify the independent and dependent variables for each equation:

31. $s = 4t + 7$
32. $V = 5t - 2$
39. $s = 3t^2 + 5t - 1$

33. $R = 0.5V$
34. $s = 65t$

35. $i = 30t - 10$
36. $E = 4V + 2$
40. $v = 2i^2 - 3i + 10$

37. $v = 50 - 6t$
38. $i = 18 - 3t$

41. The distance, s (in feet), that a body travels in t seconds is given by the equation $s = 5t + 10$. Graph the equation for positive values of s.

42. The voltage, v, in an electrical circuit varies according to the equation $v = 10t - 5$, where t is in seconds. Graph the equation for positive values of v.

43. The resistance, R, in an electrical circuit varies according to the equation $R = 1.5V$, where V is in volts. Graph the equation for positive values of R.

44. The current, I, in an electrical circuit varies according to the equation $I = 0.05V$, where V is in volts. Choose a suitable scale and graph the equation for positive values of I.

45. The voltage, v, in an electrical circuit is given by the equation $v = 60 - 5t$, where t is in μs. Graph the equation for positive values of v and t.

46. The distance, s, that a point travels in t milliseconds is given by the equation $s = 24 - 2t$. Graph the equation for positive values of s and t.

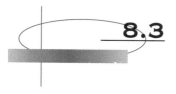

8.3 The Slope of a Line

The slope of a line or the "steepness" of a roof (see Figure 8.10) can be measured by the following ratio:

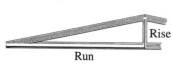

FIGURE 8.10

$$\text{slope} = \frac{\text{vertical change}}{\text{horizontal change}} = \frac{\text{rise}}{\text{run}}$$

A straight line can also be graphed using its slope and knowing one point on the line.

If two points on a line (x_1, y_1) (read "x-sub-one, y-sub-one") and (x_2, y_2) (read "x-sub-two, y-sub-two") are known (see Figure 8.11), the slope of the line is defined as follows.

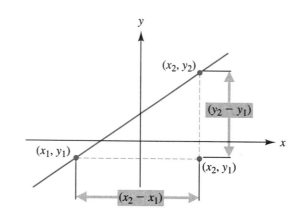

FIGURE 8.11

SLOPE OF A LINE

$$\text{slope} = m = \frac{\text{vertical change}}{\text{horizontal change}} = \frac{\text{rise}}{\text{run}} = \frac{\text{difference in } y \text{ values}}{\text{difference in } x \text{ values}} = \frac{y_2 - y_1}{x_2 - x_1}$$

• **EXAMPLE 1**   Find the slope of the line passing through the points $(-2, 3)$ and $(4, 7)$. (See Figure 8.12.)

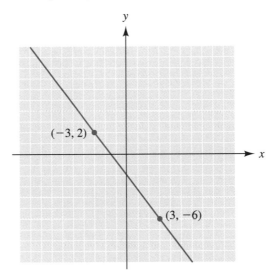

FIGURE 8.12

If we let $x_1 = -2$, $y_1 = 3$, $x_2 = 4$, and $y_2 = 7$, then

$$m = \frac{y_2 - y_1}{x_2 - x_1} = \frac{7 - 3}{4 - (-2)} = \frac{4}{6} = \frac{2}{3}$$

Note that if we reverse the order of taking the differences of the coordinates, the result is the same:

$$m = \frac{y_1 - y_2}{x_1 - x_2} = \frac{3 - 7}{-2 - 4} = \frac{-4}{-6} = \frac{2}{3}$$

• **EXAMPLE 2** Find the slope of the line passing through $(-3, 2)$ and $(3, -6)$. (See Figure 8.13.)

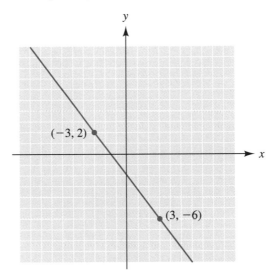

FIGURE 8.13

If we let $x_1 = -3$, $y_1 = 2$, $x_2 = 3$, and $y_2 = -6$, then

$$m = \frac{y_2 - y_1}{x_2 - x_1} = \frac{-6 - 2}{3 - (-3)} = \frac{-8}{6} = -\frac{4}{3}$$

• **EXAMPLE 3** Find the slope of the line through $(-5, 2)$ and $(3, 2)$. (See Figure 8.14.)

$$m = \frac{y_2 - y_1}{x_2 - x_1} = \frac{2 - 2}{3 - (-5)} = \frac{0}{8} = 0$$

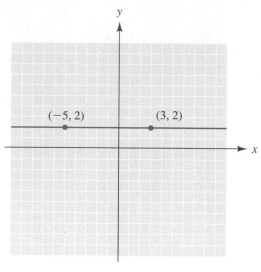

FIGURE 8.14

Note that all points on any horizontal line have the same *y* value. Therefore, *the slope of any horizontal line is 0.*

•   **EXAMPLE 4**   Find the slope of the line through (4, 2) and (4, −5). (See Figure 8.15.)

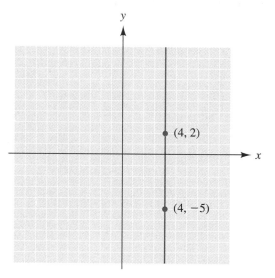

FIGURE 8.15

$$m = \frac{y_2 - y_1}{x_2 - x_1} = \frac{-5 - 2}{4 - 4} = \frac{-7}{0} \quad \text{(undefined)}$$

Division by zero is not possible, so the slope is undefined. Note that all points on any vertical line have the same *x* value. Therefore, *the slope of any vertical line is undefined.*

Note that in Example 1, the line slopes upward from left to right, whereas in Example 2 the line slopes downward. In general, the following is true.

1. If a line has positive slope, then the line slopes upward from left to right.
2. If a line has negative slope, then the line slopes downward from left to right.
3. If the line has zero slope, then the line is *horizontal*.
4. If the slope of a line is undefined, then the line is *vertical*.

The slope of a straight line can be found directly from its equation as follows:

1. Solve the equation for y.
2. The slope of the line is given by the coefficient of x.

• **EXAMPLE 5**   Find the slope of the line $4x + 6y = 15$.

First, solve the equation for y.

$$4x + 6y = 15$$
$$6y = -4x + 15 \qquad \text{Subtract } 4x \text{ from both sides.}$$
$$y = -\frac{2}{3}x + \frac{5}{2} \qquad \text{Divide both sides by 6.}$$

The slope of the line is given by the coefficient of x, or $m = -\frac{2}{3}$.

• **EXAMPLE 6**   Find the slope of the line $9x - 3y = 10$.

First, solve the equation for y.

$$9x - 3y = 10$$
$$-3y = -9x + 10 \qquad \text{Subtract } 9x \text{ from both sides.}$$
$$y = 3x - \frac{10}{3} \qquad \text{Divide both sides by } -3.$$

The slope of the line is given by the coefficient of x, or $m = 3$.

(a) Parallel lines

(b) Perpendicular lines

FIGURE 8.16

Two lines in the same plane are parallel if they do not intersect even if they are extended. (See Figure 8.16(a).) Two lines in the same plane are perpendicular if they intersect at right angles, as in Figure 8.16(b).

Since parallel lines have the same steepness, they have the same slope.

PARALLEL LINES

Two lines are parallel if either one of the following conditions holds:

1. Both lines are perpendicular to the x axis (Figure 8.17(a)), or
2. Both lines have the same slope (Figure 8.17(b))—that is, if the equations of the two lines are L_1: $y = m_1x + b_1$ and L_2: $y = m_2x + b_2$;

then

$$m_1 = m_2$$

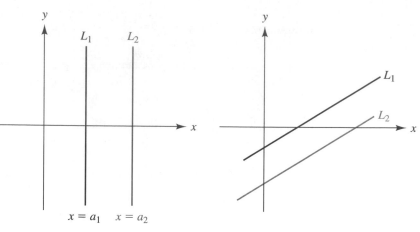

(a) Both lines are perpendicular
    to the x-axis

(b) $m_1 = m_2$

FIGURE 8.17

PERPENDICULAR LINES

Two lines are perpendicular if either one of the following conditions holds:

1. One line is vertical with equation $x = a$ and the other line is horizontal with equation $y = b$, or

2. Neither is vertical and the product of the slopes of the two lines is -1; that is, if the equations of the lines are

$$L_1: y = m_1x + b_1 \text{ and } L_2: y = m_2x + b_2$$

then

$$m_1 \cdot m_2 = -1$$

• **EXAMPLE 7** Determine whether the lines given by the equations $2x + 3y = 6$ and $6x - 4y = 9$ are parallel, perpendicular, or neither.

First, find the slope of each line by solving its equation for y.

$$
\begin{aligned}
2x + 3y &= 6 & 6x - 4y &= 9 \\
3y &= -2x + 6 & -4y &= -6x + 9 \\
y &= -\frac{2}{3}x + 2 & y &= \frac{3}{2}x - \frac{9}{4} \\
m_1 &= -\frac{2}{3} & m_2 &= \frac{3}{2}
\end{aligned}
$$

Since the slopes are not equal, the lines are not parallel. Next, find the product of the slopes.

$$m_1 \cdot m_2 = \left(-\frac{2}{3}\right)\left(\frac{3}{2}\right) = -1$$

Thus, the lines are perpendicular.

• **EXAMPLE 8** Determine whether the lines given by the equations $5x + y = 7$ and $15x + 3y = -10$ are parallel, perpendicular, or neither.

First, find the slope of each line by solving its equation for y.

$$5x + y = 7 \qquad\qquad 15x + 3y = -10$$
$$y = -5x + 7 \qquad\qquad 3y = -15x - 10$$
$$\qquad\qquad\qquad y = -5x - \frac{10}{3}$$
$$m_1 = -5 \qquad\qquad\qquad m_2 = -5$$

Since both lines have the same slope, -5, and different y intercepts, they are parallel.

• **EXAMPLE 9** Determine whether the lines given by the equations $4x + 5y = 15$ and $-3x - 2y = 12$ are parallel, perpendicular, or neither.

First, find the slope of each line by solving its equation for y.

$$4x + 5y = 15 \qquad\qquad -3x - 2y = 12$$
$$5y = -4x + 15 \qquad\qquad -2y = 3x + 12$$
$$y = -\frac{4}{5}x + 3 \qquad\qquad y = -\frac{3}{2}x - 6$$
$$m_1 = -\frac{4}{5} \qquad\qquad m_2 = -\frac{3}{2}$$

Since the slopes are not equal and do not have a product of -1, the lines are neither parallel nor perpendicular. That is, the lines intersect but not at right angles.

Exercises 8.3

Find the slope of the line passing through each pair of points:

1. $(-3, 1), (2, 6)$
2. $(4, 7), (6, 2)$
3. $(-2, 1), (-3, -5)$
4. $(5, -3), (-4, -9)$
5. $(4, 0), (0, 5)$
6. $(1, -6), (-2, 0)$
7. $(-2, 4), (5, 4)$
8. $(-3, -7), (-2, -7)$
9. $(6, 1), (6, -4)$
10. $(-8, 5), (-8, -3)$
11. $(4, -2), (-6, -8)$
12. $(-9, -1), (-3, -5)$

Find the slope of each line:

13.

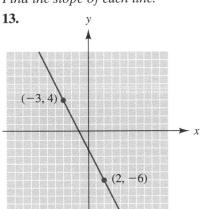

14.

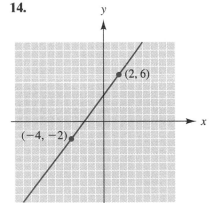

15.

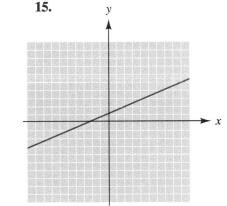

16.

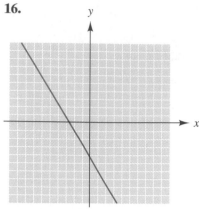

17.

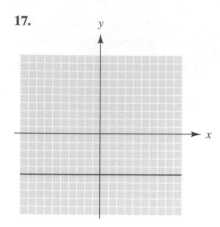

18.

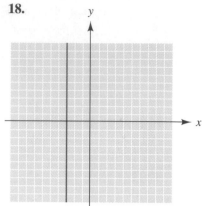

Find the slope of each line:

19. $y = 6x + 2$

20. $y = -4x + 3$

21. $y = -5x - 7$

22. $y = 9x - 13$

23. $3x + 5y = 6$

24. $9x + 12y = 8$

25. $-2x + 8y = 3$

26. $4x - 6y = 9$

27. $5x - 2y = 16$

28. $-4x - 2y = 7$

29. $x - 3 = 0$

30. $y + 15 = 0$

Determine whether the lines given by the equations are parallel, perpendicular, or neither:

31. $y = 4x - 5$

$y = 4x + 5$

32. $y = \frac{2}{3}x + 4$

$y = -\frac{3}{2}x - 5$

33. $y = \frac{3}{4}x - 2$

$y = -\frac{4}{3}x + \frac{5}{3}$

34. $y = \frac{4}{5}x - 2$

$y = -\frac{4}{5}x + \frac{2}{3}$

35. $x + 3y = 9$

$3x - y = 14$

36. $x + 2y = 11$

$2x + 4y = -5$

37. $x - 4y = 12$

$x + 4y = 16$

38. $2x + 7y = 6$

$14x - 4y = 18$

39. $y - 5x = 12$
$5x - y = -6$

40. $-3x + 9y = 20$
$x = 3y$

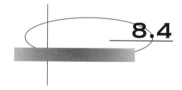

8.4   The Equation of a Line

We have learned to graph the equation of a straight line and to find the slope of a straight line given its equation or any two points on it. In this section, we will use the slope to graph the equation and to write its equation.

•   **EXAMPLE 1**   Draw the graph of the line with slope $\frac{2}{3}$ and y intercept 4.

The slope $\frac{2}{3}$ corresponds to $\frac{\text{difference in } y \text{ values}}{\text{difference in } x \text{ values}} = \frac{2}{3}$. From the y intercept 4 [the point $(0, 4)$], move 2 units up and then 3 units to the right, as shown in Figure 8.18. Then draw a straight line through $(0, 4)$ and $(3, 6)$.   ⎯⎯⎯⎯⎯⎯●

In Section 8.3, we learned how to find the slope of a line, given its equation, by solving for y. For example, for a line whose equation is $y = 3x + 5$ the slope is 3, the coefficient of x. What does the number 5 have to do with its graph? If we let $x = 0$, the equation is

$$y = 3x + 5$$
$$y = 3(0) + 5 = 0 + 5 = 5$$

The line crosses the y axis when $x = 0$. Therefore, the y intercept of the graph is 5.

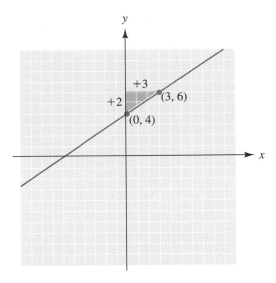

FIGURE 8.18

SLOPE-INTERCEPT FORM

When the equation of a straight line is written in the form

$$y = mx + b$$

the slope of the line is m and the y intercept is b.

• **EXAMPLE 2** Draw the graph of the equation $8x + 2y = -10$ using its slope and y intercept.

First, find the slope and y intercept by solving the equation for y as follows.

$$8x + 2y = -10$$
$$2y = -8x - 10 \qquad \text{Subtract } 8x \text{ from both sides.}$$
$$y = -4x - 5 \qquad \text{Divide both sides by 2.}$$
$$\underset{\text{slope}}{\nearrow} \qquad \underset{y \text{ intercept}}{\nwarrow}$$

The slope is -4, and the y intercept is -5. The slope -4 corresponds to

$$\frac{\text{difference in } y \text{ values}}{\text{difference in } x \text{ values}} = \frac{-4}{1}$$

When the slope is an integer, write it as a ratio with 1 in the denominator. From the y intercept -5, move 4 units down and 1 unit to the right, as shown in Figure 8.19. Draw a straight line through the points $(0, -5)$ and $(1, -9)$.

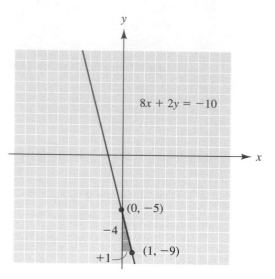

FIGURE 8.19

•   **EXAMPLE 3**   Find the equation of the line with slope $\dfrac{3}{4}$ and y intercept -2.

Use the slope-intercept form with $m = \dfrac{3}{4}$ and $b = -2$.

$$y = mx + b$$

$$y = \frac{3}{4}x + (-2)$$

$$y = \frac{3}{4}x - 2$$

$4y = 3x - 8$	Multiply both sides by 4.
$0 = 3x - 4y - 8$	Subtract $4y$ from both sides.
$8 = 3x - 4y$	Add 8 to both sides.

$$3x - 4y = 8$$

Note: Any of the last five equations in Example 3 is correct. The most common ways of writing equations are $y = mx + b$ and $cx + dy = f$.

•   **EXAMPLE 4**   Draw the graph of the straight line through the point $(-3, 6)$ with slope $-\dfrac{2}{5}$.

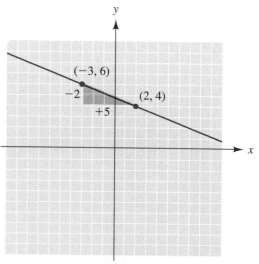

FIGURE 8.20

The slope $-\frac{2}{5}$ corresponds to $\frac{\text{difference in } y \text{ values}}{\text{difference in } x \text{ values}} = -\frac{2}{5}$. From the point $(-3, 6)$, move 2 units down and 5 units to the right, as shown in Figure 8.20. Draw a straight line through the points $(-3, 6)$ and $(2, 4)$.

Given a point on a line and its slope, we can find its equation. To show this, let m be the slope of a nonvertical straight line; let (x_1, y_1) be the coordinates of a known or given point on the line; and let (x, y) be the coordinates of any other point on the line. (See Figure 8.21.)

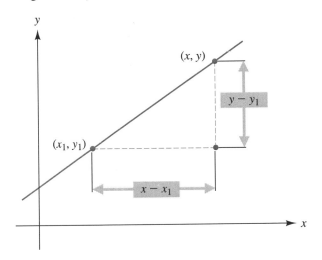

FIGURE 8.21

Then, by the definition of slope, we have

$$\frac{\text{difference in } y \text{ values}}{\text{difference in } x \text{ values}} = m$$

$$\frac{y - y_1}{x - x_1} = m$$

$$y - y_1 = m(x - x_1) \qquad \text{Multiply both sides by } (x - x_1).$$

The result is the point-slope form of the equation of a straight line.

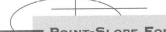

POINT-SLOPE FORM

If m is the slope and (x_1, y_1) is any point on a nonvertical straight line, its equation is
$$y - y_1 = m(x - x_1)$$

• **EXAMPLE 5**   Find the equation of the line with slope -3 that passes through the point $(-1, 4)$.

Here, $m = -3$, $x_1 = -1$, and $y_1 = 4$. Using the point-slope form, we have

$$y - y_1 = m(x - x_1)$$
$$y - 4 = -3(x - (-1))$$
$$y - 4 = -3(x + 1)$$
$$y - 4 = -3x - 3 \qquad \text{Remove parentheses.}$$
$$y = -3x + 1 \qquad \text{Add 4 to both sides.}$$

The point-slope form also can be used to find the equation of a line when two points on the line are known.

• **EXAMPLE 6**   Find the equation of the line through the points $(5, -4)$ and $(-1, 8)$.

First, find the slope.

$$m = \frac{y_2 - y_1}{x_2 - x_1} = \frac{8 - (-4)}{-1 - 5} = \frac{12}{-6} = -2$$

Substitute $m = -2$, $x_1 = 5$, and $y_1 = -4$ in the point-slope form.

$$y - y_1 = m(x - x_1)$$
$$y - (-4) = -2(x - 5)$$
$$y + 4 = -2x + 10 \qquad \text{Remove parentheses.}$$
$$y = -2x + 6 \qquad \text{Subtract 4 from both sides.}$$

We could have used the other point $(-1, 8)$, as follows:

$$y - y_1 = m(x - x_1)$$
$$y - 8 = -2(x - (-1))$$
$$y - 8 = -2(x + 1)$$
$$y - 8 = -2x - 2 \qquad \text{Remove parentheses.}$$
$$y = -2x + 6 \qquad \text{Add 8 to both sides.}$$

Exercises 8.4

Draw the graph of each line with the given slope and y intercept:

1. $m = 2, b = 5$
3. $m = -5, b = 4$
5. $m = \frac{2}{3}, b = -4$
7. $m = -\frac{2}{5}, b = -4$

2. $m = 4, b = -3$
4. $m = -1, b = 0$
6. $m = \frac{3}{4}, b = 2$
8. $m = \frac{5}{6}, b = 3$

9. $m = \frac{4}{3}, b = -1$
10. $m = -\frac{9}{4}, b = 2$

Draw the graph of each equation using the slope and y intercept:

11. $2x + y = 6$
13. $3x + 5y = 10$
15. $3x - y = 7$
17. $3x - 2y = 12$
19. $2x - 6y = 0$

12. $4x + y = -3$
14. $4x + 3y = 9$
16. $5x - y = -2$
18. $2x - 5y = 20$
20. $4x + 7y = 0$

Find the equation of the line with the given slope and y intercept:

21. $m = 2, b = 5$
23. $m = -5, b = 4$
25. $m = \frac{2}{3}, b = -4$
27. $m = -\frac{6}{5}, b = 3$

22. $m = 4, b = -3$
24. $m = -1, b = 0$
26. $m = \frac{3}{4}, b = 2$
28. $m = -\frac{12}{5}, b = -1$

29. $m = -\frac{3}{5}, b = 0$
30. $m = 0, b = 0$

Draw the graph of the line through the given point with the given slope:

31. $(3, 5), m = 2$
33. $(-5, 0), m = \frac{3}{4}$
35. $(6, -2), m = \frac{3}{2}$
37. $(-1, -1), m = -1$

32. $(-1, 4), m = -3$
34. $(0, 0), m = -\frac{1}{3}$
36. $(-3, -3), m = \frac{1}{2}$
38. $(0, -4), m = 0$

39. $(2, -7), m = 1$
40. $(8, 8), m = \frac{1}{8}$

Find the equation of the line through the given point with the given slope:

41. $(3, 5)$, $m = 2$

43. $(-5, 0)$, $m = \dfrac{3}{4}$

45. $(6, -2)$, $m = \dfrac{3}{2}$

47. $(-3, -1)$, $m = -\dfrac{10}{3}$

42. $(-1, 4)$, $m = -3$

44. $(0, 0)$, $m = -\dfrac{1}{3}$

46. $(-3, -3)$, $m = \dfrac{1}{2}$

48. $(4, 5)$, $m = -\dfrac{3}{7}$

49. $(12, -10)$, $m = -1$

50. $(15, 20)$, $m = 3$

Find the equation of the line through the given points:

51. $(2, 3)$, $(5, 1)$

53. $(-1, 4)$, $(2, -2)$

55. $(0, 1)$, $(-3, 0)$

57. $(7, 4)$, $(-1, 0)$

52. $(8, 5)$, $(2, 1)$

54. $(-1, -2)$, $(-4, 0)$

56. $(5, 5)$, $(-2, -2)$

58. $(6, -3)$, $(-3, 6)$

59. $(3, 3)$, $(1, 5)$

60. $(16, 12)$, $(4, 8)$

Chapter 8   Review

Complete the ordered pair solutions of each equation:

1. $x + 2y = 8$ $(3, \quad)$ $(0, \quad)$ $(-4, \quad)$

2. $2x - 3y = 12$ $(3, \quad)$ $(0, \quad)$ $(-3, \quad)$

3. Solve for y: $6x + y = 15$.

4. Solve for y: $3x - 5y = -10$.

Write the ordered pair corresponding to each point in Illustration 1:

5. A

6. B

7. C

8. D

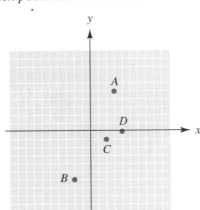

ILLUSTRATION 1

Plot each point in the number plane. Label each point by writing its ordered pair:

9. E $(3, -2)$

10. F $(-7, -4)$

11. G $(-1, 5)$

12. H $(0, -5)$

Draw the graph of each equation:

13. $x + y = 8$

15. $3x + 6y = 12$

17. $4x = 9y$

19. $x = -6$

14. $x - 2y = 5$

16. $4x - 5y = 15$

18. $y - \dfrac{1}{3}x - 4$

20. $y = 7$

Find the slope of the line passing through each pair of points:

21. $(3, -4)$, $(10, 5)$

22. $(-4, 0)$, $(2, 6)$

Find the slope of each line:

23. $y = 4x - 7$

24. $2x + 5y = 8$

25. $5x - 9y = -2$

Determine whether the lines given by the equations are parallel, perpendicular, or neither:

26. $y = 3x - 5$

$y = -\dfrac{1}{3}x - 5$

27. $3x - 4y = 12$

$8x - 6y = 15$

28. $2x + 5y = 8$

$4x + 10y = 25$

29. $x = 4$

$y = -6$

Draw the graph of each line with the given slope and y intercept:

30. $m = -2, b = 9$

31. $m = -\dfrac{2}{3}, b = -5$

Draw the graph of each equation using the slope and y intercept:

32. $3x + 5y = 20$

33. $5x - 8y = 32$

Find the equation of the line with the given slope and y intercept:

34. $m = -\dfrac{1}{2}, b = 3$

35. $m = \dfrac{8}{3}, b = 0$

36. $m = 0, b = 0$

Draw the graph of the line through the given point with the given slope:

37. $(6, -1), m = -3$

38. $(-5, -2), m = \dfrac{7}{2}$

Find the equation of the line through the given point with the given slope:

39. $(-2, 8), m = -1$

40. $(0, -5), m = -\dfrac{1}{4}$

Find the equation of the line through the given points:

41. $(2, -3), (10, 5)$

42. $(12, 0), (2, -5)$

Chapter 8   Test

Given the equation $3x - 4y = 24$, complete each ordered pair.

1. $(-4, \quad)$

2. $(0, \quad)$

3. $(4, \quad)$

4–5. Write the ordered pair corresponding to each point in Illustration 1.

6. Draw the graph of $3x + y = 3$.

7. Draw the graph of $-2x + y = 4$.

8. Draw the graph of $s = 5 + 2t$ for positive values of t.

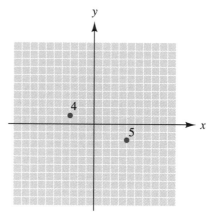

ILLUSTRATION 1

9. Find the slope of the line containing the points $(-2, 4)$ and $(5, 6)$.

Find the slope of the line with the given equation.

10. $y = 3x - 2$

11. $2x - 5y = 10$

12. Determine whether the graphs of the following pair of equations are parallel, perpendicular, or neither.

$$2x - y = 10$$
$$y = 2x - 3$$

13. Find the equation of the line having y intercept -3 and slope $\frac{1}{2}$.

14. Find the equation of the line containing the point $(-2, 3)$ and slope -2.

15. Draw the graph of the line containing the point $(3, 4)$ and having slope -3.

16. Draw the graph of the line $y = -\frac{1}{2}x + 4$, using its slope and y intercept.

Systems of Linear Equations

COMPUTER TECHNICIAN
A technician works on the mainframe computer at a college.

Systems of linear equations with two variables have many applications in the business world. Businesses can use them to predict future growth, analyze production needs, adjust marketing and sales efforts, and determine where cuts might be made.

Systems of equations are widely used for predicting the future needs of a company. This is particularly helpful for a business marketing multiple products or a corporation that owns more than one company. Employees gather information and plot the outcome; the information can then be combined into a system of linear equations to be used in the company's decision making.

9.1 Solving Pairs of Linear Equations by Graphing

Many problems can be solved by using two equations with two variables and solving them simultaneously. *To solve a pair of linear equations with two variables simultaneously, you must find an ordered pair that will make* both *equations true at the same time.*

As you know, the graph of a linear equation with two variables is a straight line. As shown in Figure 9.1, two straight lines (the graphs of *two* linear equations with two variables) in the same plane may be arranged as follows:

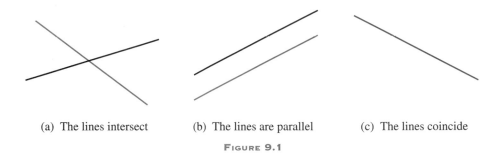

(a) The lines intersect   (b) The lines are parallel   (c) The lines coincide

FIGURE 9.1

  a. The lines may intersect. If so, they have one point in common. The equations have one common solution. The coordinates of the point of intersection define the common solution.

  b. The lines may be parallel. If so, they have no point in common. The equations have no common solution.

  c. The lines may coincide. That is, one line lies on top of the other. If so, any solution of one equation is also a solution of the other. Hence, there are infinitely many points of intersection and infinitely many common solutions. Every solution of one equation is a solution of the other.

• EXAMPLE 1   Draw the graphs of $x - y = 2$ and $x + 3y = 6$ in the same number plane. Find the common solution of the equations.

STEP 1 Draw the graph of $x - y = 2$. First, solve for y.

$$x - y = 2$$
$$x - y - x = 2 - x \qquad \text{Subtract } x \text{ from both sides.}$$
$$-y = 2 - x$$
$$\frac{-y}{-1} = \frac{2 - x}{-1} \qquad \text{Divide both sides by } -1.$$
$$y = -2 + x$$

Then find three solutions.

x	$-2 + x$	y
2	$-2 + 2$	0
0	$-2 + 0$	-2
-2	$-2 + (-2)$	-4

Three solutions are $(2, 0)$, $(0, -2)$, and $(-2, -4)$. Plot the three points that correspond to these three ordered pairs. Then draw a straight line through these three points, as in Figure 9.2.

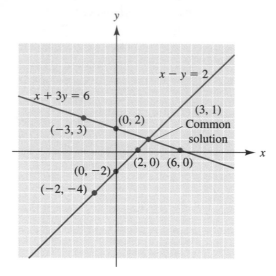

FIGURE 9.2

STEP 2 Draw the graph of $x + 3y = 6$. First, solve for y.

$$x + 3y = 6$$

$$x + 3y - x = 6 - x \qquad \text{Subtract } x \text{ from both sides.}$$

$$3y = 6 - x$$

$$\frac{3y}{3} = \frac{6 - x}{3} \qquad \text{Divide both sides by 3.}$$

$$y = \frac{6 - x}{3}$$

x	$\dfrac{6 - x}{3}$	y
6	$\dfrac{6 - 6}{3} = \dfrac{0}{3}$	0
0	$\dfrac{6 - 0}{3} = \dfrac{6}{3}$	2
-3	$\dfrac{6 - (-3)}{3} = \dfrac{9}{3}$	3

Three solutions are $(6, 0)$, $(0, 2)$, and $(-3, 3)$. Plot the three points that correspond to these three ordered pairs. Then draw a straight line through these three points, as in Figure 9.2.

The point that corresponds to $(3, 1)$ is the point of intersection. Therefore, $(3, 1)$ is the common solution. That is, $x = 3$ and $y = 1$ are the only values that satisfy both equations. The solution $(3, 1)$ should be checked in both equations.

•   **EXAMPLE 2**   Draw the graphs of $2x + 3y = 6$ and $4x + 6y = 30$ in the same number plane. Find the common solution of the equations.

STEP 1 Draw the graph of $2x + 3y = 6$. First, solve for y.

$$2x + 3y = 6$$

$2x + 3y - 2x = 6 - 2x$ Subtract $2x$ from both sides.

$$3y = 6 - 2x$$

$$\frac{3y}{3} = \frac{6 - 2x}{3}$$ Divide both sides by 3.

$$y = \frac{6 - 2x}{3}$$

Then find three solutions.

x	$\dfrac{6 - 2x}{3}$	y
3	$\dfrac{6 - 2(3)}{3} = \dfrac{0}{3}$	0
0	$\dfrac{6 - 2(0)}{3} = \dfrac{6}{3}$	2
−3	$\dfrac{6 - 2(-3)}{3} = \dfrac{12}{3}$	4

Three solutions are $(3, 0)$, $(0, 2)$, and $(-3, 4)$. Plot the three points which correspond to these three ordered pairs. Then draw a straight line through these three points, as in Figure 9.3.

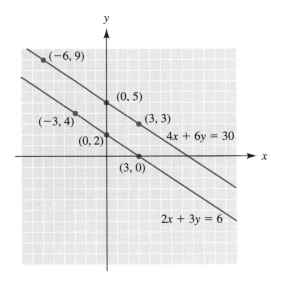

FIGURE 9.3

STEP 2 Draw the graph of $4x + 6y = 30$. First, solve for y.

$$4x + 6y = 30$$

$4x + 6y - 4x = 30 - 4x$ Subtract $4x$ from both sides.

$$6y = 30 - 4x$$

$$\frac{6y}{6} = \frac{30 - 4x}{6}$$ Divide both sides by 6.

$$y = \frac{30 - 4x}{6}$$

Then find three solutions.

x	$\dfrac{30 - 4x}{6}$	y
3	$\dfrac{30 - 4(3)}{6} = \dfrac{18}{6}$	3
0	$\dfrac{30 - 4(0)}{6} = \dfrac{30}{6}$	5
−6	$\dfrac{30 - 4(-6)}{6} = \dfrac{54}{6}$	9

Three solutions are $(3, 3)$, $(0, 5)$, and $(-6, 9)$. Plot the three points that correspond to these ordered pairs in the same number plane as the points found in Step 1. Then draw a straight line through these three points. As you see in Figure 9.3, the lines are parallel. The lines have no points in common; therefore, there is no common solution.

You may verify that the lines are parallel by showing that the slopes are equal.

—————————•

• **EXAMPLE 3**   Draw the graphs of $-2x + 8y = 24$ and $-3x + 12y = 36$ in the same number plane. Find the common solution.

STEP 1   Draw the graph of $-2x + 8y = 24$. First, solve for y.

$$-2x + 8y = 24$$
$$-2x + 8y + 2x = 24 + 2x \qquad \text{Add } 2x \text{ to both sides.}$$
$$8y = 24 + 2x$$
$$\frac{8y}{8} = \frac{24 + 2x}{8} \qquad \text{Divide both sides by 8.}$$
$$y = \frac{24 + 2x}{8}$$

Then find three solutions.

x	$\dfrac{24 + 2x}{8}$	y
−4	$\dfrac{24 + 2(-4)}{8} = \dfrac{16}{8}$	2
0	$\dfrac{24 + 2(0)}{8} = \dfrac{24}{8}$	3
8	$\dfrac{24 + 2(8)}{8} = \dfrac{40}{8}$	5

Three solutions are $(-4, 2)$, $(0, 3)$, and $(8, 5)$. Plot the three points that correspond to these three ordered pairs. Then draw a straight line through these three points.

STEP 2 Draw the graph of $-3x + 12y = 36$. First, solve for y.

$$-3x + 12y = 36$$
$$-3x + 12y + 3x = 36 + 3x \qquad \text{Add } 3x \text{ to both sides.}$$
$$12y = 36 + 3x$$
$$\frac{12y}{12} = \frac{36 + 3x}{12} \qquad \text{Divide both sides by 12.}$$
$$y = \frac{36 + 3x}{12}$$

Then find three solutions.

x	$\dfrac{36 + 3x}{12}$	y
4	$\dfrac{36 + 3(4)}{12} = \dfrac{48}{12}$	4
-8	$\dfrac{36 + 3(-8)}{12} = \dfrac{12}{12}$	1
-5	$\dfrac{36 + 3(-5)}{12} = \dfrac{21}{12}$	$\dfrac{7}{4}$ or $1\dfrac{3}{4}$

Three solutions are $(4, 4), (-8, 1), \left(-5, 1\dfrac{3}{4}\right)$. Plot the three points that correspond to these ordered pairs in the same number plane as the points found in Step 1. Then draw a straight line through these points.

Note that the lines coincide. Any solution of one equation is also a solution of the other. Hence, there are infinitely many points of intersection and infinitely many common solutions (see Figure 9.4). The solutions are the coordinates of the points on either line.

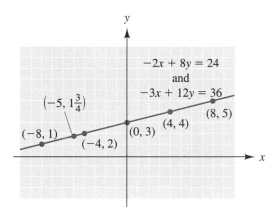

FIGURE 9.4

• **EXAMPLE 4**   The sum of two electric currents is 12 A. One current is three times the other. Find the two currents graphically.

Let x = first current
y = second current

The equations are then: $x + y = 12$

$$y = 3x$$

Draw the graph of each equation, as shown in Figure 9.5.

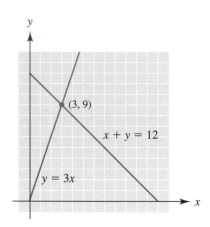

FIGURE 9.5

As you can see from Figure 9.5, the point of intersection is $(3, 9)$. Thus, $x = 3$ and $y = 9$ is the common solution; the currents are 3 A and 9 A. ────────●

Exercises | 9.1

Draw the graphs of each pair of linear equations. Find the point of intersection. If the lines do not intersect, tell whether the lines are parallel or coincide:

1. $y = 3x$
$y = x + 4$

2. $x - y = 2$
$x + 3y = 6$

3. $y = -x$
$y - x = 2$

4. $x + y = 3$
$2x + 2y = 6$

5. $x - 3y = 6$
$2x - 6y = 18$

6. $x + y = 4$
$2x + y = 5$

7. $2x - 4y = 8$
$3x - 6y = 12$

8. $4x - 3y = 7$
$6x + 5y = 8$

9. $3x - 6y = 12$
$4x - 8y = 12$

10. $2x + y = 6$
$2x - y = 6$

11. $3x + 2y = 10$
$2x - 3y = 11$

12. $5x - y = 10$
$x - 3y = -12$

13. $5x + 8y = -58$
$2x + 2y = -18$

14. $6x + 2y = 24$
$3x - 4y = 12$

15. $3x + 2y = 17$
$x = 3$

16. $5x - 4y = 28$
$y = -2$

17. $y = 2x$
$y = -x + 2$

18. $y = -5$
$y = x + 3$

19. $2x + y = 6$
$y = -2x + 1$

20. $3x + y = -5$
$2x + 5y = 1$

21. $4x + 3y = 2$
$5x - y = 12$

22. $4x - 6y = 10$
$2x - 3y = 5$

23. $2x - y = 9$
$-2x + 3y = -11$

24. $x - y = 5$
$2x - 3y = 5$

25. $8x - 3y = 0$
$4x + 3y = 0$

26. $2x + 8y = 9$
$4x + 4y = 3$

Solve Exercises 27–30 graphically:

27. The sum of two resistances is 14 Ω. Their difference is 6 Ω. Find the two resistances. If we let R_1 and R_2 be the two resistances, the equations are

$$R_1 + R_2 = 14$$
$$R_1 - R_2 = 6$$

28. A board 36 in. long is cut into two pieces so that one piece is 8 in. longer than the other. Find the

length of each piece. If we let x and y be the two lengths, the equations are

$$x + y = 36$$
$$y = x + 8$$

29. In a concrete mix, there is four times as much gravel as concrete. The total volume is 20 ft³. How much of each is in the mix? If

x = the amount of concrete

y = the amount of gravel,

the equations are

$$y = 4x$$
$$x + y = 20$$

30. An electric circuit containing two currents may be expressed by the equations

$$3i_1 + 4i_2 = 15$$
$$5i_1 - 2i_2 = -1$$

where i_1 and i_2 are the currents in microamperes (μA). Find the two currents.

 ## 9.2 Solving Pairs of Linear Equations by Addition

Exercise 13 in Section 9.1 clearly illustrates that solving linear equations by graphing gives only approximate solutions. To obtain an exact solution, use the *addition method*, outlined below.

> **SOLVING A PAIR OF LINEAR EQUATIONS BY THE ADDITION METHOD**
>
> **STEP 1** If necessary, multiply both sides of one or both equations by a number (or numbers) so that the numerical coefficients of one of the variables are negatives of each other.
>
> **STEP 2** Add the two equations from Step 1 to obtain an equation containing one variable.
>
> **STEP 3** Solve the equation from Step 2 for the one remaining variable.
>
> **STEP 4** Solve for the second variable by substituting the solution from Step 3 in either of the original equations.
>
> **STEP 5** Check your solution by substituting the ordered pair in the original equation not chosen in Step 4.

• **EXAMPLE 1** Solve the following pair of linear equations by addition. Check your solution.

$$2x - y = 6$$
$$x + y = 9$$

Step 1 of the preceding rules is unnecessary, since you can eliminate the y variable by adding the two equations as they are.

$$
\begin{array}{r}
2x - y = 6 \\
\underline{x + y = 9} \\
3x + 0 = 15 \\
3x \quad = 15 \\
x \quad = 5
\end{array}
$$

Now substitute 5 for x in either of the original equations to solve for y. (Choose the simpler equation to make the arithmetic easier.)

$$x + y = 9$$
$$5 + y = 9$$
$$y = 4$$

The solution should be $(5, 4)$.

CHECK   Use the remaining original equation, $2x - y = 6$, and substitute 5 for x and 4 for y.

$$2x - y = 6$$
$$2(5) - 4 = 6 \quad ?$$
$$10 - 4 = 6 \quad \text{True}$$

The solution checks. Thus, the solution is $(5, 4)$.

• **EXAMPLE 2**   Solve the following pair of linear equations by addition. Check your solution.

$$2x + y = 5$$
$$x + y = 4$$

First, multiply both sides of the second equation by -1 to eliminate y by addition.

$$\begin{array}{l} 2x + y = 5 \\ \underline{(-1)(x + y) = (-1)(4)} \end{array} \quad \text{or} \quad \begin{array}{l} 2x + y = 5 \\ \underline{-x - y = -4} \\ x = 1 \end{array}$$

Now substitute 1 for x in the equation $x + y = 4$ to solve for y.

$$x + y = 4$$
$$1 + y = 4$$
$$y = 3$$

The solution is $(1, 3)$. The check is left to the student.

• **EXAMPLE 3**   Solve the following pair of linear equations by addition. Check your solution.

$$4x + 2y = 2$$
$$3x - 4y = 18$$

Multiply both sides of the first equation by 2 to eliminate y by addition.

$$\begin{array}{l} 2(4x + 2y) = 2(2) \\ \underline{3x - 4y = 18} \end{array} \quad \text{or} \quad \begin{array}{l} 8x + 4y = 4 \\ \underline{3x - 4y = 18} \\ 11x = 22 \\ x = 2 \end{array}$$

Now substitute 2 for x in the equation $4x + 2y = 2$ to solve for y.

$$4x + 2y = 2$$
$$4(2) + 2y = 2$$
$$8 + 2y = 2$$
$$2y = -6$$
$$y = -3$$

The solution should be $(2, -3)$.

CHECK   Use the equation $3x - 4y = 18$.

$$3x - 4y = 18$$
$$3(2) - 4(-3) = 18 \quad ?$$
$$6 + 12 = 18 \quad \text{True}$$

The solution checks. Thus, the solution is $(2, -3)$.

• EXAMPLE 4   Solve the following pair of linear equations by addition. Check your solution.

$$3x - 4y = 11$$
$$4x - 5y = 14$$

Multiply both sides of the first equation by 4. Then multiply both sides of the second equation by -3 to eliminate x by addition.

$$\begin{array}{l} 4(3x - 4y) = 4(11) \\ -3(4x - 5y) = -3(14) \end{array} \quad \text{or} \quad \begin{array}{r} 12x - 16y = 44 \\ -12x + 15y = -42 \\ \hline 0 - \ 1y = 2 \\ y = -2 \end{array}$$

Now substitute -2 for y in the equation $3x - 4y = 11$ to solve for x.

$$3x - 4y = 11$$
$$3x - 4(-2) = 11$$
$$3x + 8 = 11$$
$$3x = 3$$
$$x = 1$$

The solution is $(1, -2)$.   ————————•

In the preceding examples, we considered only pairs of linear equations with one common solution. Thus, the graphs of these equations intersect at a point, and the ordered pair that names this point is the common solution for the pair of equations. Sometimes when solving a pair of linear equations by addition, the final statement is that two unequal numbers are equal, such as $0 = -2$. If so, the pair of equations does not have a common solution, and the graphs of these equations are parallel lines.

• EXAMPLE 5   Solve the following pair of linear equations by addition.

$$2x + 3y = 7$$
$$4x + 6y = 12$$

Multiply both sides of the first equation by -2 to eliminate x by addition.

$$\begin{array}{l} -2(2x + 3y) = -2(7) \\ 4x + 6y = 12 \end{array} \quad \text{or} \quad \begin{array}{r} -4x - 6y = -14 \\ 4x + 6y = 12 \\ \hline 0 + \ 0 = -2 \\ 0 = -2 \end{array}$$

Since $0 \neq -2$, there is no common solution, and the graphs of these two equations are parallel lines.   ————————•

If addition is used to solve a pair of linear equations and the resulting statement is $0 = 0$, then there are many common solutions. In fact, any solution of one equation is also a solution of the other. In this case, the graphs of the two equations coincide.

• EXAMPLE 6   Solve the following pair of linear equations by addition.

$$2x + \ 5y = 7$$
$$4x + 10y = 14$$

Multiply both sides of the first equation by -2 to eliminate x by addition.

$$\begin{array}{c} -2(2x + 5y) = -2(7) \\ 4x + 10y = 14 \end{array} \quad \text{or} \quad \begin{array}{r} -4x - 10y = -14 \\ 4x + 10y = 14 \\ \hline 0 + 0 = 0 \\ 0 = 0 \end{array}$$

Since $0 = 0$, there are many common solutions, and the graphs of the two equations coincide.

Note: If you multiply both sides of the first equation by 2, you obtain the second equation. Thus, the two equations are equivalent. If two equations are equivalent, they should have the same graph.

In Section 9.1, we saw that the graphs of two straight lines in the same plane may (a) intersect, (b) be parallel, or (c) coincide as shown in Figure 9.1. When using the addition method to solve a pair of linear equations, one of the same three possibilities occurs, as follows.

ADDITION METHOD POSSIBLE CASES

1. The steps of the addition method result in exactly one ordered pair, such as $x = 2$ and $y = -5$. This ordered pair is the point at which the graphs of the two linear equations intersect.

2. The steps of the addition method result in a false statement, such as $0 = 7$ or $0 = -2$. This means that there is no common solution and that the graphs of the two linear equations are parallel.

3. The steps of the addition method result in a true statement, such as $0 = 0$. This means that there are many common solutions and that the graphs of the two linear equations coincide.

Exercises 9.2

Solve each pair of linear equations by addition. If there is one common solution, give the ordered pair that names the point of intersection. If there is no one common solution, tell whether the lines are parallel or coincide:

1. $3x + y = 7$
   $x - y = 1$

2. $x + y = 8$
   $x - y = 4$

3. $2x + 5y = 18$
   $4x - 5y = 6$

4. $3x - y = 9$
   $2x + y = 6$

5. $-2x + 5y = 39$
   $2x - 3y = -25$

6. $-4x + 2y = 12$
   $-3x - 2y = 9$

7. $x + 3y = 6$
   $x - y = 2$

8. $3x - 2y = 10$
   $3x + 4y = 20$

9. $2x + 5y = 15$
   $7x + 5y = -10$

10. $4x + 5y = -17$
   $4x - y = 13$

11. $5x + 6y = 31$
   $2x + 6y = 16$

12. $6x + 7y = 0$
   $2x - 3y = 32$

13. $4x - 5y = 14$
   $2x + 3y = -4$

14. $6x - 4y = 10$
   $2x + y = 4$

15. $3x - 2y = -11$
   $7x - 10y = -47$

16. $3x + 2y = 10$
   $x + 5y = -27$

17. $x + 2y = -3$
   $2x + y = 9$

18. $5x - 2y = 6$
   $3x - 4y = 12$

19. $3x + 5y = 7$
   $2x - 7y = 15$

20. $12x + 5y = 21$
   $13x + 6y = 21$

21. $8x - 7y = -51$
   $12x + 13y = 41$

22. $5x - 7y = -20$
   $3x - 19y = -12$

23. $5x - 12y = -5$
   $9x - 16y = -2$

24. $2x + 3y = 2$
   $3x - 2y = 3$

25. $2x + 3y = 8$
$x + y = 2$

26. $4x + 7y = 9$
$12x + 21y = 12$

27. $3x - 5y = 7$
$9x - 15y = 21$

28. $2x - 3y = 8$
$4x - 3y = 0$

29. $2x + 5y = -1$
$3x - 2y = 8$

30. $3x - 7y = -9$
$2x + 14y = -6$

31. $16x - 36y = 70$
$4x - 9y = 17$

32. $8x + 12y = 36$
$16x + 15y = 45$

33. $4x + 3y = 17$
$2x - y = -4$

34. $12x + 15y = 36$
$7x - 12y = 187$

35. $2x - 5y = 8$
$4x - 10y = 16$

36. $3x - 2y = 5$
$7x + 3y = 4$

37. $5x - 8y = 10$
$-10x + 16y = 8$

38. $-3x + 2y = 5$
$-30x + y = 12$

39. $8x - 5y = 426$
$7x - 2y = 444$

40. $3x - 10y = -21$
$5x + 4y = 27$

41. $16x + 5y = 6$
$7x + \dfrac{5}{8}y = 2$

42. $\dfrac{1}{4}x - \dfrac{2}{5}y = 1$
$5x - 8y = 20$

43. $7x + 8y = 47$
$5x - 3y = 51$

44. $2x - 5y = 13$
$5x + 7y = 13$

9.3 Applications Involving Pairs of Linear Equations*

Often a technical application can be expressed mathematically as a system of linear equations. The procedure is similar to that outlined in Section 6.6, except that here you need to write two equations that express the information given in the problem and that involve both unknowns.

SOLVING APPLICATIONS INVOLVING EQUATIONS WITH TWO VARIABLES

1. Choose a different variable for each of the two unknowns you are asked to find. Write what each variable represents.
2. Write the problem as two equations using both variables. To obtain these two equations, look for two different relationships that express the two unknown quantities in equation form.
3. Solve this resulting system of equations using the methods given in this chapter.
4. Answer the question or questions asked in the problem.
5. Check your answers using the original problem.

• **EXAMPLE 1**   The sum of two voltages is 120 V. The difference between them is 24 V. Find each voltage.

Let x = large voltage
y = small voltage

The sum of two voltages is 120 V; that is,

$x + y = 120$

The difference between them is 24 V; that is,

$x - y = 24$

* In this chapter, do not use the rules for calculating with measurements.

The system of equations is

$$x + y = 120$$
$$\underline{x - y = 24}$$
$$2x = 144 \qquad \text{Add the equations.}$$
$$x = 72$$

Substitute $x = 72$ in the equation $x + y = 120$ and solve for y.

$$x + y = 120$$
$$72 + y = 120$$
$$y = 48 \qquad \text{Subtract 72 from both sides.}$$

Thus, the voltages are 72 V and 48 V.

CHECK　　The sum of the voltages, 72 V + 48 V, is 120 V. The difference between them, 72 V − 48 V, is 24 V.　　————————•

•　**EXAMPLE 2**　　How many pounds of feed mix A that is 75% corn and how many pounds of feed mix B that is 50% corn will need to be mixed to make a 400-lb mixture that is 65% corn?

Let x = number of pounds of mix A (75% corn)

y = number of pounds of mix B (50% corn)

The sum of the two mixtures is 400 lb; that is,

$$x + y = 400$$

Thus, 75% of x is corn and 50% of y is corn. Adding these amounts together results in a 400-lb final mixture that is 65% corn; that is,

$$0.75x + 0.50y = (0.65)(400)$$
or　$0.75x + 0.50y = 260$

The system of equations is

$$x + y = 400$$
$$0.75x + 0.50y = 260$$

First, let's multiply both sides of the second equation by 100 to eliminate the decimals.

$$x + y = 400$$
$$75x + 50y = 26{,}000$$

Then multiply both sides of the first equation by -50 to eliminate y by addition.

$$-50x - 50y = -20{,}000$$
$$\underline{75x + 50y = 26{,}000}$$
$$25x = 6{,}000$$
$$\frac{25x}{25} = \frac{6000}{25} \qquad \text{Divide both sides by 25.}$$
$$x = 240$$

Now substitute 240 for x in the equation $x + y = 400$ to solve for y.

$$x + y = 400$$
$$240 + y = 400$$
$$240 + y - 240 = 400 - 240$$
$$y = 160$$

Therefore, we need 240 lb of mix A and 160 lb of mix B.

Check   Use the remaining original equation, $0.75x + 0.50y = 260$, and substitute 240 for x and 160 for y.

$$0.75x + 0.50y = 260$$
$$0.75(240) + 0.50(160) = 260 \quad ?$$
$$180 + 80 = 260 \quad \text{True}$$

• **Example 3**   A company sells two grades of sand. One grade sells for 15¢/lb, and the other for 25¢/lb. How much of each grade needs to be mixed to obtain 1000 lb of a mixture worth 18¢/lb?

Let x = amount of sand selling at 15¢/lb

y = amount of sand selling at 25¢/lb

The total amount of sand is 1000 lb; that is,

$$x + y = 1000$$

One grade sells at 15¢/lb and the other sells at 25¢/lb. The two grades are mixed to obtain 1000 lb of a mixture worth 18¢/lb. Here we need to write an equation that relates the cost of the sand; that is, the cost of the sand separately equals the cost of the sand mixed.

$$15x + 25y = 18(1000)$$

That is, the cost of x pounds of sand at 15¢/lb is $15x$ cents. The cost of y pounds of sand at 25¢/lb is $25y$ cents. The cost of 1000 pounds of sand at 18¢/lb is $18(1000)$ cents. Therefore, the system of equations is

$$x + y = 1000$$
$$15x + 25y = 18{,}000$$

Multiply the first equation by -15 to eliminate x by addition.

$$\begin{array}{r} -15x - 15y = -15{,}000 \\ 15x + 25y = 18{,}000 \\ \hline 10y = 3000 \\ y = 300 \end{array}$$

Substitute $y = 300$ in the equation $x + y = 1000$ and solve for x.

$$x + y = 1000$$
$$x + 300 = 1000$$
$$x = 700$$

That is, 700 lb of sand selling at 15¢/lb and 300 lb of sand selling at 25¢/lb are needed to obtain 1000 lb of a mixture worth 18¢/lb.

Check   Left to the student.

Exercises | 9.3

1. A board 96 cm long is cut into two pieces so that one piece is 12 cm longer than the other. Find the length of each piece.

 2. Find the capacity of two trucks if 6 trips of the smaller and 4 trips of the larger make a total

haul of 36 tons, and 8 trips of the larger and 4 trips of the smaller make a total haul of 48 tons.

3. A plumbing contractor decides to field-test two new pumps. One is rated at 180 gal/h and the other at 250 gal/h. She tests one, then the other.

Over a period of 6 h, she pumps a total of 1325 gal. Assume both pumps operated as rated. How long is each in operation?

4. A bricklayer lays an average of 150 bricks per hour. During the job, he is called away and replaced by a less experienced man, who averages 120 bricks an hour. The two men laid 930 bricks in 7 h. How long did each work?

5. In a concrete mix, there is four times as much gravel as cement. The total volume of gravel and cement in the mix is 22.5 ft^3. How much of each is in the mix?

6. A contractor finds a bill for $225 for 720 ceiling tiles. She knows that there were two types of tiles used; one selling at 25¢ a tile and the other at 40¢ a tile. How many of each type were used?

7. A farmer has two types of feed. One has 5% digestible protein and the other 15% digestible protein. How much of each type will she need to mix 100 lb of 12%-digestible-protein feed?

8. A dairyman wants to make 125 lb of 12%-butterfat cream. How many pounds of 40%-butterfat cream and how many pounds of 2%-butterfat milk would he have to mix?

9. A farmer sells corn for $3.20/bu and soybeans for $5.80/bu. The entire 3150 bu brings her $11,250. How much of each does she sell?

10. A farmer has a 1.4% solution and a 2.9% solution of a pesticide. How much of each would he mix to get 2000 gal of 2% solution for his sprayer?

11. A farmer has a 6% solution and a 12% solution of pesticide. How much of each must she mix to have 300 gal of an 8% solution for her sprayer?

12. The sum of two capacitors is 85 microfarads (μF). The difference between them is 25 μF. Find the size of each capacitor.

13. Nine batteries are hooked in series to provide a 33-V power source. Some of the batteries are 3 V and some are 4.5 V. How many of each type are used?

14. In a parallel circuit, the total current is 1.25 A through the two branches. One branch has a resistance of 50 Ω, and the other has a resistance of 200 Ω. What current is flowing through each branch? *Note:* In a parallel circuit, the products of the current in amperes and the resistance in ohms are equal in all branches.

15. How much of an 8% solution and a 12% solution would you use to make 140 mL of a 9%-electrolyte solution?

16. The total current in a parallel circuit with seven branches is 1.95 A. Some of the branches have currents of 0.25 A and others 0.35 A. How many of each type of branch are in the circuit? *Note:* The total current in a parallel circuit equals the sum of the currents in each branch.

17. A small single-cylinder engine was operated on a test stand for 14 min. It was run first at 850 rpm, and then increased to 1250 rpm. A total of 15,500 revolutions was counted during the test. How long was the engine operated at each speed?

18. In testing a hybrid engine, various mixtures of gasoline and methanol are being tried. How much of a 95%-gasoline mixture and how much of an 80%-gasoline mixture would be needed to make 240 gal of a 90%-gasoline mixture?

19. An engine on a test stand was operated at two fixed settings, each with an appropriate load. At the first setting, fuel consumption was 1 gal every 12 min. At the second setting, it was 1 gal every 15 min. The test took 5 h, and 22 gal of fuel were used. How long did the engine run at each setting?

20. A mechanic stores a parts cleaner as a 65% solution, which is to be diluted to a 25% solution for use. Someone accidentally prepares a 15% solution. How much of the 65% solution and the 15% solution should be mixed to make 100 gal of the 25% solution?

21. Amy has a 3% solution and an 8% solution of a pesticide. How much of each must she mix to have 200 L of 4% solution?

22. A butcher has some 65%-lean hamburger and some 85%-lean hamburger. How much of each should be used to make 160 lb of 80%-lean hamburger?

23. When three identical compressors and five air-handling units are in operation, a total of 26.4 A are needed. When only two compressors and three air-handler units are being used, the current requirement is 17.2 A. How many amps are required **a.** by each compressor and **b.** by each fan?

24. A hospital has 35%-saline solution on hand. How much water and how much of this solution should be used to prepare 140 mL of a 20%-saline solution?

 **25.** A nurse gives 1000 cm^3 of an intravenous (I.V.) solution over a period of 8 h. At first it is given at a rate of 140 cm^3/h, then at a reduced rate of 100 cm^3/h. How long should it be given at each rate?

26. A hospital has a 4%-saline solution and an 8%-saline solution on hand. How much of each should be used to prepare 1000 cm^3 of 5%-saline solution?

27. A medication is available in 2-cm^3 vials and in 5-cm^3 vials. In a certain month, 42 vials were used, totaling 117 cm^3 of medication. How many of each type of vial were used?

28. A salesman turns in a ticket on two carpets for $2360. He sold a total of 75 yd^2 of carpet. One type was worth $27.50/yd^2, and the second was worth $36/yd^2. He neglects to note, however, how much of each type he sold. How much did he sell of each type?

29. An apartment owner rents one-bedroom apartments for $525 and two-bedroom apartments for $600. A total of 13 apartments rent for $7500 a month. How many of each type does she have?

30. A sporting goods store carries two types of snorkels. One sells for $14.95 and the other for $21.75. Records for July show that 23 snorkels were sold, for $357.45. How many of each type were sold?

9.4 Solving Pairs of Linear Equations by Substitution

For many problems, the *substitution* method is easier than the addition method for finding exact solutions. Use the substitution method when one or both equations has one variable alone as one member.

> **SOLVING A PAIR OF LINEAR EQUATIONS BY THE SUBSTITUTION METHOD**
>
> 1. From either of the two given equations, solve for one variable in terms of the other.
> 2. Substitute the result from Step 1 into the remaining equation. Note that this step eliminates one variable.
> 3. Solve the equation from Step 2 for the remaining variable.
> 4. Solve for the second variable by substituting the solution from Step 3 into the equation resulting from Step 1.
> 5. Check your solution by substituting the ordered pair in the original equation not used in Step 1.

• **EXAMPLE 1**   Solve the following pair of linear equations by substitution. Check your solution.

$$x + 3y = 15$$
$$x = 2y$$

First, substitute $2y$ for x in the first equation.

$$x + 3y = 15$$
$$2y + 3y = 15$$
$$5y = 15$$
$$y = 3$$

Now substitute 3 for y in the equation $x = 2y$ to solve for x.

$$x = 2y$$
$$x = 2(3)$$
$$x = 6$$

The solution should be $(6, 3)$.

CHECK   Use the equation $x + 3y = 15$. Substitute 6 for x and 3 for y.

$$x + 3y = 15$$
$$6 + 3(3) = 15 \quad ?$$
$$6 + 9 = 15 \quad \text{True}$$

Thus, the solution is $(6, 3)$.

The addition method is often preferred if the pair of linear equations has no numerical coefficients equal to 1. For example,

$$3x + 4y = 7$$
$$5x + 7y = 12$$

Knowing both the addition and substitution methods, you may choose the one that seems easier to you for each problem.

• EXAMPLE 2   Enclose a rectangular yard (Figure 9.6) with a fence so that the length is twice the width. The length of the 80-ft house is used to enclose part of one side of the yard. If 580 ft of fencing are used, what are the dimensions of the yard?

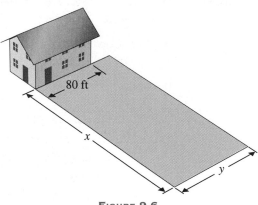

FIGURE 9.6

Let x = length of the yard
y = width of the yard

The amount of fencing used (two lengths plus two widths minus the length of the house) is 580 ft; that is,

$$2x + 2y - 80 = 580$$

or

$$2x + 2y = 660$$

The length of the yard is twice the width; that is,

$$x = 2y$$

The system of equations is

$$2x + 2y = 660$$
$$x = 2y$$

Substitute $2y$ for x in the equation $2x + 2y = 660$ and solve for y.

$$2(2y) + 2y = 660$$
$$4y + 2y = 660$$
$$6y = 660$$
$$y = 110$$

Now substitute 110 for y in the equation $x = 2y$ and solve for x.

$$x = 2y$$
$$x = 2(110)$$
$$x = 220$$

Therefore, the length is 220 ft and the width is 110 ft. (The check is left to the student.)

Exercises 9.4

Solve, using the substitution method, and check:

1. $2x + y = 12$
  $y = 3x$

2. $3x + 4y = -8$
  $x = 2y$

3. $5x - 2y = 46$
  $x = 5y$

4. $2x - y = 4$
  $y = -x$

5. $3x + 2y = 30$
  $x = y$

6. $3x - 2y = 49$
  $y = -2x$

7. $5x - y = 18$
  $y = \frac{1}{2}x$

8. $15x + 3y = 9$
  $y = -2x$

9. $x - 6y = 3$
  $3y = x$

10. $4x + 5y = 10$
  $4x = -10y$

11. $3x + y = 7$
  $4x - y = 0$

12. $5x + 2y = 1$
  $y = -3x$

13. $4x + 3y = -2$
  $x + y = 0$

14. $7x + 8y + 93 = 0$
  $y = 3x$

15. $6x - 8y = 115$
  $x = -\dfrac{y}{5}$

16. $2x + 8y = 12$
  $x = -4y$

17. $3x + 8y = 27$
  $y = 2x + 1$

18. $4x - 5y = -40$
  $x = 3 - 2y$

19. $8y - 2x = -34$
  $x = 1 - 4y$

20. $2y + 7x = 48$
  $y = 3x - 2$

21. The sum of two resistors is 550 Ω. One is 4.5 times the other. Find the size of each resistor.

22. One concrete mix contains four times as much gravel as cement. The total volume is 15 yd³. How much of each ingredient is used?

23. A wire 120 cm long is to be cut into two pieces so that one piece is three times as long as the other. Find the length of each piece.

24. The sum of two resistances is 1500 Ω. The larger is four times the smaller. Find the size of each resistance.

25. A rectangle is twice as long as it is wide. Its perimeter is 240 cm. Find the length and the width of the rectangle.

26. The sum of two voltages is 84 V. One voltage is 12 V larger than the other. Find each voltage.

27. The sum of three currents is 210 mA. Two currents are the same. The third is five times either of the other two. Find the third current.

28. Together Brenda and George earn $480 in completing a drafting project. George receives

$\frac{2}{3}$ of what Brenda does. How much does each receive?

29. The center-to-center distance between a fan and a motor shaft is 30.0 in. See Illustration 1. Pulleys with a 4.5 : 1 ratio are installed. The distance between the pulleys is 19.0 in. Find the diameter of each pulley.

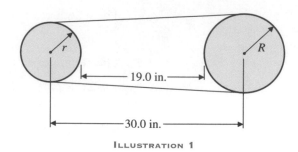

ILLUSTRATION 1

Chapter 9   Review

Draw the graphs of each pair of linear equations on the same set of coordinate axes. Find the point of intersection. If the lines do not intersect, tell whether the lines are parallel or coincide:

1. $x + y = 6$
   $2x - y = 3$

2. $y = 2x + 5$
   $y = x + 2$

3. $4x + 6y = 12$
   $6x + 9y = 18$

4. $5x - 2y = 10$
   $10x - 4y = -20$

5. $3x + 4y = -1$
   $x = -3$

6. $y = 2x$
   $y = -5$

Solve each system of equations:

7. $x + y = 7$
   $2x - y = 2$

8. $3x + 2y = 11$
   $x + 2y = 5$

9. $3x - 5y = -3$
   $2x - 3y = -1$

10. $2x - 3y = 1$
    $4x - 6y = 5$

11. $3x + 5y = 8$
    $6x - 4y = 44$

12. $5x + 7y = 22$
    $4x + 8y = 20$

13. $x + 2y = 3$
    $3x + 6y = 9$

14. $3x + 5y = 52$
    $y = 2x$

15. $5y - 4x = -6$
    $x = \frac{1}{2}y$

16. $3x - 7y = -69$
    $y = 4x + 5$

17. You can buy 20 resistors and 8 capacitors for $3.60, or 60 resistors and 40 capacitors for $14. Find the price of each.

18. The sum of the length and width of a rectangular lot is 190 ft. The lot is 75 ft longer than it is wide. Find the length and width of the lot.

19. The sum of two inductors is 90 millihenrys (mH). The larger is 3.5 times the smaller. What is the size of each inductor?

20. The sum of two lengths is 90 ft, and their difference is 20 ft. Find the two lengths.

Chapter 9   Test

Solve each system of equations by the method indicated.

1. $3x - y = 5$
   $2x - y = 0$ by graphing

2. $y = 3x - 5$
   $y = 2x - 1$ by graphing

3. $2x + 7y = -1$
   $x + 2y = 1$ by addition

4. $x - 3y = 8$
   $x + 4y = -6$ by addition

5. $y = -3x$
   $2x + 3y = 13$ by substitution

6. $x = 7y$
   $2x - 8y = 12$ by substitution

Solve each pair of linear equations. If there is one common solution, give the ordered pair that names the point of intersection. If there is no one common solution, tell whether the lines are parallel or coincide.

7. $4x - 5y = 10$
$-8x + 10y = 6$

8. $3x - y = 8$
$12x - 4y = 32$

9. $x - 3y = -8$
$2x + y = 5$

10. The perimeter of a rectangular lot is 600 m. The length is twice the width. Find the length and the width.

11. The sum of two resistances is 550 Ω. The difference between them is 250 Ω. Find the size of each resistance.

Chapters 1–9   Cumulative Review

1. Evaluate $2(6 - 5) + 3$.

2. Subtract: $+6$
          $\underline{-9}$

3. Find missing dimensions a and b in the figure in Illustration 1.

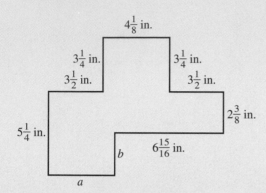

ILLUSTRATION 1

9. Use the rules for multiplication of measurements to evaluate: $1.8 \text{ m} \times 61.2 \text{ m} \times 3.2 \text{ m}$

10. Read the voltmeter scale in Illustration 3.

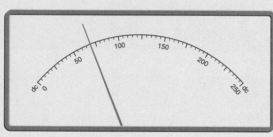

ILLUSTRATION 3

4. Add: $1.2 + 201 + 0.146 + 20.62$

5. Multiply: $(6.2 \times 10^{-3})(1.8 \times 10^{5})$

6. 61 mm = _____ m

7. Give the number of significant digits: 356,760 kg

8. Read the measurement shown on the metric micrometer in Illustration 2.

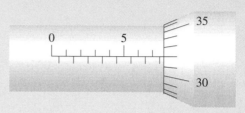

ILLUSTRATION 2

11. Simplify: $(4x - 5) - (6 - 3x)$

12. Simplify: $(-5xy^2)(8x^3y^2)$

13. Simplify: $2x(4x - 3y)$

14. Solve: $3(x - 2) + 4(3 - 2x) = 9$

15. Solve: $s = \dfrac{2V + t}{3}$ for V

Write each ratio in lowest terms:

16. 5 to 65

17. 32 in. : 3 yd

Solve each proportion (round each result to three significant digits):

18. $\dfrac{5}{13} = \dfrac{x}{156}$

19. $\dfrac{29.1}{73.8} = \dfrac{x}{104}$

20. $\dfrac{286}{x} = \dfrac{11.8}{59.7}$

21. If it costs $28.50 to repair 5 ft^2 of sidewalk, how much would it cost to repair 18 ft^2?

22. A map shows a scale of 1 in. = 40 mi. What distance is represented by $4\frac{1}{4}$ in. on the map?

23. A large gear with 16 teeth revolves at 40 rpm. It turns a small gear at 64 rpm. How many teeth does the smaller gear have?

24. Complete the ordered pair solutions of the equation:

$2x + 3y = 12$ $(3, \quad), (0, \quad), (-3, \quad)$

Solve for y:

25. $3x - y = 5$

26. $4x + 2y = 7$

27. Draw the graph of $3x - 2y = 12$.

28. Draw the graphs of $2x - y = 4$ and $x + 3y = -5$ on the same set of coordinate axes. What is the point of intersection?

Solve each system of equations:

29. $x - y = 6$
$3x + y = 2$

30. $3x - 5y = 7$
$-6x + 10y = 5$

Factoring Algebraic Expressions

AUTOMOTIVE INDUSTRY
*A rivet machine operator works
on an air bag.*

MATHEMATICS AT WORK

The concept of factoring enables us to simplify and solve a quadratic equation quickly or reduce a complicated algebraic expression into simpler terms. Engineering and science use factoring more than other professions, as they work with quadratic and higher-order degree algebraic expressions on a daily basis.

Factoring is also used extensively in calculus, in physics, and in differential equations. A sound understanding of factoring is essential to achieving a degree in any field involving mathematics or science.

10.1 Finding Monomial Factors

Factoring an algebraic expression, like finding the prime factors of a number, means *writing the expression as a product*. The prime factorization of 12 is $2 \cdot 2 \cdot 3$. Other factorizations of 12 are $2 \cdot 6$ and $4 \cdot 3$.

To factor the expression $2x + 2y$, notice that 2 is a factor common to both terms of the expression. In other words, 2 is a factor of $2x + 2y$. To find the other factor, divide by 2.

$$\frac{2x + 2y}{2} = \frac{2x}{2} + \frac{2y}{2} = x + y$$

Therefore, a factorization of $2x + 2y$ is $2(x + y)$.

When a factor such as 2 divides *each* term of an algebraic expression, it is called a *monomial factor*. (Recall that a monomial is an expression that has only one term.) When factoring any algebraic expression, *always look first for monomial factors* that are common to all terms.

• EXAMPLE 1   Factor $3a + 6b$.

First, look for a common monomial factor. Since 3 divides both $3a$ and $6b$, 3 is a common monomial factor of $3a + 6b$. Divide $3a + 6b$ by 3.

$$\frac{3a + 6b}{3} = \frac{3a}{3} + \frac{6b}{3} = a + 2b$$

Thus, $3a + 6b = 3(a + 2b)$.

Check this result by multiplication: $3(a + 2b) = 3a + 6b$.   ——————•

• EXAMPLE 2   Factor $4x^2 + 8x + 12$.

Since 4 divides each term of the expression, divide $4x^2 + 8x + 12$ by 4 to obtain the other factor.

$$\frac{4x^2 + 8x + 12}{4} = \frac{4x^2}{4} + \frac{8x}{4} + \frac{12}{4}$$
$$= x^2 + 2x + 3$$

Thus, $4x^2 + 8x + 12 = 4(x^2 + 2x + 3)$.
To check your work, multiply $4(x^2 + 2x + 3)$. Your product should be the original expression.

Note: In this example, 2 is also a common factor of each term of the expression. However, 4 is the greatest common factor. *When factoring, always choose the monomial factor that is the greatest common factor.*   ——————•

• EXAMPLE 3   Factor $15ax - 6ay$.

Note that 3 divides both $15ax$ and $6ay$, so 3 is a common factor. However, a also divides $15ax$ and $6ay$, so a is also a common factor. We are looking for the greatest common factor (GCF), which in this case is $3a$. Then we divide $15ax - 6ay$ by $3a$ to obtain the other factor.

$$\frac{15ax - 6ay}{3a} = \frac{15ax}{3a} - \frac{6ay}{3a}$$
$$= 5x - 2y$$

Thus, $15ax - 6ay = 3a(5x - 2y)$.

Note that $3(5ax - 2ay)$ or $a(15x - 6y)$ are also factored forms of $15ax - 6ay$. However, we must always use the monomial factor that is the greatest common factor.

• **EXAMPLE 4**   Factor $15xy^2 - 25x^2y + 10xy$.

The greatest common factor is $5xy$. Dividing each term by $5xy$, we have
$15xy^2 - 25x^2y + 10xy = 5xy(3y - 5x + 2)$.

Exercises | 10.1

Factor:

1. $4a + 4$

2. $3x - 6$

3. $bx + by$

4. $9 - 18y$

5. $15b - 20$

6. $12ab + 30ac$

7. $x^2 - 7x$

8. $3x^2 - 6x$

9. $a^2 - 4a$

10. $7xy - 21y$

11. $4n^2 - 8n$

12. $10x^2 + 5x$

13. $10x^2 + 25x$

14. $y^2 - 8y$

15. $3r^2 - 6r$

16. $x^3 + 13x^2 + 25x$

17. $4x^4 + 8x^3 + 12x^2$

18. $9x^4 - 15x^2 - 18x$

19. $9a^2 - 9ax^2$

20. $a - a^3$

21. $10x + 10y - 10z$

22. $2x^2 - 2x$

23. $3y - 6$

24. $y - 3y^2$

25. $14xy - 7x^2y^2$

26. $25a^2 - 25b^2$

27. $12x^2m - 7m$

28. $90r^2 - 10R^2$

29. $60ax - 12a$

30. $2x^2 - 100x^3$

31. $52m^2n^2 - 13mn$

32. $40x - 8x^3 + 4x^4$

33. $52m^2 - 14m + 2$

34. $27x^3 - 54x$

35. $36y^2 - 18y^3 + 54y^4$

36. $20y^3 - 10y^2 + 5y$

37. $6m^4 - 12m^2 + 3m$

38. $-16x^3 - 32x^2 - 16x$

39. $-4x^2y^3 - 6x^2y^4 - 10x^2y^5$

40. $18x^3y - 30x^4y + 48xy$

41. $3a^2b^2c^2 + 27a^3b^3c^2 - 81abc$

42. $15x^2yz^4 - 20x^3y^2z^2 + 25x^2y^3z^2$

43. $4x^3z^4 - 8x^2y^2z^3 + 12xyz^2$

44. $18a^2b^2c^2 + 24ab^2c^2 - 30a^2c^2$

10.2   Finding the Product of Two Binomials Mentally

In Section 5.5, you learned how to multiply two binomials such as $(2x + 3)(4x - 5)$ by the following method:

$$
\begin{array}{r}
2x + 3 \\
4x - 5 \\
\hline
-10x - 15 \\
8x^2 + 12x \\
\hline
8x^2 + 2x - 15
\end{array}
$$

This process of multiplying two binomials can be shortened by following the three steps below.

FINDING THE PRODUCT OF TWO BINOMIALS MENTALLY

1. The *first term* of the product is the product of the first terms of the binomials.

2. The *middle term* of the product is the sum of the outer product and the inner product of the binomials.

3. The *last term* of the product is the product of the last terms of the binomials.

Let's use the above steps to find the product $(2x + 3)(4x - 5)$.

Step 1 Product of the first terms: $(2x)(4x) = 8x^2$

Step 2 Outer product $= (2x)(-5) = -10x$

$$(2x + 3)(4x - 5)$$

Inner product $= (3)(4x)$ $= \underline{12x}$

Sum: $2x$

Step 3 Product of the last terms: $(3)(-5) = \underline{- 15}$

Therefore, $(2x + 3)(4x - 5)$ $= 8x^2 + 2x - 15$

Note that in each method, we found the exact same terms. The second method is much quicker, especially when you become more familiar and successful with it. The second method is used to factor polynomials. Factoring polynomials is the content of the rest of this chapter and a necessary part of the next chapter. Therefore, it is very important that you learn to find the product of two binomials, mentally, before proceeding with the next section.

This method is often called the *FOIL method*, where *F* refers to the product of the *first* terms, *O* refers to the *outer* product, *I* refers to the *inner* product, and *L* refers to the product of the *last* terms.

• EXAMPLE 1   Find the product $(2x - 7)(3x - 4)$ mentally.

Step 1 Product of the first terms: $(2x)(3x) = 6x^2$

Step 2 Outer product $= (2x)(-4) = -8x$

$$(2x - 7)(3x - 4)$$

Inner product $= (-7)(3x) = \underline{-21x}$

Sum: $-29x$

Step 3 Product of the last terms: $(-7)(-4) = \underline{28}$

Therefore, $(2x - 7)(3x - 4)$ $= 6x^2 - 29x + 28$

• EXAMPLE 2   Find the product $(x + 4)(3x + 5)$ mentally.

Step 1 Product of the first terms: $(x)(3x) = 3x^2$

Step 2 Outer product $= (x)(5)$ $= 5x$

$$(x + 4)(3x + 5)$$

Inner product $= (4)(3x) = \underline{12x}$

Sum: $17x$

Step 3 Product of the last terms: $(4)(5) = \underline{20}$

Therefore, $(x + 4)(3x + 5)$ $= 3x^2 + 17x + 20$

By now, you should be writing only the final result of each product. If you need some help, refer to the three steps at the beginning of this section and the outline shown in Examples 1 and 2.

• **EXAMPLE 3**   Find the product $(x + 8)(x + 5)$ mentally.

$$(x + 8)(x + 5) = x^2 + (5x + 8x) + 40$$
$$= x^2 + 13x + 40$$

• **EXAMPLE 4**   Find the product $(x - 6)(x - 9)$ mentally.

$$(x - 6)(x - 9) = x^2 + (-9x - 6x) + 54$$
$$= x^2 - 15x + 54$$

• **EXAMPLE 5**   Find the product $(x + 2)(x - 5)$ mentally.

$$(x + 2)(x - 5) = x^2 + (-5x + 2x) - 10$$
$$= x^2 - 3x - 10$$

• **EXAMPLE 6**   Find the product $(4x + 1)(5x + 8)$ mentally.

$$(4x + 1)(5x + 8) = 20x^2 + (32x + 5x) + 8$$
$$= 20x^2 + 37x + 8$$

• **EXAMPLE 7**   Find the product $(6x + 5)(2x - 3)$ mentally.

$$(6x + 5)(2x - 3) = 12x^2 + (-18x + 10x) - 15$$
$$= 12x^2 - 8x - 15$$

• **EXAMPLE 8**   Find the product $(4x - 5)(4x - 5)$ mentally.

$$(4x - 5)(4x - 5) = 16x^2 + (-20x - 20x) + 25$$
$$= 16x^2 - 40x + 25$$

Exercises | 10.2

Find each product mentally.

1. $(x + 5)(x + 2)$
2. $(x + 3)(2x + 7)$
3. $(2x + 3)(3x + 4)$
4. $(x + 3)(x + 18)$
5. $(x - 5)(x - 6)$
6. $(x - 9)(x - 8)$
7. $(x - 12)(x - 2)$
8. $(x - 9)(x - 4)$
9. $(x + 8)(2x + 3)$
10. $(3x - 7)(2x - 5)$

11. $(x + 6)(x - 2)$
12. $(x - 7)(x - 3)$
13. $(x - 9)(x - 10)$
14. $(x - 9)(x + 10)$
15. $(x - 12)(x + 6)$
16. $(2x + 7)(4x - 5)$
17. $(2x - 7)(4x + 5)$
18. $(2x - 5)(4x + 7)$
19. $(2x + 5)(4x - 7)$
20. $(6x + 5)(5x - 1)$

21. $(7x + 3)(2x + 5)$
22. $(5x - 7)(2x + 1)$
23. $(x - 9)(3x + 8)$
24. $(x - 8)(2x + 9)$
25. $(6x + 5)(x + 7)$
26. $(16x + 3)(x - 1)$
27. $(13x - 4)(13x - 4)$
28. $(12x + 1)(12x + 5)$
29. $(10x + 7)(12x - 3)$
30. $(10x - 7)(12x + 3)$

31. $(10x - 7)(10x - 3)$
32. $(10x + 7)(10x + 3)$
33. $(2x - 3)(2x - 5)$
34. $(2x + 3)(2x + 5)$
35. $(2x - 3)(2x + 5)$
36. $(2x + 3)(2x - 5)$
37. $(3x - 8)(2x + 7)$
38. $(3x + 8)(2x - 7)$
39. $(3x + 8)(2x + 7)$
40. $(3x - 8)(2x - 7)$

41. $(8x - 5)(2x + 3)$
42. $(x - 7)(x + 5)$
43. $(y - 7)(2y + 3)$
44. $(m - 9)(m + 2)$
45. $(3n - 6y)(2n + 5y)$
46. $(6a - b)(2a + 3b)$

47. $(4x - y)(2x + 7y)$
48. $(8x - 12)(2x + 3)$
49. $\left(\dfrac{1}{2}x - 8\right)\left(\dfrac{1}{4}x - 6\right)$
50. $\left(\dfrac{2}{3}x - 6\right)\left(\dfrac{1}{3}x + 9\right)$

10.3 Finding Binomial Factors

The factors of a trinomial are often *binomial factors*. To find these binomial factors, you must "undo" the process of multiplication as presented in Section 10.2. The following steps will enable you to undo the multiplication in a trinomial such as $x^2 + 7x + 10$.

STEP 1 Factor any common monomial. In the expression $x^2 + 7x + 10$ there is no common factor.

STEP 2 If $x^2 + 7x + 10$ can be factored, the two factors will probably be binomials. Write parentheses for the binomials.

$$x^2 + 7x + 10 = (\qquad)(\qquad)$$

STEP 3 The product of the first two terms of the binomials is the first term of the trinomial. So the first term in each binomial must be x.

$$x^2 + 7x + 10 = (x\qquad)(x\qquad)$$

STEP 4 Here, all the signs of the trinomial are positive, so the signs in the binomials are also positive.

$$x^2 + 7x + 10 = (x +\qquad)(x +\qquad)$$

STEP 5 Find the last terms of the binomials by finding two numbers that have a product of $+10$ and a sum of $+7$. The only possible factorizations of 10 are $1 \cdot 10$ and $2 \cdot 5$. The sums of the pairs of factors are $1 + 10 = 11$ and $2 + 5 = 7$. Thus, the numbers you want are 2 and 5.

$$x^2 + 7x + 10 = (x + 2)(x + 5)$$

STEP 6 Multiply the two binomials as a check, to see if their product is the same as the original trinomial.

• **EXAMPLE 1** Factor the trinomial $x^2 + 15x + 56$.

STEP 1 $x^2 + 15x + 56$ has no common monomial factor.
STEP 2 $x^2 + 15x + 56 = (\qquad)(\qquad)$
STEP 3 $x^2 + 15x + 56 = (x\qquad)(x\qquad)$
STEP 4 $x^2 + 15x + 56 = (x +\qquad)(x +\qquad)$ All the signs of the trinomial are positive.

To determine which factors of 56 to use, list all possible pairs.

$$1 \cdot 56 = 56 \qquad 1 + 56 = 57$$
$$2 \cdot 28 = 56 \qquad 2 + 28 = 30$$
$$4 \cdot 14 = 56 \qquad 4 + 14 = 18$$
$$7 \cdot \ 8 = 56 \qquad 7 + \ 8 = 15$$

Since the coefficient of x in the trinomial is 15, choose 7 and 8 for the second term of the binomial factors. There are no other pairs of positive whole numbers with a product of 56 and a sum of 15.

STEP 5 $x^2 + 15x + 56 = (x + 7)(x + 8)$

In actual work, all the above five steps are completed in one or two lines, depending on whether or not there is a common monomial.

STEP 6 *Check:* $(x + 7)(x + 8) = x^2 + 15x + 56$

• **EXAMPLE 2**   Factor the trinomial $x^2 - 13x + 36$.

Note that the only difference between this trinomial and the ones we have considered previously is the sign of the second term. Here, the sign of the second term is negative instead of positive. Thus, the steps for factoring will be the same except for Step 4.

STEP 1 $x^2 - 13x + 36$ has no common monomial factor.

STEP 2 $x^2 - 13x + 36 = x^2 + (-13)x + 36$
$$= (\quad)(\quad)$$

STEP 3 $x^2 + (-13)x + 36 = (x\quad)(x\quad)$

Note that the sign of the third term ($+36$) is positive and the coefficient of the second term ($-13x$) is negative. Since 36 is positive, the two factors of $+36$ must have like signs; and since the coefficient of $-13x$ is negative, the signs in the two factors must both be negative.

STEP 4 $x^2 + (-13)x + 36 = (x -\quad)(x -\quad)$

Find two integers whose product is 36 and whose sum is -13. Since $(-4)(-9) = 36$ and $(-4) + (-9) = -13$, these are the factors of 36 to be used.

STEP 5 $x^2 - 13x + 36 \qquad = (x - 4)(x - 9)$

STEP 6   *Check:* $(x - 4)(x - 9) = x^2 - 13x + 36$ ———————•

• **EXAMPLE 3**   Factor the trinomial $3x^2 + 12x - 36$.

STEP 1 $3x^2 + 12x - 36 = 3[x^2 + 4x - 12]$ 3 is a common factor.

STEP 2 $3[x^2 + 4x + (-12)] = 3[(\quad)(\quad)]$

STEP 3 $3[x^2 + 4x + (-12)] = 3[(x\quad)(x\quad)]$

Note that the last term of the trinomial (-12) is negative. This means that the two factors of -12 must have unlike signs since a positive number times a negative number gives a negative number.

STEP 4 $3[x^2 + 4x - 12] = 3[(x +\quad)(x -\quad)]$

Find two integers with a product of -12 and a sum of $+4$. All possible pairs of factors are shown below.

$$(-12)(+1) = -12 \qquad (-12) + (+1) = -11$$
$$(+12)(-1) = -12 \qquad (+12) + (-1) = 11$$
$$(+6)(-2) = -12 \qquad (+6) + (-2) = 4$$
$$(-6)(+2) = -12 \qquad (-6) + (+2) = -4$$
$$(-4)(+3) = -12 \qquad (-4) + (+3) = -1$$
$$(+4)(-3) = -12 \qquad (+4) + (-3) = 1$$

From these possibilities, you see that the two integers with a product of -12 and a sum of $+4$ are $+6$ and -2. Write these numbers as the last terms of the binomials.

STEP 5 $3[x^2 + 4x - 12] = 3(x + 6)(x - 2)$

STEP 6   *Check:* $3(x + 6)(x - 2) = 3[x^2 + 4x - 12]$
$$= 3x^2 + 12x - 36$$ ———————•

• **EXAMPLE 4**   Factor the trinomial $x^2 - 11x - 12$.

The signs of the factors of -12 must be different. From the list in Example 3, choose the two factors with a sum of -11.

$$x^2 - 11x - 12 = (x - 12)(x + 1)$$

CHECK $(x - 12)(x + 1) = x^2 - 11x - 12$

FACTORING TRINOMIALS

To factor a trinomial $x^2 + bx + c$, use the following steps. Assume that b and c are both positive numbers.

STEP 1 First, look for any common monomial factors.

STEP 2 a. For the trinomial $x^2 + bx + c$, use the form:

$$x^2 + bx + c = (x +\quad)(x +\quad)$$

b. For the trinomial $x^2 - bx + c$, use the form:

$$x^2 - bx + c = (x -\quad)(x -\quad)$$

c. For the trinomials $x^2 - bx - c$ and $x^2 + bx - c$, use the forms:

$$x^2 - bx - c = (x +\quad)(x -\quad)$$
$$x^2 + bx - c = (x +\quad)(x -\quad)$$

Exercises  10.3

Factor each trinomial completely:

1. $x^2 + 6x + 8$
2. $x^2 + 8x + 15$
3. $y^2 + 9y + 20$
4. $2w^2 + 20w + 32$
5. $3r^2 + 30r + 75$
6. $a^2 + 14a + 24$
7. $b^2 + 11b + 30$
8. $c^2 + 21c + 54$
9. $x^2 + 17x + 72$
10. $y^2 + 18y + 81$
11. $5a^2 + 35a + 60$
12. $r^2 + 12r + 27$
13. $x^2 - 7x + 12$
14. $y^2 - 6y + 9$
15. $2a^2 - 18a + 28$
16. $c^2 - 9c + 18$
17. $3x^2 - 30x + 63$
18. $r^2 - 12r + 35$
19. $w^2 - 13w + 42$
20. $x^2 - 14x + 49$

21. $x^2 - 19x + 90$
22. $4x^2 - 84x + 80$
23. $t^2 - 12t + 20$
24. $b^2 - 15b + 54$
25. $x^2 + 2x - 8$
26. $x^2 - 2x - 15$
27. $y^2 + y - 20$
28. $2w^2 - 12w - 32$
29. $a^2 + 5a - 24$
30. $b^2 + b - 30$
31. $c^2 - 15c - 54$
32. $b^2 - 6b - 72$
33. $3x^2 - 3x - 36$
34. $a^2 + 5a - 14$
35. $c^2 + 3c - 18$
36. $x^2 - 4x - 21$
37. $y^2 + 17y + 42$
38. $m^2 - 18m + 72$
39. $r^2 - 2r - 35$
40. $x^2 + 11x - 42$

41. $m^2 - 22m + 40$
42. $y^2 + 17y + 70$
43. $x^2 - 9x - 90$
44. $x^2 - 8x + 15$
45. $a^2 + 27a + 92$
46. $x^2 + 17x - 110$
47. $2a^2 - 12a - 110$
48. $y^2 - 14y + 40$
49. $a^2 + 29a + 100$
50. $y^2 + 14y - 120$
51. $y^2 - 14y - 95$
52. $b^2 + 20b + 36$
53. $y^2 - 18y + 32$
54. $x^2 - 8x - 128$
55. $7x^2 + 7x - 14$
56. $2x^2 - 6x - 36$
57. $6x^2 + 12x - 6$
58. $4x^2 + 16x + 16$
59. $y^2 - 12y + 35$
60. $a^2 + 16a + 63$

61. $a^2 + 2a - 63$
62. $y^2 - y - 42$
63. $x^2 + 18x + 56$
64. $x^2 + 11x - 26$
65. $2y^2 - 36y + 90$
66. $ax^2 + 2ax + a$
67. $3xy^2 - 18xy + 27x$
68. $x^2 - x - 156$
69. $x^2 + 30x + 225$
70. $x^2 - 2x - 360$
71. $x^2 - 26x + 153$
72. $x^2 + 8x - 384$
73. $x^2 + 28x + 192$
74. $x^2 + 3x - 154$
75. $x^2 + 14x - 176$
76. $x^2 - 59x + 798$
77. $2a^2b + 4ab - 48b$
78. $x^2 - 15x + 44$
79. $y^2 - y - 72$
80. $x^2 + 19x + 60$

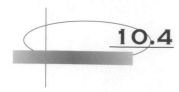

10.4 Special Products

The square of a number is the product of that number times itself. The square of 3 is $3 \cdot 3$ or 3^2 or 9. The square of a number may be found with a calculator by using the x^2 button, as shown in the examples that follow.

• **EXAMPLE 1**   Find 73.6^2 rounded to three significant digits.

Flowchart	**Buttons Pushed**	**Display**
Enter 73.6	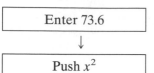	73.6
↓		
Push x^2	x^2	5416.96

Thus, $73.6^2 = 5420$ rounded to three significant digits.

• **EXAMPLE 2**   Find 0.135^2 rounded to three significant digits.

Flowchart	**Buttons Pushed**	**Display**
Enter 0.135	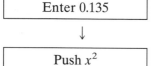	0.135
↓		
Push x^2	x^2	0.018225

Thus, $0.135^2 = 0.0182$ rounded to three significant digits.

The square of a is $a \cdot a$ or a^2 (read "a squared"). The square of the binomial $x + y$ is $(x + y)(x + y)$ or $(x + y)^2$, which is read "the quantity $x + y$ squared." When the multiplication is performed, the product is

$$(x + y)^2 = (x + y)(x + y) = x^2 + 2xy + y^2$$

A trinomial in this form is called a *perfect square trinomial*.

THE SQUARE OF A BINOMIAL

The square of the *sum* of two terms of a binomial equals the square of the first term *plus* twice the product of the two terms plus the square of the second term.

$$(a + b)(a + b) = (a + b)^2 = a^2 + 2ab + b^2$$

Similarly, the square of the *difference* of two terms of a binomial equals the square of the first term *minus* twice the product of the two terms plus the square of the second term.

$$(a - b)(a - b) = (a - b)^2 = a^2 - 2ab + b^2$$

• **EXAMPLE 3**   Find $(x + 12)^2$.

The square of the first term is x^2. Twice the product of the terms is $2(12 \cdot x)$, or $24x$. The square of the second term is 144. Thus,

$$(x + 12)^2 = x^2 + 24x + 144$$

• **EXAMPLE 4**   Find $(xy - 3)^2$.

The square of the first term is x^2y^2. Twice the product of the terms is $2(3)(xy)$, or $6xy$. The square of the second term is 9. Thus,

$$(xy - 3)^2 = x^2y^2 - 6xy + 9 \qquad\qquad\longrightarrow\bullet$$

Finding the *product of the sum and difference of two terms*, $(a + b)(a - b)$, is another special case in which the product is a binomial.

THE PRODUCT OF THE SUM AND DIFFERENCE OF TWO TERMS

This product is the difference of two squares: the square of the first term minus the square of the second term.

$$(a + b)(a - b) = a^2 - b^2$$

• **EXAMPLE 5**   Find the product $(x + 3)(x - 3)$.

The square of the first term is x^2. The square of the second term is 9. Thus,

$$(x + 3)(x - 3) = x^2 - 9 \qquad\qquad\longrightarrow\bullet$$

• **EXAMPLE 6**   Find the product $(4y + 7)(4y - 7)$.

The square of the first term is $16y^2$. The square of the second term is 49. Thus,

$$(4y + 7)(4y - 7) = 16y^2 - 49 \qquad\qquad\longrightarrow\bullet$$

• **EXAMPLE 7**   Find the product $(3x - 8y)(3x + 8y)$.

The square of the first term is $9x^2$. The square of the second term is $64y^2$. Thus,

$$(3x - 8y)(3x + 8y) = 9x^2 - 64y^2 \qquad\qquad\longrightarrow\bullet$$

Exercises  10.4

Find each product:

1. $(x + 3)(x - 3)$
2. $(x + 3)^2$
3. $(a + 5)(a - 5)$
4. $(y^2 + 9)(y^2 - 9)$
5. $(2b + 11)(2b - 11)$
6. $(x - 6)^2$
7. $(100 + 3)(100 - 3)$

8. $(90 + 2)(90 - 2)$
9. $(3y^2 + 14)(3y^2 - 14)$
10. $(y + 8)^2$
11. $(r - 12)^2$
12. $(t + 10)^2$
13. $(4y + 5)(4y - 5)$
14. $(200 + 5)(200 - 5)$

15. $(xy - 4)^2$
16. $(x^2 + y)(x^2 - y)$
17. $(ab + d)^2$
18. $(ab + c)(ab - c)$
19. $(z - 11)^2$
20. $(x^3 + 8)(x^3 - 8)$
21. $(st - 7)^2$

22. $(w + 14)(w - 14)$
23. $(x + y^2)(x - y^2)$
24. $(1 - x)^2$
25. $(x + 5)^2$
26. $(x - 6)^2$
27. $(x + 7)(x - 7)$
28. $(y - 12)(y + 12)$

29. $(x - 3)^2$
30. $(x + 4)^2$
31. $(ab + 2)(ab - 2)$
32. $(m - 3)(m + 3)$

33. $(x^2 + 2)(x^2 - 2)$
34. $(m + 15)(m - 15)$
35. $(r - 15)^2$
36. $(t + 7a)^2$

37. $(y^3 - 5)^2$
38. $(4 - x^2)^2$
39. $(10 - x)(10 + x)$
40. $(ay^2 - 3)(ay^2 + 3)$

10.5　Finding Factors of Special Products

The square root of 25 is 5 and is written as $\sqrt{25}$. The symbol $\sqrt{}$ is called a *radical*. The *square root* of a number is that positive number which, when multiplied by itself, gives the original number.

• **EXAMPLE 1**　Find the square roots of **a.** 16, **b.** 64, **c.** 100, and **d.** 144.

a. $\sqrt{16} = 4$, because $4 \times 4 = 16$
b. $\sqrt{64} = 8$, because $8 \times 8 = 64$
c. $\sqrt{100} = 10$ because $10 \times 10 = 100$
d. $\sqrt{144} = 12$ because $12 \times 12 = 144$

　　Numbers whose square roots are whole numbers are called *perfect squares*. For example, 1, 4, 9, 16, 25, 36, 49, and 64 are perfect squares.

　　The square root of a number may be found using the $\sqrt{}$ button.

• **EXAMPLE 2**　Find $\sqrt{21.4}$ rounded to three significant digits.

Flowchart	Buttons Pushed	Display
Enter 21.4	$2 \rightarrow 1 \rightarrow . \rightarrow 4$	21.4
↓		
Push square root	$\sqrt{}$	4.62601

Thus, $\sqrt{21.4} = 4.63$ rounded to three significant digits.

• **EXAMPLE 3**　Find $\sqrt{0.000594}$ rounded to three significant digits.

Flowchart	Buttons Pushed	Display
Enter 0.000594	$. \rightarrow 0 \rightarrow 0 \rightarrow 0 \rightarrow 5 \rightarrow 9 \rightarrow 4$	0.000594
↓		
Push square root	$\sqrt{}$	0.0243721

Thus, $\sqrt{0.000594} = 0.0244$ rounded to three significant digits.

　　To find the square root of a variable raised to a power, divide the exponent by 2 and use the result as the exponent of the root of the given variable.

• **EXAMPLE 4**　Assuming that x and y are positive, find the square roots of **a.** x^2, **b.** x^4, **c.** x^6, and **d.** $x^8 y^{10}$.

a. $\sqrt{x^2} = x$　　　**c.** $\sqrt{x^6} = x^3$
b. $\sqrt{x^4} = x^2$　　**d.** $\sqrt{x^8 y^{10}} = x^4 y^5$

　　To factor a trinomial, first look for a common monomial factor. Then inspect the remaining trinomial to see if it is one of the special products. If it is not a perfect square trinomial and if it can be factored, use the methods shown in Section 10.3. If it is a perfect square trinomial, it may be factored using the reverse of the rule in Section 10.4.

FACTORING PERFECT SQUARE TRINOMIALS

Each of the two factors of a perfect square trinomial with a *positive* middle term is the square root of the first term *plus* the square root of the third term. That is,

$$a^2 + 2ab + b^2 = (a + b)(a + b)$$

Similarly, each of the two factors of a perfect square trinomial with a *negative* middle term is the square root of the first term *minus* the square root of the third term. That is,

$$a^2 - 2ab + b^2 = (a - b)(a - b)$$

• **EXAMPLE 5**   Factor $9x^2 + 30x + 25$.

This perfect square trinomial has no common monomial factor. Since the middle term is positive, each of its two factors is the square root of the first term plus the square root of the third term. The square root of the first term is $3x$; the square root of the third term is 5. The sum is $3x + 5$. Therefore,

$$9x^2 + 30x + 25 = (3x + 5)(3x + 5)$$

               ————————•

• **EXAMPLE 6**   Factor $x^2 - 12x + 36$.

This perfect square trinomial has no common monomial factor. Since the middle term is negative, each of its two factors is the square root of the first term minus the square root of the third term. The square root of the first term is x; the square root of the third term is 6. The difference is $x - 6$. Therefore,

$$x^2 - 12x + 36 = (x - 6)(x - 6)$$

               ————————•

Note: If you do not recognize $x^2 - 12x + 36$ as a perfect square trinomial, you can factor it by trial and error as you would any trinomial (see Section 10.3). Your result should be the same.

• **EXAMPLE 7**   Factor $4x^2 + 24xy + 36y^2$.

First, find the common monomial factor, 4.

$$4x^2 + 24xy + 36y^2 = 4(x^2 + 6xy + 9y^2)$$

This perfect square trinomial has a positive middle term. Each of its two factors is the square root of the first term plus the square root of the third term. The square root of the first term is x; the square root of the third term is $3y$. The sum is $x + 3y$. Therefore,

$$4x^2 + 24xy + 36y^2 = 4(x + 3y)(x + 3y)$$

               ————————•

To factor a binomial that is the difference of two squares, use the reverse of the rule in Section 10.4.

Factoring the Difference of Two Squares

$$a^2 - b^2 = (a + b)(a - b)$$

Note that a is the square root of a^2 and b is the square root of b^2.

• **EXAMPLE 8**   Factor $x^2 - 4$.

First, find the square root of each term of the expression. The square root of x^2 is x, and the square root of 4 is 2. Thus, $x + 2$ is the sum of the square roots, and $x - 2$ is the difference of the square roots.

$$x^2 - 4 = (x + 2)(x - 2)$$

CHECK $(x + 2)(x - 2) = x^2 - 4$ ————————•

• **EXAMPLE 9**   Factor $1 - 36y^4$.

The square root of 1 is 1, and the square root of $36y^4$ is $6y^2$. Thus, the sum, $1 + 6y^2$, and the difference, $1 - 6y^2$, of the square roots are the factors.

$$1 - 36y^4 = (1 + 6y^2)(1 - 6y^2)$$

CHECK $(1 + 6y^2)(1 - 6y^2) = 1 - 36y^4$ ————————•

• **EXAMPLE 10**   Factor $81y^4 - 1$.

The square root of $81y^4$ is $9y^2$, and the square root of 1 is 1. The factors are the sum of the square roots, $9y^2 + 1$, and the difference of the square roots, $9y^2 - 1$.

$$81y^4 - 1 = (9y^2 + 1)(9y^2 - 1)$$

However, $9y^2 - 1$ is also the difference of two squares. Its factors are: $9y^2 - 1 = (3y + 1)(3y - 1)$. Therefore,

$$81y^4 - 1 = (9y^2 + 1)(3y + 1)(3y - 1)$$ ————————•

• **EXAMPLE 11**   Factor $2x^2 - 18$.

First, find the common monomial factor, 2.

$$2x^2 - 18 = 2(x^2 - 9)$$

Then $x^2 - 9$ is the difference of two squares whose factors are $x + 3$ and $x - 3$. Therefore,

$$2x^2 - 18 = 2(x + 3)(x - 3)$$ ————————•

Exercises | 10.5

Factor completely. Check by multiplying the factors:

1. $a^2 + 8a + 16$	**8.** $4x^2 - 1$	**15.** $49 - a^4$	**22.** $16x^2 - 1$
2. $b^2 - 2b + 1$	**9.** $y^2 - 36$	**16.** $m^2 - 2mn + n^2$	**23.** $R^2 - r^2$
3. $b^2 - c^2$	**10.** $a^2 - 64$	**17.** $49x^2 - 64y^2$	**24.** $36x^2 - 12x + 1$
4. $m^2 - 1$	**11.** $5a^2 + 10a + 5$	**18.** $x^2y^2 - 1$	**25.** $49x^2 - 25$
5. $x^2 - 4x + 4$	**12.** $9x^2 - 25$	**19.** $1 - x^2y^2$	**26.** $1 - 100y^2$
6. $2c^2 - 4c + 2$	**13.** $1 - 81y^2$	**20.** $c^2d^2 - 16$	**27.** $y^2 - 10y + 25$
7. $4 - x^2$	**14.** $16x^2 - 100$	**21.** $4x^2 - 12x + 9$	**28.** $x^2 + 6x + 9$

29. $b^2 - 9$	**33.** $4m^2 - 9$	**37.** $27x^2 - 3$
30. $16 - c^2d^2$	**34.** $16b^2 - 81$	**38.** $-9x^2 + 225x^4$
31. $m^2 + 22m + 121$	**35.** $4x^2 + 24x + 36$	**39.** $am^2 - 14am + 49a$
32. $n^2 - 30n + 225$	**36.** $-2y^2 + 12y - 18$	**40.** $-bx^2 - 12bx - 36b$

10.6 Factoring General Trinomials

In previous examples, such as $x^2 + 7x + 10 = (x + 2)(x + 5)$, there was only one possible choice for the first terms of the binomials: x and x. When the coefficient of x^2 is greater than 1, however, there may be more than one possible choice. The idea is still the same: find two binomial factors whose product equals the trinomial. Use the relationships outlined in Section 10.3 for finding the signs.

• **EXAMPLE 1**   Factor $6x^2 - x - 2$.

The first terms of the binomial factors are either $6x$ and x or $3x$ and $2x$. The $(-)$ sign of the constant term of the trinomial tells you that one sign will be $(+)$ and the other will be $(-)$ in the last terms of the binomials. The last terms are either $+2$ and -1 or -2 and $+1$.
The eight possibilities are:

1. $(6x + 2)(x - 1) = 6x^2 - 4x - 2$
2. $(6x - 1)(x + 2) = 6x^2 + 11x - 2$
3. $(6x - 2)(x + 1) = 6x^2 + 4x - 2$
4. $(6x + 1)(x - 2) = 6x^2 - 11x - 2$
5. $(3x + 2)(2x - 1) = 6x^2 + x - 2$
6. $(3x - 1)(2x + 2) = 6x^2 + 4x - 2$
7. $(3x - 2)(2x + 1) = 6x^2 - x - 2$
8. $(3x + 1)(2x - 2) = 6x^2 - 4x - 2$

Only Equation 7 gives the desired middle term, $-x$. Therefore,

$$6x^2 - x - 2 = (3x - 2)(2x + 1)$$

When factoring trinomials of this type, sometimes you may have to make several guesses or look at several combinations until you find the correct one. It is a *trial-and-error-process*. The rules for the signs, outlined in Section 10.3, simply reduce the number of possibilities you need to try.

Another way to reduce the possibilities is to eliminate any combination in which either binomial contains a common factor. In the above list, numbers 1, 3, 6, and 8 can be eliminated, since $6x + 2$, $6x - 2$, $2x + 2$, and $2x - 2$ all have the common factor 2 and cannot be correct. It is important to look for common monomial factors as the first step in factoring any trinomial.

• **EXAMPLE 2**   Factor $12x^2 - 23x + 10$.

First terms of the binomial: $12x$ and x, $6x$ and $2x$, or $4x$ and $3x$.
Signs of the last term of the binomial: both $(-)$.
Last terms of the binomial: -1 and -10 or -2 and -5.

You cannot use $12x$, $6x$, $4x$, or $2x$ with -10 or -2, since they contain the common factor 2, so the list of possible combinations is narrowed to:

1. $(12x - 1)(x - 10) = 12x^2 - 121x + 10$
2. $(12x - 5)(x - 2) = 12x^2 - 29x + 10$
3. $(4x - 5)(3x - 2) = 12x^2 - 23x + 10$
4. $(4x - 1)(3x - 10) = 12x^2 - 43x + 10$

As you can see, Equation 3 is the correct one, since it gives the desired middle term, $-23x$. Therefore,

$$12x^2 - 23x + 10 = (4x - 5)(3x - 2)$$

• **EXAMPLE 3**   Factor $12x^2 - 2x - 4$.

First look for a common factor. In this case, it is 2, so we write

$$12x^2 - 2x - 4 = 2(6x^2 - x - 2)$$

Next we try to factor the trinomial $6x^2 - x - 2$ into two binomial factors, as we did in Examples 1 and 2. Since the third term is negative (-2), we know that the signs of the second terms of the binomials are different. For the first terms of the binomials, we can try $6x$ and x or $3x$ and $2x$. The second terms can be $+2$ and -1 or -2 and $+1$. After eliminating all combinations with common factors, we have the following possibilities:

1. $(6x + 1)(x - 2) = 6x^2 - 11x - 2$
2. $(6x - 1)(x + 2) = 6x^2 + 11x - 2$
3. $(3x + 2)(2x - 1) = 6x^2 + x - 2$
4. $(3x - 2)(2x + 1) = 6x^2 - x - 2$

The correct one is Equation 4, since the middle term, $-x$, is the one we want. Therefore,

$$12x^2 - 2x - 4 = 2(3x - 2)(2x + 1)$$

Exercises   10.6

Factor completely:

1. $5x^2 - 28x - 12$
2. $4x^2 - 4x - 3$
3. $10x^2 - 29x + 21$
4. $4x^2 + 4x + 1$
5. $12x^2 - 28x + 15$
6. $9x^2 - 36x + 32$
7. $8x^2 + 26x - 45$
8. $4x^2 + 15x - 4$
9. $16x^2 - 11x - 5$

10. $6x^2 + 3x - 3$
11. $12x^2 - 16x - 16$
12. $10x^2 - 35x + 15$
13. $15y^2 - y - 6$
14. $6y^2 + y - 2$
15. $8m^2 - 10m - 3$
16. $2m^2 - 7m - 30$
17. $35a^2 - 2a - 1$
18. $12a^2 - 28a + 15$

19. $16y^2 - 8y + 1$
20. $25y^2 + 20y + 4$
21. $3x^2 + 20x - 63$
22. $4x^2 + 7x - 15$
23. $12b^2 + 5b - 2$
24. $10b^2 - 7b - 12$
25. $15y^2 - 14y - 8$
26. $5y^2 + 11y + 2$
27. $90 + 17c - 3c^2$

28. $10x^2 - x - 2$
29. $6x^2 - 13x + 5$
30. $56x^2 - 29x + 3$
31. $2y^4 + 9y^2 - 35$
32. $2y^2 + 7y - 99$
33. $4b^2 + 52b + 169$
34. $6x^2 - 19x + 15$
35. $14x^2 - 51x + 40$
36. $42x^4 - 13x^2 - 40$

37. $28x^3 + 140x^2 + 175x$
38. $-24x^3 - 54x^2 - 21x$

39. $10ab^2 - 15ab - 175a$
40. $40bx^2 - 72bx - 70b$

Chapter 10   Review

Find each product mentally:

1. $(c + d)(c - d)$
2. $(x - 6)(x + 6)$
3. $(y + 7)(y - 4)$

4. $(2x + 5)(2x - 9)$
5. $(x + 8)(x - 3)$
6. $(x - 4)(x - 9)$

7. $(x - 3)^2$
8. $(2x - 6)^2$
9. $(1 - 5x^2)^2$

Factor each expression completely:

10. $6a + 6$

11. $5x - 15$

12. $xy + 2xz$

13. $y^4 + 17y^3 - 18y^2$

14. $y^2 - 6y - 7$

15. $z^2 + 18z + 81$

16. $x^2 + 10x + 16$

17. $4a^2 + 4x^2$

18. $x^2 - 17x + 72$

19. $x^2 - 18x + 81$

20. $x^2 + 19x + 60$

21. $y^2 - 2y + 1$

22. $x^2 - 3x - 28$

23. $x^2 - 4x - 96$

24. $x^2 + x - 110$

25. $x^2 - 49$

26. $16y^2 - 9x^2$

27. $x^2 - 144$

28. $25x^2 - 81y^2$

29. $4x^2 - 24x - 364$

30. $5x^2 - 5x - 780$

31. $2x^2 + 11x + 14$

32. $12x^2 - 19x + 4$

33. $30x^2 + 7x - 15$

34. $12x^2 + 143x - 12$

35. $4x^2 - 6x + 2$

36. $36x^2 - 49y^2$

37. $28x^2 + 82x + 30$

38. $30x^2 - 27x - 21$

39. $4x^3 - 4x$

40. $25y^2 - 100$

Chapter 10   Test

Find each product mentally:

1. $(x + 8)(x - 3)$

2. $(2x - 8)(5x - 6)$

3. $(2x - 8)(2x + 8)$

4. $(3x - 5)^2$

5. $(4x - 7)(2x + 3)$

6. $(9x - 7)(5x + 4)$

Factor each expression completely:

7. $x^2 + 4x + 3$

8. $x^2 - 12x + 35$

9. $6x^2 - 7x - 90$

10. $9x^2 + 24x + 16$

11. $x^2 + 7x - 18$

12. $4x^2 - 25$

13. $6x^2 + 13x + 6$

14. $3x^2y^2 - 18x^2y + 27x^2$

15. $3x^2 - 11x - 4$

16. $15x^2 - 19x - 10$

17. $5x^2 + 7x - 6$

18. $3x^2 - 3x - 6$

19. $9x^2 - 121$

20. $9x^2 \quad 30x \mid 25$

Quadratic Equations

CHEMICAL WORK
At a desalination plant, the pH of water is checked.

MATHEMATICS AT WORK

Like linear equations, quadratic equations can be used by businesses to predict future outcomes. By plotting points into a quadratic equation, a company can see where profit is increasing or decreasing and make sound decisions to increase or decrease production in specific areas.

Quadratic equations can be used in construction trades such as landscaping and bricklaying to find maximum and minimum values. A landscaper can use quadratics to create a specified size of tract surrounding a pool or garden area. If a finish carpenter is to construct a picture frame with a fixed amount of material, he or she can use quadratics to find the maximum frame area that will require the least amount of material.

Engineers and scientists also use quadratic equations in solving higher mathematical equations involving calculus and physics.

11.1   Solving Quadratic Equations by Factoring

A *quadratic equation* in one variable is an equation in the form $ax^2 + bx + c = 0$, where $a \neq 0$.

Recall that linear equations, like $2x + 3 = 0$, have at most *one* solution. Quadratic equations have at most *two* solutions. One way to solve quadratic equations is by factoring and using the following:

> If $ab = 0$, then either $a = 0$ or $b = 0$.

That is, if you multiply two factors and the product is 0, then one or both factors are 0.

• **EXAMPLE 1**   Solve: $4(x - 2) = 0$.

If $4(x - 2) = 0$, then $4 = 0$ or $x - 2 = 0$. However, the first statement, $4 = 0$, is false; thus, the solution is $x - 2 = 0$, or $x = 2$.

• **EXAMPLE 2**   Solve: $(x - 2)(x + 3) = 0$.

If $(x - 2)(x + 3) = 0$, then either

$$x - 2 = 0 \quad \text{or} \quad x + 3 = 0$$

Therefore,

$$x = 2 \quad \text{or} \quad x = -3$$

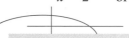

SOLVING QUADRATIC EQUATIONS BY FACTORING

1. If necessary, write an equivalent equation in the form $ax^2 + bx + c = 0$.
2. Factor the polynomial.
3. Write one or more equations by setting each factor containing a variable equal to zero.
4. Solve the two resulting first-degree equations.
5. Check.

• **EXAMPLE 3**   Solve $x^2 + 6x + 5 = 0$ for x.

STEP 1 Not needed.

STEP 2 $(x + 5)(x + 1) = 0$

STEP 3 $\quad\quad\quad x + 5 = 0 \quad\quad \text{or} \quad\quad x + 1 = 0$

STEP 4 $\quad\quad\quad\quad\quad x = -5 \quad\quad \text{or} \quad\quad\quad x = -1$

STEP 5 *Check:*

Replace x with -5.   Replace x with -1.

$$x^2 + 6x + 5 = 0 \quad\quad\quad\quad x^2 + 6x + 5 = 0$$
$$(-5)^2 + 6(-5) + 5 = 0 \text{ ?} \quad\quad (-1)^2 + 6(-1) + 5 = 0 \text{ ?}$$
$$25 - 30 + 5 = 0 \quad\quad\quad\quad 1 - 6 + 5 = 0$$
$$0 = 0 \quad\quad\quad\quad\quad\quad\quad 0 = 0$$

The roots are -5 and -1.

• **EXAMPLE 4**   Solve $x^2 + 5x = 36$ for x.

STEP 1 $x^2 + 5x - 36 = 0$

STEP 2 $(x + 9)(x - 4) = 0$

STEP 3 $x + 9 = 0$ or $x - 4 = 0$

STEP 4 $x = -9$ or $x = 4$

STEP 5   *Check:*

Replace x with -9.   Replace x with 4.

$$x^2 + 5x = 36 \qquad\qquad x^2 + 5x = 36$$

$$(-9)^2 + 5(-9) = 36? \qquad 4^2 + 5(4) = 36?$$

$$81 - 45 = 36 \qquad\qquad 16 + 20 = 36$$

$$36 = 36 \qquad\qquad\qquad 36 = 36$$

The roots are -9 and 4.

• **EXAMPLE 5**   Solve $3x^2 + 9x = 0$ for x.

STEP 1   Not needed.

STEP 2 $3x(x + 3) = 0$

STEP 3 $3x = 0$ or $x + 3 = 0$

STEP 4 $x = 0$ or $x = -3$

STEP 5   *Check:* Left to the student.

• **EXAMPLE 6**   Solve $x^2 = 4$ for x.

STEP 1 $x^2 - 4 = 0$

STEP 2 $(x + 2)(x - 2) = 0$

STEP 3 $x + 2 = 0$ or $x - 2 = 0$

STEP 4 $x = -2$ or $x = 2$

STEP 5   *Check:* Left to the student.

• **EXAMPLE 7**   Solve $6x^2 = 7x + 20$.

STEP 1 $6x^2 - 7x - 20 = 0$

STEP 2 $(3x + 4)(2x - 5) = 0$

STEP 3 $3x + 4 = 0$ or $2x - 5 = 0$

STEP 4 $3x = -4$ $2x = 5$

$$x = -\frac{4}{3} \qquad\qquad x = \frac{5}{2}$$

So the possible roots are $-\frac{4}{3}$ and $\frac{5}{2}$.

STEP 5   *Check:* Replace x with $-\frac{4}{3}$ and with $\frac{5}{2}$ in the original equation.

$$6x^2 = 7x + 20 \qquad\qquad 6x^2 = 7x + 20$$

$$6\left(-\frac{4}{3}\right)^2 = 7\left(-\frac{4}{3}\right) + 20 \qquad 6\left(\frac{5}{2}\right)^2 = 7\left(\frac{5}{2}\right) + 20$$

$$6\left(\frac{16}{9}\right) = -\frac{28}{3} + 20 \qquad 6\left(\frac{25}{4}\right) = \frac{35}{2} + 20$$

$$\frac{32}{3} = -\frac{28}{3} + \frac{60}{3} \qquad\qquad \frac{75}{2} = \frac{35}{2} + \frac{40}{2}$$

$$\frac{32}{3} = \frac{32}{3} \qquad\qquad \frac{75}{2} = \frac{75}{2}$$

So the roots are $-\frac{4}{3}$ and $\frac{5}{2}$.

Exercises  11.1

Solve each equation:

1. $x^2 + x = 12$
2. $x^2 - 3x + 2 = 0$
3. $x^2 + x - 20 = 0$
4. $d^2 + 2d - 15 = 0$
5. $x^2 - 2 = x$
6. $x^2 - 15x = -54$
7. $x^2 - 1 = 0$
8. $16n^2 = 49$
9. $x^2 - 4 = 0$
10. $4n^2 = 64$

11. $w^2 + 5w + 6 = 0$
12. $x^2 - 6x = 0$
13. $y^2 - 4y = 21$
14. $c^2 + 2 = 3c$
15. $n^2 - 6n - 40 = 0$
16. $x^2 - 17x + 16 = 0$
17. $9m = m^2$
18. $6n^2 - 15n = 0$
19. $x^2 = 108 + 3x$
20. $x^2 - x = 42$

21. $c^2 + 6c = 16$
22. $4x^2 + 4x - 3 = 0$
23. $10x^2 + 29x + 10 = 0$
24. $2x^2 = 17x - 8$
25. $4x^2 = 25$
26. $25x = x^2$
27. $9x^2 + 16 = 24x$
28. $24x^2 + 10 = 31x$
29. $3x^2 + 9x = 0$

30. A rectangle is 5 ft longer than it is wide. (See Illustration 1.) The area of the rectangle is 84 ft^2. Use a quadratic equation to find the dimensions of the rectangle.

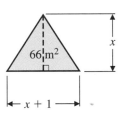

ILLUSTRATION 2

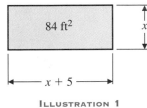

--- $x + 5$ ---

ILLUSTRATION 1

31. The area of a triangle is 66 m^2, and its base is 1 m more than the height. (See Illustration 2.) Find the base and height of the triangle. (Use a quadratic equation.)

32. A rectangle is 9 ft longer than it is wide, and its area is 360 ft^2. Use a quadratic equation to find its length and width.

33. A heating duct has a rectangular cross-section whose area is 40 in^2. If it is three inches longer than it is wide, find its length and width.

 11.2 The Quadratic Formula

Many quadratic equations cannot be solved by factoring, so let's study a method by which any quadratic equation can be solved. You may find it helpful to review the square root discussion in Section 10.5.

The roots of a quadratic equation in the form

$$ax^2 + bx + c = 0$$

may be found using the following formula:

Quadratic Formula

$$x = \frac{-b \pm \sqrt{b^2 - 4ac}}{2a}$$

where a is the coefficient of the x^2 term,

b is the coefficient of the x term, and

c is the constant term.

The symbol ($\pm$) is used to combine two expressions into one. For example, "$a \pm 4$" means "$a + 4$ or $a - 4$." Similarly,

$$x = \frac{-b \pm \sqrt{b^2 - 4ac}}{2a} \quad \text{means}$$

$$x = \frac{-b + \sqrt{b^2 - 4ac}}{2a} \quad \text{or} \quad x = \frac{-b - \sqrt{b^2 - 4ac}}{2a}$$

• **EXAMPLE 1**   In the quadratic equation $3x^2 - x - 7 = 0$, find the values of a, b, and c.

$a = 3$, $b = -1$, and $c = -7$

• **EXAMPLE 2**   Solve $x^2 + 5x - 14 = 0$ using the quadratic formula.

$$x = \frac{-b \pm \sqrt{b^2 - 4ac}}{2a}, \quad a = 1, b = 5, c = -14$$

So $x = \dfrac{-5 \pm \sqrt{5^2 - 4(1)(-14)}}{2(1)}$

$\quad = \dfrac{-5 \pm \sqrt{25 + 56}}{2}$

$\quad = \dfrac{-5 \pm \sqrt{81}}{2}$

$\quad = \dfrac{-5 \pm 9}{2}$

$\quad = \dfrac{-5 + 9}{2}$ or $\dfrac{-5 - 9}{2}$

$\quad = 2$ or -7

CHECK

| Replace x with 2. | Replace x with -7. |

$x^2 + 5x - 14 = 0$ $\qquad\qquad$ $x^2 + 5x - 14 = 0$

$2^2 + 5(2) - 14 = 0$? $\qquad$ $(-7)^2 + 5(-7) - 14 = 0$?

$4 + 10 - 14 = 0$ $\qquad\qquad$ $49 - 35 - 14 = 0$

$0 = 0$ $\qquad\qquad\qquad\qquad$ $0 = 0$

The roots are 2 and -7.

> Before using the quadratic formula, make certain the equation is written in the form $ax^2 + bx + c = 0$, such that one member is zero.

• **EXAMPLE 3**   Solve $2x^2 = x + 21$ by using the quadratic formula.

First, write the equation in the form $2x^2 - x - 21 = 0$.

$$x = \frac{-b \pm \sqrt{b^2 - 4ac}}{2a}, \quad a = 2, b = -1, c = -21$$

So $x = \dfrac{-(-1) \pm \sqrt{(-1)^2 - 4(2)(-21)}}{2(2)}$

$= \dfrac{1 \pm \sqrt{1 + 168}}{4}$

$= \dfrac{1 \pm \sqrt{169}}{4}$

$= \dfrac{1 \pm 13}{4}$

$= \dfrac{1 + 13}{4}$ or $\dfrac{1 - 13}{4}$

$= \dfrac{7}{2}$ or -3

CHECK   Left to the student. ————————•

The quantity under the radical sign, $b^2 - 4ac$, is called the *discriminant*. If the discriminant is not a perfect square, find the square root of the number by using a calculator and proceed as before. Round each final result to three significant digits.

• EXAMPLE 4   Solve $3x^2 + x - 5 = 0$ using the quadratic formula.

$x = \dfrac{-b \pm \sqrt{b^2 - 4ac}}{2a}$, $a = 3, b = 1, c = -5$

So $x = \dfrac{-1 \pm \sqrt{1^2 - 4(3)(-5)}}{2(3)}$

$= \dfrac{-1 \pm \sqrt{1 + 60}}{6}$

$= \dfrac{-1 \pm \sqrt{61}}{6}$

$= \dfrac{-1 \pm 7.81}{6}$

$= \dfrac{-1 + 7.81}{6}$ or $\dfrac{-1 - 7.81}{6}$

$= \dfrac{6.81}{6}$ or $\dfrac{-8.81}{6}$

$= 1.14$ or -1.47

The roots are 1.14 and -1.47. ————————•

The check will not work out *exactly* when the number under the radical is not a perfect square.

Exercises   11.2

Find the values of a, b, and c in each equation:

1. $x^2 - 7x + 4 = 0$ **3.** $3x^2 + 4x + 9 = 0$ **5.** $-3x^2 + 4x + 7 = 0$ **7.** $3x^2 - 14 = 0$

2. $2x^2 + x - 3 = 0$ **4.** $2x^2 - 14x + 37 = 0$ **6.** $17x^2 - x + 34 = 0$ **8.** $2x^2 + 7x = 0$

Solve each equation using the quadratic formula. Check your solutions:

9. $x^2 + x - 6 = 0$

10. $x^2 - 4x - 21 = 0$

11. $x^2 + 8x - 9 = 0$

12. $2x^2 + 5x - 12 = 0$

13. $5x^2 + 2x = 0$

14. $3x^2 - 75 = 0$

15. $48x^2 - 32x - 35 = 0$

16. $13x^2 + 178x - 56 = 0$

Solve each equation using the quadratic formula (when necessary, round results to three significant digits):

17. $2x^2 + x - 5 = 0$

18. $-3x^2 + 2x + 5 = 0$

19. $3x^2 - 5x = 0$

20. $7x^2 + 9x + 2 = 0$

21. $-2x^2 + x + 3 = 0$

22. $5x^2 - 7x + 2 = 0$

23. $6x^2 + 9x + 1 = 0$

24. $16x^2 - 25 = 0$

25. $-4x^2 = 5x + 1$

26. $9x^2 = 21x - 10$

27. $3x^2 = 17$

28. $8x^2 = 11x - 3$

29. $x^2 = 15x + 7$

30. $x^2 + x = 1$

31. $3x^2 - 31 = 5x$

32. $-3x^2 - 5 = -7x^2$

33. $52.3x = -23.8x^2 + 11.8$

34. $18.9x^2 - 44.2x = 21.5$

35. A variable voltage in an electrical circuit is given by $V = t^2 - 12t + 40$, where t is in seconds. Find the values of t when the voltage V equals **a.** 8 V, **b.** 25 V, **c.** 104 V?

36. A variable electric current is given by $i = t^2 - 7t + 12$, where t is in seconds. At what times is the current i equal to **a.** 2 A? **b.** 0 A? **c.** 4 A?

37. A rectangular piece of sheet metal is 4 ft longer than it is wide. (See Illustration 1.) The area of the piece of sheet metal is 21 ft^2. Find its length and width.

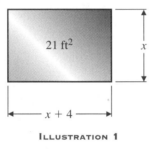

21 ft^2 x

$x + 4$

ILLUSTRATION 1

38. A square, 4 in. on a side, is cut out of each corner of a square sheet of aluminum. (See Illustration 2.) The sides are folded up to form

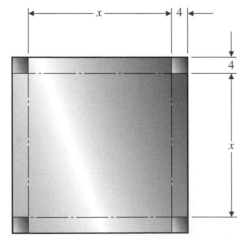

x 4

4

x

ILLUSTRATION 2

a rectangular container with no top. The volume of the resulting container is 400 in^3. What was the size of the original sheet of aluminum?

39. A square is cut out of each corner of a rectangular sheet of aluminum that is 40 cm by 60 cm. (See Illustration 3.) The sides are folded up to form a rectangular container with no top. The area of the bottom of the container is 1500 cm^2. **a.** What are the dimensions of each cut-out square? **b.** Find the volume of the container. ($V = lwh$)

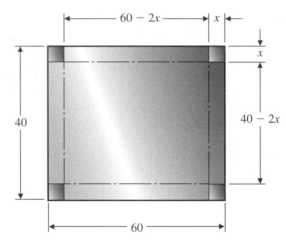

$60 - 2x$ x

x

40    $40 - 2x$

60

ILLUSTRATION 3

40. The area of a rectangular lot 80 m by 100 m is to be increased by 4000 m^2. (See Illustration 4.) The length and the width will be increased by the same amount. What are the dimensions of the larger lot?

41. A border of uniform width is built around a rectangular garden that measures 16 ft by 20 ft. (See Illustration 5.) The area of the border is 160 ft^2. Find the width of the border.

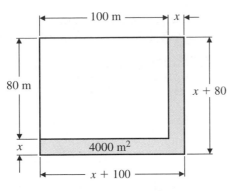

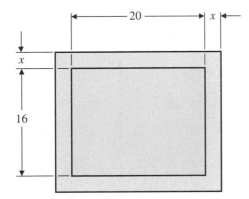

ILLUSTRATION 4

42. A border of uniform width is printed on a page measuring 11 in. by 14 in. (See Illustration 6.) The area of the border is 66 in². Find the width of the border.

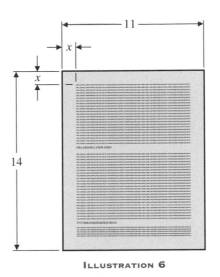

ILLUSTRATION 6

ILLUSTRATION 5

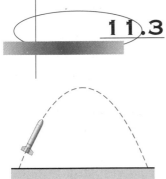

FIGURE 11.1

11.3 Graphs of Quadratic Equations

Many physical phenomena follow a path called a *parabola*. The trajectory of a rocket (Figure 11.1) is one such phenomenon. The parabolic curve is really the graph of a quadratic equation in two variables. This relationship is very important; it allows us to use computers in analyzing rocket launches and to test the launches against the planned flight path.

The quadratic equation that represents a parabola is written in the form

$$y = ax^2 + bx + c$$

where a, b, and c are real numbers and $a \neq 0$. The quadratic equation in the form $x = ay^2 + by + c$ also represents a parabola, but we will not work with this equation in this book.

To draw the graph of a parabola, find points whose ordered pairs satisfy the equation (that is, make the equation a true statement). Since this graph is not a straight line, you will need to find many points to get an accurate graph of the curve. A table is helpful for listing these ordered pairs.

• **EXAMPLE 1**   Graph the equation $y = x^2$.

First, set up a table as follows.

x	-4	-3	-2	-1	0	1	2	3	4
$y = x^2$	16	9	4	1	0	1	4	9	16

Then, using a rectangular coordinate system, plot these points. Notice that with only the points shown in Figure 11.2, there isn't a definite outline of the curve. So let's look closer at values of x between 0 and 1.

x	$\frac{1}{6}$	$\frac{1}{5}$	$\frac{1}{4}$	$\frac{1}{3}$	$\frac{2}{5}$	$\frac{1}{2}$	$\frac{3}{5}$	$\frac{3}{4}$	$\frac{2}{3}$	$\frac{4}{5}$	$\frac{5}{6}$
$y = x^2$	$\frac{1}{36}$	$\frac{1}{25}$	$\frac{1}{16}$	$\frac{1}{9}$	$\frac{4}{25}$	$\frac{1}{4}$	$\frac{9}{25}$	$\frac{9}{16}$	$\frac{4}{9}$	$\frac{16}{25}$	$\frac{25}{36}$

Plotting these additional ordered pairs, we get a better graph (Figure 11.3). If we were to continue to choose more and more values, the graph would appear as a solid line.

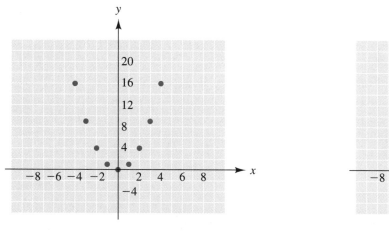

FIGURE 11.2

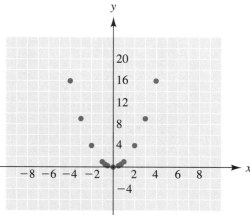

FIGURE 11.3

Since it is impossible to find *all* the ordered pairs that satisfy the equation, we will assume that all the points between any two of the ordered pairs already located could be found, and they do lie on the graph. Thus, assume that the graph of $y = x^2$ looks like the graph in Figure 11.4.

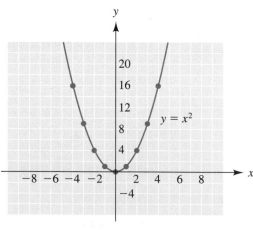

FIGURE 11.4

In summary, to draw the graph of a quadratic equation, form a table to find many ordered pairs that satisfy the equation. Then plot these ordered pairs and connect them with a smooth curved line.

• **EXAMPLE 2**   Graph the equation $y = 2x^2 - 4x + 5$.

Let $x = -7$; then $y = 2(-7)^2 - 4(-7) + 5 = 131$.
Let $x = -6$; then $y = 2(-6)^2 - 4(-6) + 5 = 101$.
Let $x = -5$; then $y = 2(-5)^2 - 4(-5) + 5 = 75$.
Let $x = -4$; then $y = 2(-4)^2 - 4(-4) + 5 = 53$.
Let $x = -3$; then $y = 2(-3)^2 - 4(-3) + 5 = 35$.
Let $x = -2$; then $y = 2(-2)^2 - 4(-2) + 5 = 21$.
Let $x = -1$; then $y = 2(-1)^2 - 4(-1) + 5 = 11$.
Let $x = 0$; then $y = 2(0)^2 - 4(0) + 5 = 5$.
Let $x = 1$; then $y = 2(1)^2 - 4(1) + 5 = 3$.
Let $x = 2$; then $y = 2(2)^2 - 4(2) + 5 = 5$.
Let $x = 3$; then $y = 2(3)^2 - 4(3) + 5 = 11$.
Let $x = 4$; then $y = 2(4)^2 - 4(4) + 5 = 21$.
Let $x = 5$; then $y = 2(5)^2 - 4(5) + 5 = 35$.
Let $x = 6$; then $y = 2(6)^2 - 4(6) + 5 = 53$.
Let $x = 7$; then $y = 2(7)^2 - 4(7) + 5 = 75$.
Let $x = 8$; then $y = 2(8)^2 - 4(8) + 5 = 101$.
Let $x = 9$; then $y = 2(9)^2 - 4(9) + 5 = 131$.

x	-7	-6	-5	-4	-3	-2	-1	0	1	2	3	4	5	6	7	8	9
y	131	101	75	53	35	21	11	5	3	5	11	21	35	53	75	101	131

Then plot the points from the table. Disregard those coordinates that cannot be plotted. Connect the points with a smooth curved line, as shown in Figure 11.5.

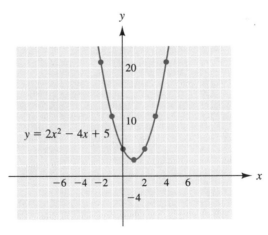

$y = 2x^2 - 4x + 5$

FIGURE 11.5

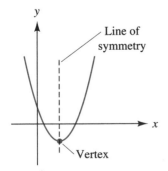

FIGURE 11.6

All parabolas have a property called *symmetry,* which means that a line can be drawn through a parabola dividing it into two parts that are mirror images of each other. (See Figure 11.6.)

The intersection of this line of symmetry and the parabola is called the *vertex.* In this section, the vertex is either the highest or lowest point of the parabola. Locating the vertex is most helpful in drawing the graph of a parabola.

The vertex of a parabola whose equation is in the form $y = ax^2 + bx + c$ may be found as follows.

VERTEX OF A PARABOLA

1. The x coordinate is the value of $-\dfrac{b}{2a}$.

2. The y coordinate is found by substituting the x coordinate from Step 1 into the parabola equation and solving for y.

EXAMPLE 3   Find the vertex of the parabola $y = 2x^2 - 4x + 5$ in Example 2.

Note that $a = 2$ and $b = -4$.

1. The x coordinate is $-\dfrac{b}{2a} = -\dfrac{-4}{2(2)} = 1$.

2. Substitute $x = 1$ into $y = 2x^2 - 4x + 5$ and solve for y.

$$y = 2(1)^2 - 4(1) + 5$$
$$= 2 - 4 + 5 = 3$$

Thus, the vertex is $(1, 3)$.

EXAMPLE 4   Graph the equation $y = -x^2 + 6x$.

First, find the vertex. The x-coordinate is $\dfrac{-b}{2a} = \dfrac{-6}{2(-1)} = \dfrac{-6}{-2} = 3$. Then substitute $x = 3$ into $y = -x^2 + 6x$.

$$y = -(3)^2 + 6(3)$$
$$y = -9 + 18 = 9$$

The vertex is $(3, 9)$. Graph the vertex in Figure 11.7.

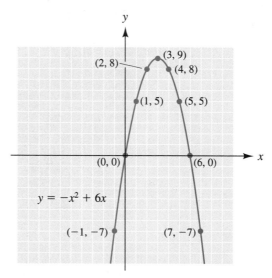

FIGURE 11.7

To find other points, let

$x = 4$, then $y = -(4)^2 + 6(4) = -16 + 24 = 8$; graph $(4, 8)$.
$x = 5$, then $y = -(5)^2 + 6(5) = -25 + 30 = 5$; graph $(5, 5)$.
$x = 6$, then $y = -(6)^2 + 6(6) = -36 + 36 = 0$; graph $(6, 0)$.
$x = 7$, then $y = -(7)^2 + 6(7) = -49 + 42 = -7$; graph $(7, -7)$.

From symmetry, do you see that you can graph the points $(2, 8)$, $(1, 5)$, $(0, 0)$, and $(-1, -7)$ in Figure 11.7 without calculation? If not, let $x = 2$ and solve for y.

You may also note that the parabola in Figure 11.7 opens down. In general, the graph of $y = ax^2 + bx + c$

opens up with $a > 0$ and
opens down when $a < 0$.

Exercises  **11.3**

Draw the graph of each equation and label each vertex:

1. $y = 2x^2$

2. $y = -2x^2$

3. $y = \frac{1}{2}x^2$

4. $y = -\frac{1}{2}x^2$

5. $y = x^2 + 3$

6. $y = x^2 - 4$

7. $y = 2(x - 3)^2$

8. $y = -(x + 2)^2$

9. $y = x^2 - 2x + 1$

10. $y = 2(x + 1)^2 - 3$

11. $y = 2x^2 - 5$

12. $y = -3x^2 - 2x$

13. $y = x^2 - 2x - 5$

14. $y = -3x^2 + 6x + 15$

15. $y = x^2 - 2x - 15$

16. $y = 2x^2 - x - 15$

17. $y = -4x^2 - 5x + 9$

18. $y = 4x^2 - 12x + 9$

19. $y = \frac{1}{5}x^2 - \frac{2}{5}x + 4$

20. $y = -0.4x^2 + 2.4x + 0.7$

 **11.4** Imaginary Numbers

What is the meaning of $\sqrt{-4}$? What number squared is -4? Try to find its value on your calculator.

As you can see, this is a different kind of number. Up to now, we have considered only real numbers. The number $\sqrt{-4}$ is not a real number. The square root of a negative number is called an *imaginary number*. The imaginary unit is defined as $\sqrt{-1}$ and in many mathematics texts is given by the symbol i. However, in technical work, i is commonly used for current. To avoid confusion, many technical books use j for $\sqrt{-1}$, which is what we use in this book.

> **Imaginary Unit**
> $$\sqrt{-1} = j$$

Then $\sqrt{-4} = \sqrt{(-1)(4)} = (\sqrt{-1})(\sqrt{4}) = (j)(2)$ or $2j$.

• **EXAMPLE 1**   Express each number in terms of j: **a.** $\sqrt{-25}$, **b.** $\sqrt{-45}$, **c.** $\sqrt{-183}$.

a. $\sqrt{-25}$

$= \sqrt{(-1)(25)}$

$= (\sqrt{-1})(\sqrt{25})$

$= (j)(5)$ or $5j$

b. $\sqrt{-45}$

$= \sqrt{(-1)(45)}$

$= (\sqrt{-1})(\sqrt{45})$

$= (j)(6.71)$ or $6.71\,j$

c. $\sqrt{-183}$

$= \sqrt{(-1)(183)}$

$= (\sqrt{-1})(\sqrt{183})$

$= (j)(13.5)$ or $13.5\,j$

Now let's consider powers of j or $\sqrt{-1}$. Using the rules of exponents and the definition of j, carefully study the following powers of j:

$$j = j$$
$$j^2 = (\sqrt{-1})^2 = -1$$
$$j^3 = (j^2)(j) = (-1)(j) = -j$$
$$j^4 = (j^2)(j^2) = (-1)(-1) = 1$$
$$j^5 = (j^4)(j) = (1)(j) = j$$
$$j^6 = (j^4)(j^2) = (1)(-1) = -1$$
$$j^7 = (j^4)(j^3) = (1)(-j) = -j$$
$$j^8 = (j^4)^2 = 1^2 = 1$$
$$j^9 = (j^8)(j) = (1)(j) = j$$
$$j^{10} = (j^8)(j^2) = (1)(-1) = -1$$

As you can see, the values of j to a power repeat in the order of j, -1, $-j$, 1, j, -1, $-j$, 1, . . . Also, j to any power divisible by 4 equals 1.

• **EXAMPLE 2**   Simplify **a.** j^{15}, **b.** j^{21}, **c.** j^{72}.

a. $j^{15} = (j^{12})(j^3)$

$= (1)(-j) = -j$

b. $j^{21} = (j^{20})(j)$

$= (1)(j) = j$

c. $j^{72} = 1$

When solving the quadratic equation $ax^2 + bx + c = 0$, its solutions are given by the quadratic formula

$$x = \frac{-b \pm \sqrt{b^2 - 4ac}}{2a}$$

The part under the radical sign, $b^2 - 4ac$, is called the *discriminant*. The value of $b^2 - 4ac$ determines what kind of solutions the quadratic equation has, and how many solutions it has when a, b, and c are integers.

If $b^2 - 4ac$ is	Roots
positive and a perfect square,	both roots are rational.
positive and not a perfect square,	both roots are irrational.
zero,	there is only one rational root.
negative,	both roots are imaginary.

The relationship between the graph of $y = ax^2 + bx + c$ and the value of the discriminant is shown in Figure 11.8.

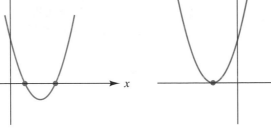

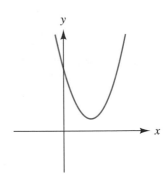

(a)	(b)	(c)
$b^2 - 4ac > 0$	$b^2 - 4ac = 0$	$b^2 - 4ac < 0$
Two solutions, as indicated by the two points of intersection on the x-axis	One solution, as indicated by the one point of intersection on the x-axis	The graph does not cross the x-axis; both roots are imaginary

FIGURE 11.8

• **EXAMPLE 3**   Determine the nature of the roots of $3x^2 + 5x - 2 = 0$ without solving the equation.

$$a = 3, b = 5, c = -2$$

The value of the discriminant is

$$b^2 - 4ac = (5)^2 - 4(3)(-2)$$
$$= 25 + 24 = 49$$

Since 49 is a perfect square, both roots are rational.   ⎯⎯⎯⎯•

• **EXAMPLE 4**   Determine the nature of the roots of $4x^2 - 12x + 9 = 0$ without solving the equation.

$$a = 4, b = -12, c = 9$$

The value of the discriminant is

$$b^2 - 4ac = (-12)^2 - 4(4)(9)$$
$$= 144 - 144 = 0$$

Therefore, there is only one rational root.   ⎯⎯⎯⎯•

• **EXAMPLE 5**   Determine the nature of the roots of $x^2 - 3x + 8 = 0$ without solving the equation.

$$a = 1, b = -3, c = 8$$

The value of the discriminant is

$$b^2 - 4ac = (-3)^2 - 4(1)(8)$$
$$= 9 - 32 = -23$$

Since -23 is negative, both roots are imaginary.   ⎯⎯⎯⎯•

• **EXAMPLE 6**   Solve $4x^2 - 6x + 5 = 0$ using the quadratic formula.

$$x = \frac{-b \pm \sqrt{b^2 - 4ac}}{2a}, \quad a = 4, b = -6, c = 5$$

$$x = \frac{-(-6) \pm \sqrt{(-6)^2 - 4(4)(5)}}{2(4)} = \frac{6 \pm \sqrt{36 - 80}}{8}$$

$$= \frac{6 \pm \sqrt{-44}}{8} = \frac{6 \pm 6.63j}{8} \qquad \sqrt{-44} = (\sqrt{-1})(\sqrt{44}) = 6.63j$$

$$= \frac{6 + 6.63j}{8} \quad \text{or} \quad \frac{6 - 6.63j}{8}$$

$$= 0.75 + 0.829j \quad \text{or} \quad 0.75 - 0.829j$$

The roots are $0.75 + 0.829j$ and $0.75 - 0.829j$.

Exercises 11.4

Express each number in terms of j (when necessary, round the result to three significant digits):

1. $\sqrt{-49}$ **3.** $\sqrt{-14}$ **5.** $\sqrt{-2}$ **7.** $\sqrt{-56}$ **9.** $\sqrt{-169}$ **11.** $\sqrt{-27}$

2. $\sqrt{-64}$ **4.** $\sqrt{-5}$ **6.** $\sqrt{-3}$ **8.** $\sqrt{-121}$ **10.** $\sqrt{-60}$ **12.** $\sqrt{-40}$

Simplify:

13. j^3 **15.** j^{13} **17.** j^{19} **19.** j^{24} **21.** j^{38} **23.** $\dfrac{1}{j}$

14. j^6 **16.** j^{16} **18.** j^{31} **20.** j^{26} **22.** j^{81} **24.** $\dfrac{1}{j^6}$

Determine the nature of the roots of each quadratic equation without solving it:

25. $x^2 + 3x - 10 = 0$ **29.** $3x + 1 = 2x^2$ **33.** $x^2 + 25 = 0$

26. $2x^2 - 7x + 3 = 0$ **30.** $3x^2 = 4x - 8$ **34.** $x^2 - 4 = 0$

27. $5x^2 + 4x + 1 = 0$ **31.** $2x^2 + 6 = x$

28. $9x^2 + 12x + 4 = 0$ **32.** $2x^2 + 7x = 4$

Solve each quadratic equation using the quadratic formula (when necessary, round results to three significant digits):

35. $x^2 - 6x + 10 = 0$ **41.** $6x^2 + 5x + 8 = 0$ **47.** $5x^2 + 14x - 3 = 0$

36. $x^2 - x + 2 = 0$ **42.** $4x^2 + 3x - 1 = 0$ **48.** $2x^2 - x + 1 = 0$

37. $x^2 - 14x + 53 = 0$ **43.** $3x^2 = 6x - 7$ **49.** $x^2 + x + 1 = 0$

38. $x^2 + 10x + 34 = 0$ **44.** $5x^2 + 2x = -3$ **50.** $12x^2 + 23x + 10 = 0$

39. $x^2 + 8x + 41 = 0$ **45.** $5x^2 + 8x + 4 = 0$

40. $x^2 - 6x + 13 = 0$ **46.** $2x^2 + x + 3 = 0$

Chapter 11   Review

1. If $ab = 0$, what is known about either a or b?   **2.** Solve for x: $3x(x - 2) = 0$.

Solve each equation by factoring:

3. $x^2 - 4 = 0$ **5.** $5x^2 - 6x = 0$ **7.** $x^2 - 14x = -45$ **9.** $3x^2 + 20x + 32 = 0$

4. $x^2 - x = 6$ **6.** $x^2 - 3x - 28 = 0$ **8.** $x^2 - 18 - 3x = 0$

Solve each equation using the quadratic formula (when necessary, round results to three significant digits):

10. $3x^2 - 16x - 12 = 0$ **12.** $2x^2 + x = 15$ **14.** $3x^2 - 4x = 5$

11. $x^2 + 7x - 5 = 0$ **13.** $x^2 - 4x = 2$

15. The area of a piece of plywood is 36 ft². Its length is 5 ft more than its width. Find its length and width.

16. A variable electric current is given by the formula $i = t^2 - 12t + 36$, where t is in μs. At what times is the current i equal to **a.** 4 A? **b.** 0 A? **c.** 10 A?

Draw the graph of each equation and label each vertex:

17. $y = x^2 - x - 6$

18. $y = -3x^2 + 2$

Express each number in terms of j:

19. $\sqrt{-36}$

20. $\sqrt{-73}$

Simplify:

21. j^{12}

22. j^{27}

Determine the nature of the roots of each quadratic equation without solving it:

23. $9x^2 + 30x + 25 = 0$

24. $3x^2 - 2x + 4 = 0$

Solve each equation using the quadratic formula (when necessary, round results to three significant digits):

25. $x^2 - 4x + 5 = 0$

26. $5x^2 - 6x + 4 = 0$

27. A solar-heated house has a rectangular heat collector with a length 1 ft more than three times its width. The area of the collector is 21.25 ft². Find its length and width.

28. A rectangular opening is 15 in. wide and 26 in. long. (See Illustration 1.) A strip of constant width is to be removed from around the opening to increase the area to 672 in². How wide must the strip be?

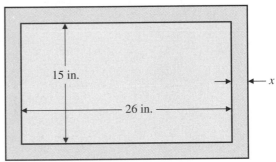

15 in.

26 in.

x

ILLUSTRATION 1

Chapter 11   Test

Solve each equation.

1. $x^2 = 64$

3. $x^2 + 9x - 36 = 0$

2. $x^2 - 8x = 0$

4. $12x^2 + 4x = 1$

Solve each equation using the quadratic formula (when necessary, round results to three significant digits.)

5. $5x^2 + 6x - 10 = 0$

6. $3x^2 - 4x - 9 = 0$

Solve each equation. (When necessary, round results to three significant digits.)

7. $21x^2 - 29x - 10 = 0$

9. $3x^2 - 39x + 90 = 0$

8. $5x^2 - 7x - 2 = 0$

10. $6x^2 - 8x - 5 = 0$

11. Draw the graph of $y = -x^2 - 8x - 15$ and label the vertex.

12. Draw the graph of $y = 2x^2 + 8x + 11$ and label the vertex. Express each number in terms of j.

Express each number in terms of j:

13. $\sqrt{-16}$

14. $\sqrt{-29}$

Simplify:

15. j^9

16. j^{28}

17. Determine the nature of the roots of $3x^2 - x + 4 = 0$ without solving it.

18. One side of a rectangle is 5 cm more than another. Its area is 204 cm². Find its length and width.

Geometry

CARPENTRY
Finish carpenters and woodworkers must have a sound knowledge of geometry.

MATHEMATICS AT WORK

Geometry is a widely used tool in many occupations. Carpenters use the Pythagorean theorem to be sure footings and foundations are square and to predetermine the length of rafters without having to climb on the wall and measure the distance with a tape measure. Bricklayers use theorems dealing with circles to calculate the arc length of a curved-shaped doorway or fireplace.

Drafters and machinists use geometric theorems and postulates in preparing blueprints for machining and manufacturing. Geometry is also used with trigonometric equations for milling parts that are beveled or curved.

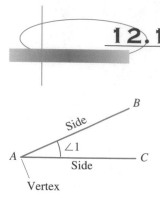

FIGURE 12.1
Basic parts of an angle

12.1   Angles and Polygons

Some fundamentals of geometry must be understood in order to solve many technical applications. Geometry is also needed to follow some of the mathematical developments in technical mathematics courses, in technical support courses, and in on-the-job training programs. Here, we will cover the most basic and most often used geometric terms and relationships.

An *angle* is determined by the parts of two lines that have a common endpoint. The common endpoint is called the *vertex* of the angle. The parts of the lines are called the *sides* of the angle. An angle is designated either by a number, by a single letter, or by three letters. For example, the angle in Figure 12.1 is referred to as $\angle A$ or $\angle BAC$. The middle letter of the three letters is always the one at the vertex.

The measure of an angle is the amount of rotation needed to make one side coincide with the other side. The measure can be expressed in any one of many different units. The standard metric unit of plane angles is the radian (rad). While the radian is the metric unit of angle measurement, many ordinary measurements continue to be made in degrees (°). Although some trades subdivide the degree into the traditional minutes and seconds, most others use tenths and hundredths of degrees. Radian measure is developed in Section 12.6.

One degree is $\frac{1}{360}$ of one complete revolution; that is, $360° =$ one revolution. The protractor in Figure 12.2 is an instrument, marked in degrees, used to measure angles.

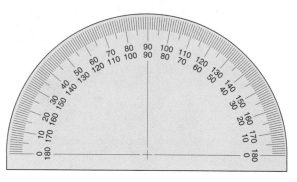

FIGURE 12.2
Protractor

USING A PROTRACTOR

STEP 1 Place the protractor so that the center mark on its base coincides with the vertex of the angle, and so that the 0° mark is on one side of the angle.

STEP 2 Read the mark on the protractor that is on the other side of the angle (extended, if necessary).

    a. If the side of the angle under the 0° mark extends to the *right* from the vertex, read the inner scale to find the degree measure.

    b. If the side of the angle under the 0° mark extends to the *left* from the vertex, read the outer scale to find the degree measure.

•   **EXAMPLE 1**   Find the degree measure of the angle in Figure 12.3.

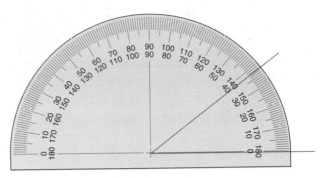

FIGURE 12.3

The measure of the angle is 38°, using Step 2(a).

• **EXAMPLE 2** Find the degree measure of the angle in Figure 12.4.

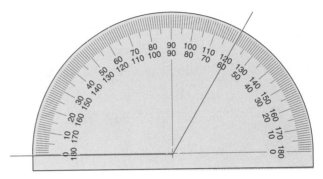

FIGURE 12.4

The measure of the angle is 120°, using Step 2(b).

Angles are often classified by degree measure. A *right* angle is an angle with a measure of 90°. In a sketch or diagram, a 90° angle is denoted by placing ⌐ or ⌐ in the angle, as shown in Figure 12.5. An *acute* angle is an angle with a measure less than 90°. An *obtuse* angle is an angle with a measure greater than 90° but less than 180°.

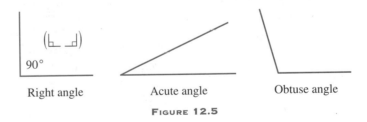

Right angle   Acute angle   Obtuse angle

FIGURE 12.5

We will first study some geometric relationships of angles and lines in the same plane. Two lines *intersect* if they have only one point in common. (See Figure 12.6.)

Two lines in the same plane are *parallel* (∥) if they do not intersect even when extended. (See Figure 12.7.)

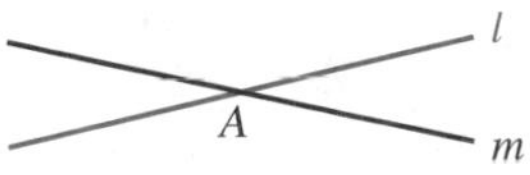

FIGURE 12.6
Lines *l* and *m* intersect at point *A*.

FIGURE 12.7
Lines *l* and *m* are parallel.

Two angles are *adjacent* if they have a common vertex and a common side between them, with no common interior points. See Figure 12.8, where $\angle 1$ and $\angle 2$ are adjacent angles because both have a common vertex B and a common side BD between them.

Two lines in the same plane are *perpendicular* ($\perp$) if they intersect and form equal adjacent angles. Each of these equal adjacent angles is a right angle. See Figure 12.9, where $l \perp m$ because $\angle 1 = \angle 2$. Angles 1 and 2 are right angles.

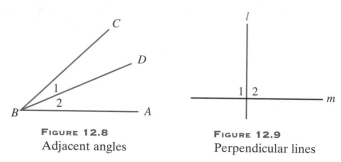

FIGURE 12.8
Adjacent angles

FIGURE 12.9
Perpendicular lines

Two angles are *complementary* if the sum of their measures is 90°. (See Figure 12.10.) Angles A and B in Figure 12.10(a) are complementary:

$$52° + 38° = 90°$$

Angles LMN and NMP in Figure 12.10(b) are complementary:

$$25° + 65° = 90°$$

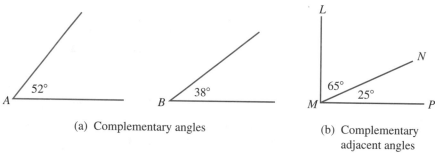

(a) Complementary angles

(b) Complementary adjacent angles

FIGURE 12.10

Two angles are *supplementary* if the sum of their measures is 180°. (See Figure 12.11.) Angles C and D in Figure 12.11(a) are supplementary: $70° + 110° = 180°$.

Two adjacent angles with their exterior sides in a straight line are *supplementary*. (See Figure 12.11(b).) Angles 1 and 2 have their exterior sides in a straight line, so they are supplementary: $\angle 1 + \angle 2 = 180°$.

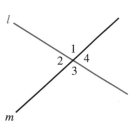

FIGURE 12.12
Vertical angles

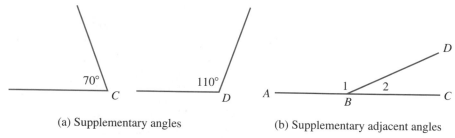

(a) Supplementary angles

(b) Supplementary adjacent angles

FIGURE 12.11

When two lines intersect, the angles opposite each other are called *vertical angles*. (See Figure 12.12.) If two straight lines intersect, the vertical angles formed

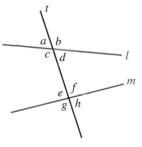

FIGURE 12.13
Line *t* is a transversal of lines *l* and *m*.

are equal. In Figure 12.12, angles 1 and 3 are vertical angles, so $\angle 1 = \angle 3$. Angles 2 and 4 are vertical angles, so $\angle 2 = \angle 4$.

A *transversal* is a line that intersects two or more lines in different points in the same plane (Figure 12.13). Angles formed between *l* and *m* are called *interior angles*. Those angles outside *l* and *m* are called *exterior angles*.

Interior angles: $\angle c, \angle d, \angle e, \angle f$

Exterior angles: $\angle a, \angle b, \angle g, \angle h$

Exterior-interior angles on the same side of the transversal are *corresponding angles*. For example, $\angle c$ and $\angle g$ in Figure 12.13 are corresponding angles. Angles *b* and *f* are also corresponding angles.

Angles on opposite sides of the transversal with different vertices are *alternate angles*. Angles *a* and *f* in Figure 12.13 are alternate angles. Angles *a* and *h* are also alternate angles.

If two *parallel* lines are cut by a transversal, then

- *corresponding* angles are *equal.*
- *alternate-interior* angles are *equal.*
- *alternate-exterior* angles are *equal.*
- *interior* angles on the same side of the transversal are *supplementary.*

In Figure 12.14, lines *l* and *m* are parallel and *t* is a transversal. The corresponding angles are equal. That is, $\angle 1 = \angle 5$, $\angle 2 = \angle 6$, $\angle 3 = \angle 7$, and $\angle 4 = \angle 8$.

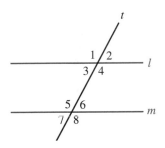

FIGURE 12.14

The alternate-interior angles are equal. That is, $\angle 3 = \angle 6$ and $\angle 4 = \angle 5$.

The alternate-exterior angles are equal. That is, $\angle 1 = \angle 8$ and $\angle 2 = \angle 7$.

The interior angles on the same side of the transversal are supplementary. That is, $\angle 3 + \angle 5 = 180°$ and $\angle 4 + \angle 6 = 180°$.

• **EXAMPLE 3**

In Figure 12.15, lines *l* and *m* are parallel and line *t* is a transversal. The measure of $\angle 2$ is 65°. Find the measure of $\angle 5$.

There are many ways of finding $\angle 5$. We show two ways.

Method 1: Angles 2 and 4 are supplementary, so

$$\angle 4 = 180° - 65° = 115°$$

Angles 4 and 5 are alternate-interior angles, so

$$\angle 4 = \angle 5 = 115°$$

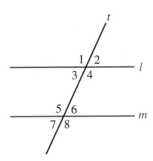

FIGURE 12.15

Method 2: Angles 2 and 3 are vertical angles, so

$$\angle 2 = \angle 3 = 65°$$

Angles 3 and 5 are interior angles on the same side of the transversal; therefore they are supplementary. $\angle 5 = 180° - 65° = 115°$ ————•

• **EXAMPLE 4**

In Figure 12.16, lines *l* and *m* are parallel and line *t* is a transversal. Given $\angle 2 = 2x + 10$ and $\angle 3 = 3x - 5$, find the measure of $\angle 1$.

Angles 2 and 3 are alternate-interior angles, so they are equal. That is,

$$2x + 10 = 3x - 5$$
$$15 = x$$

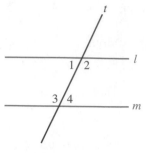

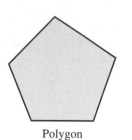

FIGURE 12.16

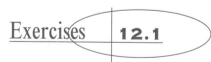

Polygon

FIGURE 12.17

Then $\angle 2 = 2x + 10 = 2(15) + 10 = 40°$. Since $\angle 1$ and $\angle 2$ are supplementary, $\angle 1 = 180° - 40° = 140°$.

Note: We let $\overline{AB}$ be the line segment with endpoints at A and B:

$$\bullet\!\!-\!\!-\!\!-\!\!-\!\!-\!\!\bullet$$
$A \qquad B$

Let $\overleftrightarrow{AB}$ be the line containing A and B:

$A \qquad B$

And let AB be the length of $\overline{AB}$.

A *polygon* is a closed figure whose sides are straight lines. A polygon is shown in Figure 12.17. Polygons are named according to the number of sides they have (see Figure 12.18). A *triangle* is a polygon with three sides. A *quadrilateral* is a polygon with four sides. A *pentagon* is a polygon with five sides. A *regular* polygon has all its sides and angles equal.

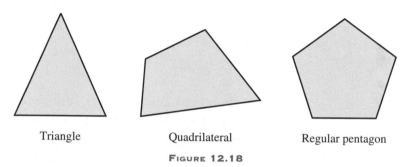

Triangle   Quadrilateral   Regular pentagon

FIGURE 12.18

Some polygons with more than five sides are named as follows.

Number of sides	Name of polygon
6	Hexagon
7	Heptagon
8	Octagon
9	Nonagon

Exercises   12.1

Use a protractor to measure each angle. Classify each angle as right, acute, or obtuse:

1.

3.

5.

7.

2.

4.

6.

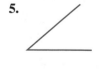

8.

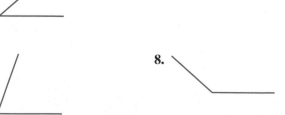

9. In Illustration 1, line *l* intersects line *m* and forms a right angle. Then ∠1 is a ___?___ angle. Lines *l* and *m* are ___?___.

10. Suppose *l* ∥ *m* and *t* ⊥ *l*. Is t ⊥ *m*? Why or why not?

11. In Illustration 2, **a.** name the pairs of adjacent angles; **b.** name the pairs of vertical angles.

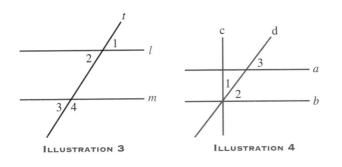

| ILLUSTRATION 1 | ILLUSTRATION 2 |

12. In Illustration 2, suppose ∠3 = 40° and ∠7 = 97°. Find the measures of the other angles.

13. In Illustration 3, suppose *l* ∥ *m* and ∠1 = 57°. What are the measures of the other angles?

14. In Illustration 4, suppose *a* ∥ *b*, *a* ⊥ *c*, and ∠1 = 37°. Find the measures of angles 2 and 3.

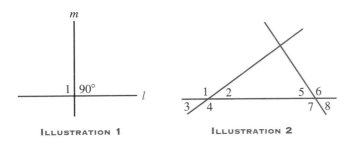

| ILLUSTRATION 3 | ILLUSTRATION 4 |

15. In Illustration 5, suppose $\overleftrightarrow{AOB}$ is a straight line and ∠AOC = 119°. What is the measure of ∠COB?

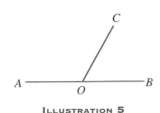

ILLUSTRATION 5

16. Suppose angles 1 and 2 are supplementary and ∠1 = 63°. Then ∠2 = ?

17. Suppose angles 3 and 4 are complementary and ∠3 = 38°. Then ∠4 = ?

18. In Illustration 6, suppose *l* ∥ *m*, $\overleftrightarrow{AOB}$ is a straight line, and ∠3 = ∠6 = 68°. Find the measure of each of the other angles.

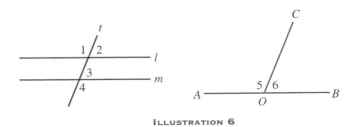

ILLUSTRATION 6

19. In Illustration 7, suppose *a* ∥ *b*, *t* ∥ *x*, ∠3 = 38°, and ∠1 = 52°.
a. Is *x* ⊥ *a*? **b.** Is *x* ⊥ *b*?

20. Suppose angles 1 and 2 are complementary and ∠1 = ∠2. Find the measure of each angle.

21. In Illustration 8, suppose ∠1 and ∠3 are supplementary. Find the measure of each angle.

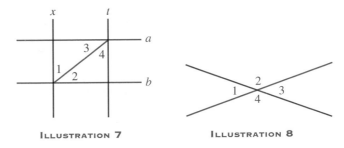

| ILLUSTRATION 7 | ILLUSTRATION 8 |

22. In Illustration 9, suppose *l* ∥ *m*, ∠1 = 3x − 50, and ∠2 = x + 60. Find the value of *x*.

23. In Illustration 9, suppose *l* ∥ *m*, ∠1 = 4x + 55, and ∠3 = 10x − 85. Find the value of *x*.

24. In Illustration 9, suppose *l* ∥ *m*, ∠1 = 8x + 60 and ∠4 = 3x + 10. Find the value of *x*.

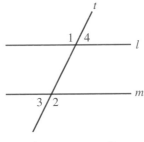

ILLUSTRATION 9

25. A plumber wishes to add a pipe parallel to an existing pipe as shown in Illustration 10. Find angle *x*.

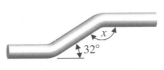

ILLUSTRATION 10

26. A machinist needs to weld a piece of iron parallel to an existing piece of iron as shown in Illustation 11. What is angle *y*?

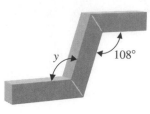

ILLUSTRATION 11

28. Given $\overline{AB} \parallel \overline{CD}$ in Illustration 13, find the measure of **a.** angle 1, **b.** angle 2, **c.** angle 3.

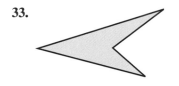

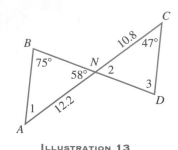

ILLUSTRATION 12   ILLUSTRATION 13

27. In Illustration 12, find angle *z* if *m* ∥ *n*.

Name each polygon:

29.

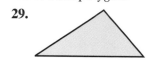

31.

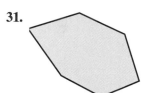

33.

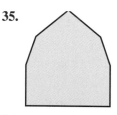

35.

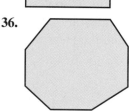

30.

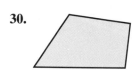

32.

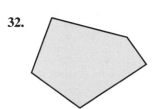

34.

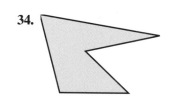

36.

12.2   Quadrilaterals

A *parallelogram* is a quadrilateral with opposite sides parallel. In Figure 12.19, sides $\overline{AB}$ and $\overline{CD}$ are parallel, and sides $\overline{AD}$ and $\overline{BC}$ are parallel. Polygon *ABCD* is therefore a parallelogram.

Figure 12.20(a) shows the same parallelogram with a perpendicular line segment drawn from point *D* to side $\overline{AB}$. This line segment is an *altitude*.

Figure 12.20(b) shows the result of removing the triangle at the left side of the parallelogram and placing it at the right side. You now have a rectangle with sides of lengths *b* and *h*. Note that the area of this rectangle, *bh* square units, is the same as the area of the parallelogram. So the area of a parallelogram is given by the formula $A = bh$, where *b* is the length of the base and *h* is the length of the altitude drawn to that base. The perimeter is $2a + 2b$ or $2(a + b)$.

FIGURE 12.19

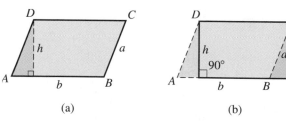

(a)   (b)

FIGURE 12.20

A *rectangle* is a parallelogram with four right angles. The area of the rectangle with sides of lengths *b* and *h* is given by the formula $A = bh$. (See Figure 12.21.)

Another way to find the area of a rectangle is to count the number of square units in it. In Figure 12.22, there are 15 squares in the rectangle, so the area is 15 square units.

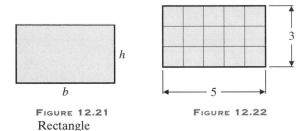

FIGURE 12.21
Rectangle

FIGURE 12.22

The formula for the area of each of the following quadrilaterals follows from the formula for the area of a rectangle.

A *square* (Figure 12.23) is a rectangle with the lengths of all four sides equal. Its area is given by the formula $A = b \cdot b = b^2$. The perimeter is $b + b + b + b$, or $4b$. Note that the length of the altitude is also b.

FIGURE 12.23
Square

A *rhombus* (Figure 12.24) is a parallelogram with the lengths of all four sides equal. Its area is given by the formula $A = bh$. The perimeter is $b + b + b + b$, or $4b$.

A *trapezoid* (Figure 12.25) is a quadrilateral with only two sides parallel. Its area is given by the formula $A = \left(\dfrac{a + b}{2}\right)h$. The perimeter is $a + b + c + d$.

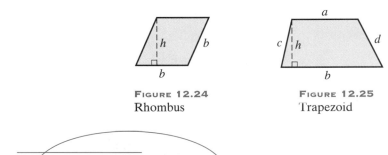

FIGURE 12.24
Rhombus

FIGURE 12.25
Trapezoid

SUMMARY OF FORMULAS FOR AREA AND PERIMETER OF QUADRILATERALS		
Quadrilateral	Area	Perimeter
Rectangle	$A = bh$	$P = 2(b + h)$
Square	$A = b^2$	$P = 4b$
Parallelogram	$A = bh$	$P = 2(a + b)$
Rhombus	$A = bh$	$P = 4b$
Trapezoid	$A = \left(\dfrac{a + b}{2}\right)h$	$P = a + b + c + d$

Note: Follow the rules for working with measurements in the rest of this chapter.

• **EXAMPLE 1**   Find the area and the perimeter of the parallelogram shown in Figure 12.26.

The formula for the area of a parallelogram is

$$A = bh$$

So $A = (27.2 \text{ m})(15.5 \text{ m})$

$= 422 \text{ m}^2$

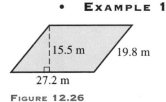

FIGURE 12.26

The formula for the perimeter of a parallelogram is

$$P = 2(a + b)$$
$$\text{So} \quad P = 2(19.8\text{ m} + 27.2\text{ m})$$
$$= 2(47.0\text{ m})$$
$$= 94.0\text{ m}$$

• **EXAMPLE 2**   A rectangular lot 121.5 ft by 98.7 ft must be fenced. (See Figure 12.27.) A fence is installed for $5.50 per running foot. Find the cost of fencing the lot.

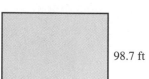

98.7 ft

121.5 ft

FIGURE 12.27

The length of fencing needed equals the perimeter of the rectangle. The formula for the perimeter is

$$P = 2(b + h)$$
$$\text{So} \quad P = 2(121.5\text{ ft} + 98.7\text{ ft})$$
$$= 2(220.2\text{ ft})$$
$$= 440.4\text{ ft}$$

$$\text{Cost} = \frac{5.50}{1\text{ ft}} \times 440.4\text{ ft} = \$2422.20$$

• **EXAMPLE 3**   Find the cost of the fertilizer needed for the lawn in Example 2. One bag covers 2500 ft^2 and costs $12.95.

First, find the area of the rectangle. The formula for the area is

$$A = bh$$
$$\text{So} \quad A = (121.5\text{ ft})(98.7\text{ ft})$$
$$= 12{,}\overline{0}00\text{ ft}^2$$

The amount of fertilizer needed is found by dividing the total area by the area covered by one bag:

$$\frac{12{,}000\text{ ft}^2}{2500\text{ ft}^2} = 4.8\text{ bags, or 5 bags.}$$

$$\text{Cost} = 5\text{ bags} \times \frac{\$12.95}{1\text{ bag}}$$
$$= \$64.75$$

Exercises   12.2

Find the perimeter and the area of each quadrilateral:

1.

15.0 cm

15.0 cm

2.

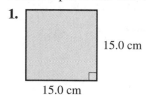

8.00 cm

10.0 cm

3.

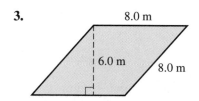

8.0 m

6.0 m

8.0 m

4.

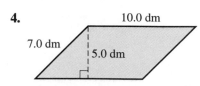

10.0 dm

7.0 dm

5.0 dm

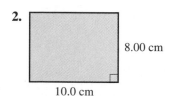

5.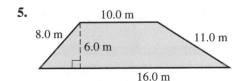

6.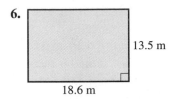

7.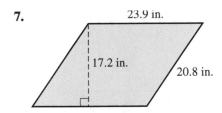

8.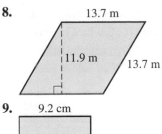

9.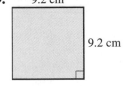

10.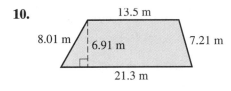

In Exercises 11–15, use the formula $A = bh$:

11. $A = 24\overline{0}$ cm^2, $b = 10.0$ cm; find h.

12. $A = 792$ m^2, $h = 25.0$ m; find b.

13. $h = 17.8$ mm, $A = 207$ mm^2; find b.

14. $A = 179.3$ cm^2, $b = 13.24$ cm; find h.

15. $A = 29.6$ ft^2, $b = 5.04$ ft; find h.

16. The area of a rectangle is 280 cm^2. Its width is 14 cm. Find its length.

17. The area of a parallelogram is 486 ft^2. The length of its base is 36.2 ft. Find its height.

18. The area of a parallelogram is 273 in^2. The length of its base is 17.6 in. Find its height.

19. The area of a rectangle is 315 m^2. Its height is 38.7 m. Find its width.

20. The area of a rectangle is 1087 cm^2. Its length is 204.2 cm. Find its width.

21. A parallelogram has an area of 2014 cm^2. The length of its base is 52.62 cm. Find its height.

22. A rectangle has a length of 79.2 m. Its area is 1050 m^2. Find its width.

23. The area of a rectangle is 2056.7 km^2. The length of its base is 45.147 km. Find its height.

24. A rectangular X-ray film measures 15 cm by 32 cm. What is its area?

25. Each hospital bed and its accessories use 96 ft^2 of floor space. How many beds can be placed in a ward 24 ft by 36 ft?

26. A respirator unit needs a rectangular floor space of 0.79 m by 1.2 m. How many units could be placed in a storeroom having $2\overline{0}$ m^2 of floor space?

27. A 108 ft^2 roll of fiberglass is 36 in. wide. What is its length in feet?

28. The cost of the fiberglass in Exercise 27 is $9.16/yd^2. How much would this roll cost?

29. How many pieces of fiberglass, 36 in. wide and 72 in. long, can be cut from the roll in Exercise 27?

30. A mechanic plans to build a storage garage for 85 cars. Each car needs a space of 15.0 ft by 10.0 ft. Find the floor area of the garage.

31. At a cost of $14/ft^2, find the cost of the garage in Exercise 30.

32. A machinist plans to build a screen around his shop area. The area is rectangular and measures 16.2 ft by 20.7 ft. **a.** How many linear feet of screen will be needed? **b.** If the screen is to be 8.0 ft high, how many square feet of screen will be needed?

33. A piece of sheet metal in the shape of a parallelogram has a rectangular hole in it, as shown in Illustration 1. Find **a.** the area of the piece that was punched out, and **b.** the area of the metal that is left.

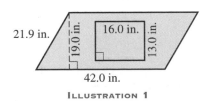

ILLUSTRATION 1

34. A rectangular piece of sheet metal has an area of 10,680 in². Its length is 72.0 in. Find its width.

35. What is the acreage of a ranch that is a square 25.0 mi on a side?

36. A rectangular field of corn is averaging 97 bu/acre. The field measures 1020 yd by 928 yd. How many bushels of corn will there be?

37. Find the amount of sheathing needed for the roof in Illustration 2. How many squares of shingles must be purchased? (1 square = 100 ft²)

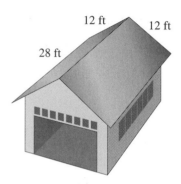

ILLUSTRATION 2

38. A ceiling is 12 ft by 15 ft. How many 1-ft-by-3-ft suspension panels are needed to cover the ceiling?

39. The Smith family plan to paint their home (shown in Illustration 3). The area of the openings not to be painted is 325 ft². The cost per square foot is $0.85. Find the cost of painting the house.

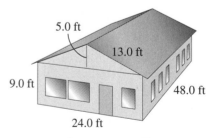

ILLUSTRATION 3

40. An 8-in.-thick wall uses 15 standard bricks (8 in. by $2\frac{1}{4}$ in. by $3\frac{3}{4}$ in.) for one square foot. Find the number of bricks needed for a wall 8 in. thick, 18 ft long, and 8.5 ft high.

41. What is the display floor space of a parallelogram-shaped space that is 29.0 ft long and 8.7 ft deep?

42. Canvas that costs $\frac{3}{4}$¢/in² is used to make golf bags. Find the cost of 200 rectangular pieces of canvas, each 8 in. by 40 in.

43. By law, all businesses outside the Parkville city limits must fence their lots. How many feet of fence will be needed to fence the parallelogram-shaped lot shown in Illustration 4.

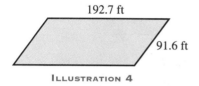

ILLUSTRATION 4

44. Find the area of the three trapezoid-shaped display floor spaces of the stores shown in Illustration 5.

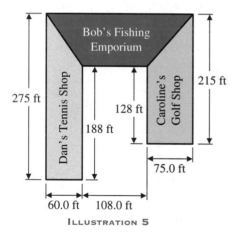

ILLUSTRATION 5

12.3 Triangles

Triangles are often classified in two ways:

1. by the number of equal sides
2. by the measures of the angles of the triangle

Triangles may be classified or named by the relative lengths of their sides. In each triangle in Figure 12.28, the lengths of the sides are represented by *a*, *b*, and *c*.

An *equilateral triangle* has all three sides equal. All three angles are also equal. An *isosceles triangle* has two sides equal. The angles opposite these two sides are also equal. A *scalene triangle* has no sides equal. No angles are equal either.

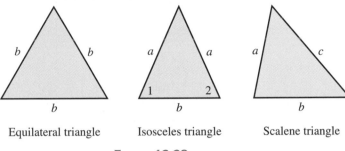

Equilateral triangle   Isosceles triangle   Scalene triangle

FIGURE 12.28
Triangles named by sides

Triangles may also be classified or named in terms of the measures of their angles (see Figure 12.29). A *right triangle* has one right angle. The two sides of the right angle are called *legs* of the right triangle. The side opposite the right angle is called the *hypotenuse*. An *acute triangle* has three acute angles. An *obtuse triangle* has one obtuse angle.

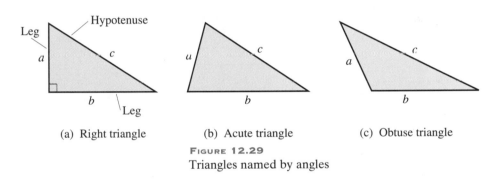

(a) Right triangle   (b) Acute triangle   (c) Obtuse triangle

FIGURE 12.29
Triangles named by angles

In a right triangle, the side opposite the right angle is called the *hypotenuse*, which we label c. The other two sides are called *legs*, which we label a and b. (See Figure 12.29(a).) The *Pythagorean theorem* relates the lengths of the sides of any right triangle.

Pythagorean Theorem

$$c^2 = a^2 + b^2 \text{ or } c = \sqrt{a^2 + b^2}$$

The Pythagorean theorem states that *the square of the hypotenuse is equal to the sum of the squares of the two legs.* Alternate forms of the Pythagorean theorem are:

$$a^2 = c^2 - b^2 \text{ or } a = \sqrt{c^2 - b^2}$$
and $\quad b^2 = c^2 - a^2 \text{ or } b = \sqrt{c^2 - a^2}$

Before we use the Pythagorean theorem, you may wish to review square roots, in Section 10.5.

• **EXAMPLE 1**   Find the length of the hypotenuse of the triangle in Figure 12.30.

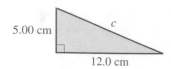

FIGURE 12.30

Substitute 5.00 cm for a and 12.0 cm for b in the formula:

$$c = \sqrt{a^2 + b^2}$$
$$c = \sqrt{(5.00 \text{ cm})^2 + (12.0 \text{ cm})^2}$$
$$= \sqrt{25.0 \text{ cm}^2 + 144 \text{ cm}^2}$$
$$= \sqrt{169 \text{ cm}^2} = 13.0 \text{ cm}$$

Flowchart	Buttons Pushed	Display
Enter 5	5	5
↓		
Push square	x^2	25
↓		
Push plus	+	25
↓		
Enter 12	1 → 2	12
↓		
Push square	x^2	144
↓		
Push equals	=	169
↓		
Push square root	√	13

•   **EXAMPLE 2**   Find the length of the hypotenuse of the triangle in Figure 12.31.

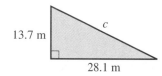

FIGURE 12.31

$$c = \sqrt{a^2 + b^2}$$
$$c = \sqrt{(13.7 \text{ m})^2 + (28.1 \text{ m})^2}$$
$$= 31.3 \text{ m}$$

•   **EXAMPLE 3**   Find the length of side b of the triangle in Figure 12.32.

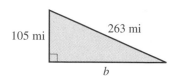

FIGURE 12.32

$$b = \sqrt{c^2 - a^2}$$
$$b = \sqrt{(263 \text{ mi})^2 - (105 \text{ mi})^2}$$
$$= 241 \text{ mi}$$

•   **EXAMPLE 4**   The right triangle in Figure 12.33 gives the relationship in a circuit among the applied voltage, the voltage across a resistance, and the voltage across a coil. The voltage across the resistance is 79 V. The voltage across the coil is 82 V. Find the applied voltage.

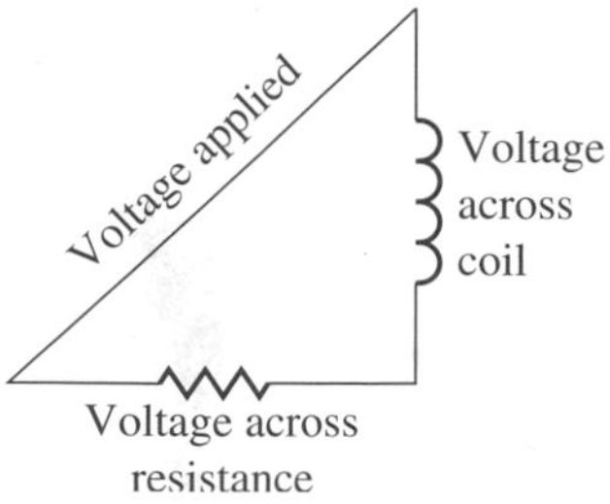

FIGURE 12.33

Using the Pythagorean theorem:

$$\text{Voltage applied} = \sqrt{(\text{voltage across coil})^2 + (\text{voltage across resistance})^2}$$
$$= \sqrt{(82 \text{ V})^2 + (79 \text{ V})^2}$$
$$= 110 \text{ V}$$

Perimeter and Area

To find the perimeter of a triangle, find the sum of the lengths of the three sides. The formula is $P = a + b + c$, where P is the perimeter and a, b, and c are the lengths of the sides.

An *altitude* of a triangle is a line segment drawn perpendicular from one vertex to the opposite side. Sometimes this opposite side must be extended. See Figure 12.34.

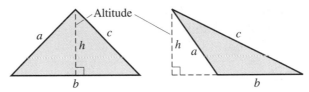

FIGURE 12.34

Look closely at a parallelogram (Figure 12.35(a)) to find the formula for the area of a triangle. Remember that the area of a parallelogram with sides of lengths a and b is given by $A = bh$. In this formula, b is the length of the base of the parallelogram, and h is the length of the altitude drawn to that base.

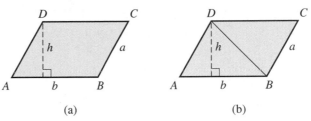

(a)   (b)

FIGURE 12.35

Next, draw a line segment from B to D in the parallelogram as in Figure 12.35(b). Two triangles are formed. We know from geometry that these two triangles have equal areas. Since the area of the parallelogram is bh square units, the area of one triangle is one-half the area of the parallelogram. So the formula for the area of a triangle is

---

Area of Triangle

$$A = \frac{1}{2}bh$$

---

where b is the length of the base of the triangle (the side to which the altitude is drawn) and h is the length of the altitude.

• **EXAMPLE 5**   The length of the base of a triangle is 10.0 cm. The length of the altitude to that base is 6.00 cm. Find the area of the triangle. (See Figure 12.36.)

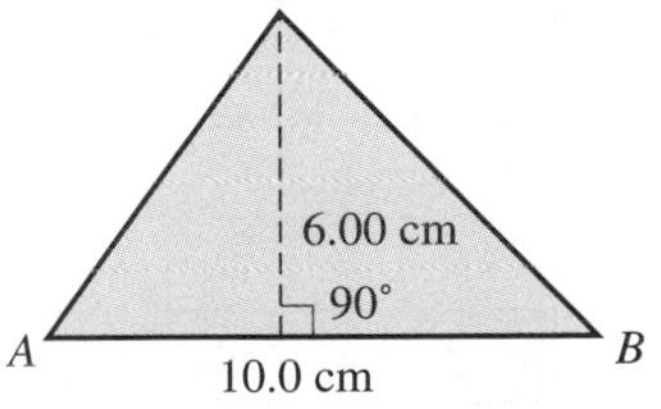

$$A = \frac{1}{2}bh$$

$$A = \frac{1}{2}(10.0 \text{ cm})(6.00 \text{ cm})$$

$$= 30.0 \text{ cm}^2$$

The area is 30.0 cm².

FIGURE 12.36

If only the lengths of the three sides are known, the area of a triangle is found by the following formula (called Heron's formula):

$$A = s\sqrt{(s-a)(s-b)(s-c)}$$

where a, b, and c are the lengths of the three sides and $s = \frac{1}{2}(a + b + c)$.

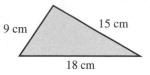

FIGURE 12.37

9 cm   15 cm   18 cm

EXAMPLE 6   Find the perimeter and the area (rounded to three significant digits) of the triangle in Figure 12.37.

$$P = a + b + c$$
$$P = 9 \text{ cm} + 15 \text{ cm} + 18 \text{ cm} = 42 \text{ cm}$$

To find the area, first find s.

$$s = \frac{1}{2}(a + b + c)$$

$$s = \frac{1}{2}(9 + 15 + 18) = \frac{1}{2}(42) = 21$$

$$A = \sqrt{s(s-a)(s-b)(s-c)}$$
$$A = \sqrt{21(21-9)(21-15)(21-18)}$$
$$= \sqrt{21(12)(6)(3)}$$
$$= \sqrt{4536}$$
$$= 67.3 \text{ cm}^2$$

Flowchart	Buttons Pushed	Display
Enter 21	2 → 1	21
Push times	×	21
Left parenthesis	(	21
Enter 21	2 → 1	21
Push minus	−	21
Enter 9	9	9
Right parenthesis	)	12
Push times	×	252
Left parenthesis	(	252

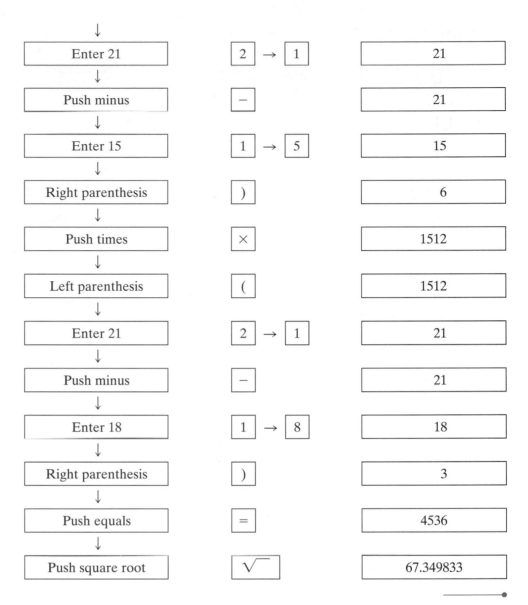

Enter 21	2 → 1	21
Push minus	−	21
Enter 15	1 → 5	15
Right parenthesis	)	6
Push times	×	1512
Left parenthesis	(	1512
Enter 21	2 → 1	21
Push minus	−	21
Enter 18	1 → 8	18
Right parenthesis	)	3
Push equals	=	4536
Push square root	√	67.349833

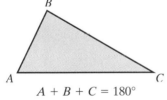

$$A + B + C = 180°$$

FIGURE 12.38

The following relationship is often used in geometry and trigonometry.

The sum of the measures of the angles of any triangle is 180° (Figure 12.38).

• **EXAMPLE 7** Two angles of a triangle have measures 80° and 40°. (See Figure 12.39.) Find the measure of the third angle of the triangle.

Since the sum of the measures of the angles of any triangle is 180°, we know that:

$$40° + 80° + x = 180°$$
$$120° + x = 180°$$
$$x = 60°$$

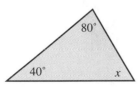

FIGURE 12.39

So the measure of the missing angle is 60°.

Exercises 12.3

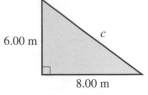

Use the rules for working with measurements.

Find the length of the hypotenuse in each triangle:

1.

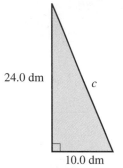

6.00 m, c, 8.00 m

3.

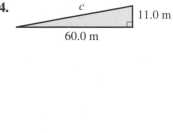

24.0 m, 7.00 m, c

5.

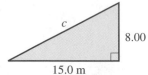

c, 8.00, 15.0 m

2.

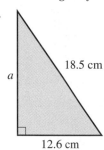

24.0 dm, c, 10.0 dm

4.

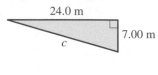

c, 11.0 m, 60.0 m

6.

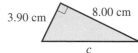

3.90 cm, 8.00 cm, c

Find the length of the missing side in each triangle:

7.

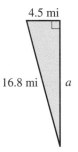

18.5 cm, a, 12.6 cm

9.

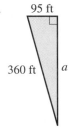

b, 1980 km, 2460 km

11.

95 ft, 360 ft, a

8.

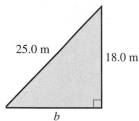

25.0 m, 18.0 m, b

10.

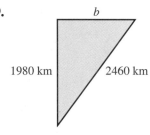

4.5 mi, 16.8 mi, a

12.

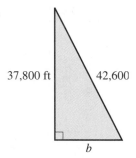

37,800 ft, 42,600, b

13. Find the length of the braces needed for the rectangular supports shown in Illustration 1.

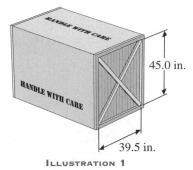

HANDLE WITH CARE

45.0 in.

39.5 in.

ILLUSTRATION 1

14. Find the center-to-center distance between the two holes in Illustration 2.

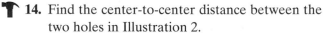

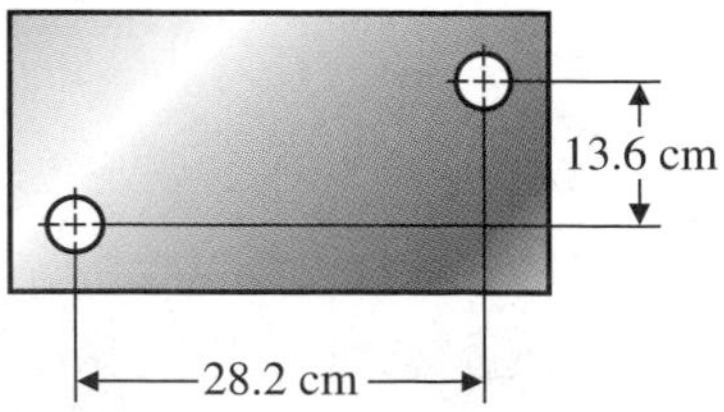

13.6 cm

28.2 cm

ILLUSTRATION 2

15. Find the length of the brace needed in Illustration 3.

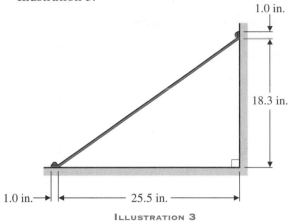

ILLUSTRATION 3

16. Often a machinist must cut a keyway in a shaft. The total depth of cut equals the keyway depth plus the height of a circular segment. The height of a circular segment is found by applying the Pythagorean theorem or by using the formula

$$h = r - \sqrt{r^2 - \left(\frac{l}{2}\right)^2}$$

where h is the height of the segment, r is the radius of the shaft, and l is the length of the chord (or the width of the keyway). Find the total depth of cut shown in Illustration 4.

17. A piece of 4.00-in.-diameter round stock is to be milled into a square piece of stock with the largest dimensions possible. (See Illustration 5.) What will be the length of the side of the square?

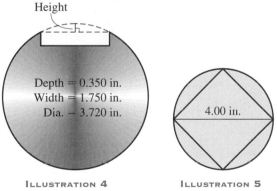

ILLUSTRATION 4   ILLUSTRATION 5

18. Find the length of the rafter in Illustration 6.

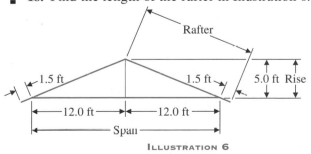

ILLUSTRATION 6

19. Find the offset distance x (rounded to nearest tenth of an inch) of the 6-ft length of pipe shown in Illustration 7.

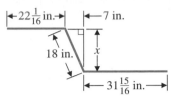

ILLUSTRATION 7

20. A conduit is run in a building (see Illustration 8). **a.** Find the length of the conduit from A_1 to A_6. **b.** Find the straight-line distance from A_1 to A_6.

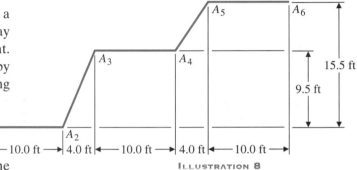

ILLUSTRATION 8

21. The voltage across a resistance is 85.2 V. The voltage across a coil is 78.4 V. Find the voltage applied in the circuit. (See Illustration 9.)

22. The voltage across a coil is 362 V. The voltage applied is 537 V. Find the voltage across the resistance. (See Illustration 9.)

23. The resistor current is 24 A. The total current is 32 A. Find the coil current. (See Illustration 10.)

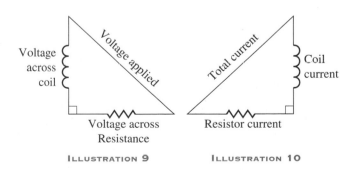

ILLUSTRATION 9   ILLUSTRATION 10

24. The resistor current is 50.2 A. The coil current is 65.3 A. Find the total current. (See Illustration 10.)

In Exercises 25–27, see Illustration 11:

25. Find the reactance of a circuit with impedance 165 Ω and resistance 105 Ω.

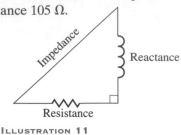

Impedance

Reactance

Resistance

ILLUSTRATION 11

26. Find the impedance of a circuit with reactance 20.2 Ω and resistance 38.3 Ω.

27. Find the resistance of a circuit with impedance 4.5 Ω and reactance 3.7 Ω.

28. The base of a window is 7.2 m above the ground. The lower end of a ladder is 3.1 m from the side of the house. How long must a ladder be to reach the base of the window?

Find the area and perimeter of each isosceles triangle:

29.

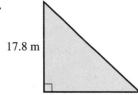

17.8 m

30.

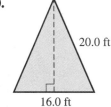

20.0 ft

16.0 ft

31. Find the area and perimeter of an equilateral triangle with one side 6.00 cm long.

32. Find the area and perimeter of an equilateral triangle with one side 18.0 m long.

Find the area and perimeter of each triangle:

33.

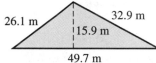

26.1 m   32.9 m

15.9 m

49.7 m

37.

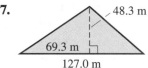

48.3 m

69.3 m

127.0 m

34.

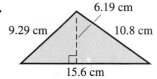

6.19 cm

9.29 cm   10.8 cm

15.6 cm

38.

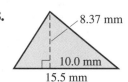

8.37 mm

10.0 mm

15.5 mm

35.

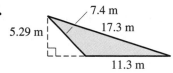

7.4 m

5.29 m   17.3 m

11.3 m

39.

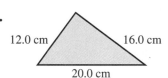

12.0 cm   16.0 cm

20.0 cm

36.

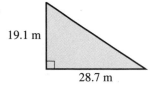

19.1 m

28.7 m

40.

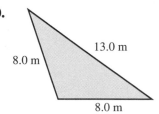

13.0 m

8.0 m

8.0 m

41. A triangular machine guard is to be made of steel mesh. The machine has a base of 12.5 ft and a height of 8.3 ft. How many square feet of mesh are needed?

42. A triangular piece of metal stock weighs 38 lb. The length of the base is 4.0 ft and the altitude is 2.5 ft. What is the weight per square foot of the metal?

43. A steel plate is punched with a triangular hole as shown in Illustration 12. Find the area of the hole.

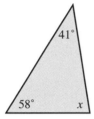

28.6 in.

11.0 in.

26.4 in.

ILLUSTRATION 12

44. A square hole is cut from the equilateral triangle in Illustration 13. Find the area remaining in the triangle.

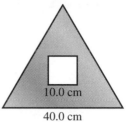

10.0 cm

40.0 cm

ILLUSTRATION 13

Find the measure of the missing angle in each triangle (do not use a protractor):

45.

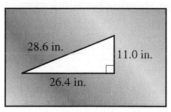

41°

58°   x

47.

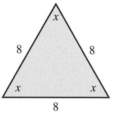

x

8   8

x   x

8

49.

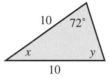

10   72°

x   y

10

46.

x

62°

48.

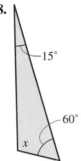

15°

60°

x

50.

x

62.1°   19.7°

12.4   Similar Polygons

Polygons with the same shape are called *similar* polygons. Polygons are similar when the *corresponding angles are equal*. In Figure 12.40, polygon *ABCDE* is similar to polygon *A'B'C'D'E'* because the corresponding angles are equal. That is, $\angle A = \angle A'$, $\angle B = \angle B'$, $\angle C = \angle C'$, $\angle D = \angle D'$, and $\angle E = \angle E'$.

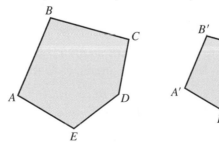

FIGURE 12.40
Similar polygons

When two polygons are similar, the lengths of the *corresponding sides are proportional.* That is,

$$\frac{AB}{A'B'} = \frac{BC}{B'C'} = \frac{CD}{C'D'} = \frac{DE}{D'E'} = \frac{EA}{E'A'}$$

• **EXAMPLE 1**   The polygons in Figure 12.40 are similar and $AB = 12$, $DE = 3$, $A'B' = 8$. Find $D'E'$.

Since the polygons are similar,

$$\frac{AB}{A'B'} = \frac{DE}{D'E'}$$

$$\frac{12}{8} = \frac{3}{D'E'}$$

$$12(D'E') = (8)(3)$$

$$D'E' = \frac{24}{12} = 2$$

Two triangles are similar when two pairs of corresponding angles are equal, as in Figure 12.41. (If two pairs of corresponding angles are equal, then the third pair of corresponding angles must also be equal. Why?)

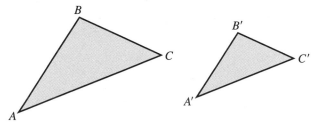

FIGURE 12.41
Similar triangles

Triangle ABC is similar to triangle $A'B'C'$ because $\angle A = \angle A'$, $\angle B = \angle B'$, and $\angle C = \angle C'$.

When two triangles are similar, the lengths of the corresponding sides are proportional. That is,

$$\frac{AB}{A'B'} = \frac{BC}{B'C'} = \frac{CA}{C'A'}$$

Or

$$\frac{AB}{A'B'} = \frac{BC}{B'C'} \qquad \frac{BC}{B'C'} = \frac{CA}{C'A'} \quad \text{and} \quad \frac{AB}{A'B'} = \frac{CA}{C'A'}$$

• **EXAMPLE 2**   Find DE and AE in Figure 12.42.

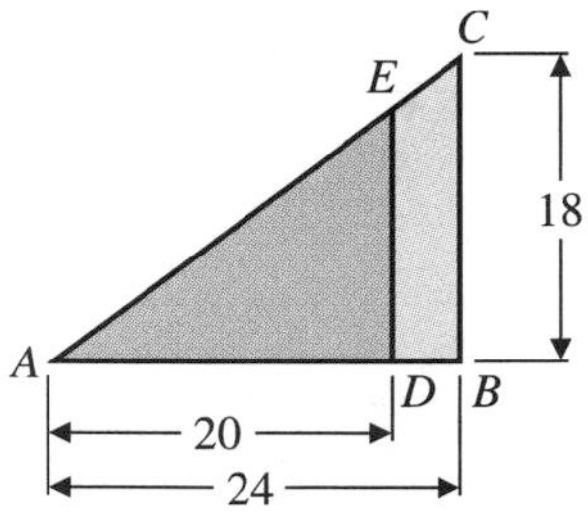

Triangles ADE and ABC are similar, because $\angle A$ is common to both and each triangle has a right angle. So the lengths of the corresponding sides are proportional.

$$\frac{AB}{AD} = \frac{BC}{DE}$$

$$\frac{24}{20} = \frac{18}{DE}$$

$$24(DE) = (20)(18)$$

$$DE = \frac{360}{24} = 15$$

FIGURE 12.42

Use the Pythagorean theorem to find AE.

$$AE = \sqrt{(AD)^2 + (DE)^2}$$
$$AE = \sqrt{(20)^2 + (15)^2}$$
$$= \sqrt{400 + 225} = \sqrt{625} = 25$$

Exercises | **12.4**

Follow the rules for working with measurements beginning with Exercise 6:

1. In Illustration 1, suppose $\overline{DE} \parallel \overline{BC}$. Find DE.

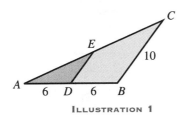

ILLUSTRATION 1

2. In Illustration 2, polygon $ABCD$ is similar to polygon $FGHI$. Find **a.** $\angle H$ **b.** FI **c.** IH **d.** BC.

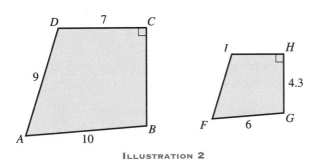

ILLUSTRATION 2

3. In Illustration 3, $\overline{AB} \parallel \overline{CD}$. Is triangle ABO similar to triangle CDO? Why or why not?

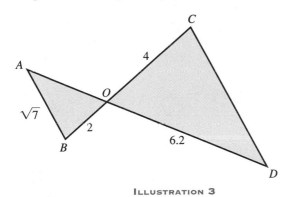

ILLUSTRATION 3

4. Find the lengths of $\overline{AO}$ and $\overline{CD}$ in Illustration 3.

5. In Illustration 4, quadrilaterals $ABCD$ and $XYZW$ are similar rectangles. $AB = 12$, $BC = 8$, and $XY = 8$. Find YZ.

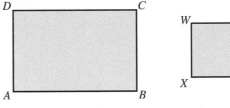

ILLUSTRATION 4

6. An inclined ramp is to be built so that it reaches a height of 6.00 ft over a 15.00 ft run. (See Illustration 5.) Braces are placed every 5.00 ft. Find the height of braces x and y.

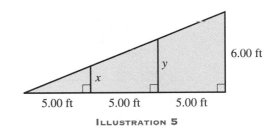

ILLUSTRATION 5

7. A tree casts a shadow $8\bar{0}$ ft long when a vertical rod 6.0 ft high casts a shadow 4.0 ft long. (See Illustration 6.) How tall is the tree?

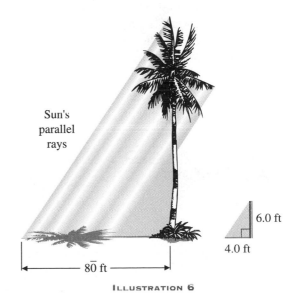

Sun's parallel rays

6.0 ft

4.0 ft

$8\bar{0}$ ft

ILLUSTRATION 6

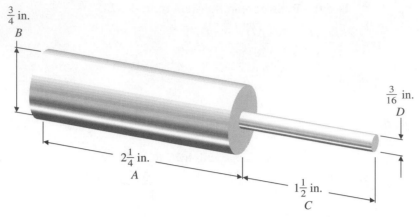

ILLUSTRATION 7

8. A machinist must follow a part drawing with scale 1 to 16. Find the dimensions of the finished stock shown in Illustration 7. That is, find lengths *A*, *B*, *C*, and *D*.

9. Find the length of *DE* in Illustration 8.

10. Find the length of *BC* in Illustration 8.

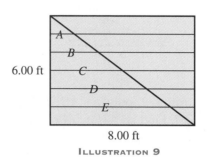

ILLUSTRATION 9

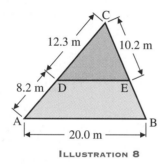

ILLUSTRATION 8

11. A 6.00-ft-by-8.00-ft bookcase is to be built. It has horizontal shelves every foot. A support is to be notched in the shelves diagonally from one corner to the opposite corner. At what point should each of the shelves be notched? That is, find lengths *A*, *B*, *C*, *D,* and *E* in Illustration 9. How long is the crosspiece?

12. A vertical tower 132 ft high is anchored to the ground by guy wires as shown in Illustration 10. How long is each guy wire?

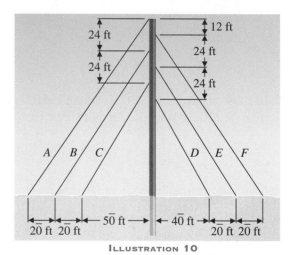

ILLUSTRATION 10

12.5   Circles

A *circle* is a plane curve consisting of all points at a given distance (called the *radius*, *r*) from a fixed point in the plane, called the *center*. (See Figure 12.43.) The *diameter*, *d*, of the circle is a line segment through the center of the circle with endpoints on the circle. Note that the length of the diameter equals the length of two radii—that is, $d = 2r$.

The *circumference* of a circle is the distance around the circle. The ratio of the circumference of a circle to the length of its diameter is a constant called π (pi). The

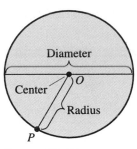

FIGURE 12.43
Circle

number π cannot be written exactly as a decimal. Decimal approximations for π are 3.14 or 3.1416. When solving problems with π, use the π button on your calculator.

The following formulas are used to find the circumference and the area of a circle. C is the circumference and A is the area of a circle; d is the length of the diameter and r is the length of the radius.

CIRCUMFERENCE OF CIRCLE:

$$C = \pi d$$
$$C = 2\pi r$$

AREA OF CIRCLE:

$$A = \pi r^2$$
$$A = \frac{\pi d^2}{4}$$

• **EXAMPLE 1** Find the area and circumference of the circle shown in Figure 12.44.

The formula for the area of a circle given the radius is

$$A = \pi r^2$$
$$A = \pi(16.0 \text{ cm})^2$$
$$= 804 \text{ cm}^2$$

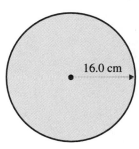

FIGURE 12.44

The formula for the circumference of a circle given the radius is

$$C = 2\pi r$$
$$C = 2\pi(16.0 \text{ cm})$$
$$= 101 \text{ cm}$$

• **EXAMPLE 2** The area of a circle is 576 m². Find the radius.

The formula for the area of a circle in terms of the radius is

$$A = \pi r^2$$
$$576 \text{ m}^2 = \pi r^2$$
$$\frac{576 \text{ m}^2}{\pi} = r^2 \qquad \text{Divide both sides by } \pi.$$
$$\sqrt{\frac{576 \text{ m}^2}{\pi}} = r \qquad \text{Take the square root of both sides.}$$
$$13.5 \text{ m} = r$$

• **EXAMPLE 3** The circumference of a circle is 28.2 cm. Find the radius.

The formula for the circumference of a circle in terms of the radius is

$$C = 2\pi r$$
$$28.2 \text{ cm} = 2\pi r$$
$$\frac{28.2 \text{ cm}}{2\pi} = r \qquad \text{Divide both sides by } 2\pi.$$
$$4.49 \text{ cm} = r$$

An angle whose vertex is at the center of a circle is called a *central* angle. Angle *A* in Figure 12.45 is a central angle.

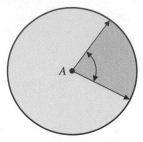

FIGURE 12.45
Central angle

•   **EXAMPLE 4**   Measure the central angles in Figure 12.46 and find their sum.

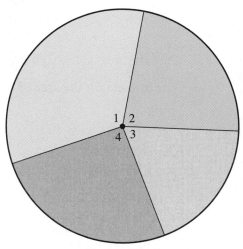

FIGURE 12.46

Measure of angle 1 = 117°
Measure of angle 2 =   83°
Measure of angle 3 =   68°
Measure of angle 4 =   92°
Sum: 360°

The sum of the measures of all the central angles of any circle is 360°.

Common Terms and Relationships of a Circle

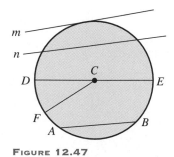

FIGURE 12.47

A *chord* is a line segment that has its endpoints on the circle.

A *secant* is any line that has a chord in it.

A *tangent* is a line (or a line segment) that has only one point in common with a circle and lies totally outside the circle. In Figure 12.47, *C* is the center. $\overline{AB}$ is a chord. Line *n* is a secant. Line *m* is a tangent. $\overline{DE}$ is a diameter. $\overline{CF}$ is a radius.

An *inscribed angle* is an angle whose vertex is on the circle and whose sides are chords.

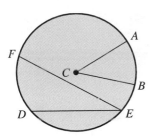

FIGURE 12.48

Arcs

That part of the circle between the two sides of an inscribed or central angle is called the *intercepted arc.* In Figure 12.48, C is the center and $\angle ACB$ is a central angle. $\angle DEF$ is an inscribed angle. $\overset{\frown}{AB}$ is the intercepted arc of $\angle ACB$. $\overset{\frown}{DF}$ is the intercepted arc of $\angle DEF$.

The number of degrees in a central angle is equal to the number of arc degrees in its intercepted arc. (See Figure 12.49.) The measure of an inscribed angle is equal in degrees to one-half the number of arc degrees in its intercepted arc.

$\overset{\frown}{AB}$ = 40 arc degrees
$\angle ACB = 40°$
$\overset{\frown}{MN}$ = 50 arc degrees
$\angle MAN = 25°$

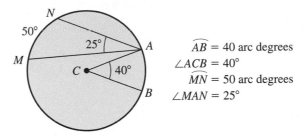

FIGURE 12.49

The measure of an angle formed by two intersecting chords is equal in number to one-half the sum of the measures of the intercepted arcs. (See Figure 12.50.)

$\overset{\frown}{AC}$ = 38 arc degrees
$\overset{\frown}{DB}$ = 22 arc degrees
$\angle AEC = \frac{1}{2}(38° + 22°) = 30°$

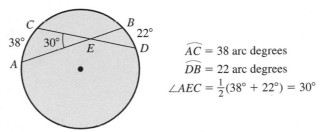

FIGURE 12.50

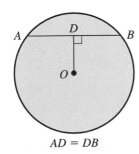

$AD = DB$

FIGURE 12.51

Other Chords and Tangents

A diameter that is perpendicular to a chord bisects the chord. (See Figure 12.51.)

A line segment from the center of a circle to the point of tangency is perpendicular to the tangent. (See Figure 12.52.)

Two tangents drawn from a point outside a circle to the circle are equal. The line segment drawn from the center of the circle to this point outside the circle bisects the angle formed by the tangents. (See Figure 12.53.)

$CE \perp m$

FIGURE 12.52

$AM = AN$
$\angle CAM = \angle CAN$

FIGURE 12.53

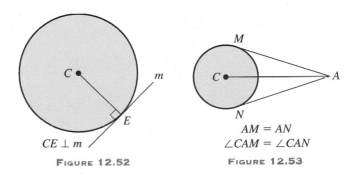

Exercises 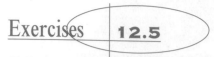 12.5

Follow the rules for working with measurements.

Find **a.** *the circumference and* **b.** *the area of each circle:*

1.

5.00 in.

3.

9.21 mm

5.

56.1 mi

2.

20.0 m

4.

2.70 cm

6.

39.8 mm

Find the measure of each unknown angle:

7.

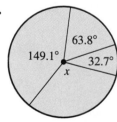

97°
92° *x*

9.

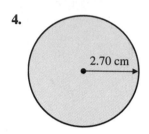

x 31.8°
111.1°   29.8°
143.9°

8.

63.8°
149.1°   32.7°
x

10.

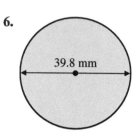

30.8°
x
130.8°   117.9°
28.3°

11. The area of a circle is 28.2 cm². Find its radius.

12. The area of a circle is 214 ft². Find its radius.

13. The area of a circle is 21.2 in². Find its radius.

14. The area of a circle is 792 m². Find its radius.

15. The circumference of a circle is 62.9 m. Find its radius.

16. The circumference of a circle is 17.2 in. Find its radius.

17. The circumference of a circle is 202 ft. Find its radius.

18. The circumference of a circle is 2.38 mm. Find its radius.

19. How many degrees are in a central angle whose arc is $\frac{1}{4}$ of a circle?

20. How many degrees are in a central angle whose arc is $\frac{2}{3}$ of a circle?

21. A wheel of radius 1.80 ft is used to measure a field. The wheel rotates 236 times while going the length of the field. How long is the field?

22. Find the length of the diameter of a circular silo with circumference of 52.0 ft.

23. A rectangular piece of insulation is to be wrapped around a pipe 4.25 in. in diameter. (See Illustration 1.) How wide does the rectangular piece need to be?

Insulation

4.25 in.

ILLUSTRATION 1

24. How many 1.5-in.-diameter pipes have approximately the same total cross-sectional area as one whose diameter is 5.0 in.?

25. A manifold is being designed to carry compressed gas from a tank to four processing stations where the gas is being used. (See Illustration 2.) The main line from the tank is 2.50 in. in diameter. The same equivalent cross-sectional area must be maintained in the four outlet pipes as the cross-sectional area of the main line. For simplicity, we will not consider flow restriction due to friction, turbulence, or bends in the cylindrical lines. What diameter manifold discharge pipes are required?

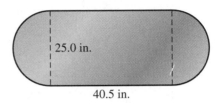

Output manifold

Input

ILLUSTRATION 2

26. A pipe has a 3.50-in. outside diameter and a 3.25-in. inside diameter. (See Illustration 3.) Find the area of its cross section.

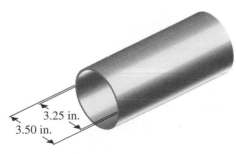

3.25 in.

3.50 in.

ILLUSTRATION 3

27. In Illustration 4, find the area of the rectangular piece of metal after the two circles are removed.

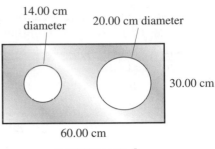

14.00 cm diameter   20.00 cm diameter

30.00 cm

60.00 cm

ILLUSTRATION 4

28. Find the area and perimeter of the figure in Illustration 5.

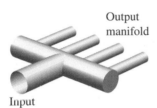

25.0 in.

40.5 in.

ILLUSTRATION 5

29. Find the length of strapping needed for the pipe in Illustration 6.

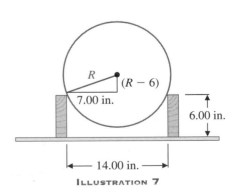

1.50 in.

12.00 in. diameter pipe

1.50 in.

ILLUSTRATION 6

30. See Illustration 7. Find the smallest circular duct that will rest on the ceiling joists without touching the ceiling.

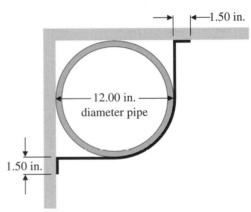

R $(R - 6)$

7.00 in.

6.00 in.

14.00 in.

ILLUSTRATION 7

31. A boiler 5.00 ft in diameter is to be placed in a corner of a room shown in Illustration 8.
  a. How far from corner C are points A and B of the boiler?
  b. How long is a pipe from C to the center of the boiler M?

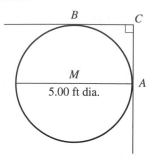

ILLUSTRATION 8

32. A pulley is connected to a spindle of a wheel by a belt. The distance from the spindle to the center of the pulley is 15.0 in. The diameter of the pulley is 15.0 in. The belt is in contact with the pulley for 31.4 in. What is the length of the belt? (See Illustration 9.)

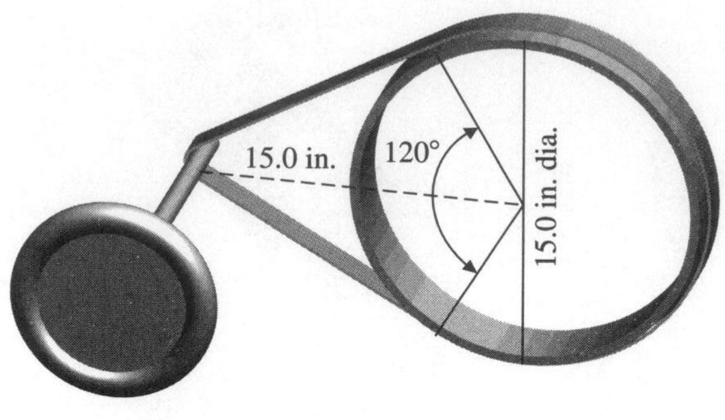

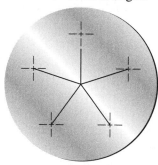

ILLUSTRATION 9

33. Mary needs to punch 5 equally spaced holes in a circular metal plate (see Illustration 10). Find the measure of each central angle.

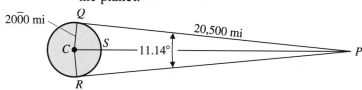

ILLUSTRATION 10

For Exercises 34–36, see Illustration 11:

34. Find the measure of $\angle 1$, where C is the center.

35. Find the measure of $\angle 2$.

36. Find the measure of $\angle 3$, given that $\overline{AC} \parallel \overline{DB}$.

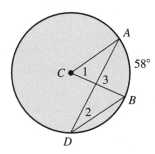

ILLUSTRATION 11

37. Find the measure of $\angle 1$ in Illustration 12.

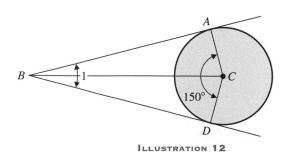

ILLUSTRATION 12

38. In Illustration 12, the length of AC is 5 and the distance between B and center C is 13. Find the length of $\overline{AB}$.

39. Illustration 13 shows a satellite at position P relative to a strange planet of radius $20\overline{0}0$ miles. The angle between the tangent lines is 11.14°. The distance from the satellite to Q is 20,500 miles. Find the altitude $\overline{SP}$ of the satellite above the planet.

2000 mi · 20,500 mi · 11.14°

ILLUSTRATION 13

40. In Illustration 14, $CP = 12.2$ m and $PB = 10.8$ m. Find the radius of the circle, where C is the center.

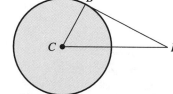

ILLUSTRATION 14

Exercises 41–44 refer to Illustration 15. $\overrightarrow{AB}$ and $\overrightarrow{AC}$ are secants; $\overline{CD}$ and $\overline{BF}$ are chords.

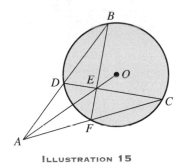

ILLUSTRATION 15

41. Suppose $\overset{\frown}{BD}$ = 85 arc degrees and $\angle DEF = 52°$. Find the number of arc degrees in $\overset{\frown}{CF}$.

42. Suppose $\overset{\frown}{BC}$ = 100 arc degrees and $\overset{\frown}{DF}$ = 40 arc degrees. Find the number of degrees in $\angle BAC$.

43. Suppose $\angle BEC = 78°$ and $\overset{\frown}{BC}$ = 142 arc degrees. Find the number of arc degrees in $\overset{\frown}{DF}$.

44. Suppose $\angle DCF = 30°$, $\angle EFC = 52°$, and $\overset{\frown}{CF}$ = 110 arc degrees. Find the number of arc degrees in $\overset{\frown}{BD}$.

45. Inscribe an equilateral triangle in a circle. **a.** How many arc degrees are contained in each arc? **b.** How many degrees are contained in each inscribed angle? **c.** Draw a central angle to each arc. How many degrees are contained in each central angle?

46. Inscribe a square in a circle. **a.** How many arc degrees are contained in each arc? **b.** How many degrees are contained in each inscribed angle? **c.** Draw a central angle to each arc. How many degrees are contained in each central angle?

47. Inscribe a regular hexagon in a circle. **a.** How many arc degrees are contained in each arc? **b.** How many degrees are contained in each inscribed angle? **c.** Draw a central angle to each arc. How many degrees are contained in each central angle?

48. An arc of a circle is doubled. Is its central angle doubled? Is its chord doubled?

49. In designing a bracket for use in a satellite, weight is of major importance. (See Illustration 16.) Find **a.** the area of the part in in^2, **b.** the overall length of the part, **c.** the overall height of the part, and **d.** its total weight.

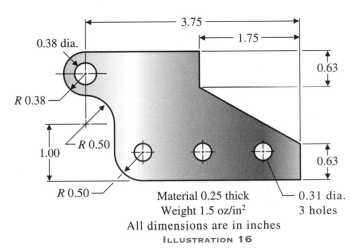

Material 0.25 thick
Weight 1.5 oz/in^2
All dimensions are in inches
ILLUSTRATION 16

50. A piece of aluminum flat bar stock for an equipment bracket on an aircraft is to be bent to the shape shown in Illustration 17. Disregarding material consumed in cutting and squaring ends, what is the total length of the material required? *Note:* In computing material length, the measurement to the mean thickness of material is used for greater accuracy.

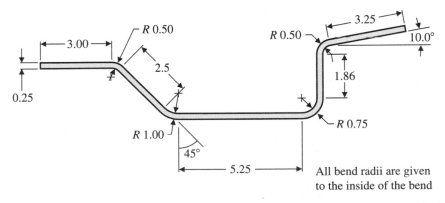

All bend radii are given
to the inside of the bend

All dimensions are in inches

ILLUSTRATION 17

12.6 Radian Measure

Radian measure, the metric unit of angle measure, is used in many applications, such as arc length and rotary motion. The *radian* (rad) unit is defined as the measure of an angle with its vertex at the center of a circle and with an intercepted arc on the circle equal in length to the radius. (In Figure 12.54, $\angle AOB$ forms the intercepted arc AB on the circle.)

In general, the radian is defined as the ratio of the length of arc that an angle intercepts on a circle to the length of its radius. In a complete circle or one complete revolution, the circumference $C = 2\pi r$. This means that for any circle the ratio of the circumference to the radius is constant (2π) because $\frac{C}{r} = 2\pi$. That is, the radian measure of one complete revolution is 2π rad.

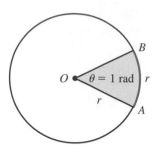

Radian

FIGURE 12.54

What is the relationship between radians and degrees?

One complete revolution = 360°

One complete revolution = 2π rad

Therefore,

$$360° = 2\pi \text{ rad}$$

$$180° = \pi \text{ rad}$$

This gives us the conversion factors as follows.

$$\frac{\pi \text{ rad}}{180°} = 1 \text{ and } \frac{180°}{\pi \text{ rad}} = 1$$

For comparison purposes,

$$1 \text{ rad} = \frac{180°}{\pi} = 57.3°$$

$$1° = \frac{\pi \text{ rad}}{180°} = 0.01745 \text{ rad}$$

EXAMPLE 1   How many degrees are in an angle that measures $\frac{\pi}{2}$ rad?

Use the conversion factor $\frac{180°}{\pi \text{ rad}}$.

$$\frac{\pi}{2} \text{ rad} \times \frac{180°}{\pi \text{ rad}} = \frac{180°}{2} = 90°$$

EXAMPLE 2   How many radians are in an angle that measures 30°?

Use the conversion factor $\frac{\pi \text{ rad}}{180°}$.

$$30° \times \frac{\pi \text{ rad}}{180°} = \frac{\pi}{6} \text{ rad} \quad \text{or} \quad 0.524 \text{ rad}$$

As a wheel rolls along a surface, the distance s that a point on the wheel travels equals the product of the radius r and angle θ, measured in radians, through which the wheel turns. (See Figure 12.55.)

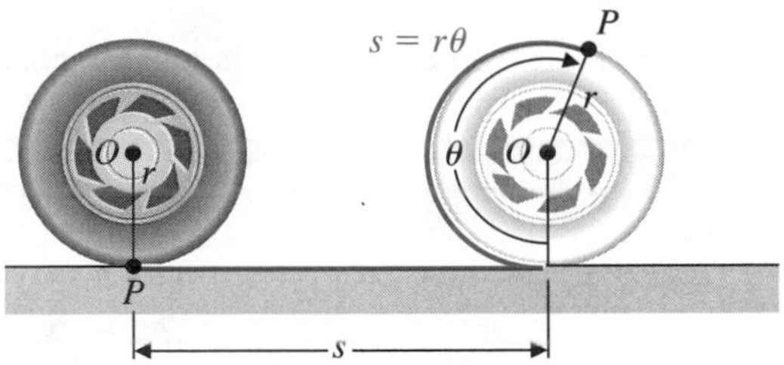

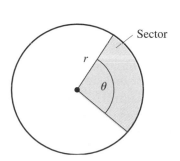

FIGURE 12.55

$$s = r\theta \qquad (\theta \text{ in rad})$$

• **EXAMPLE 3**   Find the distance a point on the surface of a pulley travels if its radius is 10.0 cm and the angle of turn is $\frac{5}{4}$ rad.

$$s = r\theta$$

$$s = 10.0 \text{ cm} \times \frac{5}{4}$$

$$= 12.5 \text{ cm}$$

• **EXAMPLE 4**   Find the distance a point on the surface of a gear travels if its radius is 15 cm and the angle of the turn is 420°.

$$s = r\theta$$

Since the angle is given in degrees, you must change 420° to radians. Use the conversion factor $\frac{\pi \text{ rad}}{180°}$.

$$\overset{7}{\cancel{420°}} \times \frac{\pi \text{ rad}}{\underset{3}{\cancel{180°}}} = \frac{7\pi}{3} \text{ rad}$$

$$s = \overset{5}{\cancel{15}} \text{ cm} \times \frac{7\pi}{\cancel{3}} = 35\pi \text{ cm or 110 cm}$$

• **EXAMPLE 5**   A wheel with radius 5.4 cm travels a distance of 21 cm. Find angle θ **a.** in radians, and **b.** in degrees that the wheel turns.

a. $s = r\theta$

Solve for θ:

$$\frac{s}{r} = \theta$$

$$\frac{21 \text{ cm}}{5.4 \text{ cm}} = \theta$$

$$3.9 \text{ rad} = \theta$$

b. $3.9 \text{ rad} \times \dfrac{180°}{\pi \text{ rad}} = 223°$

A *sector* of a circle is the region bounded by two radii of a circle and the arc intercepted by them. (See Figure 12.56.) The area of a given sector is proportional to the area of the circle itself; the area of the sector is a fraction of the area of the whole circle.

FIGURE 12.56
Sector of a circle

If the central angle of a sector is measured in degrees, the ratio of the measure of the central angle to 360° specifies the fraction of the area of the circle contained in the sector as follows.

AREA OF A SECTOR OF A CIRCLE (WITH THE CENTRAL ANGLE MEASURED IN DEGREES)

$$A = \frac{\theta}{360°} \pi r^2$$

where θ is the measure of the central angle in degrees and r is the radius.

If the central angle is measured in radians, the ratio of the measure of the central angle to 2π specifies the fraction of the area of the circle contained in the sector as follows:

$$A = \frac{\theta}{2\pi} \cdot \pi r^2 = \frac{1}{2} r^2 \theta$$

That is,

AREA OF A SECTOR OF A CIRCLE (WITH THE CENTRAL ANGLE MEASURED IN RADIANS)

$$A = \frac{1}{2} r^2 \theta$$

where θ is the measure of the central angle in radians and r is the radius of the circle.

• **EXAMPLE 6**

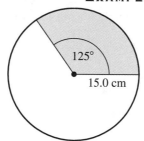

FIGURE 12.57

Find the area of the sector of a circle of radius 15.0 cm with a central angle of 125°. (See Figure 12.57.)

$$A = \frac{\theta}{360°} \cdot \pi r^2$$

$$A = \frac{125°}{360°} \cdot \pi (15.0 \text{ cm})^2$$

$$= 245 \text{ cm}^2$$

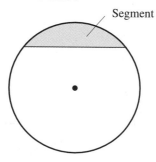

FIGURE 12.58
Segment of a circle

A segment of a circle is the region between a chord and an arc subtended by the chord. (See Figure 12.58.) The area of a segment may be found by finding the area of the sector and subtracting the area of the isosceles triangle formed by the chord and two radii. Draw the radii to the ends of the chord as in Figure 12.59. Next, draw altitude h from the center, perpendicular to the chord. The area of the isosceles triangle is $A = \frac{1}{2}ch$, where c is the length of the chord. Using the Pythagorean theorem in $\triangle OMB$, we have

$$(OM)^2 + (MB)^2 = (OB)^2 \qquad \text{Pythagorean theorem}$$

$$h^2 + \left(\frac{c}{2}\right)^2 = r^2$$

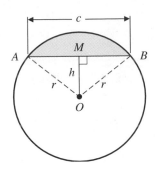

FIGURE 12.59

$$h^2 = r^2 - \frac{c^2}{4} \qquad \text{Subtract } \frac{c^2}{4} \text{ from both sides.}$$

$$h^2 = \frac{4r^2 - c^2}{4} \qquad \text{Write with LCD = 4.}$$

$$h = \frac{\sqrt{4r^2 - c^2}}{2} \qquad \text{Take the square root of both sides.}$$

The area of the isosceles triangle is then

$$A = \frac{1}{2}ch$$

$$= \frac{1}{2}c\left(\frac{\sqrt{4r^2 - c^2}}{2}\right)$$

$$= \frac{c\sqrt{4r^2 - c^2}}{4}$$

The area of the segment is the area of the sector minus the area of the isosceles triangle:

$$A = \frac{1}{2}r^2\theta - \frac{c\sqrt{4r^2 - c^2}}{4}$$

where r is the radius, θ is the measure of the central angle, and c is the length of the chord.

If only the length of the chord and the radius of the circle are known, trigonometry is required. This application is treated in Chapter 13.

• **EXAMPLE 7**   The chord in Figure 12.60 has a length of 26.6 cm. The radius of the circle is 15.0 cm. The measure of the central angle is 125°, as in Example 6. Find the area of the segment.

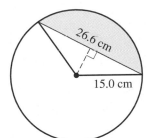

FIGURE 12.60

The area of the isosceles triangle is

$$A = \frac{c\sqrt{4r^2 - c^2}}{4}$$

$$A = \frac{(26.6)\sqrt{4(15.0)^2 - (26.6)^2}}{4}$$

$$= 92.3 \text{ cm}^2$$

Using the result from Example 6, the area of the segment is then

$$245 \text{ cm}^2 - 92.3 \text{ cm}^2 = 153 \text{ cm}^2$$

Exercises | 12.6

1. π rad _____°

2. 1.7 rad = _____°

3. 21° = _____ rad

4. 45° = _____ rad

5. Change $\frac{\pi}{3}$ rad to degrees.

6. Change 150° to radians.

7. Change 135° to radians.

8. Change $\frac{\pi}{12}$ rad to degrees.

9. How many radians are contained in a central angle which is $\frac{2}{3}$ of a circle?

10. What percent of 2π rad is $\frac{\pi}{2}$ rad?

11. Find the number of radians in a central angle whose arc is $\frac{2}{5}$ of a circle.

12. What percent of 2 rad is $\frac{\pi}{12}$ rad?

Complete the table using the formula s = rθ (θ in radians):

	Radius, *r*	Angle, θ	Distance, *s*
13.	25.0 cm	$\frac{2\pi}{5}$ rad	
14.	30.0 cm	$\frac{4\pi}{3}$ rad	
15.	6.0 cm	45°	
16.	172 mm	$\frac{\pi}{4}$ rad	
17.	18.0 cm	330°	
18.	3.0 m	250°	
19.	40.0 cm	rad	112 cm
20.	0.0081 mm	rad	0.011 mm
21.	0.500 m	°	0.860 m
22.	0.027 m	°	0.0283 m
23.		$\frac{2\pi}{3}$ rad	18.5 cm
24.		315°	106 m

25. A pulley is turning at an angular velocity of 10.0 rad per second. How many revolutions is the pulley making each second? *Hint:* One revolution equals 2π rad.)

26. The radius of a wheel is 20.0 in. It turns through an angle of 2.75 rad. What is the distance a point travels on the surface of the wheel?

27. The radius of a gear is 22.0 cm. It turns through an angle of 240°. What is the distance a point travels on the surface of the gear?

28. A wheel of diameter 6.00 m travels a distance of 31.6 m. Find the angle θ (in radians) that the wheel turns.

29. A wheel of diameter 15.2 cm turns through an angle of 3.40 rad. Find the distance a point travels on the surface of the wheel.

30. In Illustration 1, find:
    a. the length of arc *s*
    b. the area of the sector
    c. the area of the segment

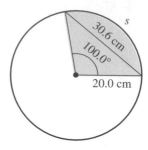

ILLUSTRATION 1

31. In Illustration 2, find:
    a. the length of arc *s*
    b. the area of the sector
    c. the area of the segment

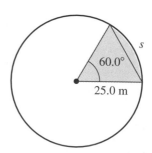

ILLUSTRATION 2

32. Given two concentric circles (circles with the same center) with central angle 45.0°, r_1 = 4.00 m, and r_2 = 8.00 m, find the shaded area in Illustration 3.

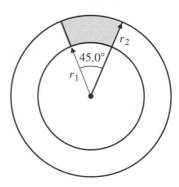

ILLUSTRATION 3

12.7 Prisms

Up to now, you have studied the geometry of two dimensions. *Solid geometry* is the geometry of three dimensions—length, width, and depth.

A *prism* is a solid whose sides are parallelograms and whose bases are one pair of polygons that are parallel and congruent (same size and shape). (See Figure

12.61.) (Recall that a closed figure made up of straight lines in two dimensions is called a *polygon*.) The two parallel polygons (which may be any type of polygon) are called the *bases* of the prism. The remaining polygons will be parallelograms and are called *lateral faces*. A *right prism* has lateral faces that are rectangles and are therefore perpendicular to the bases.

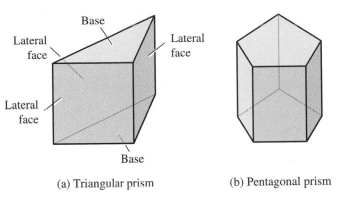

(a) Triangular prism   (b) Pentagonal prism

FIGURE 12.61

The name of the polygon used as the base names the type of prism. For example, a prism with bases that are triangles is called a *triangular* prism (see Figure 12.61(a)). A prism with bases that are pentagons is called a *pentagonal* prism (see Figure 12.61(b)), and so on.

> The lateral surface area of a prism is the sum of the areas of the lateral faces of the prism.
>
> The total surface area of a prism is the sum of the areas of the lateral faces and the areas of the bases of the prism.

• **EXAMPLE 1**   Find the lateral surface area of the triangular right prism in Figure 12.62.

To find the lateral surface area, find the area of each lateral face. Then find the sum of these areas.

The area of the rectangular face located on the right front side (see Figure 12.63(a)) is

$$A = lw$$
$$A = (12.0 \text{ cm})(5.0 \text{ cm})$$
$$= 6\overline{0} \text{ cm}^2$$

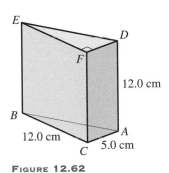

FIGURE 12.62

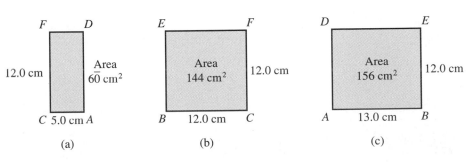

FIGURE 12.63

The area of the square face on the left front side (see Figure 12.63(b)) is

$$A = s^2$$
$$A = (12.0 \text{ cm})^2$$
$$= 144 \text{ cm}^2$$

To find the area of the third face (the back side of the prism), first find length AB. Since AB is also the hypotenuse of the right triangle ABC, use the Pythagorean theorem as follows:

$$c = \sqrt{a^2 + b^2}$$
$$c = \sqrt{(12.0 \text{ cm})^2 + (5.0 \text{ cm})^2}$$
$$= \sqrt{144 \text{ cm}^2 + 25 \text{ cm}^2}$$
$$= \sqrt{169 \text{ cm}^2}$$
$$= 13.0 \text{ cm}$$

The area of the third face (see Figure 12.63(c)) is

$$A = lw$$
$$A = (13.0 \text{ cm})(12.0 \text{ cm})$$
$$= 156 \text{ cm}^2$$

Therefore, the lateral surface area is

$$6\overline{0} \text{ cm}^2 + 144 \text{ cm}^2 + 156 \text{ cm}^2 = 36\overline{0} \text{ cm}^2$$

• **EXAMPLE 2**   Find the total surface area of the prism in Example 1.

To find the total surface area, first find the area of the bases. Then add this result to the lateral surface area from Example 1. The bases have the same size and shape, so just find the area of one base and then double it.

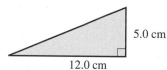

5.0 cm

12.0 cm

FIGURE 12.64

The area of one base as shown in Figure 12.64 is

$$B = A = \frac{1}{2}bh$$

$$B = \frac{1}{2}(12.0 \text{ cm})(5.0 \text{ cm})$$

$$= 3\overline{0} \text{ cm}^2$$

Double this to find the area of both the bases.

$$2(3\overline{0} \text{ cm}^2) = 6\overline{0} \text{ cm}^2$$

Add this area to the lateral surface area to find the total area.

$$36\overline{0} \text{ cm}^2 + 6\overline{0} \text{ cm}^2 = 42\overline{0} \text{ cm}^2$$

So the total surface area is $42\overline{0}$ cm^2.

The volume of a prism is found by the following formula.

Volume of prism

$$V = Bh$$

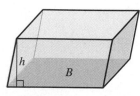

h

B

FIGURE 12.65

where B is the area of one of the bases and h is the altitude (perpendicular distance between the parallel bases). See Figure 12.65.

• **EXAMPLE 3** Find the volume of the prism in Example 1.

To find the volume of the prism, use the formula $V = Bh$. B is the area of the base which we found to be $3\overline{0}$ cm². The altitude of a lateral face is 12.0 cm.

$$V = Bh$$
$$V = (3\overline{0} \text{ cm}^2)(12.0 \text{ cm})$$
$$= 360 \text{ cm}^3$$

• **EXAMPLE 4** Find the volume of the prism in Figure 12.66.

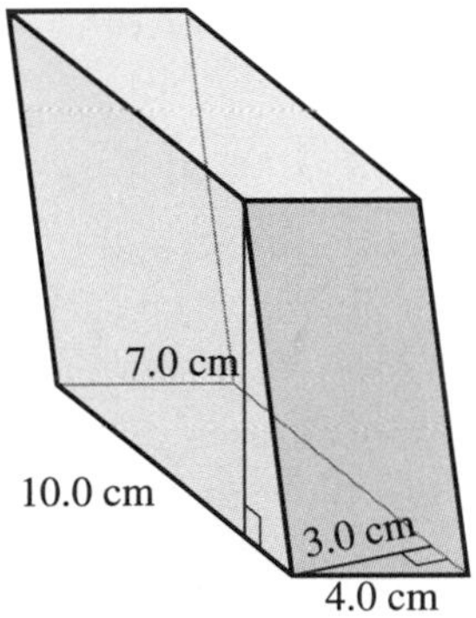

Use the formula $V = Bh$. The base is a parallelogram with sides of length 10.0 cm and 4.0 cm. The altitude of the base is 3.0 cm. First, find B, the area of the base.

$$B = bh$$
$$B = (10.0 \text{ cm})(3.0 \text{ cm})$$
$$= 3\overline{0} \text{ cm}^2$$

The altitude of a lateral face of the prism is 7.0 cm. Therefore,

$$V = Bh$$
$$V = (3\overline{0} \text{ cm}^2)(7.0 \text{ cm})$$
$$= 210 \text{ cm}^3$$

7.0 cm

10.0 cm 3.0 cm
4.0 cm

FIGURE 12.66

• **EXAMPLE 5** A rectangular piece of steel is 24.1 in. by 13.2 in. by 8.20 in. (see Figure 12.67). Steel weighs 0.28 lb/in³. Find its weight, in pounds.

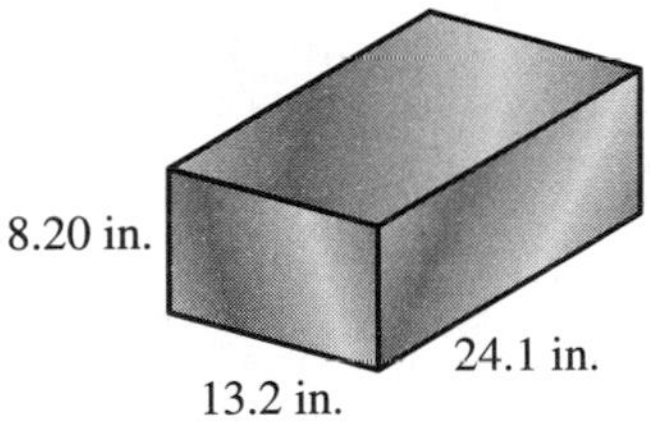

Find the volume using the formula for the volume of a prism, $V = Bh$. First, find B, the area of the base of the prism.

$$B = lw$$
$$B = (13.2 \text{ in.})(24.1 \text{ in.})$$

The volume is then

$$V = Bh$$
$$V = [(13.2 \text{ in.})(24.1 \text{ in.})](8.20 \text{ in.})$$
$$= 2610 \text{ in}^3$$

Since steel weighs 0.28 lb/in³, the total weight is

$$2610 \text{ in}^3 \times \frac{0.28 \text{ lb}}{1 \text{ in}^3} = 730 \text{ lb}$$

8.20 in.

24.1 in.
13.2 in.

FIGURE 12.67

Exercises | 12.7

Follow the rules for working with measurements:

1. **a.** Find the lateral surface area of the prism shown In Illustration 1.
   b. Find the total surface area of the prism.
   c. Find the volume of the prism.

2. **a.** Find the lateral surface area of the rectangular prism shown in Illustration 2.
   b. Find the total surface area of the prism.
   c. Find the volume of the prism.
   d. What is the name given to this geometric solid?

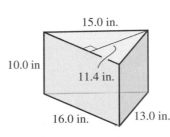

15.0 in.

10.0 in

11.4 in.

16.0 in. 13.0 in.

ILLUSTRATION 1

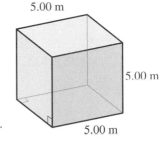

5.00 m

5.00 m

5.00 m

ILLUSTRATION 2

3. a. What is the area of the lateral surface to be painted in Illustration 3? (Assume no windows.)
  b. What is the area of roof to be covered with shingles?
  c. What is the volume of concrete needed to pour a floor 16 cm deep?
  d. What is the total surface area of the figure? (Include painted surface, roof, and floor.)

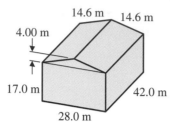

ILLUSTRATION 3

4. Find the volume of the wagon box in Illustration 4.

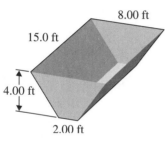

ILLUSTRATION 4

5. Find the volume of the gravity bin in Illustration 5.

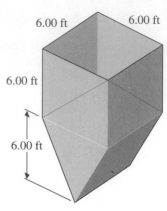

ILLUSTRATION 5

6. Steel weighs 0.28 lb/in^3. What is the weight of a rectangular piece of steel 0.3125 in. by 12.0 in. by 20.0 in.?

7. A steel rod of cross-sectional area 5.0 in^2 weighs 42.0 lb. Find its length. (Steel weighs 0.28 lb/in^3.)

8. The rectangular lead sleeve shown in Illustration 6 has a cored hole 2.0 in. by 3.0 in. How many cubic inches of lead are in this sleeve?

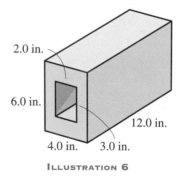

ILLUSTRATION 6

12.8 Cylinders

A *cylinder* is a geometric solid with a curved lateral surface. The *axis* of a cylinder is the line segment between the centers of the bases. The altitude, *h*, is the shortest (perpendicular) distance between the bases. If the axis is perpendicular to the bases, and the bases are circles, the cylinder is called a *right circular cylinder*, and the axis is the same length as the altitude. See Figure 12.68.

The volume of a right circular cylinder is found by the formula $V = Bh$, where B is the area of the base. The base is a circle with area $B = \pi r^2$. Therefore, the formula for the volume of a cylinder is written as follows.

Volume of cylinder

$$V = \pi r^2 h$$

where *r* is the radius of the base and *h* is the altitude.

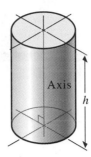

(a) Cylinder   (b) Right circular cylinder

FIGURE 12.68

• **EXAMPLE 1** Find the volume of the right circular cylinder in Figure 12.69.

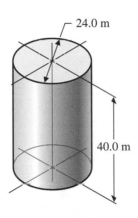

24.0 m

40.0 m

FIGURE 12.69

The diameter is 24.0 m, so the radius is 12.0 m.

$$V = \pi r^2 h$$
$$V = \pi (12.0 \text{ m})^2 (40.0 \text{ m})$$
$$= 18,100 \text{ m}^3$$

• **EXAMPLE 2** Find the diameter of a cylindrical tank 23.8 ft high with a capacity of 136,000 gallons (1 ft³ = 7.48 gal).

First, find the volume of the cylinder in ft³.

$$136,000 \text{ gal} \times \frac{1 \text{ ft}^3}{7.48 \text{ gal}} = 18,200 \text{ ft}^3$$

Since V and h are known, find r using the formula

$$V = \pi r^2 h$$

$$r^2 = \frac{V}{\pi h} \qquad \text{Divide both sides by } \pi h.$$

$$r = \sqrt{\frac{V}{\pi h}} \qquad \text{Take the square root of both sides.}$$

$$r = \sqrt{\frac{18,200 \text{ ft}^3}{\pi (23.8 \text{ ft})}}$$

$$= 15.6 \text{ ft}$$

Diameter is $2r$. So the diameter is 2(15.6 ft) = 31.2 ft.

The lateral surface area of a right circular cylinder can be visualized as a can without ends. Cut through the side of the can and then flatten it out, as shown in Figure 12.70.

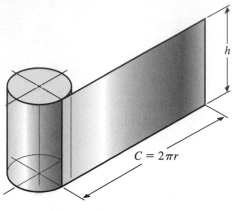

FIGURE 12.70

The lateral surface area of a right cylinder is a rectangle with the base $2\pi r$ and the altitude h. The formula for the lateral surface area is

Lateral surface area of cylinder

$$A = 2\pi rh$$

The total surface area of a cylinder is the area of the bases plus the lateral surface area.

•   **EXAMPLE 3**   Find the total surface area of the right circular cylinder in Figure 12.71.

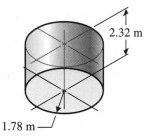

FIGURE 12.71

The area of one base is

$$A = \pi r^2 = \pi(1.78 \text{ m})^2$$
$$= 9.95 \text{ m}^2$$

The area of both bases, then, is

$$2(9.95 \text{ m}^2) = 19.9 \text{ m}^2$$

The lateral surface area $= 2\pi rh$
$$= 2\pi(1.78 \text{ m})(2.32 \text{ m})$$
$$= 25.9 \text{ m}^2$$

The total surface area $= 19.9 \text{ m}^2 + 25.9 \text{ m}^2$
$$= 45.8 \text{ m}^2$$

Exercises | 12.8

Follow the rules for working with measurements:

Find the volume of each cylinder:

1.

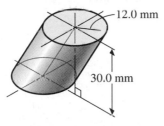

12.0 mm

30.0 mm

2.

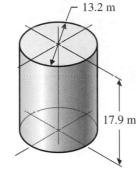

13.2 m

17.9 m

3. How many litres does a cylindrical tank of height 39.2 m with radius 8.20 m hold? (1 m³ = 1000 L)

4. A welder is assigned to fabricate the sheet metal water trough shown in Illustration 1. **a.** Find its cubic foot capacity. **b.** What is the length of side y?

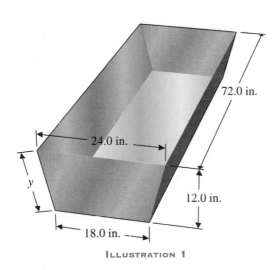

72.0 in.

24.0 in.

y

12.0 in.

18.0 in.

ILLUSTRATION 1

5. A technician draws plans for a 40$\overline{0}$,000-gallon cylindrical tank with radius 20.0 ft. What should the height be? (1 ft³ = 7.48 gal)

6. The compression ratio, C, of a cylinder is the ratio of the volume of the cylinder when the piston is at the bottom of its stroke (V_B) to the volume of the cylinder when the piston is at the top of the stroke (V_T). That is, $C = \frac{V_B}{V_T}$. The diameter of a cylinder is 5.500 in. The height of the cylinder with the piston at the bottom of the stroke is 5.250 in. The height at the top of the

stroke is 0.750 in. **a.** Find C. **b.** Note that the ratio of the heights gives the same result. Why?

7. Use the formula in Exercise 6 to find the compression ratio of a cylinder of diameter 8.250 in., a height of 7.00 in. with piston at the bottom of the stroke, and a height of 0.500 in. with piston at the top of the stroke.

8. An engine has 8 cylinders. Each cylinder has a bore of 4.70 in. diameter and a stroke of 5.25 in. Find its piston displacement.

9. A cylindrical tank is 25 ft 9 in. long and 7 ft 6 in. in diameter. How many cubic feet does it hold?

10. A 3.0-in.-diameter cylindrical rod is 16 in. long. Find its volume.

11. A cylindrical piece of steel is 10.0 in. long. Its volume is 25.3 in³. Find its diameter.

12. A 2.50-in.-diameter cylindrical rod has a volume of 15.6 in³. Find its length.

13. Copper tubing $\frac{1}{2}$ in. I.D. (inside diameter) is 12.0 ft long. What is the volume of the refrigerant contained in the tubing?

14. A steel I-beam with center thickness 0.75 in. has five 2.00-in. holes drilled as shown in Illustration 2. What is the total volume of steel removed from the beam?

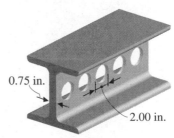

0.75 in.

2.00 in.

ILLUSTRATION 2

15. Find **a.** the lateral surface area and **b.** the total surface area of the right circular cylinder shown in Illustration 3.

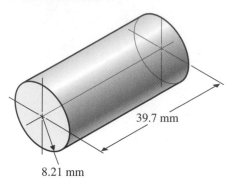

39.7 mm

8.21 mm

ILLUSTRATION 3

16. Find **a.** the lateral surface area and **b.** the total surface area of the cylinder shown in Illustration 4.

17. Find the total amount (area) of paper used for labels for 1000 cans like the one shown in Illustration 5.

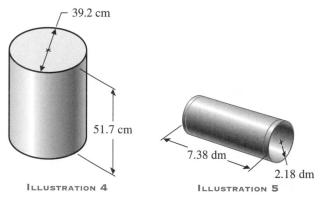

39.2 cm

51.7 cm

7.38 dm

2.18 dm

ILLUSTRATION 4   ILLUSTRATION 5

18. How many square feet of sheet metal are needed for the sides of the cylindrical tank shown in Illustration 6? (Allow 2.0 in. for seam overlap.)

2.0 ft

10.5 ft

ILLUSTRATION 6

19. A cylindrical piece of stock is turned on a lathe from 3.10 in. down to 2.24 in. in diameter. The cut is 5.00 in. long. What is the volume of the metal removed?

20. What is the volume of lead in the "pig" shown in Illustration 7? What is the volume of the mold?

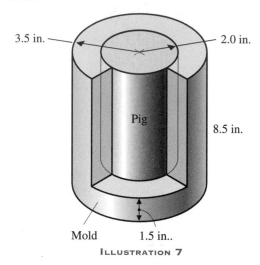

3.5 in.   2.0 in.

Pig

8.5 in.

Mold   1.5 in..

ILLUSTRATION 7

21. A cylinder bore is increased in diameter from 2.78 in. to 2.86 in. The cylinder is 5.50 in. high. How much has the surface area of the walls been increased?

22. Each cylinder bore of a 6-cylinder engine has a diameter of 2.50 in. and a height of 4.90 in. What is the lateral surface area of the six cylinder bores?

23. The sides of a cylindrical silo 15 ft in diameter and 26 ft high are to be painted. Each gallon of paint will cover $2\overline{0}0$ ft^2. How many gallons of paint will be needed?

24. How many square feet of sheet metal are needed to form the trough shown in Illustration 8?

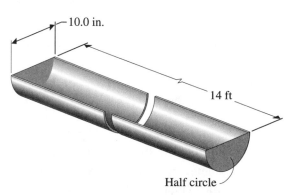

10.0 in.

14 ft

Half circle

ILLUSTRATION 8

25. Find the number of kilograms of metal needed for 2,700,000 cans with ends of the type shown in Illustration 9. The metal has a density of 0.000147 g/cm².

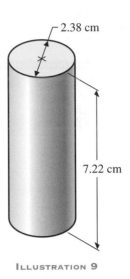

2.38 cm

7.22 cm

ILLUSTRATION 9

12.9 Pyramids and Cones

A *pyramid* is a geometric solid whose base is a polygon and whose lateral faces are triangles with a common vertex. The common vertex is called the *apex* of the pyramid. If a pyramid has a base that is a triangle, then it is called a *triangular pyramid*. If a pyramid has a base that is a hexagon, then it is called a *hexagonal pyramid*. In general, a pyramid is named by the shape of its base. Two types of pyramids are shown in Figure 12.72.

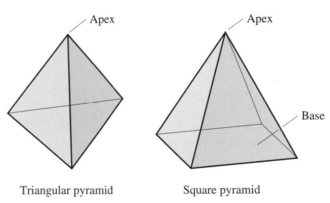

Apex   Apex

Base

Triangular pyramid   Square pyramid

FIGURE 12.72

The volume of a pyramid is found using the following formula.

Volume of pyramid

$$V = \frac{1}{3}Bh$$

where B is the area of the base and h is the height of the pyramid. The height of a pyramid is the shortest (perpendicular) distance between the apex and the base of the pyramid.

• **EXAMPLE 1**   Find the volume of the pyramid in Figure 12.73.

The base is a right triangle with legs 6.0 in. and 8.0 in. Therefore,

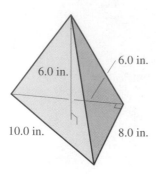

FIGURE 12.73

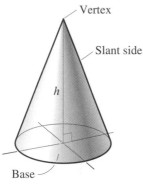

FIGURE 12.74

$$B = \frac{1}{2}(6.0 \text{ in.})(8.0 \text{ in.}) = 24 \text{ in}^2$$

The height is 6.0 in. Therefore,

$$V = \frac{1}{3}Bh = \frac{1}{3}(24 \text{ in}^2)(6.0 \text{ in.}) = 48 \text{ in}^3$$

A *cone* is a geometric solid whose base is a circle (see Figure 12.74). It has a curved lateral surface that comes to a point called a *vertex*. The *axis* of a cone is a line segment from the vertex to the center of the base.

The height, *h*, of a cone is the shortest (perpendicular) distance between the vertex and the base. A *right circular cone* is a cone in which the height is the distance from the vertex to the center of the base. The *slant height* of a right circular cone is the length of a line segment that joins the vertex to any point on the circle that forms the base of the cone.

The volume of a circular cone is given by the formula $V = \frac{1}{3}Bh$. Since the base, *B*, is always a circle, its area is πr^2, where *r* is the radius of the base. Thus, the formula for the volume of a right circular cone is usually written as follows.

> **Volume of cone**
>
> $$V = \frac{1}{3}\pi r^2 h$$

The lateral surface area of a right circular cone is found using the following formula:

> **Lateral surface area**
>
> $$A = \pi rs$$

where *s* is the slant height of the cone.

•   **EXAMPLE 2**   Find the volume of the right circular cone in Figure 12.75.

$$V = \frac{1}{3}\pi r^2 h$$

$$V = \frac{1}{3}\pi(6.1 \text{ m})^2(17.2 \text{ m})$$

$$= 670 \text{ m}^3$$

•   **EXAMPLE 3**   Find the lateral surface area of the right circular cone in Figure 12.75.

The formula for the lateral surface area is $A = \pi rs$. The slant height is not given. However, a right triangle is formed by the axis, the radius, and the slant height. Therefore, to find the slant height, *s*, use the formula

$$c = \sqrt{a^2 + b^2}$$

Then $s = \sqrt{(6.1 \text{ m})^2 + (17.2 \text{ m})^2}$

$$= 18.2 \text{ m}$$

The lateral surface area can then be found as follows.

$$A = \pi rs$$

$$= \pi(6.1 \text{ m})(18.2 \text{ m})$$

$$= 350 \text{ m}^2$$

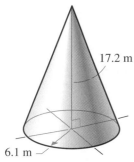

FIGURE 12.75

The frustum of a pyramid is the section of a pyramid between the base and a plane parallel to the base, as shown in Figure 12.76.

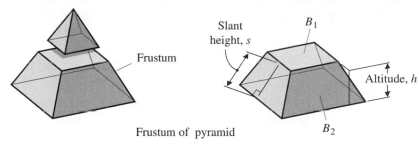

Frustum

Slant height, s

B_1

Altitude, h

B_2

Frustum of pyramid

FIGURE 12.76

The altitude of the frustum is the perpendicular distance between the two bases. The volume of the frustum of a pyramid is

$$V = \frac{1}{3}h(B_1 + B_2 + \sqrt{B_1 B_2})$$

where h is the altitude and B_1 and B_2 are the areas of the bases.

The lateral surface area of the frustum of a pyramid is

$$A = \frac{1}{2}s(P_1 + P_2)$$

where s is the slant height and P_1 and P_2 are the perimeters of the bases.

• **EXAMPLE 4**   Find the lateral surface area and the volume of the frustum in Figure 12.77.

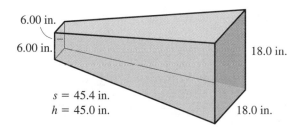

6.00 in.

6.00 in.

18.0 in.

$s = 45.4$ in.
$h = 45.0$ in.

18.0 in.

FIGURE 12.77
Air duct connection

$P_1 = 4(6.00 \text{ in.}) = 24.0 \text{ in.}$

$P_2 = 4(18.0 \text{ in.}) = 72.0 \text{ in.}$

$A = \frac{1}{2}s(P_1 + P_2)$

$A = \frac{1}{2}(45.4 \text{ in.})(24.0 \text{ in.} + 72.0 \text{ in.})$

$\quad = 2180 \text{ in}^2$

$B_1 = (6.00 \text{ in.})^2 = 36.0 \text{ in}^2$

$B_2 = (18.0 \text{ in.})^2 = 324 \text{ in}^2$

$V = \frac{1}{3}h(B_1 + B_2 + \sqrt{B_1 B_2})$

$V = \frac{1}{3}(45.0 \text{ in.})(36.0 \text{ in}^2 + 324 \text{ in}^2 + \sqrt{(36.0 \text{ in}^2)(324 \text{ in}^2)})$

$\quad = 7020 \text{ in}^3$

The frustum of a cone is the section of the cone between the base and a plane parallel to the base, as shown in Figure 12.78.

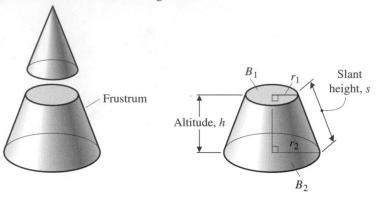

Frustrum

FIGURE 12.78

The altitude of the frustum is the perpendicular distance between the two bases. The volume of the frustum of a cone is

$$V = \frac{1}{3}h(B_1 + B_2 + \sqrt{B_1 B_2})$$

where h is the altitude and B_1 and B_2 are the areas of the bases.

The lateral surface area of a frustum of a right circular cone (area of the curved surface) is

$$A = \pi s(r_1 + r_2)$$

where s is the slant height and r_1 and r_2 are the radii of the bases.

Exercises | 12.9

Follow the rules for working with measurements:

Find the volume of each figure in Exercises 1–10:

1.

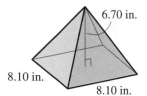

6.70 in.
8.10 in.
8.10 in.

2. 22.6 cm   42.5 cm

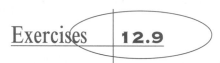

42.0 cm

3.

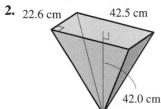

16.2 m
10.8 m
18.8 m

4.

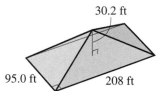

30.2 ft
95.0 ft   208 ft

5.

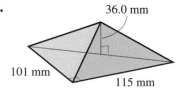

36.0 mm
101 mm
115 mm

6.

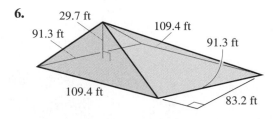

29.7 ft
91.3 ft   109.4 ft
91.3 ft
109.4 ft
83.2 ft

7.

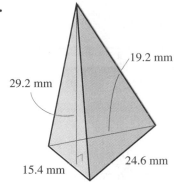

19.2 mm

29.2 mm

15.4 mm

24.6 mm

Hint: To find *B*, use

$$B = A = \sqrt{s(s-a)(s-b)(s-c)}$$

8.

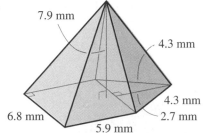

7.9 mm

4.3 mm

4.3 mm

6.8 mm   2.7 mm

5.9 mm

Hint: To find the area of the base of the pyramid, find the sum of the areas of the rectangle and the triangle.

9.

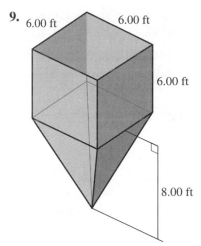

6.00 ft   6.00 ft

6.00 ft

8.00 ft

10.

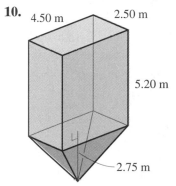

4.50 m   2.50 m

5.20 m

2.75 m

Find **a.** *the volume and* **b.** *the lateral surface area of each right circular cone in Exercises 11–12:*

11.

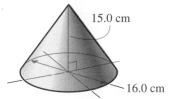

15.0 cm

16.0 cm

12.

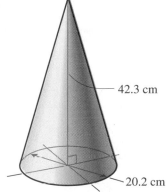

42.3 cm

20.2 cm

13. A loading chute in a flour mill goes directly into a feeding bin. The feeding bin is in the shape of an inverted right circular cone, as shown in Illustration 1. How many bushels of wheat can be placed in the feeding bin? (0.804 bu $= 1$ ft^3)

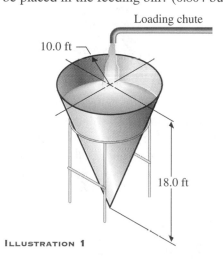

Loading chute

10.0 ft

18.0 ft

ILLUSTRATION 1

14. The circular tank in Illustration 2 is made of $\frac{1}{2}$-in. steel weighing 19.8 lb/ft^2. **a.** What is the total weight of the top? **b.** What is the total weight of the top, sides, and bottom of the tank?

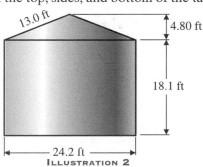

13.0 ft   4.80 ft

18.1 ft

24.2 ft

ILLUSTRATION 2

15. Find **a.** the volume and **b.** the lateral surface area of the right circular cone shown in Illustration 3.

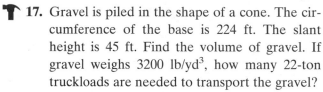

ILLUSTRATION 3

16. The volume of a right circular cone is 238 cm³. The radius of its base is 5.82 cm. Find its height.

17. Gravel is piled in the shape of a cone. The circumference of the base is 224 ft. The slant height is 45 ft. Find the volume of gravel. If gravel weighs 3200 lb/yd³, how many 22-ton truckloads are needed to transport the gravel?

18. Find the weight of the display model shown in Illustration 4. The model is made of pine. Pine weighs 31.2 lb/ft³.

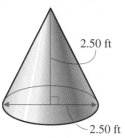

2.50 ft

2.50 ft

ILLUSTRATION 4

19. Find the volume of the frustum of the pyramid shown in Illustration 5.

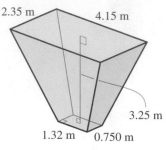

2.35 m   4.15 m

3.25 m

1.32 m   0.750 m

ILLUSTRATION 5

20. Find the volume and lateral surface area of the frustum of the cone shown in Illustration 6.

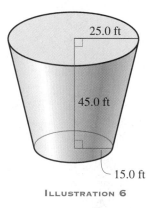

25.0 ft

45.0 ft

15.0 ft

ILLUSTRATION 6

Find the volume and the lateral surface area of each storage bin:

21.

15.0 ft

22.9 ft   25.0 ft

40.0 ft

25.0 ft

22.

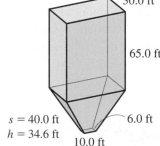

50.0 ft   30.0 ft

65.0 ft

$s = 40.0$ ft   6.0 ft
$h = 34.6$ ft   10.0 ft

23. A hopper must be designed to contain plastic resin pellets for an injection mold machine. (See Illustration 7.) The cylindrical portion of the tank is 18.0 in. in diameter. The spout is a conical frustum. For the hopper to hold 5.00 ft³ of resin pellets, find the length L of the cylindrical part.

24. A piece of one-inch (diameter) round stock is tapered so that its tip is a cone. (See Illustration 8.) If the taper begins 3.00 in. from the end of the stock, find the volume of stock that was removed in order to produce the tapered end.

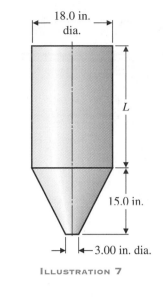

18.0 in.
dia.

L

15.0 in.

3.00 in. dia.

ILLUSTRATION 7

1.00 in.

3.00 in.

ILLUSTRATION 8

25. A welder is assigned to fabricate luggage storage compartments to fit in the luggage storage area of an aircraft. What is the cubic foot displacement of the compartment shown in Illustration 9?

26. Find the area of the pieces of metal to fabricate each compartment in Illustration 9.

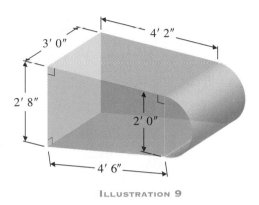

4′ 2″

3′ 0″

2′ 8″

2′ 0″

4′ 6″

ILLUSTRATION 9

12.10 Spheres

A *sphere* (Figure 12.79) is a geometric solid formed by a closed curved surface, with all points on the surface the same distance from a given point (the center). The given distance from a point on the surface to the center is called the *radius*.

The volume of a sphere is found by using the following formula.

Volume of sphere

$$V = \frac{4}{3}\pi r^3$$

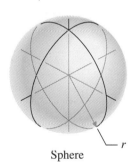

Sphere

FIGURE 12.79

where r is the radius of the sphere.

The surface area of a sphere is found by using the formula:

Surface area of sphere

$$A = 4\pi r^2$$

• **EXAMPLE 1** Find the surface area of a sphere of radius 2.80 cm. (See Figure 12.80.)

$A = 4\pi r^2$

$A = 4\pi(2.80 \text{ cm})^2$

$= 4\pi(2.80)^2 \text{ cm}^2$

$= 98.5 \text{ cm}^2$

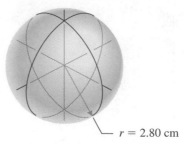

$r = 2.80$ cm

FIGURE 12.80

•   **EXAMPLE 2**   Find the volume of the sphere in Example 1. The formula for the volume of a sphere is

$$V = \frac{4}{3}\pi r^3$$

$$V = \frac{4}{3}\pi(2.80 \text{ cm})^3$$

$$= \frac{4}{3}\pi(2.80)^3 \text{ cm}^3$$

$$= 92.0 \text{ cm}^3$$

Exercises   **12.10**

Follow the rules for working with measurements.

*Find **a.** the surface area and **b.** the volume of each sphere:*

1.

8.00 m

3.

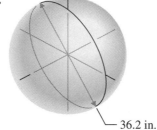

36.2 in.

2.

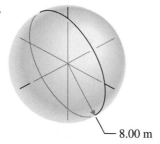

18.7 cm

4.

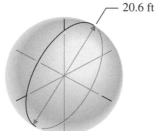

20.6 ft

5. A balloon 30.1 m in radius is to be filled with helium. How many m³ of helium are needed to fill it?

6. An experimental balloon is to have a diameter of 5.72 m. How much material is needed for this balloon?

7. How many gallons of water can be stored in the spherical portion of the water tank shown in Illustration 1? (7.48 gal = 1 ft³)

8. A city drains 150,000 gal of water from a full spherical tank with a radius of 26 ft. How many

ILLUSTRATION 1

gallons of water are left in the tank? ($1 \text{ ft}^3 =$ 7.48 gal)

🔨 **9.** A spherical tank for liquefied petroleum is 16.0 in. in diameter. **a.** What is the ratio of surface area to the volume of the tank? **b.** Find the same ratio for a tank 24.0 in. in diameter.

10. Find the volume of the cylindrical-shaped silo with a hemispherical top shown in Illustration 2.

ILLUSTRATION 2

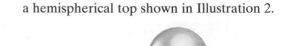

Chapter 12   Review

Use a protractor to measure each angle:

1.

2.

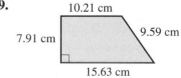

For Exercises 3–5, see Illustration 1:

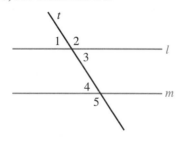

ILLUSTRATION 1

3. In the figure, *l* is parallel to *m* and $\angle 5 = 121°$. Find the measure of each angle.

4. In the figure, $\angle 4$ and $\angle 5$ are called __?__ angles.

5. Suppose $\angle 1 = 4x + 5$ and $\angle 2 = 2x + 55$. Find the value of *x*.

6. Name the polygon which has **a.** 4 sides; **b.** 5 sides; **c.** 6 sides; **d.** 3 sides; and **e.** 8 sides.

Find the perimeter and the area of each quadrilateral:

7.

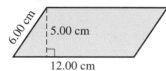

8.

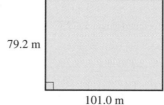

9.

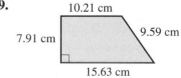

10. The area of a rectangle is 79.6 m². The length is 10.3 m. What is the width?

11. The area of a parallelogram is 2.53 cm². Find its height if the base is 1.76 cm.

Find the area and the perimeter of each triangle:

12.

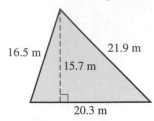

16.5 m 21.9 m 15.7 m 20.3 m

13.

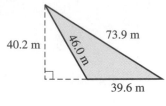

40.2 m 46.0 m 73.9 m 39.6 m

14.

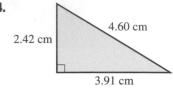

2.42 cm 4.60 cm 3.91 cm

Find the length of the hypotenuse of each triangle:

15.

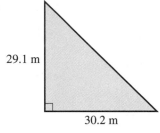

29.1 m 30.2 m

16.

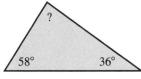

1.72 cm 3.44 cm

17. Find the measure of the missing angle in Illustration 2.

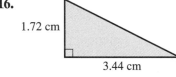

? 58° 36°

ILLUSTRATION 2

18. In Illustration 3, suppose $\overline{DE} \parallel \overline{BC}$. Find length BC.

19. Find the area and circumference of the circle in Illustration 4.

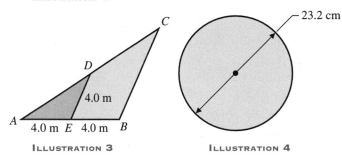

C D 4.0 m A 4.0 m E 4.0 m B

ILLUSTRATION 3

23.2 cm

ILLUSTRATION 4

20. The area of a circle is 462 cm². Find its radius.

21. How many degrees are in a central angle whose arc is $\frac{3}{5}$ of a circle?

22. Change 24° to radians.

23. Change $\frac{\pi}{18}$ rad to degrees.

24. The radius of a wheel is 75.3 cm. The wheel turns 0.561 rad. Find the distance the wheel travels.

25. A wheel of diameter 25.8 cm travels a distance of 20.0 cm. Find the angle θ (in radians) that the wheel turns.

26. A wheel of radius 16.2 cm turns an angle of 1028°. Find the distance a point travels on the surface of the wheel.

For Exercises 27–28, see Illustration 5:

27. Find **a.** the lateral surface area and **b.** the total surface area of the prism.

28. Find the volume of the prism.

29. Find the volume of the right circular cylinder shown in Illustration 6.

30. A metallurgist needs to cast a molten alloy in the

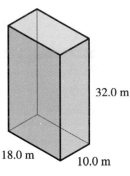

32.0 m
18.0 m
10.0 m

Illustration 5

5.62 cm
12.7 cm

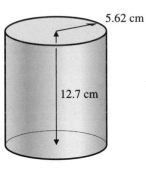

Illustration 6

shape of a right circular cylinder. The dimensions of the mold are as shown in Illustration 7. Find the amount (volume) of molten alloy needed.

31. Find the lateral surface area and the total surface area of the cylinder that was cast in Exercise 30.

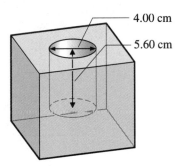

4.00 cm
5.60 cm

Illustration 7

32. Find the volume of the pyramid shown in Illustration 8.

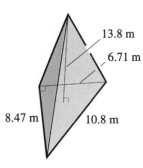

13.8 m
6.71 m
8.47 m
10.8 m

Illustration 8

33. Find **a.** the volume and **b.** the lateral surface area of the right circular cone shown in Illustration 9.

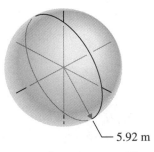

38.7 m
37.6 m

Illustration 9

34. Find **a.** the volume and **b.** the surface area of the sphere shown in Illustration 10.

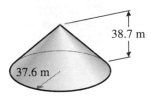

5.92 m

Illustration 10

35. Find the volume and lateral surface area of the frustum of the cone shown in Illustration 11.

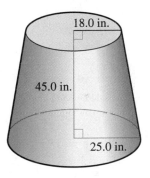

18.0 in.
45.0 in.
25.0 in.

Illustration 11

Chapter 12 Test

1. Find the area of a rectangle 18.0 ft long and 6.00 ft wide.

2. Find the perimeter of a square lot 160 m on a side.

Given the trapezoid in Illustration 1, find:

3. its area   **4.** its perimeter

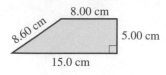

8.00 cm

8.60 cm   5.00 cm

15.0 cm

ILLUSTRATION 1

5. Find the length of side *a* in Illustration 2.

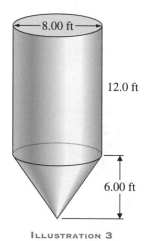

B

30.0 km

a

C   24.0 km   *A*

ILLUSTRATION 2

Given a circle of radius 20.0 cm, find:

6. its area   **7.** its circumference

8. Change 240° to radians.

9. Change $\frac{7\pi}{4}$ rad to degrees.

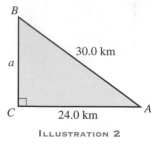

8.00 ft

12.0 ft

6.00 ft

10. Find the volume of a rectangular box
12.0 ft × 8.00 ft × 9.00 ft.

11. Find the total surface area of the box in Exercise 10.

ILLUSTRATION 3

12. Find the volume of a cylindrical tank 20.0 m in
diameter and 30.0 m high.

13. Find **a.** the volume and **b.** the lateral surface area
of the bin in Illustration 3.

7.60 m

8.90 m

5.20 m

14. Find the volume of the frustum of the cone in
Illustration 4.

ILLUSTRATION 4

Cumulative Review   Chapters 1–12

1. Add $-8 + (+7) + (-3)$.

2. Given $P = 2(l + w)$, where $l = 4\frac{1}{8}$ in. and $w = 2\frac{3}{4}$ in., find P.

3. The mass of a full-size automobile is **a.** 100 kg **b.** 1500 kg **c.** 10 kg or **d.** 15,000 kg.

4. Find **a.** precision and **b.** greatest possible error: 20,400 L

5. Simplify: $\dfrac{14x^3 - 56x^2 - 28x}{7x}$

Find each product mentally:

11. $(3x - 5)(2x + 7)$

Factor each expression completely:

13. $5x^3 - 15x$

14. $x^2 - 3x - 28$

15. $x^2 - 4$

Solve each equation using the quadratic formula (when necessary, round the results to three significant digits):

18. $5x^2 + 13x - 6 = 0$

19. $4x^2 - 10x - 29 = 0$

20. Draw the graph of $y = 2x^2 - 3x - 2$ and label the vertex.

21. Simplify j^{33}.

22. In Illustration 1, $m \parallel n$ and $\angle 4 = 82°$. Find the measure of the other angles.

6. Solve: $4 - 2x = 18$

7. $E = mv^2$; find m if $E = 952$ and $v = 7.00$.

8. A 160-lb object, 28.0 in. from the fulcrum of a lever, balances a second object 80.0 in. from the other end of the lever. What is the weight of the second object?

9. Solve for y: $5x - 8y = 10$

10. Solve: $5x - y = 12$
    $\qquad\quad y = 2x$

12. $(4x - 3)^2$

16. Solve: $10x^2 - 5x = 105$

17. Solve: $2x^2 - x = 3$

23. In Illustration 1, if $\angle 4 = 2x - 3$ and $\angle 5 = x + 6$, find the value of x.

24. Find the perimeter and area of the quadrilateral in Illustration 2.

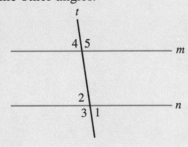

ILLUSTRATION 1

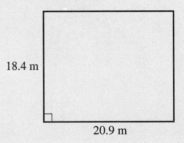

ILLUSTRATION 2

Find the area and perimeter of each triangle:

25.

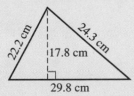

26.

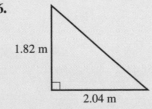

27. In Illustration 3, find the measure of the angle whose measure is not given.

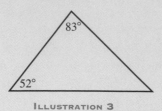

ILLUSTRATION 3

28. Find the area and the circumference of the circle in Illustration 4.

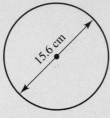

ILLUSTRATION 4

29. The area of a circle is 168 cm^2. Find its radius.

30. Find the volume and total surface area of the cylinder shown in Illustration 5.

ILLUSTRATION 5

Right Triangle Trigonometry

MEDICAL RESEARCH
At the New York State Department of Health, a research worker runs radiobiological tests.

MATHEMATICS AT WORK

Trigonometry is a mathematical tool that is used in many trades, such as engineering, drafting, machining, welding, sheetmetal work, plumbing, and manufacturing processes that require a high degree of accuracy.

Trigonometry is also used in aviation, aeronautics, and surveying, to find the precise range and bearing from point to another. Geologists use trig to survey and plot grids when recording geological sites.

Trigonometry is an important tool for anyone planning to pursue a profession that requires the use of higher-order mathematics such as calculus and physics. A firm understanding of trigonometry can open many doors to success in high-paying jobs.

13.1   Trigonometric Ratios

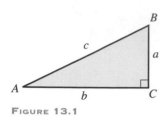

FIGURE 13.1

Many applications in science and technology require the use of triangles and trigonometry. Early applications of trigonometry, beginning in the second century B.C., were in astronomy, surveying, and navigation. Applications that you may study include electronics, the motion of projectiles, light refraction in optics, and sound.

In this chapter, we consider only right triangles. A right triangle has one right angle, two acute angles, a hypotenuse, and two legs. The right angle, as shown in Figure 13.1, is usually labeled with the capital letter C. The vertices of the two acute angles are usually labeled with the capital letters A and B. The hypotenuse is the side opposite the right angle, the longest side of a right triangle, and is usually labeled with the lowercase letter c. The legs are the sides opposite the acute angles. The leg (side) opposite angle A is labeled a, and the leg opposite angle B is labeled b. Note that each side of the triangle is labeled with the lowercase of the letter of the angle opposite that side.

The two legs are also named as the side *opposite* angle A and the side *adjacent to* (or next to) angle A; or as the side opposite angle B and the side adjacent to angle B. See Figure 13.2.

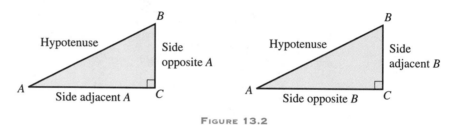

FIGURE 13.2

PYTHAGOREAN THEOREM

In any right triangle,

$$c^2 = a^2 + b^2$$

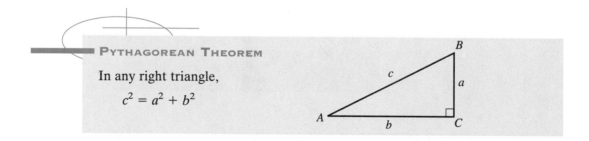

That is, the square of the length of the hypotenuse is equal to the sum of the squares of the lengths of the legs. The following equivalent formulas are often more useful:

$c = \sqrt{a^2 + b^2}$ used to find the length of the hypotenuse

$a = \sqrt{c^2 - b^2}$ used to find the length of leg a

$b = \sqrt{c^2 - a^2}$ used to find the length of leg b

Recall that the Pythagorean theorem was developed in detail in Section 12.3.

• EXAMPLE 1

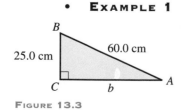

FIGURE 13.3

Find the length of side b in Figure 13.3.

Using the formula to find the length of leg b, we have

$$b = \sqrt{c^2 - a^2}$$
$$b = \sqrt{(60.0 \text{ cm})^2 - (25.0 \text{ cm})^2}$$
$$= 54.5 \text{ cm}$$

• **EXAMPLE 2**   Find the length of side c in Figure 13.4.

Using the formula to find the hypotenuse c, we have

$$c = \sqrt{a^2 + b^2}$$
$$c = \sqrt{(29.7 \text{ m})^2 + (34.2 \text{ m})^2}$$
$$= 45.3 \text{ m}$$

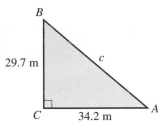

FIGURE 13.4

A *ratio* is the comparison of two quantities by division. The ratios of the sides of a right triangle can be used to find an unknown part—or parts—of that right triangle. Such a ratio is called a *trigonometric ratio* and expresses the relationship between an acute angle and the lengths of two of the sides of a right triangle.

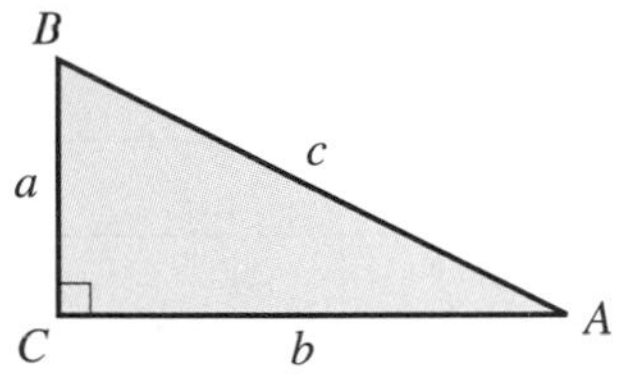

The *sine* of angle A, abbreviated "sin A," equals the ratio of the length of the side opposite angle A, which is a, to the length of the hypotenuse, c.

The *cosine* of angle A, abbreviated "cos A," equals the ratio of the length of the side adjacent to angle A, which is b, to the length of the hypotenuse, c.

The *tangent* of angle A, abbreviated "tan A," equals the ratio of the length of the side opposite angle A, which is a, to the length of the side adjacent to angle A, which is b.

That is, in a right triangle (Figure 13.5), we have the following ratios.

FIGURE 13.5

TRIGONOMETRIC RATIOS

$$\sin A = \frac{\text{length of side opposite angle } A}{\text{length of hypotenuse}} = \frac{a}{c}$$

$$\cos A = \frac{\text{length of side adjacent to angle } A}{\text{length of hypotenuse}} = \frac{b}{c}$$

$$\tan A = \frac{\text{length of side opposite angle } A}{\text{length of side adjacent to angle } A} = \frac{a}{b}$$

Similarly, the ratios can be defined for angle B.

$$\sin B = \frac{\text{length of side opposite angle } B}{\text{length of hypotenuse}} = \frac{b}{c}$$

$$\cos B = \frac{\text{length of side adjacent to angle } B}{\text{length of hypotenuse}} = \frac{a}{c}$$

$$\tan B = \frac{\text{length of side opposite angle } B}{\text{length of side adjacent to angle } B} = \frac{b}{a}$$

• **EXAMPLE 3**   Find the three trigonometric ratios for angle A in the triangle in Figure 13.6.

$$\sin A = \frac{\text{length of side opposite angle } A}{\text{length of hypotenuse}} = \frac{a}{c} = \frac{144 \text{ m}}{156 \text{ m}} = 0.9231$$

$$\cos A = \frac{\text{length of side adjacent to angle } A}{\text{length of hypotenuse}} = \frac{b}{c} = \frac{60.0 \text{ m}}{156 \text{ m}} = 0.3846$$

$$\tan A = \frac{\text{length of side opposite angle } A}{\text{length of side adjacent to angle } A} = \frac{a}{b} = \frac{144 \text{ m}}{60.0 \text{ m}} = 2.400$$

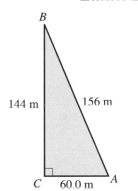

FIGURE 13.6

The values of the trigonometric ratios of various angles are found in tables and from calculators. We shall use calculators. You will need a calculator that has sin, cos, and tan buttons.

• **EXAMPLE 4**   Find sin 37.5° rounded to four significant digits.

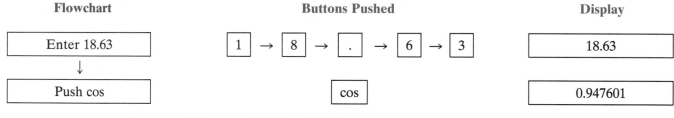

Thus, sin 37.5° = 0.6088 rounded to four significant digits.

• **EXAMPLE 5**   Find cos 18.63° rounded to four significant digits.

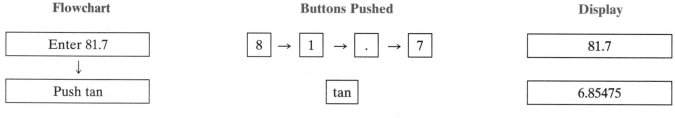

Thus, cos 18.63° = 0.9476 rounded to four significant digits.

• **EXAMPLE 6**   Find tan 81.7° rounded to four significant digits.

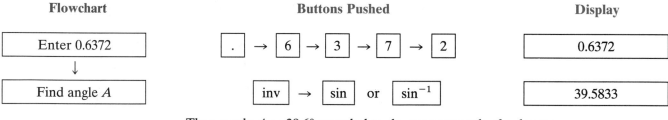

Thus, tan 81.7° = 6.855 rounded to four significant digits.

A calculator may also be used to find the *angle* when the value of the trigonometric ratio is known. The procedure is shown in the examples below.

• **EXAMPLE 7**   Find angle A to the nearest tenth of a degree when sin A = 0.6372.

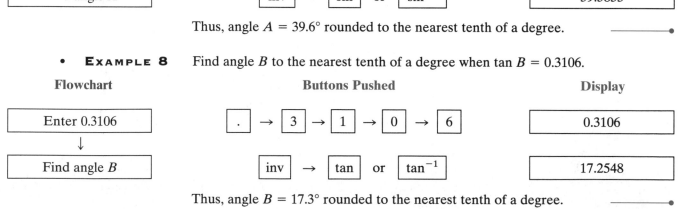

Thus, angle A = 39.6° rounded to the nearest tenth of a degree.

• **EXAMPLE 8**   Find angle B to the nearest tenth of a degree when tan B = 0.3106.

Thus, angle B = 17.3° rounded to the nearest tenth of a degree.

• **EXAMPLE 9**   Find angle A to the nearest hundredth of a degree when $\cos A = 0.4165$.

Flowchart   **Buttons Pushed**   **Display**

| Enter 0.4165 |

| 0.4165 |

↓

| Find Angle A |

| inv | → | cos | or | $\cos^{-1}$ |

| 65.3862 |

Thus, angle $A = 65.39°$ rounded to the nearest hundredth of a degree.

Exercises   **13.1**

Refer to right triangle ABC in Illustration 1 for Exercises 1–10:

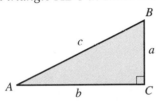

ILLUSTRATION 1

1. The side opposite angle A is ___?___.
2. The side opposite angle B is ___?___.

3. The hypotenuse is ___?___.
4. The side adjacent to angle A is ___?___.
5. The side adjacent to angle B is ___?___.
6. The angle opposite side a is ___?___.
7. The angle opposite side b is ___?___.
8. The angle opposite side c is ___?___.
9. The angle adjacent to side a is ___?___.
10. The angle adjacent to side b is ___?___.

Use right triangle ABC in Illustration 1 and the Pythagorean theorem to find each unknown side, rounded to three significant digits:

11. $c = 75.0$ m, $a = 45.0$ m
12. $a = 25.0$ cm, $b = 60.0$ cm
13. $a = 29.0$ mi, $b = 47.0$ mi
14. $a = 12.0$ km, $c = 61.0$ km
15. $c = 18.9$ cm, $a = 6.71$ cm
16. $a = 20.2$ mi, $b = 19.3$ mi
17. $a = 171$ ft, $b = 203$ ft

18. $c = 35.3$ m, $b = 25.0$ m
19. $a = 202$ m, $c = 404$ m
20. $a = 1.91$ km, $b = 3.32$ km
21. $b = 1520$ km, $c = 2160$ km
22. $a = 203,000$ ft, $c = 521,000$ ft
23. $a = 45,800$ m, $b = 38,600$ m
24. $c = 3960$ m, $b = 3540$ m

Use the triangle in Illustration 2 for Exercises 25–30:

25. Find $\sin A$.
26. Find $\cos A$.
27. Find $\tan A$.

28. Find $\sin B$.
29. Find $\cos B$.
30. Find $\tan B$.

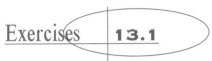

ILLUSTRATION 2

Find the value of each trigonometric ratio rounded to four significant digits:

31. $\sin 49.6°$
32. $\cos 55.2°$
33. $\tan 65.3°$

34. $\sin 69.7°$
35. $\cos 29.7°$
36. $\tan 14.6°$

37. $\sin 31.64°$
38. $\tan 13.25°$
39. $\cos 75.31°$

40. $\cos 84.83°$
41. $\tan 3.05°$
42. $\sin 6.74°$

43. $\sin 37.62°$
44. $\cos 18.94°$
45. $\tan 21.45°$

46. $\sin 11.31°$
47. $\cos 47.16°$
48. $\tan 81.85°$

Find each angle rounded to the nearest tenth of a degree:

49. $\sin A = 0.7941$
50. $\tan A = 0.2962$
51. $\cos B = 0.4602$

52. $\cos A = 0.1876$
53. $\tan B = 1.386$
54. $\sin B = 0.3040$

55. $\sin B = 0.1592$
56. $\tan B = 2.316$
57. $\cos A = 0.8592$

58. $\cos B = 0.3666$
59. $\tan A = 0.8644$
60. $\sin A = 0.5831$

Find each angle rounded to the nearest hundredth of a degree:

61. $\tan A = 0.1941$ **64.** $\cos B = 0.2597$ **67.** $\tan B = 3.806$ **70.** $\cos A = 0.6427$

62. $\sin B = 0.9324$ **65.** $\sin A = 0.1506$ **68.** $\sin A = 0.4232$ **71.** $\sin B = 0.3441$

63. $\cos A = 0.3572$ **66.** $\tan B = 2.500$ **69.** $\cos B = 0.7311$ **72.** $\tan A = 0.5536$

73. In Exercises 25–30, there are two pairs of ratios that are equal. Name them.

13.2 Using Trigonometric Ratios to Find Angles

A trigonometric ratio may be used to find an angle of a right triangle, given the lengths of any two sides.

•   **EXAMPLE 1**

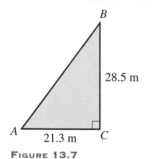

FIGURE 13.7

In Figure 13.7, find angle B using a calculator, as follows.

We know the sides opposite and adjacent to angle A. So we use the tangent ratio:

$$\tan A = \frac{\text{length of side opposite angle } A}{\text{length of side adjacent to angle } A}$$

$$\tan A = \frac{28.5 \text{ m}}{21.3 \text{ m}} = 1.338$$

Next, find angle A to the nearest tenth of a degree when $\tan A = 1.338$. The complete set of operations on a calculator follows.

Flowchart	Buttons Pushed	Display
Enter 28.5	2 → 8 → . → 5	28.5
Push divide	÷	28.5
Enter 21.3	2 → 1 → . → 3	21.3
Push equals	=	1.33803
Find angle A	inv → tan or tan⁻¹	53.2267

Thus, angle $A = 53.2°$ rounded to the nearest tenth of a degree. ————————●

When calculations involve a trigonometric ratio, we shall use the following rule for significant digits.

Angles expressed to the nearest	The length of each side of the triangle contains
1°	Two significant digits
0.1°	Three significant digits
0.01°	Four significant digits

An example of each case is shown in Figure 13.8.

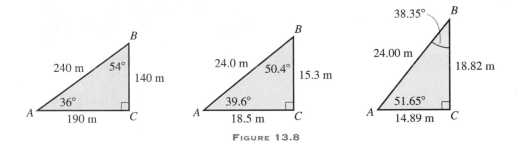

FIGURE 13.8

• **EXAMPLE 2**   Find angle B in the triangle in Figure 13.9.

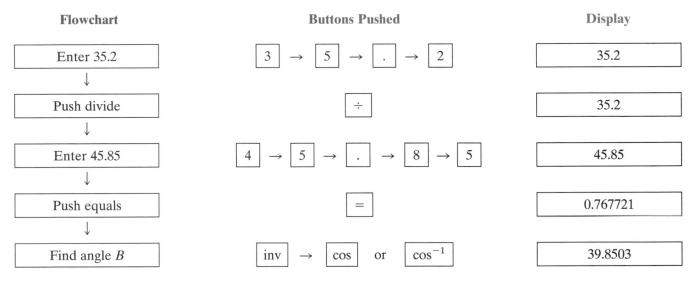

FIGURE 13.9

We know the hypotenuse and the side adjacent to angle B. So let's use the cosine ratio.

$$\cos B = \frac{\text{length of side adjacent to angle } B}{\text{length of hypotenuse}}$$

$$\cos B = \frac{35.20 \text{ cm}}{45.85 \text{ cm}}$$

Find angle B using a calculator as follows.

Flowchart	Buttons Pushed	Display
Enter 35.2	$3 \rightarrow 5 \rightarrow . \rightarrow 2$	35.2
Push divide	$\div$	35.2
Enter 45.85	$4 \rightarrow 5 \rightarrow . \rightarrow 8 \rightarrow 5$	45.85
Push equals	$=$	0.767721
Find angle B	inv $\rightarrow$ cos or $\cos^{-1}$	39.8503

Thus, angle $B = 39.85°$ rounded to the nearest hundredth of a degree.

The question is often raised, "Which of the three trig ratios do I use?" First, notice that each trigonometric ratio consists of two sides and one angle, or three quantities in all. To find the solution to such an equation, two of the quantities must be known. We will answer the question in two parts.

WHICH TRIG RATIO TO USE

1. If you are finding an angle, two sides must be known. Label these two known sides as *side opposite* the angle you are finding, *side adjacent* to the angle you are finding, or *hypotenuse*. Then choose the trig ratio that has these two sides.

2. If you are finding a side, one side and one angle must be known. Label the known side and the unknown side as *side opposite* the known angle, *side adjacent* to the known angle, or *hypotenuse*. Then choose the trig ratio that has these two sides.

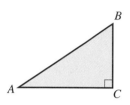

FIGURE 13.10

A useful and time-saving fact about right triangles (Figure 13.10) is that the sum of the acute angles of any right triangle is 90°.

$$A + B = 90°$$

Why? We know that the sum of the interior angles of any triangle is 180°. A right triangle, by definition, contains a right angle, whose measure is 90°. That leaves 90° to be divided between the two acute angles.

Note, then, that if one acute angle is given or known, the other acute angle may be found by subtracting the known angle from 90°. That is,

$$A = 90° - B$$

$$B = 90° - A$$

• **EXAMPLE 3** Find angle A and angle B in the triangle in Figure 13.11.

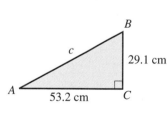

FIGURE 13.11

$$\tan A = \frac{\text{length of side opposite angle } A}{\text{length of side adjacent to angle } A}$$

$$\tan A = \frac{29.1 \text{ cm}}{53.2 \text{ cm}} = 0.5470$$

$$A = 28.7°$$

Angle $B = 90° - 28.7° = 61.3°$.

Exercises | 13.2

Using Illustration 1, find the measure of each acute angle for each right triangle:

1. $a = 36.0$ m, $b = 50.9$ m
2. $a = 72.0$ cm, $c = 144$ cm
3. $b = 39.7$ cm, $c = 43.6$ cm
4. $a = 171$ km, $b = 695$ km
5. $b = 13.6$ m, $c = 18.7$ m
6. $b = 409$ km, $c = 612$ km
7. $a = 29.7$ m, $b = 29.7$ m, $c = 42.0$ m

8. $a = 36.2$ mm, $b = 62.7$ mm, $c = 72.4$ mm
9. $a = 2902$ km, $b = 1412$ km
10. $b = 21.34$ m, $c = 47.65$ m
11. $a = 0.6341$ cm, $c = 0.7982$ cm
12. $b = 4.372$ m, $c = 5.806$ m
13. $b = 1455$ ft, $c = 1895$ ft
14. $a = 25.45$ in., $c = 41.25$ in.

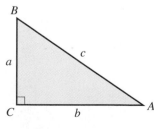

ILLUSTRATION 1

15. $a = 243.2$ km, $b = 271.5$ km

16. $a = 351.6$ m, $b = 493.0$ m

17. $a = 16.7$ m, $c = 81.4$ m

18. $a = 847$ m, $b = 105$ m

19. $b = 1185$ ft, $c = 1384$ ft

20. $a = 48.7$ cm, $c = 59.5$ cm

21. $a = 845$ km, $b = 2960$ km

22. $b = 2450$ km, $c = 3570$ km

23. $a = 8897$ m, $c = 9845$ m

24. $a = 58.44$ mi, $b = 98.86$ mi

13.3 Using Trigonometric Ratios to Find Sides

We also use a trigonometric ratio to find a side of a right triangle, given one side and the measure of one of the acute angles.

• **EXAMPLE 1**

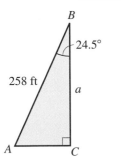

FIGURE 13.12

Find side a in the triangle in Figure 13.12.

With respect to the known angle B, we know the hypotenuse and are finding the adjacent side. So we use the cosine ratio.

$$\cos B = \frac{\text{length of side adjacent to angle } B}{\text{length of hypotenuse}}$$

$$\cos 24.5° = \frac{a}{258 \text{ ft}}$$

$$a = (\cos 24.5°)(258 \text{ ft}) \qquad \text{Multiply both sides by 258 ft.}$$

Side a can be found using a calculator as follows.

Flowchart	Buttons Pushed	Display
Enter 24.5	2 → 4 → . → 5	24.5
Push cos	cos	0.909961
Push times	×	0.909961
Enter 258	2 → 5 → 8	258
Push equals	=	234.770

Thus, side $a = 235$ ft rounded to three significant digits.

• **EXAMPLE 2**

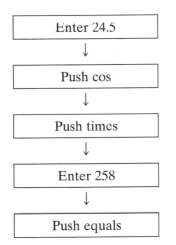

FIGURE 13.13

Find the sides b and c in the triangle in Figure 13.13.

If we find side b first, we are looking for the side adjacent to angle A, the known angle. We are given the side opposite angle A. Thus, we should use the tangent ratio.

$$\tan A = \frac{\text{length of side opposite angle } A}{\text{length of side adjacent to angle } A}$$

$$\tan 52.3° = \frac{29.7 \text{ m}}{b}$$

$$b(\tan 52.3°) = 29.7 \text{ m} \qquad \text{Multiply both sides by } b.$$

$$b = \frac{29.7 \text{ m}}{\tan 52.3°} \qquad \text{Divide both sides by } \tan 52.3°.$$

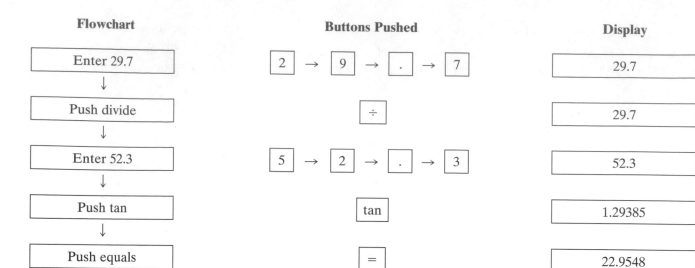

Flowchart	Buttons Pushed	Display

Thus, side $b = 23.0$ m rounded to three significant digits.

To find side c, we are looking for the hypotenuse, and we have the opposite side given. Thus, we should use the sine ratio.

$$\sin A = \frac{\text{length of side opposite angle } A}{\text{length of hypotenuse}}$$

$$\sin 52.3° = \frac{29.7 \text{ m}}{c}$$

$c(\sin 52.3°) = 29.7$ m   Multiply both sides by c.

$c = \dfrac{29.7 \text{ m}}{\sin 52.3°}$ Divide both sides by $\sin 52.3°$.

$= 37.5$ m

The Pythagorean theorem may be used to check your work.

Exercises 13.3

Find the unknown sides of each right triangle (see Illustration 1):

1. $a = 36.7$ m, $A = 42.1°$
2. $b = 73.6$ cm, $B = 19.0°$
3. $a = 236$ km, $B = 49.7°$
4. $b = 28.9$ ft, $A = 65.2°$
5. $c = 49.1$ cm, $A = 36.7°$
6. $c = 236$ m, $A = 12.9°$
7. $b = 23.7$ cm, $A = 23.7°$
8. $a = 19.2$ km, $B = 63.2°$
9. $b = 29,200$ km, $A = 12.9°$
10. $c = 36.7$ mi, $B = 68.3°$

11. $a = 19.72$ m, $A = 19.75°$
12. $b = 125.3$ cm, $B = 23.34°$
13. $c = 255.6$ mi, $A = 39.25°$
14. $c = 7.363$ km, $B = 14.80°$
15. $b = 12,350$ m, $B = 69.72°$
16. $a = 3678$ m, $B = 10.04°$
17. $a = 1980$ m, $A = 18.4°$
18. $a = 9820$ ft, $B = 35.7°$
19. $b = 841.6$ km, $A = 18.91°$
20. $c = 289.5$ cm, $A = 24.63°$

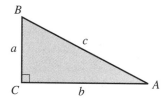

ILLUSTRATION 1

21. $c = 185.6$ m, $B = 61.45°$
22. $b = 21.63$ km, $B = 82.06°$
23. $c = 256$ cm, $A = 25.6°$
24. $a = 18.3$ mi, $A = 71.2°$

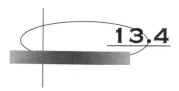

13.4 Solving Right Triangles

To *solve* a triangle means to find the measures of the various parts of a triangle that are not given. We proceed as we did in the last two sections.

• **EXAMPLE 1**

FIGURE 13.14

Solve the right triangle in Figure 13.14.

We are given the measure of one acute angle and the length of one leg.

$$A = 90° - B$$
$$A = 90° - 36.7° = 53.3°$$

We then can use either the sine or the cosine ratio to find side c.

$$\sin B = \frac{\text{length of side opposite angle } B}{\text{length of hypotenuse}}$$

$$\sin 36.7° = \frac{19.2 \text{ m}}{c}$$

$c(\sin 36.7°) = 19.2 \text{ m}$ Multiply both sides by c.

$c = \dfrac{19.2 \text{ m}}{\sin 36.7°}$ Divide both sides by $\sin 36.7°$.

$\quad = 32.1 \text{ m}$

Now we may use either a trigonometric ratio or the Pythagorean theorem to find side a.

Solution by a Trigonometric Ratio:

$$\tan B = \frac{\text{length of side opposite angle } B}{\text{length of side adjacent to angle } B}$$

$$\tan 36.7° = \frac{19.2 \text{ m}}{a}$$

$a(\tan 36.7°) = 19.2 \text{ m}$ Multiply both sides by a.

$a = \dfrac{19.2 \text{ m}}{\tan 36.7°}$ Divide both sides by $\tan 36.7°$.

$\quad = 25.8 \text{ m}$

Solution by the Pythagorean Theorem:

$$a = \sqrt{c^2 - b^2}$$
$$a = \sqrt{(32.1 \text{ m})^2 - (19.2 \text{ m})^2}$$
$$\quad = 25.7 \text{ m}$$

Can you explain the difference in these two results? —————•

• **EXAMPLE 2**

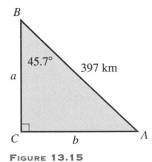

FIGURE 13.15

Solve the right triangle in Figure 13.15.

We are given the measure of one acute angle and the length of the hypotenuse.

$$A = 90° - B$$
$$A = 90° - 45.7° = 44.3°$$

To find side b, we must use the sine or the cosine ratio, since the hypotenuse is given.

$$\sin B = \frac{\text{length of side opposite angle } B}{\text{length of hypotenuse}}$$

$$\sin 45.7° = \frac{b}{397 \text{ km}}$$

$(\sin 45.7°)(397 \text{ km}) = b$ Multiply both sides by 397 km.

$284 \text{ km} = b$

Again, we can use either a trigonometric ratio or the Pythagorean theorem to find side a.

Solution by a Trigonometric Ratio:

$$\cos B = \frac{\text{length of side adjacent to angle } B}{\text{length of hypotenuse}}$$

$$\cos 45.7° = \frac{a}{397 \text{ km}}$$

$(\cos 45.7°)(397 \text{ km}) = a$ Multiply both sides by 397 km.

$$277 \text{ km} = a$$

Solution by the Pythagorean Theorem:

$$a = \sqrt{c^2 - b^2}$$
$$a = \sqrt{(397 \text{ km})^2 - (284 \text{ km})^2}$$
$$= 277 \text{ km}$$

• **EXAMPLE 3**   Solve the right triangle in Figure 13.16.

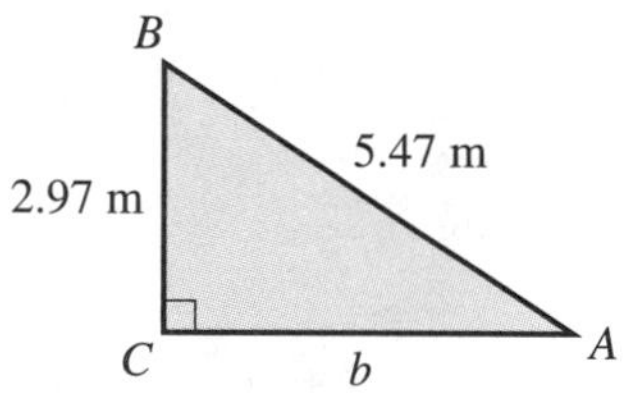

FIGURE 13.16

We are given two sides of the right triangle.

To find angle A or angle B, we could use either the sine or cosine, since the hypotenuse is given.

$$\sin A = \frac{\text{length of side opposite angle } A}{\text{length of hypotenuse}}$$

$$\sin A = \frac{2.97 \text{ m}}{5.47 \text{ m}} = 0.5430$$

$$A = 32.9°$$

Then

$$B = 90° - A$$
$$B = 90° - 32.9° = 57.1°$$

The unknown side b can be found using the Pythagorean theorem.

$$b = \sqrt{c^2 - a^2}$$
$$b = \sqrt{(5.47 \text{ m})^2 - (2.97 \text{ m})^2}$$
$$= 4.59 \text{ m}$$

Exercises | 13.4

Using Illustration 1, solve each right triangle:

1. $A = 50.6°$, $c = 49.0$ m
2. $a = 30.0$ cm, $b = 40.0$ cm
3. $B = 41.2°$, $a = 267$ ft
4. $A = 39.7°$, $b = 49.6$ km
5. $b = 72.0$ mi, $c = 78.0$ mi
6. $B = 22.4°$, $c = 46.0$ mi
7. $A = 52.1°$, $a = 72.0$ mm
8. $B = 42.3°$, $b = 637$ m

9. $A = 68.8°$, $c = 39.4$ m
10. $a = 13.6$ cm, $b = 13.6$ cm
11. $a = 12.00$ m, $b = 24.55$ m
12. $B = 38.52°$, $a = 4315$ m
13. $A = 29.19°$, $c = 2975$ ft
14. $B = 29.86°$, $a = 72.62$ m
15. $a = 46.72$ m, $b = 19.26$ m
16. $a = 2436$ ft, $c = 4195$ ft

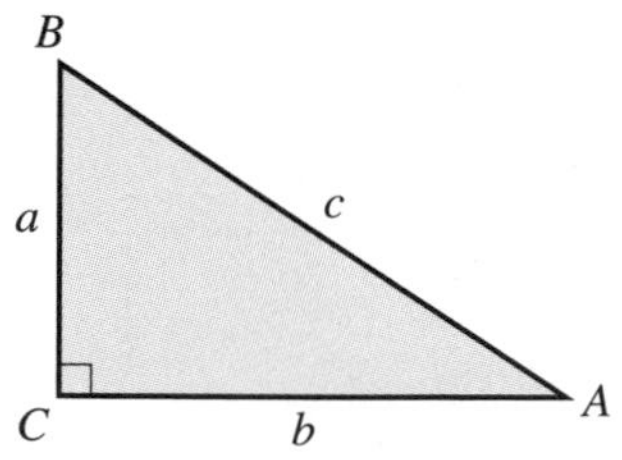

ILLUSTRATION 1

17. $A = 41.1°$, $c = 485$ m
18. $a = 1250$ km, $b = 1650$ km

19. $B = 9.45°$, $a = 1585$ ft

20. $A = 14.60°$, $b = 135.7$ cm

21. $b = 269.5$ m, $c = 380.5$ m

22. $B = 75.65°$, $c = 92.75$ km

23. $B = 81.5°$, $b = 9370$ ft

24. $a = 14.6$ mi, $c = 31.2$ mi

13.5 Applications Involving Trigonometric Ratios

Trigonometric ratios can be used to solve many applications similar to those problems we solved in the last sections. However, instead of having to find all the parts of a right triangle, we usually need to find only one.

• **EXAMPLE 1**   The roof in Figure 13.17 has a rise of 7.50 ft and a run of 18.0 ft. Find angle A.

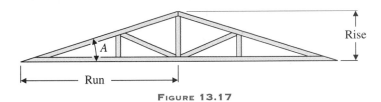

Rise

A

Run

FIGURE 13.17

We know the length of the side opposite angle A and the length of the side adjacent to angle A. So we use the tangent ratio.

$$\tan A = \frac{\text{length of side opposite angle } A}{\text{length of side adjacent to angle } A}$$

$$\tan A = \frac{7.50 \text{ ft}}{18.0 \text{ ft}} = 0.4167$$

$$A = 22.6°$$

The *angle of depression* is the angle between the horizontal and the line of sight to an object that is *below* the horizontal. The *angle of elevation* is the angle between the horizontal and the line of sight to an object that is *above* the horizontal.

In Figure 13.18, angle A is the angle of depression for an observer in the helicopter sighting down to the building on the ground, and angle B is the angle of elevation for an observer in the building sighting up to the helicopter.

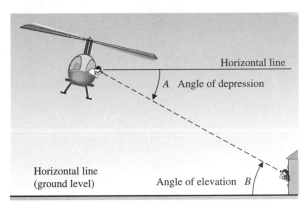

Horizontal line

A Angle of depression

Horizontal line
(ground level)

Angle of elevation B

FIGURE 13.18

• **EXAMPLE 2**   A ship's navigator measures the angle of elevation to the beacon of a lighthouse to be 10.1°. He knows that this particular beacon is 225 m above sea level. How far is the ship from the lighthouse?

First, you should sketch the problem, as in Figure 13.19. Since this problem involves finding the length of the side adjacent to an angle when the opposite side is known, we use the tangent ratio.

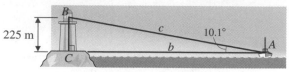

FIGURE 13.19

$$\tan A = \frac{\text{length of side opposite angle } A}{\text{length of side adjacent to angle } A}$$

$$\tan 10.1° = \frac{225 \text{ m}}{b}$$

$b(\tan 10.1°) = 225 \text{ m}$ Multiply both sides by b.

$$b = \frac{225 \text{ m}}{\tan 10.1°}$$ Divide both sides by $\tan 10.1°$.

$$= 1260 \text{ m}$$

• **EXAMPLE 3**   In ac (alternating current) circuits, the relationship between impedance Z (in ohms), the resistance R (in ohms), and the phase angle θ is shown by the right triangle in Figure 13.20(a). If the resistance is 250 Ω and the phase angle is 41°, find the impedance.

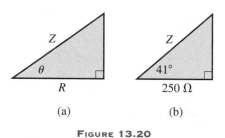

(a)   (b)

FIGURE 13.20

Here, we know the adjacent side and the angle and wish to find the hypotenuse (see Figure 13.20(b)). So we use the cosine ratio.

$$\cos \theta = \frac{\text{length of side adjacent to angle } \theta}{}$$

$$\cos 41° = \frac{250 \text{ } \Omega}{Z}$$

$Z(\cos 41°) = 250 \text{ } \Omega$ Multiply both sides by Z.

$$Z = \frac{250 \text{ } \Omega}{\cos 41°}$$ Divide both sides by $\cos 41°$.

$$= 330 \text{ } \Omega$$

Exercises | **13.5**

⌐ 1. A conveyor is used to lift paper to a shredder. The most efficient operating angle of elevation for the conveyor is 35.8°. The paper is to be elevated 11.0 m. What length of conveyor is needed?

⌐ 2. Maria is to weld a support for a 23-m conveyor

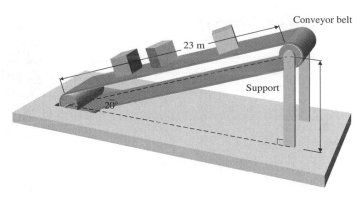

ILLUSTRATION 1

so that it will operate at a $2\overline{0}°$ angle. What is the length of the support? See Illustration 1.

3. A bullet is found embedded in the wall of a room 2.3 m above the floor. The bullet entered the wall going upward at an angle of 12°. How far from the wall was the bullet fired if the gun was held 1.2 m above the floor?

4. The recommended safety angle of a ladder against a building is 78°. A $1\overline{0}$-m ladder will be used. How high up on the side of the building will the ladder safely reach? (See Illustration 2.)

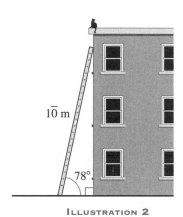

$1\overline{0}$ m

78°

ILLUSTRATION 2

5. A piece of conduit 38.0 ft long cuts across the corner of a room, as shown in Illustration 3. Find length x and angle A.

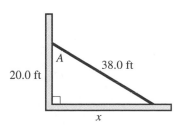

A

20.0 ft   38.0 ft

x

ILLUSTRATION 3

6. Find the width of the river in Illustration 4.

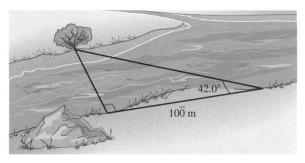

42.0°

100 m

ILLUSTRATION 4

7. A roadbed rises 220 ft for each 2300 ft of horizontal. (See Illustration 5.) Find the angle of inclination of the roadbed. (This is usually referred to as *% of grade*.)

220 ft

2300 ft

ILLUSTRATION 5

8. A smokestack is $18\overline{0}$ ft high. A guy wire must be fastened to the stack 20.0 ft from the top. The guy wire makes an angle of 40.0° with the ground. Find the length of the guy wire.

9. A railroad track has an angle of elevation of 1.0°. What is the difference in height (in feet) of two points on the track 1.00 mi apart?

10. A machinist needs to drill four holes 1.00 in. apart in a straight line in a metal plate, as shown in Illustration 6. If the first hole is placed at the origin and the line forms an angle of 32.0°

with the vertical axis, find the coordinates of the other three holes (*A*, *B*, and *C*).

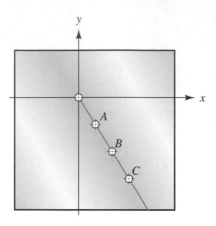

ILLUSTRATION 6

11. A draftsperson has to draft a triangular roof to a house. (See Illustration 7.) The roof is 30.0 ft wide. If the rafters are 17.0 ft long, at what angle will the rafters be laid at the eaves? Assume no overhang.

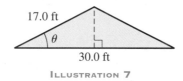

17.0 ft
θ
30.0 ft

ILLUSTRATION 7

12. A draftsperson needs to find the height of a cone-shaped hill. (See Illustration 8.) The diameter measures 280 ft. From a point on the circumference of the base, the angle of elevation measures 43°. What is the height?

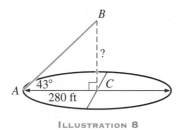

B
?
43°
A — 280 ft — *C*

ILLUSTRATION 8

13. A gauge is used to check the diameter of a crankshaft journal. It is constructed to make measurements on the basis of a right triangle with a 60.0° angle. The distance *AB* in Illustration 9 is 11.4 cm. Find the radius, *BC*, of the journal.

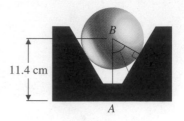

11.4 cm

B

A

ILLUSTRATION 9

14. A television station has cables cut in 18-m lengths. These are to be fastened to the top of their relay antennae and to the ground. The cables make an angle of 25° with level ground. Allow 1 m for fastening. Find the height of the relay antennae.

15. The cables attached to a TV relay tower are $11\overline{0}$ m long. They meet level ground at an angle of 60.0°, as in Illustration 10. Find the height of the tower.

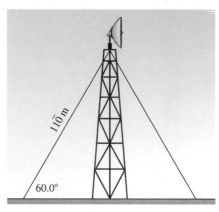

$11\overline{0}$ m

60.0°

ILLUSTRATION 10

16. A lunar module is resting on the moon's surface directly below a spaceship in its orbit, 12.0 km above the moon. (See Illustration 11.) Two lunar explorers find that the angle from their position to that of the spaceship is 82.9°. What distance are they from the lunar module?

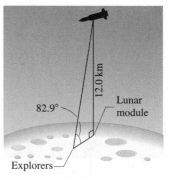

82.9°
12.0 km
Lunar module
Explorers

ILLUSTRATION 11

17. In ac (alternating current) circuits, the relationship between the impedance Z (in ohms), the resistance R (in ohms), the phase angle θ, and the reactance X (in ohms) is shown by the right triangle in Illustration 12.

   a. Find the impedance if the resistance is 350 Ω and the phase angle is 35°.

   b. Suppose the resistance is 550 Ω and the impedance is $7\overline{0}0$ Ω. What is the phase angle?

   c. Suppose the reactance is 182 Ω and the resistance is 240 Ω. Find the impedance and the phase angle.

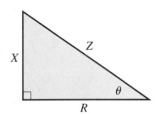

ILLUSTRATION 12

18. A right circular conical tank with its point down (Illustration 13) has a height of 4.00 m and a radius of 1.20 m. The tank is filled to a height of 3.70 m with liquid. How many litres of liquid are in the tank? (1000 litres $= 1$ m^3)

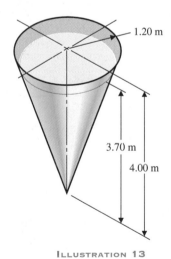

ILLUSTRATION 13

19. Use the right triangle in Illustration 14.

   a. Find the voltage applied if the voltage across the coil is 35.6 V and the voltage across the resistance is 40.2 V.

   b. Find the voltage across the resistance if the voltage applied is 378 V and the voltage across the coil is 268 V.

   c. Find the voltage across the coil if the voltage applied is 448 V and the voltage across the resistance is 381 V.

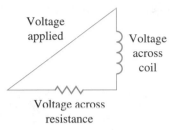

ILLUSTRATION 14

20. Machinists often use a coordinate system to drill holes by placing the origin at the most convenient location. A bolt circle is the circle formed by completing an arc through the centers of the bolt holes in a piece of metal. Find the coordinates of the centers of eight equally spaced $\frac{1}{4}$-in. holes on a bolt circle of radius 6.00 in., as shown in Illustration 15.

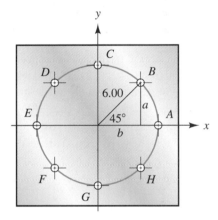

ILLUSTRATION 15

21. Twelve equally spaced holes must be drilled on a 14.500-in.-diameter bolt circle. (See Illustration 16.) What is the straight-line center-to-center distance between two adjacent holes?

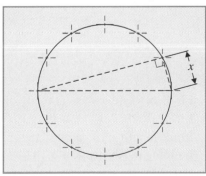

ILLUSTRATION 16

22. Dimension x in the dovetail shown in Illustration 17 is a check dimension. Precision steel balls of diameter 0.1875 in. are used in this procedure. What should this check dimension be?

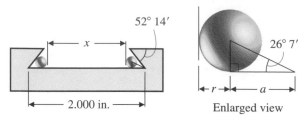

ILLUSTRATION 17

23. Find angle θ of the taper in Illustration 18.

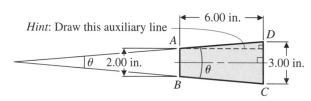

ILLUSTRATION 18

24. You need to use a metal screw with a head angle of angle A, which is not less than 65 degrees and no larger than 70 degrees. The team leader wants you to find angle A from the sketch shown in Illustration 19 and determine if the head angle will be satisfactory. Find the head angle A.

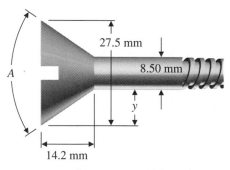

ILLUSTRATION 19

25. Find **a.** distance x and **b.** distance BD in Illustration 20. Length $BC = 5.50$ in.

26. Find length AB along the roofline of the building in Illustration 21. The slope of the roof is 45.0°.

27. Find length x and angle A in Illustration 22.

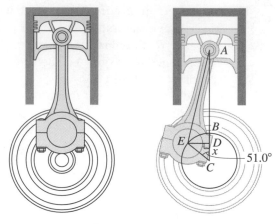

ILLUSTRATION 20

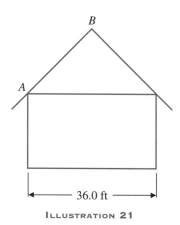

ILLUSTRATION 21

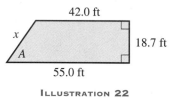

ILLUSTRATION 22

28. From the base of a building, measure out a horizontal distance of 215 ft along the ground. The angle of elevation to the top of the building is 63.0°. Find the height of the building.

29. A mechanical draftsperson needs to find the distance across the corners of a hex-bolt. See Illustration 23. If the distance across the flats is 2.25 cm, find the distance across the corners.

30. A hydraulic control valve has two parallel angular passages that must connect to two threaded ports, as shown in Illustration 24. What are the missing dimensions necessary for the location of the two ports?

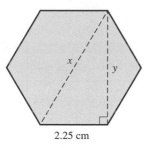

2.25 cm

ILLUSTRATION 23

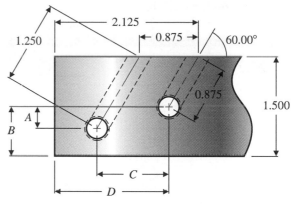

ILLUSTRATION 24

Chapter 13   Review

For Exercises 1–7, see Illustration 1:

1. What is the length of the side opposite angle A in the right triangle?

2. What is the angle adjacent to the side whose length is 29.7 m?

3. The side of the triangle denoted by c is known as the _____?_____.

4. What is the length of the side denoted by c?

5. $\dfrac{\text{length of side opposite angle } A}{\text{length of hypotenuse}}$ is what ratio?

6. $\cos A = \dfrac{?}{\text{length of hypotenuse}}$

7. $\tan B = \dfrac{?}{?}$

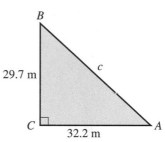

ILLUSTRATION 1

Find the value of each trigonometric ratio rounded to four significant digits:

8. $\cos 36.2°$

9. $\tan 48.7°$

10. $\sin 23.72°$

Find each angle rounded to the nearest tenth of a degree:

11. $\sin A = 0.7136$

12. $\tan B = 0.1835$

13. $\cos A = 0.4104$

14. Find angle A in Illustration 2.

15. Find angle B in Illustration 2.

16. Find side b in Illustration 3.

17. Find side c in Illustration 3.

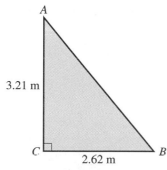

ILLUSTRATION 2

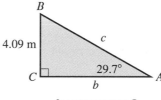

ILLUSTRATION 3

Solve each right triangle:

18.

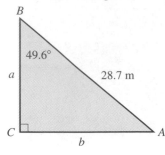

19.

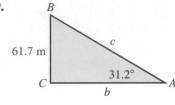

20.

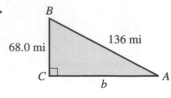

21. A satellite is directly overhead one observer station when it is at an angle of 68.0° from another observer station. (See Illustration 4.) The distance between the two stations is 2000 m. What is the height of the satellite?

ILLUSTRATION 4

22. A ranger at the top of a fire tower observes the angle of depression to a fire on level ground to be 3.0°. If the tower is 275 ft tall, what is the ground distance from the base of the tower to the fire?

23. Find the angle of slope of the symmetrical roof in Illustration 5.

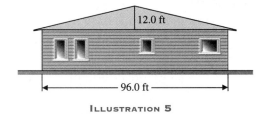

ILLUSTRATION 5

Chapter 13   Test

Find the value of each trigonometric ratio rounded to four significant digits:

1. sin 35.5°

2. cos 16.9°

3. tan 57.1°

Find each angle rounded to the nearest tenth of a degree:

4. cos A = 0.5577

5. tan B = 0.8888

6. sin A = 0.4166

7. Find angle B in Illustration 1.

8. Find side a in Illustration 1.

9. Find side c in Illustration 1.

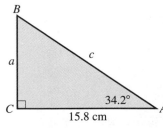

ILLUSTRATION 1

10. Find angle A in Illustration 2.

11. Find angle B in Illustration 2.

12. Find side b in Illustration 2.

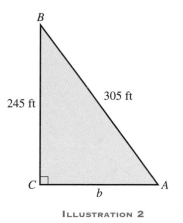

ILLUSTRATION 2

13. A tower 50.0 ft high has a guy wire that is attached to its top and anchored in the ground 15.0 ft from its base. Find the length of the guy wire.

14. Find length x in the retaining wall in Illustration 3.

15. Find angle A in the retaining wall in Illustration 3.

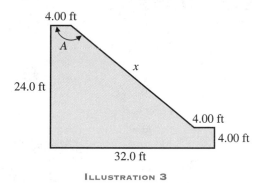

ILLUSTRATION 3

Trigonometry with Any Angle

CHAPTER 14

CONSTRUCTION
A worker uses trigonometry in checking dimensions on a construction project.

MATHEMATICS AT WORK

The law of sines and the law of cosines are used to find angles and sides of oblique triangles. For example, in machining, drafting, and engineering, these laws come in handy when we do not have enough information to work with right triangles.

Aviators and oceanographers use the laws of sines and cosines to find their range and bearing from one point to another. Geologists, archeologists, and surveyors use the laws when surveying, measuring, and mapping triangular sections of land.

14.1   Sine and Cosine Graphs

Up to this point, we have considered only the trigonometric ratios of angles between 0° and 90°, because we were working only with right triangles. For many applications, we need to consider values greater than 90°. In this section, we use a calculator to find the values of the sine and cosine ratios of angles greater than 90°. Then we use these values to construct various sine and cosine graphs.

The procedure for finding the value of the sine or cosine of an angle greater than 90° is the same as for angles between 0° and 90°, as shown in Section 13.1.

• **EXAMPLE 1**   Find sin 255° rounded to four significant digits.

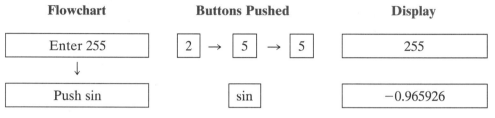

Flowchart	Buttons Pushed	Display
Enter 255	2 → 5 → 5	255
↓		
Push sin	sin	−0.965926

Thus, sin 255° = −0.9659 rounded to four significant digits. ————————•

Let's graph $y = \sin x$ for values of x between 0° and 360°. First, find a large number of values of x and y that satisfy the equation and plot them in the xy plane. For convenience, we will choose values of x in multiples of 30° and round the values of y to two significant digits.

x	0°	30°	60°	90°	120°	150°	180°	210°	240°	270°	300°	330°	360°
y	0	0.50	0.87	1.0	0.87	0.50	0	−0.50	−0.87	−1.0	−0.87	−0.50	0

Now choose a convenient scale for the x axis so that one unit equals 30°, and mark the x axis between 0° and 360°. Then choose a convenient scale for the y axis so that one unit equals 0.1, and mark the y axis between +1.0 and −1.0. Plot the points corresponding to the ordered pairs (x, y) from the table above. Then connect the points with a smooth, continuous curve. The graph is shown in Figure 14.1.

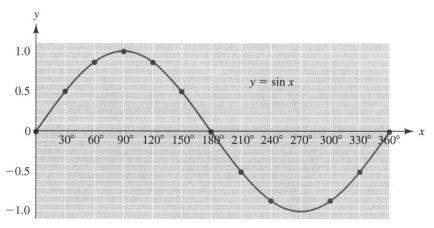

FIGURE 14.1

The graph of $y = \cos x$ for values of x between 0° and 360° can be found in a similar manner and is shown on page 428.

x	0°	30°	60°	90°	120°	150°	180°	210°	240°	270°	300°	330°	360°
y	1.0	0.87	0.50	0	−0.50	−0.87	−1	−0.87	−0.50	0	0.50	0.87	1.0

Plot the points corresponding to these ordered pairs and connect them with a smooth curve, as shown in Figure 14.2.

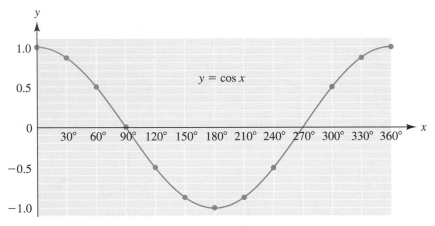

FIGURE 14.2

•   **EXAMPLE 2**   Graph $y = 4 \sin x$ for values of x between 0° and 360°.

Prepare a table listing values of x in multiples of 30°. To find each value of y, find the sine of the angle, multiply this value by 4, and round to two significant digits.

x	0°	30°	60°	90°	120°	150°	180°	210°	240°	270°	300°	330°	360°
y	0	2.0	3.5	4	3.5	2.0	0	−2.0	−3.5	−4	−3.5	−2.0	0

Here, let's choose the scale for the y axis so that one unit equals 0.5 and mark the y axis between +4.0 and −4.0. Plot the points corresponding to these ordered pairs in the table and then connect them with a smooth curve, as shown in Figure 14.3.

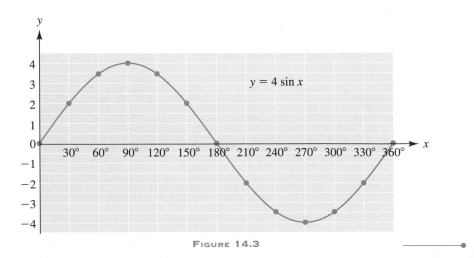

FIGURE 14.3

Note that the graphs for $y = \sin x$ and $y = 4 \sin x$ are similar. That is, each starts at $(0°, 0)$, reaches its maximum at $x = 90°$, crosses the x axis at $(180°, 0)$, reaches its minimum at $x = 270°$, and meets the x axis at $(360°, 0)$. In general, the graphs of equations in the form

$$y = A \sin x \quad \text{and} \quad y = A \cos x, \quad \text{where } A > 0$$

reach a maximum value of A and a minimum value of $-A$. The value of A is usually called the *amplitude*.

One of the most common applications of waves is in alternating current. In a generator, a coil of wire is rotated in a magnetic field, which produces an electric current. See Figure 14.4.

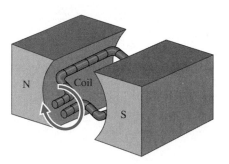

FIGURE 14.4

In a simple generator, the current i changes as the coil rotates according to the equation

$$i = I \sin x$$

where I is the maximum current and x is the angle through which the coil rotates.

Similarly, the voltage v also changes as the coil rotates according to the equation

$$v = V \sin x$$

where V is the maximum voltage and x is the angle through which the coil rotates.

• **EXAMPLE 3**   The maximum voltage V in a simple generator is 25 V. The changing voltage v as the coil rotates is given by

$$v = 25 \sin x$$

Graph the equation in multiples of 30° for one complete revolution of the coil.

First, prepare a table. To find each value of y, find the sine of the angle, multiply this value by 25, and round to two significant digits.

x	0°	30°	60°	90°	120°	150°	180°	210°	240°	270°	300°	330°	360°
y	0	13	22	25	22	13	0	−13	−22	−25	−22	−13	0

Let's choose the scale for the y axis so that one unit equals 5 V. Mark the y axis between +25 V and −25 V. Plot the points corresponding to the ordered pairs in the table. Then connect these with a smooth curve, as shown in Figure 14.5.

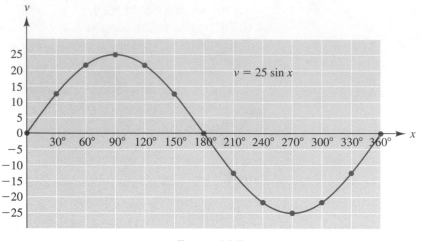

FIGURE 14.5

If you were to continue finding ordered pairs in the previous tables by choosing values of *x* greater than 360° and less than 0°, you would find that the *y* values repeat themselves and that the graphs form *waves*, as shown in Figure 14.6.

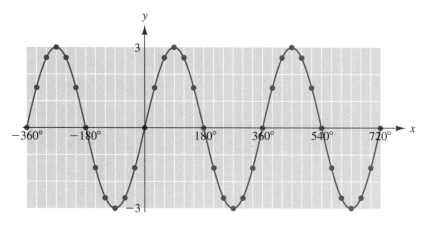

(a) $y = 3 \sin x$ for *x* between −360° and 720°

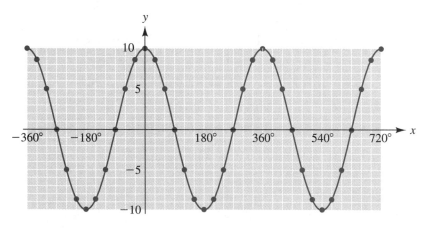

(b) $y = 10 \cos x$ for *x* between −360° and 720°

FIGURE 14.6

In general, if the *x*-axis variable is *distance*, the length of one complete wave is called the *wavelength* and is given by the symbol λ, the Greek letter lambda. See Figure 14.7.

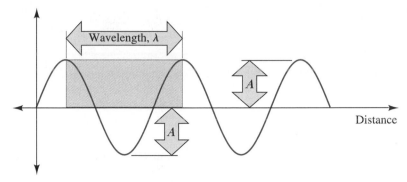

FIGURE 14.7

If the *x*-axis variable is *time*, the time required for one complete wave to pass a given point is called the *period*, *T*. See Figure 14.8.

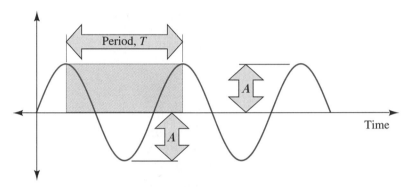

FIGURE 14.8

The frequency *f* is the number of waves that pass a given point each second. That is,

$$f = \frac{1}{T}$$

The unit of frequency is the hertz (Hz), where

1 Hz = 1 wave/s   or   1 cycle/s

Common multiples of the hertz are the kilohertz (kHz, 10^3 Hz) and the megahertz (MHz, 10^6 Hz).

• **EXAMPLE 4** Find the period of a wave whose frequency is 250 Hz.

Given $f = \dfrac{1}{T}$

$\quad fT = 1$ Multiply both sides by *T*.

$\quad T = \dfrac{1}{f}$ Divide both sides by *f*.

$\quad T = \dfrac{1}{250 \text{ Hz}}$ Substitute *f* = 250 Hz.

$\quad = \dfrac{1}{250} \cdot \dfrac{1}{\text{Hz}}$

$$= \frac{1}{250} \cdot \frac{\frac{1}{\text{wave}}}{\text{s}} \qquad \frac{\frac{1}{\text{wave}}}{\text{s}} = 1 \div \frac{\text{wave}}{\text{s}} = 1 \cdot \frac{\text{s}}{\text{wave}} = \frac{\text{s}}{\text{wave}}$$

$$= \frac{1}{250} \cdot \frac{\text{s}}{\text{wave}}$$

$$= 4.0 \times 10^{-3} \text{ s}$$ That is, one wave passes a given point each 4.0×10^{-3} s or 4.0 ms.

Frequency and wavelength are related to wave velocity by the formula

$$v = \lambda f$$

where v is the wave velocity, λ is the wavelength, and f is the frequency.

•   **EXAMPLE 5**   The FM band of a radio station is 90.9 MHz (megahertz). The speed of a radio wave is the same as the speed of light, which is 3.00×10^8 m/s. Find its wavelength.

$$v = \lambda f$$

$$\lambda = \frac{v}{f} \qquad \text{Divide both sides by } f.$$

$$\lambda = \frac{3.00 \times 10^8 \text{ m/s}}{90.9 \text{ MHz}} \qquad \text{Mega} = 10^6$$

$$= \frac{3.00 \times 10^8 \text{ m/s}}{90.9 \times 10^6 \text{ Hz}} \qquad 1 \text{ Hz} = 1 \text{ cycle/s} = 1 \text{ wave/s}$$

$$= \frac{3.00 \times 10^8 \text{ m/s}}{90.9 \times 10^6 \text{ waves/s}} \qquad \frac{\text{m}}{\text{s}} \div \frac{\text{waves}}{\text{s}} = \frac{\text{m}}{\text{s}} \cdot \frac{\text{s}}{\text{waves}} = \frac{\text{m}}{\text{wave}}$$

$$= 3.30 \text{ m} \qquad \text{That is, the length of each wave is 3.30 m.}$$

Exercises | 14.1

Find each value rounded to four significant digits:

1. sin 137°   **4.** cos 295°   **7.** cos 166.5°   **10.** tan 125.5°

2. sin 318°   **5.** sin 205.8°   **8.** cos 348.2°   **11.** tan 156.3°

3. cos 246°   **6.** sin 106.3°   **9.** tan 217.6°   **12.** tan 418.5°

Graph each equation for values of x between 0° and 360° in multiples of 30°:

13. $y = 6 \sin x$ **14.** $y = 2 \sin x$ **15.** $y = 5 \cos x$ **16.** $y = 4 \cos x$

Graph each equation for values of x between 0° and 360° in multiples of 15°:

17. $y = \sin 2x$ **18.** $y = \cos 2x$

Graph each equation for values of x between 0° and 360° in multiples of 10°:

19. $y = 4 \cos 3x$ **20.** $y = 2 \sin 3x$

The maximum voltage in a simple generator is V. The changing voltage v as the coil rotates is given by

$$v = V \sin x$$

Graph this equation in multiples of 30° for one complete revolution of the coil for each value of V.

21. $V = 36$ V   **22.** $V = 48$ V

The maximum current in a simple generator is I. *The changing current i as the coil rotates is given by*

$$i = I \sin x$$

Graph this equation in multiples of 30° for one complete revolution of the coil for each value of I:

23. $I = 5.0$ A

24. $I = 7.5$ A

25. From the graph in Exercise 21, estimate the value of v at $x = 45°$ and $x = 295°$.

26. From the graph in Exercise 22, estimate the value of v at $x = 135°$ and $x = 225°$.

27. From the graph in Exercise 23, estimate the value of i at $x = 135°$ and $x = 225°$.

28. From the graph in Exercise 24, estimate the value of i at $x = 45°$ and $x = 190°$.

29. Find the period of a wave whose frequency is 5.0 kHz.

30. Find the period of a wave whose frequency is 1.1 MHz.

31. Find the frequency of a wave whose period is 0.56 s.

32. Find the frequency of a wave whose period is 25 μs.

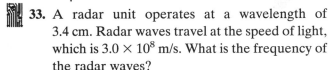 **33.** A radar unit operates at a wavelength of 3.4 cm. Radar waves travel at the speed of light, which is 3.0×10^8 m/s. What is the frequency of the radar waves?

34. A local AM radio station broadcasts at 1400 kHz. What is the wavelength of its radio waves? (They travel at the speed of light, which is 3.0×10^8 m/s.)

35. Find the speed of a wave having frequency 4.50/s and wavelength 0.500 m.

36. Find the wavelength of water waves with frequency 0.55/s and speed 1.4 m/s.

 ## 14.2 Period and Phase Shift

As we saw in Figure 14.8, the period is the length of one complete cycle of the sine or cosine graph. In general, the period for $y = A \sin Bx$ and for $y = A \cos Bx$ may be found by the formula

$$P = \frac{360°}{B}$$

• **EXAMPLE 1** Find the period and amplitude of $y = 2 \cos 3x$ and draw its graph.

$$P = \frac{360°}{3} = 120°$$

Draw the cosine graph with amplitude 2 and period 120°, as in Figure 14.9.

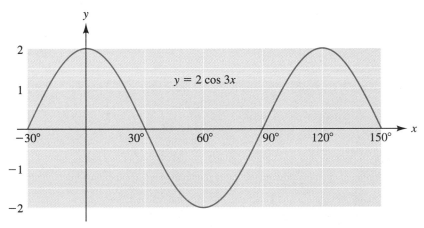

FIGURE 14.9

•   **EXAMPLE 2**   Find the period and amplitude of $y = 5 \sin 4x$ and draw one period of its graph.

$$P = \frac{360°}{4} = 90°$$

Draw the sine graph with amplitude 5 and period 90°, as in Figure 14.10.

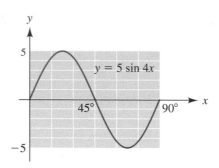

FIGURE 14.10

If the graph of a sine curve does not pass through the origin (0°, 0) or if the graph of a cosine curve does not pass through the point (0°, A), where A is the amplitude, the curve is *out of phase*. If the curve is out of phase, the *phase shift* is the distance between two successive corresponding points of the curve $y = A \sin Bx$ or $y = A \cos Bx$ and the out-of-phase curve. (See Figure 14.11.)

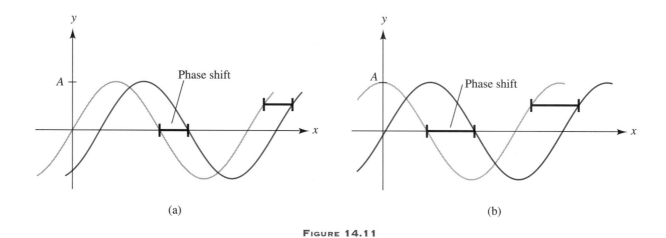

(a)   (b)

FIGURE 14.11

•   **EXAMPLE 3**   Graph $y = 2 \sin x$ and $y = 2 \sin (x - 45°)$ on the same set of coordinate axes.
For $y = 2 \sin (x - 45°)$,

x	−45°	0°	45°	90°	135°	180°	225°	270°	315°	360°	405°	450°
y	−2	−1.4	0	1.4	2	1.4	0	−1.4	−2	−1.4	0	1.4

Graph both equations, as in Figure 14.12.

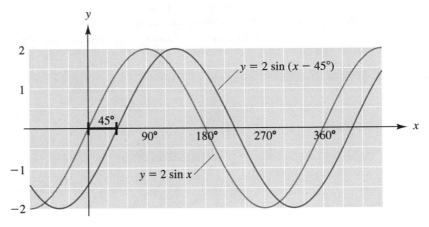

FIGURE 14.12

Graphing equations in the form $y = A \sin (Bx + C)$ or $y = A \cos (Bx + C)$ involves a phase shift.

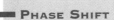

PHASE SHIFT

The effect of C in each equation is to shift the curve $y = A \sin Bx$ or $y = A \cos Bx$

1. to the *left* $\frac{C}{B}$ units if $\frac{C}{B}$ is positive.
2. to the *right* $\frac{C}{B}$ units if $\frac{C}{B}$ is negative.

EXAMPLE 4 Graph $y = 4 \sin (3x + 60°)$.

The amplitude is 4. The period is $P = \frac{360°}{3} = 120°$. The phase shift is $\frac{C}{B} = \frac{60°}{3} = 20°$, or 20° to the *left*.

Graph $y = 4 \sin 3x$ and shift the curve 20° to the left, as shown in Figure 14.13.

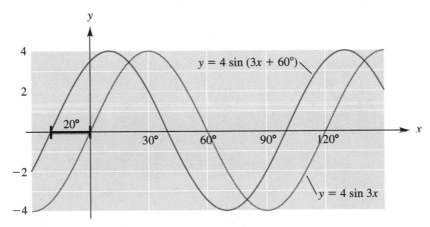

FIGURE 14.13

•   **EXAMPLE 5**   Graph $y = 6 \cos (2x - 90°)$.

The amplitude is 6. The period is $\frac{360°}{2} = 180°$. The phase shift is $\frac{C}{B} = \frac{-90°}{2} = -45°$, or 45° to the *right*.

Graph $y = 6 \cos 2x$ and shift the curve 45° to the right, as shown in Figure 14.14.

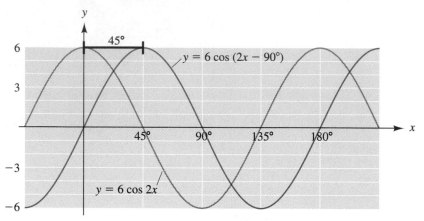

FIGURE 14.14

Note that the graph of $y = 6 \cos (2x - 90°)$ is the same as the graph of $y = 6 \sin 2x$. Each sine or cosine graph may be expressed in terms of the other trigonometric function with the appropriate phase shift.

Exercises   14.2

Find the period and amplitude and graph at least two periods of each equation:

1. $y = 3 \sin 3x$

2. $y = 7 \cos 4x$

3. $y = 8 \cos 6x$

4. $y = 9 \sin 5x$

5. $y = 10 \sin 9x$

6. $y = 15 \cos 10x$

7. $y = 6 \cos \frac{1}{2}x$

8. $y = 4 \sin \frac{1}{3}x$

9. $y = 3.5 \sin \frac{2}{3}x$

10. $y = 1.8 \cos \frac{3}{4}x$

11. $y = 4 \sin \frac{5}{2}x$

12. $y = 6 \cos \frac{4}{3}x$

Find the period, amplitude, and phase shift, and graph at least two periods of each equation:

13. $y = \sin (x + 30°)$

14. $y = \cos (x + 45°)$

15. $y = 2 \cos (x - 60°)$

16. $y = 3 \sin (x - 120°)$

17. $y = 4 \sin (3x + 180°)$

18. $y = 5 \cos (2x + 60°)$

19. $y = 10 \sin (4x - 120°)$

20. $y = 12 \cos (4x + 180°)$

21. $y = 5 \sin (\frac{1}{2}x + 90°)$

22. $y = 6 \cos (\frac{3}{4}x - 240°)$

23. $y = 10 \cos(\frac{1}{4}x + 180°)$

24. $y = 15 \sin (\frac{2}{3}x - 120°)$

14.3   Solving Oblique Triangles: Law of Sines

An *oblique triangle* is a triangle with no right angle. We use the common notation of labeling vertices of a triangle by the capital letters A, B, and C and using the small letters a, b, and c as the sides opposite angles A, B, and C, respectively.

Recall that an *acute* angle is an angle that has a measure less than 90°. An *obtuse* angle is an angle that has a measure greater than 90° but less than 180°.

The trigonometric ratios used in Chapter 13 apply *only* to right triangles. So we must use other ways to solve oblique triangles. One law that we use to solve oblique triangles is the law of sines.

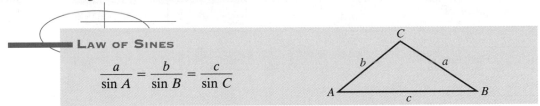

LAW OF SINES

$$\frac{a}{\sin A} = \frac{b}{\sin B} = \frac{c}{\sin C}$$

That is, *for any triangle, the ratio of the length of any side to the sine of the opposite angle equals the ratio of the length of any other side to the sine of its opposite angle.*

When using this law, you must form a proportion by choosing two of the three ratios in which three of the four terms are known. In order to use the law of sines, you must know:

a. two angles and a side opposite one of them (actually, knowing two angles and any side is enough, because in knowing two angles, the third is easily found), or

b. two sides and an angle opposite one of them.

• **EXAMPLE 1**   If $C = 28.0°$, $c = 46.8$ cm, and $B = 101.5°$, solve the triangle.*

First, find side b, in Figure 14.15.

$$\frac{c}{\sin C} = \frac{b}{\sin B}$$

$$\frac{46.8 \text{ cm}}{\sin 28.0°} = \frac{b}{\sin 101.5°}$$

$b(\sin 28.0°) = (\sin 101.5°)(46.8 \text{ cm})$ Multiply both sides by the LCD.

$$b = \frac{(\sin 101.5°)(46.8 \text{ cm})}{\sin 28.0°}$$ Divide both sides by $\sin 28.0°$.

$$= 97.7 \text{ cm}$$

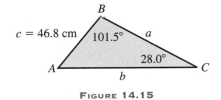

FIGURE 14.15

You may use a calculator to do this calculation as follows:

Flowchart	Buttons Pushed	Display
Enter 101.5	$1 \rightarrow 0 \rightarrow 1 \rightarrow . \rightarrow 5$	101.5
Push sin	sin	0.979925
Push times	×	0.979925
Enter 46.8	$4 \rightarrow 6 \rightarrow . \rightarrow 8$	46.8

*As in previous sections, round sides to three significant digits and angles to the nearest tenth of a degree.

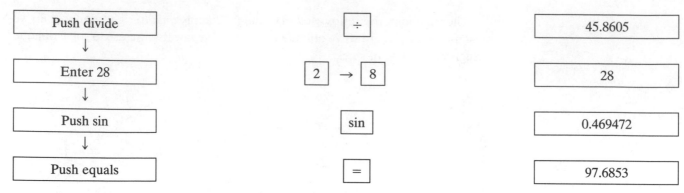

$$A = 180° - B - C = 180° - 101.5° - 28.0° = 50.5°$$

To find side a,

$$\frac{c}{\sin C} = \frac{a}{\sin A}$$

$$\frac{46.8 \text{ cm}}{\sin 28.0°} = \frac{a}{\sin 50.5°}$$

$a(\sin 28.0°) = (\sin 50.5°)(46.8 \text{ cm})$ Multiply both sides by the LCD.

$a = \dfrac{(\sin 50.5°)(46.8 \text{ cm})}{\sin 28.0°}$ Divide both sides by sin 28.0°.

$$= 76.9 \text{ cm}$$

The solution is $a = 76.9$ cm, $b = 97.7$ cm, and $A = 50.5°$. ————•

A wide variety of applications may be solved using the law of sines.

• **EXAMPLE 2**   Find the lengths of rafters AC and BC for the roofline shown in Figure 14.16.

First, find angle C:

$$C = 180° - A - B = 180° - 35.0° - 65.0° = 80.0°$$

To find side AC,

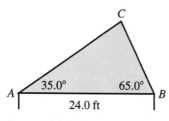

FIGURE 14.16

$$\frac{AC}{\sin B} = \frac{AB}{\sin C}$$

$$\frac{AC}{\sin 65.0°} = \frac{24.0 \text{ ft}}{\sin 80.0°}$$

$AC(\sin 80.0°) = (\sin 65.0°)(24.0 \text{ ft})$ Multiply both sides by the LCD.

$AC = \dfrac{(\sin 65.0°)(24.0 \text{ ft})}{\sin 80.0°}$ Divide both sides by sin 80.0°.

$$= 22.1 \text{ ft}$$

To find side BC,

$$\frac{BC}{\sin A} = \frac{AB}{\sin C}$$

$$\frac{BC}{\sin 35.0°} = \frac{24.0 \text{ ft}}{\sin 80.0°}$$

$BC(\sin 80.0°) = (\sin 35.0°)(24.0 \text{ ft})$ Multiply both sides by the LCD.

$BC = \dfrac{(\sin 35.0°)(24.0 \text{ ft})}{\sin 80.0°}$ Divide both sides by sin 80.0°.

$$= 14.0 \text{ ft}$$

————•

Exercises 14.3

Solve each triangle using the labels as shown in Illustration 1 (round lengths of sides to three significant digits and angles to the nearest tenth of a degree):

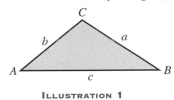

ILLUSTRATION 1

1. $A = 68.0°$, $a = 24.5$ m, $b = 17.5$ m
2. $C = 56.3°$, $c = 142$ cm, $b = 155$ cm
3. $A = 61.5°$, $B = 75.6°$, $b = 255$ ft
4. $B = 41.8°$, $C = 59.3°$, $c = 24.7$ km
5. $A = 14.6°$, $B = 35.1°$, $c = 43.7$ cm

6. $B = 24.7°$, $C = 136.1°$, $a = 342$ m
7. $A = 54.0°$, $C = 43.1°$, $a = 26.5$ m
8. $B = 64.3°$, $b = 135$ m, $c = 118$ m
9. $A = 20.1°$, $a = 47.5$ mi, $c = 35.6$ mi
10. $B = 75.2°$, $A = 65.1°$, $b = 305$ ft
11. $C = 48.7°$, $B = 56.4°$, $b = 5960$ m
12. $A = 118.0°$, $a = 5750$ m, $b = 4750$ m
13. $B = 105.5°$, $c = 11.3$ km, $b = 31.4$ km
14. $A = 58.2°$, $a = 39.7$ mi, $c = 27.5$ mi
15. $A = 16.5°$, $a = 206$ ft, $b = 189$ ft
16. $A = 35.0°$, $B = 49.3°$, $a = 48.7$ m

17. Find the distance *AC* across the river shown in Illustration 2.

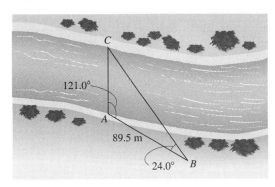

ILLUSTRATION 2

18. Find the lengths of rafters *AC* and *BC* of the roof shown in Illustration 3.

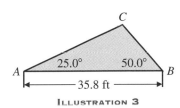

ILLUSTRATION 3

19. Find the distance *AB* between the ships shown in Illustration 4.

20. Find the height of the cliff shown in Illustration 5.

21. A contractor needs to grade the slope of a subdivision lot to place a house on level ground. (See Illustration 6.) The present slope of the lot is 12.5°. The contractor needs a level lot that is 105 ft deep. To control erosion, the back of the lot must be cut to a

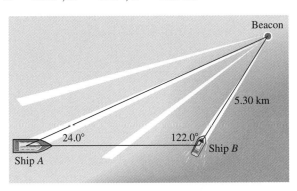

ILLUSTRATION 4

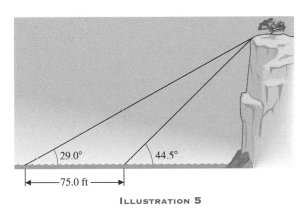

ILLUSTRATION 5

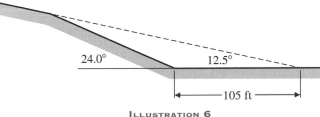

ILLUSTRATION 6

slope of 24.0°. How far from the street, measured along the present slope, will the excavation extend?

22. A weather balloon is sighted from points A and B, which are 4.00 km apart on level ground. The angle of elevation of the balloon from point A is 29.0°. Its angle of elevation from point B is 48.0°.

a. Find the height (in m) of the balloon if it is between A and B.

b. Find its height (in m) if point B is between point A and the weather balloon.

14.4 Law of Sines: The Ambiguous Case

The solution of a triangle in which two sides and an angle opposite one of the sides are given needs special care. In this case, there may be one, two, or no triangles formed from the given information. Let's look at the possibilities.

• **EXAMPLE 1** Construct a triangle given that $A = 32°$, $a = 18$ cm, and $b = 24$ cm.

As you can see from Figure 14.17, there are two triangles that satisfy these conditions: triangles ABC and $AB'C$. In one case, angle B is acute. In the other, angle B' is obtuse.

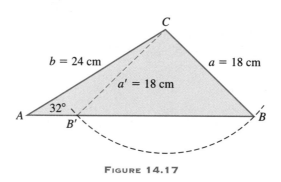

FIGURE 14.17

• **EXAMPLE 2** Construct a triangle given that $A = 40°$, $a = 12$ cm, and $b = 24$ cm.

As you can see from Figure 14.18, there is no triangle that satisfies these conditions. Side a is just not long enough to reach AB.

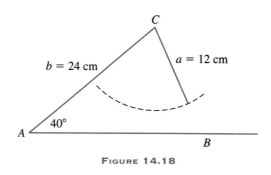

FIGURE 14.18

• **EXAMPLE 3** Construct a triangle given that $A = 50°$, $a = 12$ cm, and $b = 8$ cm.

As you can see from Figure 14.19, there is only one triangle that satisfies these conditions. Side a is too long for two solutions.

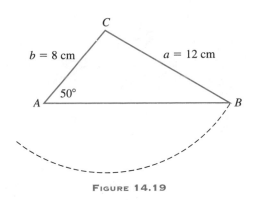

FIGURE 14.19

Let's summarize the possible cases when two sides and an angle opposite one of the sides are given. Assume that angle A and adjacent side b are given. From these two parts, the altitude ($h = b \sin A$) is determined and fixed.

If angle A is *acute*, we have four possible cases as shown in Figure 14.20.

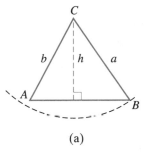

(a)

When $h < b < a$, we have only one possible triangle. That is, when the side opposite the given acute angle is greater than the known adjacent side, there is only one possible triangle.

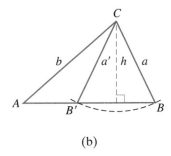

(b)

When $h < a < b$, we have two possible triangles. That is, when the side opposite the given acute angle is less than the known adjacent side but greater than the altitude, there are two possible triangles.

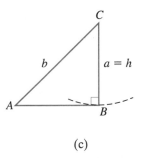

(c)

When $a = h$, we have one possible triangle. That is, when the side opposite the given acute angle equals the altitude, there is only one possible triangle—a right triangle.

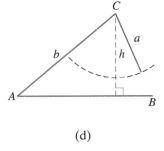

(d)

When $a < h$, there is no possible triangle. That is, when the side opposite the given acute angle is less than the altitude, there is no possible triangle.

FIGURE 14.20

If angle A is *obtuse*, we have two possible cases, as shown in Figure 14.21.

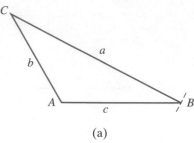

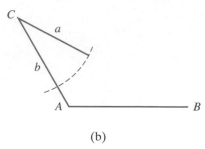

(a)	(b)
When $a > b$, we have one possible triangle. That is, when the side opposite the given obtuse angle is greater than the known adjacent side, there is only one possible triangle.	When $a \leq b$, there is no possible triangle. That is, when the side opposite the given obtuse angle is less than or equal to the known adjacent side, there is no possible triangle.

FIGURE 14.21

The table below summarizes these possibilities for the ambiguous case of the law of sines. Assume that the given parts are angle A, the side opposite, a, and the side adjacent, b.

If angle A is acute:		If angle A is obtuse:	
a. $h < b < a$	One triangle	**a.** $a > b$	One triangle
b. $h < a < b$	Two triangles	**b.** $a \leq b$	No triangle
c. $a = h$	One right triangle		
d. $a < h$	No triangle		

Note: If the given parts are not angle A, side opposite, a, and side adjacent, b, then you must substitute the given angle for A, the side opposite for a, and the side adjacent for b in this table. This is why it is important to understand the general cases described above.

•   **EXAMPLE 4**   If $A = 25.0°$, $a = 50.0$ m, and $b = 80.0$ m, solve the triangle.

First, find h.

$$h = b \sin A = (80.0 \text{ m})(\sin 25.0°) = 33.8 \text{ m}$$

Since $h < a < b$, we have two solutions. First, let's find *acute* angle B in Figure 14.22.

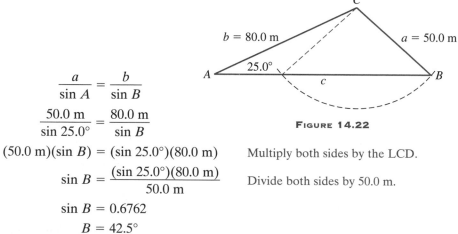

FIGURE 14.22

$$\frac{a}{\sin A} = \frac{b}{\sin B}$$

$$\frac{50.0 \text{ m}}{\sin 25.0°} = \frac{80.0 \text{ m}}{\sin B}$$

$(50.0 \text{ m})(\sin B) = (\sin 25.0°)(80.0 \text{ m})$ Multiply both sides by the LCD.

$$\sin B = \frac{(\sin 25.0°)(80.0 \text{ m})}{50.0 \text{ m}}$$ Divide both sides by 50.0 m.

$$\sin B = 0.6762$$

$$B = 42.5°$$

You may use a calculator to do this calculation as follows.

Flowchart	Buttons Pushed	Display
Enter 25	2 → 5	25
Push sin	sin	0.422618
Push times	×	0.422618
Enter 80	8 → 0	80
Push divide	÷	33.8095
Enter 50	5 → 0	50
Push equals	=	0.676189
Find angle B	inv → sin or sin⁻¹	42.5466

$C = 180° \quad A \quad B = 180° \quad 25.0° \quad 42.5° = 112.5°$

To find side c,

$$\frac{c}{\sin C} = \frac{a}{\sin A}$$

$$\frac{c}{\sin 112.5°} = \frac{50.0 \text{ m}}{\sin 25.0°}$$

$c(\sin 25.0°) = (\sin 112.5°)(50.0 \text{ m})$ Multiply both sides by the LCD.

$$c = \frac{(\sin 112.5°)(50.0 \text{ m})}{\sin 25.0°}$$ Divide both sides by 25.0°.

$$= 109 \text{ m}$$

Next, let's find *obtuse* angle B in Figure 14.23.

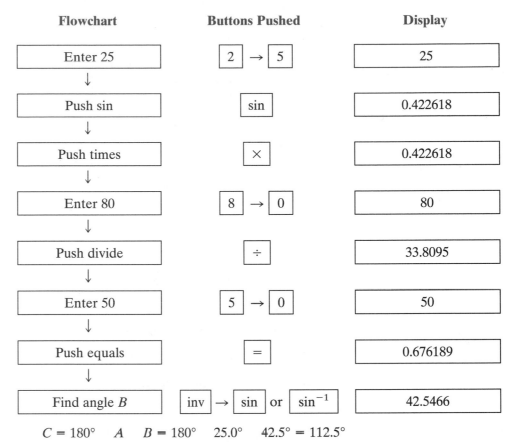

$$\frac{a}{\sin A} = \frac{b}{\sin B}$$

$$\frac{50.0 \text{ m}}{\sin 25.0°} = \frac{80.0 \text{ m}}{\sin B}$$

$(50.0 \text{ m})(\sin B) = (\sin 25.0°)(80.0 \text{ m})$

$$\sin B = \frac{(\sin 25.0°)(80.0 \text{ m})}{50.0 \text{ m}}$$

$$\sin B = 0.6762$$

$$B = 180° - 42.5° = 137.5°$$

FIGURE 14.23

Note: If B is acute, B is [inv] → [sin] of 0.6762.

If B is obtuse, B is $180° - ($ [inv] → [sin] $)$ of 0.6762.

$$C = 180° - A - B = 180° - 25.0° - 137.5° = 17.5°$$

To find side c,

$$\frac{c}{\sin C} = \frac{a}{\sin A}$$

$$\frac{c}{\sin 17.5°} = \frac{50.0 \text{ m}}{\sin 25.0°}$$

$$c(\sin 25.0°) = (\sin 17.5°)(50.0 \text{ m})$$

$$c = \frac{(\sin 17.5°)(50.0 \text{ m})}{\sin 25.0°}$$

$$= 35.6 \text{ m}$$

The two solutions are $c = 109$ m, $B = 42.5°$, $C = 112.5°$ and $c = 35.6$ m, $B = 137.5°$, $C = 17.5°$.

• **EXAMPLE 5**   If $A = 59.0°$, $a = 205$ m, and $b = 465$ m, solve the triangle.

First, find h.

$$h = b \sin A = (465 \text{ m})(\sin 59.0°) = 399 \text{ m}$$

Since $a < h$, there is no possible triangle.

What would happen if you tried to apply the law of sines anyway?

$$\frac{a}{\sin A} = \frac{b}{\sin B}$$

$$\frac{205 \text{ m}}{\sin 59.0°} = \frac{465 \text{ m}}{\sin B}$$

$$(205 \text{ m})(\sin B) = (\sin 59.0°)(465 \text{ m})$$

$$\sin B = \frac{(\sin 59.0°)(465 \text{ m})}{205 \text{ m}} = 1.944$$

Note: $\sin B = 1.944$ is impossible, because $-1 \leq \sin B \leq 1$. Recall that the graph of $y = \sin x$ has an amplitude of 1, which means that the values of $\sin x$ vary between 1 and -1. Your calculator will also indicate an error when you try to find angle B.

As a final check to make certain that your solution is correct, verify that the following geometric triangle property is satisfied: In any triangle, the largest side is opposite the largest angle and the smallest side is opposite the smallest angle.

Exercises | 14.4

*For each general triangle, **a.** determine the number of solutions and **b.** solve the triangle, if possible, using the labels as shown in Illustration 1 (round lengths to three significant digits and angles to the nearest tenth of a degree):*

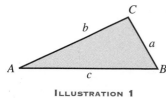

ILLUSTRATION 1

1. $A = 38.0°$, $a = 42.3$ m, $b = 32.5$ m

2. $C = 47.6°$, $a = 85.2$ cm, $c = 96.1$ cm

3. $A = 25.6°$, $b = 306$ m, $a = 275$ m

4. $B = 41.2°$, $c = 1860$ ft, $b = 1540$ ft

5. $A = 71.6°$, $b = 48.5$ m, $a = 15.7$ m

6. $B = 40.3°$, $b = 161$ cm, $c = 288$ cm

7. $C = 71.2°$, $a = 245$ cm, $c = 238$ cm

8. $A = 36.1°$, $b = 14.5$ m, $a = 12.5$ m

9. $B = 105.0°$, $b = 33.0$ mi, $a = 24.0$ mi

10. $A = 98.3°$, $a = 1420$ ft, $b = 1170$ ft

11. $A = 31.5°$, $a = 376$ m, $c = 406$ m

12. $B = 50.0°$, $b = 4130$ ft, $c = 4560$ ft

13. $C = 60.0°$, $c = 151$ m, $b = 181$ m

14. $A = 30.0°$, $a = 4850$ mi, $c = 3650$ mi

15. $B = 8.0°$, $b = 451$ m, $c = 855$ m

16. $C = 8.7°$, $c = 89.3$ mi, $b = 61.9$ mi

17. The owner of a triangular lot wishes to fence it along the lot lines. Lot markers at A and B have been located, but the lot marker at C cannot be found. The owner's attorney gives the following information by phone: $AB = 355$ ft, $BC = 295$ ft, and $A = 36.0°$. What is the length of AC?

18. The average distance from the sun to earth is 1.5×10^8 km and from the sun to Venus is 1.1×10^8 km. Find the distance between earth and Venus when the angle between earth and the sun and earth and Venus is $24.7°$. (Assume that earth and Venus have circular orbits around the sun.)

14.5 Solving Oblique Triangles: Law of Cosines

A second law used to solve oblique triangles is the law of cosines.

LAW OF COSINES

$$a^2 = b^2 + c^2 - 2bc \cos A$$
$$b^2 = a^2 + c^2 - 2ac \cos B$$
$$c^2 = a^2 + b^2 - 2ab \cos C$$

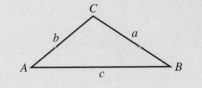

That is, *for any triangle, the square of any side equals the sum of the squares of the other two sides minus twice the product of these two sides and the cosine of their included angle.*

In order to use the law of cosines, you must know:

a. two sides and the included angle, or

b. all three sides.

• **EXAMPLE 1** If $A = 115.2°$, $b = 18.5$ m, and $c = 21.7$ m, solve the triangle. (See Figure 14.24.)

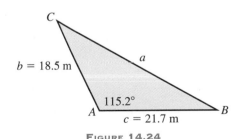

FIGURE 14.24

To find side a,

$$a^2 = b^2 + c^2 - 2bc \cos A$$
$$a^2 = (18.5 \text{ m})^2 + (21.7 \text{ m})^2 - 2(18.5 \text{ m})(21.7 \text{ m})(\cos 115.2°)$$
$$a = 34.0 \text{ m}$$

You may use a calculator to do this calculation as follows.

Flowchart	**Buttons Pushed**	**Display**
Enter 115.2	$\boxed{1} \rightarrow \boxed{1} \rightarrow \boxed{5} \rightarrow \boxed{.} \rightarrow \boxed{2}$	115.2
Push cos	$\boxed{\text{cos}}$	−0.425779
Push times	$\boxed{\times}$	−0.425779
Enter 21.7	$\boxed{2} \rightarrow \boxed{1} \rightarrow \boxed{.} \rightarrow \boxed{7}$	21.7
Push times	$\boxed{\times}$	−9.23941
Enter 18.5	$\boxed{1} \rightarrow \boxed{8} \rightarrow \boxed{.} \rightarrow \boxed{5}$	18.5
Push times	$\boxed{\times}$	−170.929
Enter −2	$\boxed{2} \rightarrow \boxed{+/-}$	−2
Push plus	$\boxed{+}$	341.858
Enter 21.7	$\boxed{2} \rightarrow \boxed{1} \rightarrow \boxed{.} \rightarrow \boxed{7}$	21.7
Push square	$\boxed{x^2}$	470.89
Push plus	$\boxed{+}$	812.748
Enter 18.5	$\boxed{1} \rightarrow \boxed{8} \rightarrow \boxed{.} \rightarrow \boxed{5}$	18.5
Push square	$\boxed{x^2}$	342.25
Push equals	$\boxed{=}$	1154.998
Push square root	$\boxed{\sqrt{}}$	33.9853

To find angle B, use the law of sines, as it requires less computation.

$$\frac{a}{\sin A} = \frac{b}{\sin B}$$

$$\frac{34.0 \text{ m}}{\sin 115.2°} = \frac{18.5 \text{ m}}{\sin B}$$

$(34.0 \text{ m})(\sin B) = (\sin 115.2°)(18.5 \text{ m})$ Multiply both sides by the LCD.

$$\sin B = \frac{(\sin 115.2°)(18.5 \text{ m})}{34.0 \text{ m}}$$ Divide both sides by 34.0 m.

$$\sin B = 0.4923$$

$$B = 29.5°$$

$$C = 180° - A - B = 180° - 115.2° - 29.5° = 35.3°$$

• **EXAMPLE 2**   If $a = 125$ cm, $b = 285$ cm, and $c = 382$ cm, solve the triangle. (See Figure 14.25.)

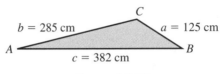

$b = 285$ cm C $a = 125$ cm

A $c = 382$ cm B

FIGURE 14.25

When three sides are given, you are advised to find the angle opposite the largest side first. Why?

To find angle C,

$$c^2 = a^2 + b^2 - 2ab \cos C$$

$$(382 \text{ cm})^2 = (125 \text{ cm})^2 + (285 \text{ cm})^2 - 2(125 \text{ cm})(285 \text{ cm}) \cos C$$

$$(382 \text{ cm})^2 - (125 \text{ cm})^2 - (285 \text{ cm})^2 = -2(125 \text{ cm})(285 \text{ cm}) \cos C$$

$$\frac{(382 \text{ cm})^2 - (125 \text{ cm})^2 - (285 \text{ cm})^2}{-2(125 \text{ cm})(285 \text{ cm})} = \cos C$$

$$-0.6888 = \cos C$$

$$133.5° = C$$

You may use a calculator to do this calculation as follows.

Flowchart	Buttons Pushed	Display
Enter 382	$3 \rightarrow 8 \rightarrow 2$	382
Push square	x^2	145924
Push minus	$-$	145924
Enter 125	$1 \rightarrow 2 \rightarrow 5$	125

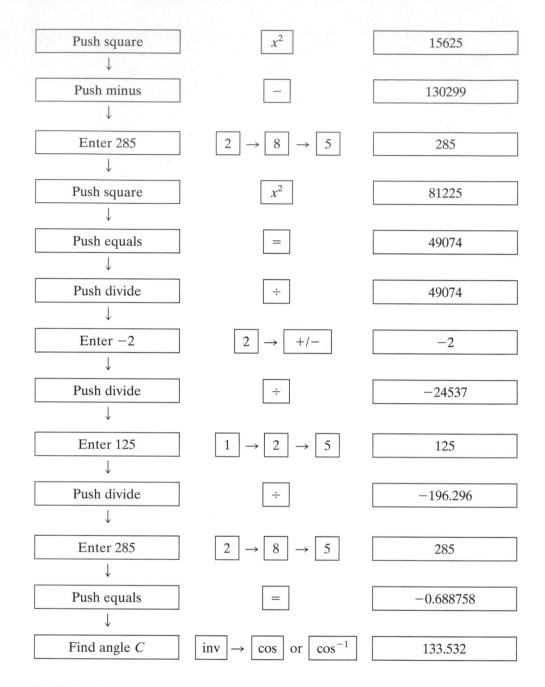

To find angle A, let's use the law of sines.

$$\frac{c}{\sin C} = \frac{a}{\sin A}$$

$$\frac{382 \text{ cm}}{\sin 133.5°} = \frac{125 \text{ cm}}{\sin A}$$

$$(382 \text{ cm})(\sin A) = (\sin 133.5°)(125 \text{ cm})$$

$$\sin A = \frac{(\sin 133.5°)(125 \text{ cm})}{382 \text{ cm}}$$

$$\sin A = 0.2374$$

$$A = 13.7°$$

$$B = 180° - A - C = 180° - 13.7° - 133.5° = 32.8°$$

EXAMPLE 3 Find the lengths of guy wires AC and BC for a tower located on a hillside, as shown in Figure 14.26(a). The height of the tower is 50.0 m; $\angle ADC = 120.0°$; $AD = 20.0$ m; $BD = 15.0$ m.

First, let's use triangle ACD in Figure 14.26(b) to find length AC. Using the law of cosines,

$$(AC)^2 = (AD)^2 + (DC)^2 - 2(AD)(DC) \cos ADC$$
$$(AC)^2 = (20.0 \text{ m})^2 + (50.0 \text{ m})^2 - 2(20.0 \text{ m})(50.0 \text{ m}) \cos 120.0°$$
$$AC = 62.4 \text{ m}$$

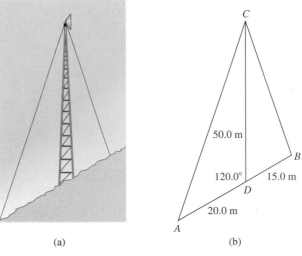

(a)   (b)

FIGURE 14.26

Next, use triangle CDB and the law of cosines to find length BC. Note that $\angle CDB = 180° - 120.0° = 60.0°$.

$$(BC)^2 = (BD)^2 + (DC)^2 - 2(BD)(DC) \cos CDB$$
$$(BC)^2 = (15.0 \text{ m})^2 + (50.0 \text{ m})^2 - 2(15.0 \text{ m})(50.0 \text{ m}) \cos 60.0°$$
$$BC = 44.4 \text{ m}$$

Exercises 14.5

Solve each triangle using the labels as shown in Illustration 1 (round lengths of sides to three significant digits and angles to the nearest tenth of a degree):

1. $A = 55.0°$, $b = 21.2$ m, $c = 24.0$ m

2. $B = 14.5°$, $a = 37.6$ cm, $c = 48.2$ cm

3. $C = 115.0°$, $a = 247$ ft, $b = 316$ ft

4. $A = 130.0°$, $b = 15.2$ km, $c = 9.50$ km

5. $a = 136$ m, $b = 155$ m, $c = 168$ m

6. $a = 3.96$ in., $b = 4.81$ in., $c = 6.45$ in.

7. $C = 71.6°$, $a = 5.42$ km, $b = 6.03$ km

8. $B = 96.1°$, $a = 452$ yd, $c = 365$ yd

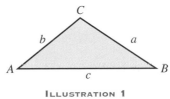

ILLUSTRATION 1

9. $a = 38{,}500$ mi, $b = 67{,}500$ mi, $c = 47{,}200$ mi

10. $a = 146$ cm, $b = 271$ cm, $c = 205$ cm

11. $B = 19.3°$, $a = 4820$ ft, $c = 1930$ ft

12. $C = 108.5°$, $a = 415$ m, $b = 325$ m

13. $a = 19.5$ m, $b = 36.5$ m, $c = 25.6$ m

14. $a = 207$ mi, $b = 106$ mi, $c = 142$ mi

15. Find the distance a across the pond in Illustration 2.

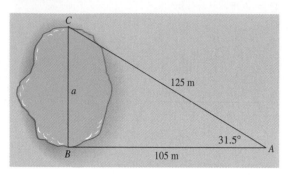

ILLUSTRATION 2

16. Find the length of rafter AC in Illustration 3.

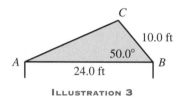

ILLUSTRATION 3

17. Find angles A and C in the roof in Illustration 4.

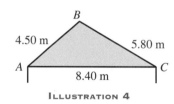

ILLUSTRATION 4

18. a. Find the angles A and ACB in the roof in Illustration 5. $AC = BC$
   b. Find length CD.

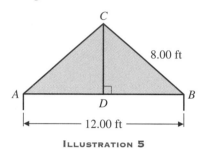

ILLUSTRATION 5

19. In the framework shown in Illustration 6, we know that $AE = CD$, $AB = BC$, $BD = BE$, and $\overline{AC} \parallel \overline{ED}$. Find **a.** $\angle BEA$, **b.** $\angle A$, **c.** length BE, **d.** length DE.

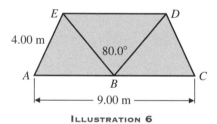

ILLUSTRATION 6

20. In the framework shown in Illustration 7, we know that $AB = DE$, $BC = CD$, $AH = FE$, $HG = GF$. Find **a.** length HB, **b.** $\angle AHB$, **c.** length GC, and **d.** length AG.

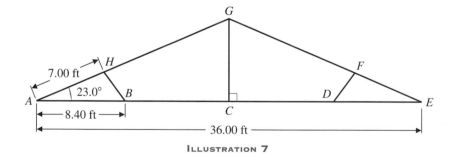

ILLUSTRATION 7

21. A triangular lot has sides 1580 ft, 2860 ft, and 1820 ft long. Find its largest angle.

22. A ship starts at point A and travels 125 mi northeast. It then travels 150 mi due east and arrives at point B. If the ship had sailed directly from A to B, what distance would it have traveled?

23. See Illustration 8. Deloney and Jackson Streets meet at a 45° angle. A lot extends $5\overline{0}$ yards along Jackson and $4\overline{0}$ yards along Deloney. Find the length of the back border.

24. See Illustration 9. From its home port (H), a ship sails 52 miles at a bearing of 147° to point A, and then sails 78 miles at a bearing of 213° to point *B*. How far is point *B* from home port, and at what bearing will it return?

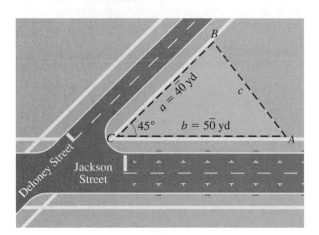

ILLUSTRATION 8

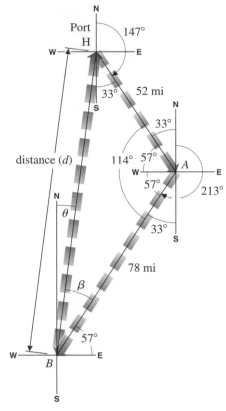

ILLUSTRATION 9

Chapter 14   Review

Find each value rounded to four significant digits:

1. tan 143°   **2.** sin 209.8°   **3.** cos 317.4°

Graph each equation for values of x between 0° and 360° in multiples of 15°:

4. $y = 6 \cos x$ **5.** $y = 3 \sin 2x$

Find the period and amplitude, and graph at least two periods of each equation:

6. $y = 5 \sin 3x$ **7.** $y = 3 \cos 4x$

Find the period, amplitude, and phase shift, and graph at least two periods of each equation:

8. $y = 4 \cos (x + 60°)$ **9.** $y = 6 \sin (2x - 180°)$

Solve each triangle using the labels as shown in Illustration 1 (round lengths of sides to three significant digits and angles to the nearest tenth of a degree):

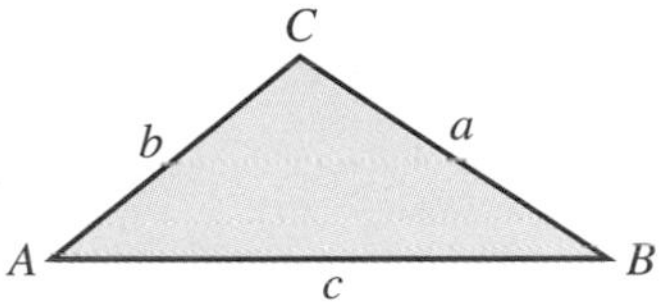

10. $B = 52.7°$, $b = 206$ m, $a = 175$ m

11. $A = 61.2°$, $C = 75.6°$, $c = 88.0$ cm

12. $B = 17.5°$, $a = 345$ m, $c = 405$ m

13. $a = 48.6$ cm, $b = 31.2$ cm, $c = 51.5$ cm

14. $A = 29.5°$, $b = 20.5$ m, $a = 18.5$ m

15. $B = 18.5°$, $a = 1680$ m, $b = 1520$ m

16. $a = 575$ ft, $b = 1080$ ft, $c = 1250$ ft

17. $C = 73.5°$, $c = 58.2$ ft, $b = 81.2$ ft

ILLUSTRATION 1

18. Find **a.** angle *B* and **b.** length *x* in Illustration 2.

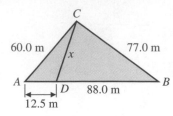

ILLUSTRATION 2

19. The centers of five holes are equally spaced around a circle of diameter 16.00 in. Find the distance between the centers of two successive holes.

20. In the roof truss in Illustration 3, *AB* = *DE*, *BC* = *CD*, and *AE* = 36.0 m. Find the lengths: **a.** *AF*, **b.** *BF*, **c.** *CF*, and **d.** *BC*.

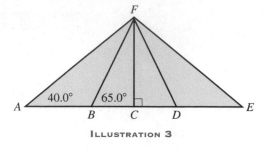

ILLUSTRATION 3

Chapter 14   Test

Find each value rounded to four significant digits:

1. cos 182.9°

2. tan 261°

3. Find the period, amplitude, and phase shift, and draw at least two periods of $y = 2 \sin (3x + 45°)$.

4. Find angle *B* in Illustration 1.

5. Find angle *C* in Illustration 1.

6. Find side *c* in Illustration 1.

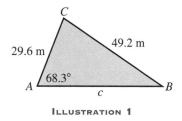

ILLUSTRATION 1

7. Find angle *C* in Illustration 2.

8. Find angle *B* in Illustration 2.

9. Find angle *A* in Illustration 2.

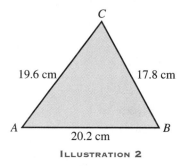

ILLUSTRATION 2

10. Find the length of the rafter shown in Illustration 3.

11. Find angle *A* in Illustration 3.

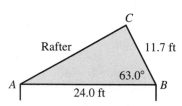

Basic Statistics

CHAPTER 15

QUALITY CONTROL
A worker takes a sample measurement of the density of a concrete pad.

MATHEMATICS AT WORK

Basic statistics play an important role in our daily lives. The government has statisticians working around the clock to compute the expected average number of people who will be on welfare, unemployment compensation, and Medicare, as well as the average number of highway fatalities, the expected number of homicides, etc.

State education systems rely on statistics to determine the average number of graduates and dropouts, or to predict the number of students expected to enroll in each district in a given year. The results of such statistical analysis are used to determine budgets and the amount of government funding to expect.

In manufacturing, statistics are used to calculate the average number of parts being produced, which then determines production quotas. Statistics are also used to determine the average number of defective parts to expect during a particular run, so as to minimize defective parts shipped to customers.

15.1   Bar Graphs

Statistics is the branch of mathematics that deals with the collection, analysis, interpretation, and presentation of masses of numerical data. In this chapter, we will first study the different ways that data can be presented using graphs. Then, in the later sections, we will study some of the different ways to describe small sets of data. We will study only the most basic parts to help you read and better understand newspapers, magazines, and some of the technical reports in your field of interest. The chapter concludes with an examination of statistical process control, a technique widely used in manufacturing.

A graph is a picture that shows the relationship between several types of collected information. A graph is very useful when there are large quantities of information to analyze. There are many ways of graphing. One common way is the *bar graph*. It is made of several parallel bars of lengths proportional to the quantities shown by each bar. Each bar stands for a fixed category, and the length of the bar shows the amount or value of that category. Look closely at the bar graph in Figure 15.1.

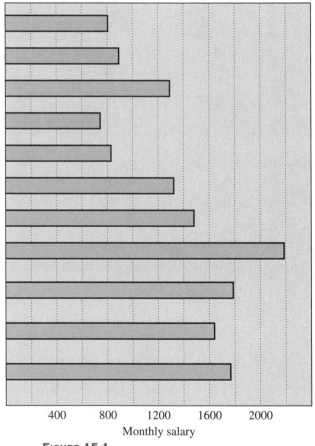

Classification of workers

Unskilled work, high school graduate

Semiskilled work with experience, high school graduate

Skilled technician, community college graduate

Clerk in retail store, high school graduate

Office worker and typist, high school graduate

Semiprofessional technician, community college graduate

Semiprofessional technician, after 5 years of experience

Senior engineering technician, after 10 years of experience

Data-processing technician, community college graduate with 2 years experience

Architectural draftsperson, community college graduate with 2 years experience

Auto mechanic, community college graduate with 5 years experience

400   800   1200   1600   2000

Monthly salary

FIGURE 15.1

• **EXAMPLE 1**   What are the monthly earnings of a data-processing technician with two years of experience?

Find the data-processing technician in the "Classification of workers" column. Read the right end of the bar on the "Monthly salary" scale: $1800.   ———•

Exercises | 15.1

Find the monthly earnings of the workers listed below from the bar graph in Figure 15.1:

1. Auto mechanic
2. Office worker and typist
3. Clerk in retail store
4. Skilled technician, community college graduate
5. Unskilled worker
6. Semiskilled worker
7. Architectural draftsperson
8. Semiprofessional technician, community college graduate
9. Semiprofessional technician after 5 years' experience
10. Senior engineering technician after 10 years' experience

Find the following information from Illustration 1.

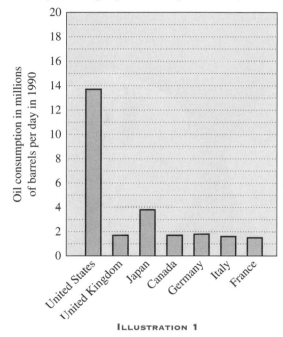

ILLUSTRATION 1

11. How many barrels per day were used by France?
12. How many barrels per day were used by Japan?
13. What country used the most barrels per day?
14. What country used the fewest barrels per day?
15. How many barrels per day were used by Canada?
16. How many barrels per day were used by Italy?
17. How many barrels per day were used by the United States?
18. How many barrels per day were used by the United Kingdom?
19. How many barrels per day were used by Germany?
20. What was the total number of barrels per day used by all the countries listed?

Using the bar graph in Illustration 2, find the number of deaths per 100,000 for each cause listed below:

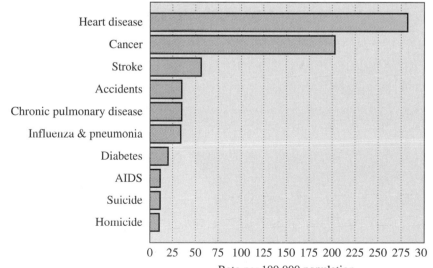

Rate per 100,000 population
Ten leading causes of death in United States in 1990

ILLUSTRATION 2

21. Heart disease

22. Accidents

23. Influenza and pneumonia

24. Suicide

25. Diabetes

26. Chronic pulmonary disease

27. Cancer

28. Homicide

29. AIDS

30. Stroke

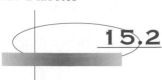

15.2   Circle Graphs

Another type of graph used quite often to give results of surveys is the *circle graph* (see Figure 15.2). The circle graph is used to show the relationship between the parts and the whole.

To make a circle graph with data given in percents, first draw the circle. Since there are 360 degrees in a circle, multiply the percent of an item by 360 to find what part of the circle is used by that item.

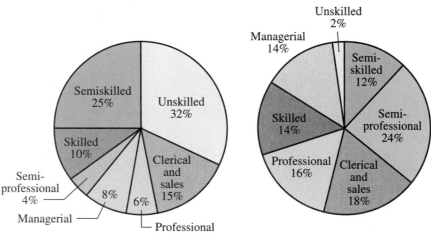

U.S. labor force in 1930   U.S. labor force in 1990

FIGURE 15.2

• **EXAMPLE 1**   Draw a circle graph with the following data.

In 1930, 58% of the people working had a grade school education or less, 32% had a high school education, and 10% had a college education.

58% of 360° = 0.58 × 360° = 208.8°, or about 209°

32% of 360° = 0.32 × 360° = 115.2°, or about 115°

10% of 360° = 0.10 × 360° = 36°

With a protractor draw central angles of 209°, 115°, and 36°. Then label the sections (see Figure 15.3).

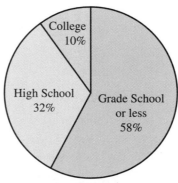

FIGURE 15.3

Sometimes data are not written in percent form. When a circle graph is to be drawn from data that are not in percent form, the data must be converted to percents. Once the data are in this form, the steps in drawing the graph are the same as those already given.

•   EXAMPLE 2   Draw a circle graph with the following data.
Suggested semester credit-hour requirements for a community college curriculum in engineering technology are as follows:

Course	Semester Hours
Mathematics (technical)	10
Applied science	10
Technical courses in major	34
General education courses	12
	66

Write each area of study as a percent of the whole program. Then draw the central angles and label the sections (see Figure 15.4).

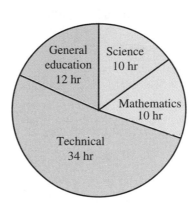

FIGURE 15.4

Mathematics:
$$\frac{10}{66} = \frac{r}{100}$$
$$66r = 1000$$
$$r = 15.2\%$$
$$15.2\% \times 360° = 0.152 \times 360°$$
$$= 55° \quad \text{(rounded to nearest whole degree)}$$

Science: Same as mathematics: 55°

Technical courses:
$$\frac{34}{66} = \frac{r}{100}$$
$$66r = 3400$$
$$r = 51.5\%$$
$$51.5\% \times 360° = 0.515 \times 360° = 185°$$

General education:
$$\frac{12}{66} = \frac{r}{100}$$
$$66r = 1200$$
$$r = 18.2\%$$
$$18.2\% \times 360° = 0.182 \times 360° = 66°$$

Exercises  15.2

1. Find 26% of 360°.
2. Find 52% of 360°.
3. Find 15.2% of 360°.
4. Find 37.1% of 360°.
5. Find 75% of 360°.
6. Find 47.7% of 360°.

7. Of 744 students, 452 are taking mathematics. What angle of a circle would show the percent of students taking mathematics?

8. Of 2017 students, 189 are taking technical physics. What angle of a circle would show the percent of students taking technical physics?

9. Of 5020 TV sets, 208 are found to be defective. What angle of a circle would show the percent of defective TV sets?

10. Candidate A was one of four candidates in an election. Of 29,106 votes cast, 4060 were for Candidate A. What angle of a circle would show the percent of votes *not* cast for Candidate A?

11. A department spends $16,192 of its $182,100 budget for supplies. What angle of a circle would show the percent of money the department spends on things other than supplies?

12. In one month, the sales of calculators were as follows:

    Brand A: 29   Brand D: 75
    Brand B: 52   Brand E: 43
    Brand C: 15

    What central angle of a circle graph would show Brand B's sales as a percent of the total sales for the month?

13. Draw a circle graph using the following data. In the United States for 1991, the percentage of national household income by quintiles is as follows: 1st, 4.4%; 2nd, 10.7%; 3rd, 16.6%; 4th, 24.1%; 5th, 44.2%.

14. Draw a circle graph with the following data. In 1991 the United States population by age was as follows: under 15, 21.9%; 15–29, 22.7%; 30–44, 24.4%; 45–59, 14.3%; 60–74, 11.4%; 75 and over, 5.3%.

15. Draw a circle graph to depict the suggested semester credit-hour requirements for a community college curriculum in industrial technology, as shown in Illustration 1.

Course	Semester Hours
Mathematics	6
Applied science	8
Technical specialties	34
General education courses	12
	60

ILLUSTRATION 1

16. A company interviewed its 473 employees to find the toughest day to work of a five-day work week. Draw a graph using the data in Illustration 2.

Day	Number
Monday	251
Tuesday	33
Wednesday	57
Thursday	43
Friday	89

ILLUSTRATION 2

17. Draw a circle graph with the following data. In 1992, total cotton production, in millions of 480-lb bales, was as follows: Western Hemisphere, 25.3;

Europe, 1.5; former Soviet republics, 11.3; Africa, 11.3; Asia and Oceania, 52.5.

18. Draw a circle graph with the following data. United States wholesale sales, in millions of dollars, for durable goods in 1991 were as follows: automotive, 156,200; electrical, 113,700; furniture and home furnishings, 28,100; hardware, 41,300; lumber, 52,600; machinery, 846,500; metals, 76,300; miscellaneous, 128,600.

19. Draw a circle graph with the following data. In

1993, federal budget receipts for the United States, in millions of dollars, were as follows: individual income taxes, 507,024; corporate income taxes, 111,638; social security taxes, 444,228; other taxes, 100,009.

20. Draw a circle graph with the following data. The number of health personnel in the United States during 1991 was as follows: physicians, 612,000; dentists, 160,000; nurses, 1,648,000; pharmacists, 174,000; midwives, 3000.

15.3 Line Graphs

The line graph is used to show changing conditions, often over a certain time interval.

• **EXAMPLE 1** An industrial technician must keep a chemical at a temperature below 60°F. He must also keep an hourly record of its temperature and record each day's temperatures on a line graph. The following table shows the data he collected.

Time	8	9	10	11	12	1	2	3	4	5
Temp. (°F)	60°	58°	54°	51°	52°	57°	54°	52°	54°	58°

When drawing line graphs, (a) use graph paper, because it is already subdivided both vertically and horizontally; (b) choose horizontal and vertical scales so that the line uses up most of the space allowed for the graph; (c) name and label each scale so that all marks on the scale are the same distance apart and show equal intervals; (d) plot the points from the given data; (e) connect each pair of points in order by a straight line. When you have taken all these steps, you will have a line graph (see Figure 15.5).

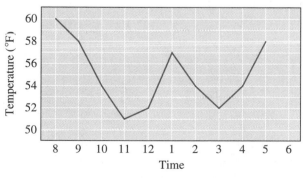

FIGURE 15.5

Exercises   15.3

1. The data in Illustration 1 below are from the records of the industrial technician in Example 1. These data were recorded on the following day. Draw a line graph for them.

2. An inspector recorded the number of faulty calculators and the hour in which they passed by his station, as shown in Illustration 2. Draw a line graph for these data.

Time	8	9	10	11	12	1	2	3	4	5
Temp. (°F)	59°	57°	55°	54°	53°	57°	55°	53°	56°	59°

ILLUSTRATION 1

Time	7–8	8–9	9–10	10–11	11–12	1–2	2–3	3–4	4–5	5–6
Number of faulty calculators	1	2	2	3	6	2	4	4	7	10

ILLUSTRATION 2

3. A survey of 100 families was taken to find the number of times the families had gone out to eat in the past month. The data are given in Illustration 3. Draw a line graph for this survey.

4. Illustration 4 lists the winning batting average for the years 1981 through 1988 for the major leagues. Draw a line graph for these data.

5. Illustration 5 shows the average test scores on chapter tests given in a mathematics class. Draw a line graph for these scores.

Times Out in Past Month	Number of Families
0	2
1	15
2	51
3	17
4	10
5 or more	5

ILLUSTRATION 3

Year	1981	1982	1983	1984	1985	1986	1987	1988
Batting average	0.341	0.332	0.361	0.351	0.368	0.334	0.370	0.313

ILLUSTRATION 4

Chapter	1	2	3	4	5	6	7	8	9	10	11	12	13	14
Score	78	81	75	77	84	81	79	70	72	73	75	69	81	72

ILLUSTRATION 5

6. Illustration 6 gives the average wage per week in 1992 for workers in certain industries. Draw a line graph for these data.

7. Illustration 7 lists the male life expectancy for certain countries. Draw a line graph for these data.

8. Illustration 8 gives the female life expectancy for certain countries. Draw a line graph for these data.

Industry	Metal Mining	Coal Mining	Oil and Gas	Contract Construction	Transpor-tation	Health Service	Retail Trade	Hotel
Wage (in $)	658	700	607	547	526	375	208	227

ILLUSTRATION 6

Country	Egypt	Nigeria	Israel	Japan	France	Sweden	Canada	Mexico	U.S.	Russia
Years	59.0	50.8	73.9	75.9	72.7	74.4	73.3	67.8	72.0	63.9

ILLUSTRATION 7

Country	Egypt	Nigeria	Israel	Japan	France	Sweden	Canada	Mexico	U.S.	Russia
Years	60.0	54.9	77.5	81.8	80.9	80.2	80.0	73.9	78.8	74.3

ILLUSTRATION 8

A technician is often asked to read graphs drawn by a machine. The machine records measurements by inking them on a graph. Any field in which quality control or continuous information is needed might use this way of recording measurements. Figure 15.6 shows a microbarograph used by the weather service to record atmospheric pressure in inches. For example, the reading on Monday at 8:00 P.M. was 29.34 in.

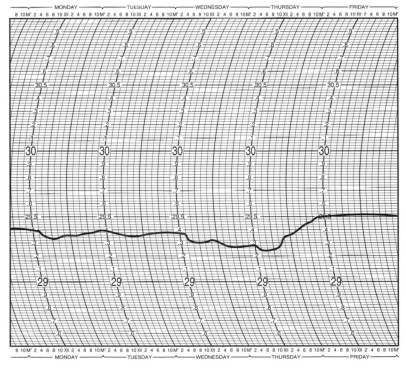

FIGURE 15.6

Use the microbarograph in Figure 15.6 to answer the following questions:

9. What was the atmospheric pressure recorded for Tuesday at 2:00 P.M.?

10. What was the highest atmospheric pressure recorded? When was it recorded?

11. What was the lowest atmospheric pressure recorded?

12. What was the atmospheric pressure recorded for Thursday at 10:00 P.M.?

13. What was the atmospheric pressure recorded for Monday at noon?

A hygrothermograph is used by the weather services to record temperature and relative humidity (see Figure 15.7). The lower part of the graph is used to measure relative humidity from 0% to 100%. The upper part of the graph is used to measure temperature from 10°F to 110°F. For example, at 8:00 P.M., the temperature was 86°F and the relative humidity 82%.

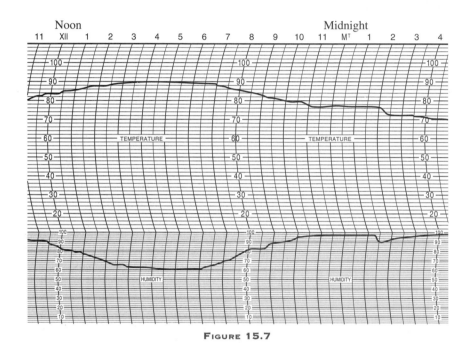

FIGURE 15.7

Use the hygrothermograph in Figure 15.7 to answer the following questions:

14. What was the relative humidity at 12:00 midnight?

15. What was the temperature at 3:30 A.M.?

16. What was the highest temperature recorded?

17. What was the lowest temperature recorded?

18. What was the relative humidity at 2:00 A.M.?

19. What was the lowest relative humidity recorded?

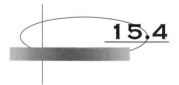

15.4 Other Graphs

A graph can be a curved line, as shown in Figure 15.8. This graph shows typical power gain for class B push-pull amplifiers with 9-volt power supply.

• **EXAMPLE 1**   What is the power output in Figure 15.8 when the gain is 22 decibels (db)?

Find 22 on the horizontal axis and read up until you meet the graph. Read left to the vertical axis and read 160 milliwatts (mW).

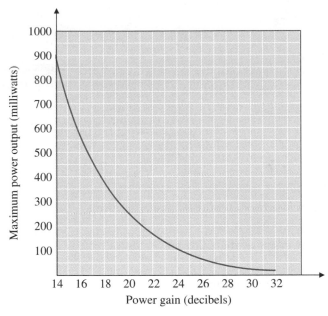

FIGURE 15.8

One way to avoid having to use a curved line for a graph is to use *semilogarithmic* graph paper. It has a logarithmic (nonuniform) scale for one axis and a uniform scale for the other axis. Figure 15.9 shows the data from Figure 15.8 plotted on semilogarithmic graph paper.

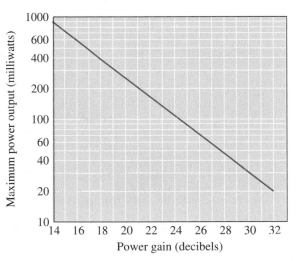

FIGURE 15.9

• **EXAMPLE 2**   Find the power gain in Figure 15.9 when the power output is 60 mW.

Find 60 on the vertical axis, which is the first mark above 60. Read across until you meet the graph; read down to the horizontal axis and read 27 db.   ———————•

We can see that each type of graph has advantages. The least and greatest changes are easier to read from the curved-line graph, but it is easier to read specific values from the straight-line graph.

Exercises   15.4

Use Figure 15.8 to find the answers for Exercises 1–5:

1. What is the highest power output? What is the power gain at the highest power output?
2. What is the power gain when the power output is 600 mW?
3. What is the power output when the power gain is 25 db?

4. Between what two db readings is the greatest change in power output found?
5. Between what two db readings is the least change in power output found?

Use Figure 15.9 to find the answers for Exercises 6–10:

6. What is the highest power output? What is the power gain at the highest power output?
7. What is the power gain when the power output is 600 mW?
8. What is the power output when the power gain is 25 db?

9. Between what two db readings is the greatest change in power output found?
10. Between what two db readings is the least change in power output found?

15.5   Mean Measurement

We have already seen in other chapters that with each technical measurement, a certain amount of error is made. One way that a technician can offset this error is to use what is called the *mean* of the measurements or the *mean measurement*. To find the mean measurement, the technician takes several measurements. The mean measurement is then found by taking the sum of these measurements and dividing this sum by the number of measurements taken.

$$\text{mean measurement} = \frac{\text{sum of the measurements}}{\text{number of measurements}}$$

•   **EXAMPLE**   A machinist measured the thickness of a metal disc with a micrometer at four different places. She found the following values:

2.147 in., 2.143 in., 2.151 in., 2.148 in.

Find the mean measurement.

STEP 1:   Add the measurements:

2.147 in.
2.143 in.
2.151 in.
2.148 in.
8.589 in.

STEP 2:   Divide the sum of the measurements by the number of measurements.

$$\text{mean measurement} = \frac{\text{sum of measurements}}{\text{number of measurements}}$$
$$= \frac{8.589 \text{ in.}}{4}$$
$$= 2.14725 \text{ in.}$$

So the mean measurement is 2.147 in.

Note that the mean measurement is written so that it has the same precision as each of the measurements.

Find the mean measurement for each set of measurements:

1. 47.61 cm; 48.23 cm; 47.92 cm; 47.81 cm
2. 9234 m; 9228 m; 9237 m; 9235 m; 9231 m
3. 0.2617 in.; 0.2614 in.; 0.2624 in.; 0.2620 in.; 0.2619 in.; 0.2617 in.
4. 6.643 mm; 6.644 mm; 6.647 mm; 6.645 mm; 6.650 mm
5. 25,740 mi; 25,780 mi; 25,790 mi; 25,810 mi; 25,720 mi; 25,760 mi
6. 3414 mg; 3433 mg; 3431 mg; 3419 mg; 3441 mg; 3417 mg; 3427 mg; 3434 mg; 3435 mg; 3432 mg
7. 2018 km; 2100̄ km; 2005 km; 2025 km; 2030̄ km
8. 69°; 81°; 74°; 83°; 67°; 71°; 75°; 63°
9. 728 lb; 475 lb; 803 lb; 915 lb; 1002 lb; 256 lb; 781 lb
10. 108 kW; 210̄ kW; 175 kW; 160̄ kW; 190̄ kW; 120̄ kW
11. 6091; 5050̄; 7102; 4111; 6060̄; 5910̄; 7112; 5855; 6280̄; 10,171; 9020̄; 10,172
12. 2.7; 8.1; 9.3; 7.2; 10.6; 11.4; 12.9; 13.5; 16.1; 10.9; 12.7; 15.9; 20.7; 21.9; 30.6; 42.9
13. 2050̄; 1951; 2132; 2232; 2147; 1867; 1996; 1785
14. 0.018; 0.115; 0.052; 0.198; 0.222; 0.189; 0.228; 0.346; 0.196; 0.258; 0.337; 0.532
15. 1.005; 1.102; 1.112; 1.058; 1.068; 1.115; 1.213
16. 248; 625; 324; 125; 762; 951; 843; 62; 853; 192; 346; 367; 484; 281; 628; 733; 801; 97; 218
17. 21; 53; 78; 42; 63; 28; 57; 83; 91; 32; 18
18. 0.82; 0.31; 1.63; 0.79; 1.08; 0.78; 1.14; 1.93; 0.068
19. 1.69; 2.38; 4.17; 7.13; 3.68; 2.83; 4.17; 8.29; 4.73; 3.68; 6.18; 1.86; 6.32; 4.17; 2.83; 1.08; 9.62; 7.71
20. 3182; 4440̄; 2967; 7632; 1188; 6653; 2161; 8197; 5108; 9668; 5108; 6203; 1988; 4033; 1204; 3206; 4699; 3307; 7226

15.6   Other Average Measurements and Percentiles

There are other procedures to determine an average measurement besides finding the mean measurement. The *median measurement* is the value that falls in the middle of a group of measurements arranged in order of size. That is, one-half of the measurements will be larger than or equal to the median, and one-half of the measurements will be less than or equal to the median.

• **EXAMPLE 1**   Find the median of the following set of measurements.

2.151 mm, 2.148 mm, 2.146 mm, 2.143 mm, 2.149 mm

STEP 1: Arrange the measurements in order of size.

2.151 mm
2.149 mm
2.148 mm
2.146 mm
2.143 mm

STEP 2: Find the middle measurement.

Since there are five measurements, the third measurement, 2.148 mm, is the median.

In Example 1, there was an odd number of measurements. When there is an even number of measurements, there is no one middle measurement. In this case, the median is found by taking the mean of the two middle measurements.

•   **EXAMPLE 2**   Find the median measurement of the following set of measurements.

54°, 57°, 59°, 55°, 53°, 57°, 50°, 56°

STEP 1: Arrange the measurements in order of size.

59°
57°
57°
56°
55°
54°
53°
50°

STEP 2: Since there are eight measurements, find the mean of the two middle measurements.

$$\frac{55° + 56°}{2} = \frac{111°}{2} = 55.5°$$

So the median measurement is 55.5°.   —————•

Another kind of average often used is the mode. The *mode* is the value that appears most often. In Example 2 above, 57° is the mode, because two of the measurements have this value. However, the mode can present problems. There can be more than one mode, and the mode may or may not be near the middle.

•   **EXAMPLE 3**   Find the mode of the following set of measurements.

3.8 cm, 3.2 cm, 3.7 cm, 3.5 cm, 3.8 cm, 3.9 cm, 3.5 cm, 3.1 cm

The measurements 3.5 cm and 3.8 cm are both modes because each appears most often—twice.   —————•

Related to averages is the idea of measuring the position of a piece of data relative to the rest of the data. Percentiles are numbers that divide the data into 100 equal parts. The *nth percentile* is the number P_n such that n percent of the data (ranked from smallest to largest) is smaller than P_n. For example, if you score in the 64th percentile on some standardized test, this means that you scored higher than 64% of those who took the test and you scored lower than 36% of those who took the test.

•   **EXAMPLE 4**   The list below gives ranked data (ranked from smallest to largest). Note that there are 50 pieces of data.

Ranked Data				
16	49	82	121	147
19	50	88	125	148
23	51	89	126	150
27	52	99	129	155
31	57	101	130	156
32	64	103	131	161
32	71	104	138	163
39	72	107	142	169
43	78	118	143	172
47	79	120	145	179

   a. Find the 98th percentile.
   b. Find the 75th percentile.
   c. Find the 26th percentile.

Solution:

   a. The 98th percentile is 172 (the 49th piece of data: $0.98 \times 50 = 49$). Ninety-eight percent of the data is smaller in value than 172.

   b. The 75th percentile is 142 (the 38th piece of data: $0.75 \times 50 = 37.5$ or 38). Seventy-five percent of the data is smaller in value than 142.

   c. The 26th percentile is 51 (the 13th piece of data: $0.26 \times 50 = 13$). Twenty-six percent of the data is smaller in value than 51.

Exercises 15.6

1–20. *Find the median measurement for each set of measurements in Exercises 1–20 of Exercises 15.5.*

Find the following percentiles for the data listed in Example 4:

21. 94th percentile
22. 80th percentile

23. 55th percentile
24. 12th percentile

25. 5th percentile
26. 50th percentile

15.7 Grouped Data

Finding the mean of a large number of measurements can take much time and can be subject to mistakes. Grouping the measurements (the data) can make the work in finding the mean much easier.

   Grouping data means arranging the data in groups. The groups are determined by setting up intervals. An *interval* contains all numbers between two given numbers a and b. We will show such an interval here by a–b. For example, 2–8 stands for all numbers between 2 and 8.

   The number a is called the *lower limit* and b is called the *upper limit* of the interval. The number midway between a and b, $\frac{a+b}{2}$, is called the *midpoint* of the interval. In the above example, the lower limit is 2, the upper limit is 8, and the midpoint is $\frac{2+8}{2} = 5$.

While there are no given rules for choosing these intervals, the following general rules are helpful.

1. The number of intervals chosen should be between 6 and 20.

2. The length of all intervals should be the same and should always be an odd number.

3. The midpoint of each interval should have the same number of digits as each of the measurements that fall within that interval. The lower limit and the upper limit of each interval will have one more digit than the measurements within the interval. In this way, no actual measurement will have exactly the same value as any of these limits. It will therefore be clear to which interval each measurement belongs.

4. The lower limit of the first interval should be lower than the lowest measurement value, and the upper limit of the last interval should be higher than the highest measurement value.

Once the intervals have been chosen, form a frequency distribution. A *frequency distribution* lists each interval, its midpoint, and the number of measurements that lie in that interval (frequency).

EXAMPLE 1   Make a frequency distribution for the recorded high temperatures for the days from November 1 to January 31 as given in Table 15.1.

TABLE 15.1

November	High Temperature (°F)	December	High Temperature (°F)	January	High Temperature (°F)
1	42	1	20	1	29
2	45	2	27	2	29
3	36	3	32	3	30
4	41	4	45	4	26
5	29	5	26	5	2
6	40	6	24	6	45
7	29	7	28	7	41
8	18	8	45	8	12
9	45	9	13	9	31
10	49	10	32	10	26
11	30	11	41	11	25
12	38	12	49	12	15
13	20	13	32	13	52
14	41	14	23	14	42
15	26	15	46	15	22
16	15	16	31	16	30
17	46	17	12	17	19
18	50	18	31	18	19

TABLE 15.1 CONT.

November	High Temperature (°F)	December	High Temperature (°F)	January	High Temperature (°F)
19	31	19	40	19	19
20	36	20	9	20	55
21	31	21	42	21	23
22	38	22	40	22	17
23	22	23	15	23	26
24	29	24	24	24	12
25	39	25	28	25	16
26	52	26	27	26	21
27	29	27	29	27	39
28	25	28	8	28	20
29	30	29	36	29	23
30	36	30	45	30	22
		31	12	31	9

First, choose the number and size of the group intervals to be used. We must have enough group intervals to cover the range of the data (the difference between the highest and the lowest values). Here, the range is $55° - 2° = 53°$. Since 53 is close to 54, let us choose the odd number 9 as the interval length. This means that we will need $54 \div 9 = 6$ group intervals. This satisfies our general rule for the number of intervals.

Our first interval is 1.5–10.5 with a midpoint of 6. Here, 1.5 is the lower limit and 10.5 is the upper limit of the interval. We then make the frequency distribution as shown in Table 15.2.

TABLE 15.2

Temperature (°F)	Midpoint x	Tally	Frequency f
1.5–10.5	6	////	4
10.5–19.5	15	卌 卌 ////	14
19.5–28.5	24	卌 卌 卌 卌 ///	23
28.5–37.5	33	卌 卌 卌 卌 ///	23
37.5–46.5	42	卌 卌 卌 卌 //	22
46.5–55.5	51	卌 /	6
			92

To find the mean from the frequency distribution, (a) multiply the frequency of each interval by the midpoint of that interval, xf; (b) add the products, xf; and (c) divide by the number of data, sum of f.

$$\text{mean} = \frac{\text{sum of } xf}{\text{sum of } f}$$

• **EXAMPLE 2**   Find the mean of the data given in Example 1.

A frequency distribution table (Table 15.3) gives the information for finding the mean.

TABLE 15.3

Temperature (°F)	Midpoint x	Frequency f	Product xf
1.5–10.5	6	4	24
10.5–19.5	15	14	210
19.5–28.5	24	23	552
28.5–37.5	33	23	759
37.5–46.5	42	22	924
46.5–55.5	51	6	306
		92	2775

The mean temperature is found as follows.

$$\text{mean} = \frac{\text{sum of } xf}{\text{sum of } f}$$

$$\text{mean} = \frac{2775}{92} = 30.2°F$$

Note: If the mean of the data in Example 2 were found by summing the actual temperatures and dividing by the number of temperatures, the value of the mean would be 29.86°F, or 29.9°F. There is a small difference between the two calculated means. This is because we are using the midpoints of the intervals rather than the actual data. However, since the mean is easier to find by this method, the small difference in values is acceptable.

• **EXAMPLE 3**   Find the mean of the data given in Example 1, this time using an interval length of 5.

The range of the data is 53, which is close to 55, a number divisible by 5. Since $55 \div 5 = 11$, we will use 11 intervals, each of length 5. Now make a frequency distribution with 1.5–6.5 as the first interval using 4 as the first midpoint. The frequency distribution then becomes as shown in Table 15.4.

TABLE 15.4

Temperature (°F)	Midpoint x	Frequency f	Product xf
1.5–6.5	4	1	4
6.5–11.5	9	3	27
11.5–16.5	14	9	126
16.5–21.5	19	9	171
21.5–26.5	24	15	360
26.5–31.5	29	20	580
31.5–36.5	34	7	238
36.5–41.5	39	11	429
41.5–46.5	44	11	484
46.5–51.5	49	3	147
51.5–56.5	54	3	162
		92	2728

Find the mean temperature:

$$\text{mean} = \frac{\text{sum of } xf}{\text{sum of } f}$$

$$\text{mean} = \frac{2728}{92}$$

$$= 29.7°F$$

Exercises | 15.7

1. From the following grouped data, find the mean.

Interval	Midpoint x	Frequency f	Product xf
41.5–48.5		12	
48.5–55.5		15	
55.5–62.5		20	
62.5–69.5		25	
69.5–76.5		4	
76.5–83.5		2	

2. The following are scores from a mathematics test. Make a frequency distribution and use it to find the mean score.

85, 73, 74, 69, 87, 81, 68, 76, 78, 75, 88, 85, 67, 83, 82, 95, 63, 84, 94, 66, 84, 78, 96, 67, 63, 59, 100, 90, 100, 94, 79, 79, 74

3. A laboratory technician records the life span (in months) of rats treated at birth with a fertility hormone. From the frequency distribution below, find the mean life span.

Life Span (months)	Midpoint x	Frequency f	Product xf
−0.5–2.5		12	
2.5–5.5		18	
5.5–8.5		22	
8.5–11.5		30	
11.5–14.5		18	

4. The life expectancy of a fluorescent light bulb is given by the number of hours that it will burn. From the frequency distribution on the next page, find the mean life of this type of bulb.

Life of Bulb (hours)	Midpoint x	Frequency f	Product xf
−0.5–499.5		2	
499.5–999.5		12	
999.5–1499.5		14	
1499.5–1999.5		17	
1999.5–2499.5		28	
2499.5–2999.5		33	
2999.5–3499.5		14	
3499.5–3999.5		5	

5. The shipment times in hours for a load of goods from a factory to market are tabulated in the frequency distribution below. Find the mean shipment time.

Shipment Time (hours)	Midpoint x	Frequency f	Product xf
22.5–27.5		2	
27.5–32.5		41	
32.5–37.5		79	
37.5–42.5		28	
42.5–47.5		15	
47.5–52.5		6	

6. The cost of goods stolen from a department store during the month of December has been tabulated by dollar amounts in the frequency distribution below. Find the mean cost of the thefts.

Cost ($)	Midpoint x	Frequency f	Product xf
−0.5–24.5		2	
24.5–49.5		17	
49.5–74.5		25	
74.5–99.5		51	
99.5–124.5		38	
124.5–149.5		32	

7. The number of passengers and their luggage weight in pounds on flight 2102 have been tabulated in

the frequency distribution below. Find the mean luggage weight.

Weight (lb)	Midpoint x	Frequency f	Product xf
0.5–9.5		1	
9.5–18.5		3	
18.5–27.5		22	
27.5–36.5		37	
36.5–45.5		56	
45.5–54.5		19	
54.5–63.5		17	
63.5–72.5		10	
72.5–81.5		5	
81.5–90.5		2	

8. The income of the residents in a neighborhood was tabulated. The results are shown in the frequency distribution below. Find the mean income.

Income ($)	Midpoint x	Frequency f	Product xf
2,500–12,500		1	
12,500–22,500		2	
22,500–32,500		15	
32,500–42,500		25	
42,500–52,500		8	

9. The number of defective parts per shipment has been tabulated in the frequency distribution below. Find the mean of defective parts per shipment.

Number of Defective Parts per Shipment	Midpoint x	Frequency f	Product xf
0.5–3.5		1	
3.5–6.5		•7	
6.5–9.5		20	
9.5–12.5		9	
12.5–15.5		32	
15.5–18.5		3	

10. The following are the dollar amounts of the traffic fines collected in one day in a city. Make a frequency distribution and use it to find the mean amount of the fines.

$30, $28, $15, $14, $32, $67, $45, $30, $17, $25,
$30, $19, $27, $32, $51, $45, $36, $42, $72, $50,
$18, $41, $23, $32, $35, $46, $50, $61, $82, $78,
$39, $42, $27, $20

11. The length of hospital stays for patients at a local hospital has been tabulated, and the results are shown in the frequency distribution below. Find the mean length for a hospital stay.

Length of Stay (days)	Frequency
0.5–1.5	50
1.5–2.5	32
2.5–3.5	18
3.5–4.5	10
4.5–5.5	8
5.5–6.5	5
6.5–7.5	26
7.5–8.5	17
8.5–9.5	22

12. The frequency of repair for the trucks owned by a trucking firm over a five-year period has been tabulated. The results are shown in the frequency distribution below. Find the mean number of repairs over the five-year period.

Times Repaired	Frequency
1.5–2.5	22
2.5–3.5	53
3.5–4.5	71
4.5–5.5	108
5.5–6.5	102
6.5–7.5	120
7.5–8.5	146
8.5–9.5	135
9.5–10.5	98
10.5–11.5	84
11.5–12.5	42
12.5–13.5	12
13.5–14.5	8

13. The scores that golfers shot on 18 holes at a local course were tabulated. The results are shown in the frequency distribution below. Find the mean score of the golfers.

Score	Frequency
68.5–73.5	5
73.5–78.5	7
78.5–83.5	10
83.5–88.5	12
88.5–93.5	20
93.5–98.5	22
98.5–103.5	25
103.5–108.5	32
108.5–113.5	17
113.5–118.5	12
118.5–123.5	9

14. The number of bushels of corn per acre for a certain hybrid planted by farmers during the year was tabulated. The results are shown in the frequency distribution below. Find the mean number of bushels per acre.

Yield (bu/acre)	Frequency
45.5–54.5	2
54.5–63.5	1
63.5–72.5	3
72.5–81.5	6
81.5–90.5	27
90.5–99.5	43
99.5–108.5	201
108.5–117.5	197
117.5–126.5	483
126.5–135.5	332
135.5–144.5	962
144.5–153.5	481
153.5–162.5	512
162.5–171.5	193
171.5–180.5	185
180.5–189.5	92
189.5–198.5	87
198.5–207.5	53
207.5–216.5	38

15. The following are the squad sizes of the football teams in a regional area. Make a frequency distribution and use it to find the mean.

> 108, 115, 97, 68, 72, 63, 18, 24, 202, 38, 43, 52, 83, 74, 39, 40, 51, 22, 37, 43, 48, 19, 23, 56, 72, 63, 23, 31, 43

16. The number of miles traveled by an experimental tire before it became unfit for use is recorded below. Make a frequency distribution and use it to find the mean.

> 8,457; 22,180; 15,036; 32,168; 9,168; 25,068; 32,192; 38,163; 18,132; 34,186; 36,192; 37,072; 14,183; 42,183; 19,182; 33,337; 38,162; 28,048; 20,208; 34,408; 35,108; 40,002; 29,208; 32,225; 33,207

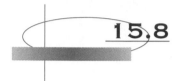

15.8   Variance and Standard Deviation

The mean measurement gives the technician the average value of a group of measurements, but the mean does not give any information about how the actual data vary in values. Some type of measurement that gives the amount of variation is often helpful in analyzing a set of data.

One way of describing the variation in the data is to find the range. The *range* is the difference between the highest value and the lowest value of the data.

• **EXAMPLE 1**   Find the range of the following measurements:

$$54°, 57°, 59°, 55°, 53°, 57°, 50°, 56°$$

The range is the difference between the highest value, 59°, and the lowest value, 50°. The range is $59° − 50° = 9°$. ————•

The range gives us an idea of how much the data are spread out, but another measure, the standard deviation, is often more helpful. The *standard deviation* tells how the data typically vary from the mean.

To find the standard deviation, we first need to find the variance. The *variance* is the mean of the squares of the differences between each measurement and the mean of the measurements.

$$\text{variance} = \frac{\text{sum of (measurement − mean)}^2}{\text{number of measurements}}$$

The unit of measurement for the variance is the square of the unit of measurement. Rather than work with the square of the units, we can find the standard deviation. The standard deviation is the square root of the variance. It therefore has the same unit of measurement as the original data.

$$\text{standard deviation} = \sqrt{\text{variance}}$$

• **EXAMPLE 2**   Find the standard deviation for the data given in Example 1.

STEP 1   Find the mean.

$$\text{mean} = \frac{\text{sum of measurements}}{\text{number of measurements}}$$

$$\text{mean} = \frac{54° + 57° + 59° + 55° + 53° + 57° + 50° + 56°}{}$$

$$= \frac{441°}{8} = 55.1°$$

STEP 2 Find the difference between each piece of data and the mean.

$$54 - 55.1 = -1.1$$
$$57 - 55.1 = 1.9$$
$$59 - 55.1 = 3.9$$
$$55 - 55.1 = -0.1$$
$$53 - 55.1 = -2.1$$
$$57 - 55.1 = 1.9$$
$$50 - 55.1 = -5.1$$
$$56 - 55.1 = 0.9$$

STEP 3 Square each difference and find the sum of the squared amounts.

$$(-1.1)^2 = 1.21$$
$$(1.9)^2 = 3.61$$
$$(3.9)^2 = 15.21$$
$$(-0.1)^2 = 0.01$$
$$(-2.1)^2 = 4.41$$
$$(1.9)^2 = 3.61$$
$$(-5.1)^2 = 26.01$$
$$(0.9)^2 = \underline{0.81}$$
$$54.88$$

STEP 4 Divide the sum in Step 3 by the number of measurements to find the variance.

$$\text{variance} = \frac{54.88}{8} = 6.86 \text{ or } 6.9$$

STEP 5 To find the standard deviation, find the square root of the variance from Step 4.

$$\text{standard deviation} = \sqrt{6.9} = 2.6°$$

For grouped data, the variance and standard deviation are found in a way similar to the way the mean is found. A frequency table is used. Columns are added to show the difference, D, between the midpoints and the mean ($D = x - $ mean); the square of D, D^2; and the frequency times D^2, $D^2 f$. The following formulas give the variance and standard deviation for grouped data.

$$\text{variance} = \frac{\text{sum of } D^2 f}{n}$$

$$\text{standard deviation} = \sqrt{\frac{\text{sum of } D^2 f}{n}}$$

where n is the number of pieces of data.

• **EXAMPLE 3** Given the grouped data in Table 15.5, find **a.** the mean and **b.** the standard deviation.

TABLE 15.5

Interval (cm)	Frequency f
3.5–12.5	3
12.5–21.5	3
21.5–30.5	7
30.5–39.5	6
39.5–48.5	4
48.5–57.5	1

First, we add the following columns to the frequency distribution: x, xf, D, D^2, and D^2f. See Table 15.6.

TABLE 15.6

Interval (cm)	Midpoint x	Frequency f	Product xf	$x - $ mean D	D^2	D^2f
3.5–12.5	8	3	24	−21	441	1323
12.5–21.5	17	3	51	−12	144	432
21.5–30.5	26	7	182	−3	9	63
30.5–39.5	35	6	210	6	36	216
39.5–48.5	44	4	176	15	225	900
48.5–57.5	53	1	53	24	576	576
		$n = 24$	696			3510

a. The mean $= \dfrac{\text{sum of } xf}{n} = \dfrac{696}{24} = 29.0$ cm.

b. The variance $= \dfrac{\text{sum of } D^2f}{n} = \dfrac{3510}{24} = 146.25$ cm^2.

The standard deviation $= \sqrt{\text{variance}} = \sqrt{146.25 \text{ cm}^2} = 12.1$ cm.

So the average measurement is 29.0 cm, and the data typically tend to vary from the mean by 12.1 cm.

Many scientific calculators have statistical functions. These can be used to find the mean (usually denoted by $\bar{x}$), variance (denoted by Σx^2), and the standard variation (denoted by σ). If you have a calculator with these functions, you should read its manual.

Exercises   15.8

1–20. Find the standard deviation for each set of measurements in Exercises 1–20 of Exercises 15.5.

21–36. Find the standard deviation for each set of data in Exercises 1–16 of Exercises 15.7.

15.9   Statistical Process Control

One of the many uses of statistics is in random sampling of processed goods. By watching the process, technicians can make changes early rather than waiting until a large number of defective goods have been produced.

Control charts are used to help find the information that the technician desires. Different kinds of control charts give different information; three types of control charts are listed below.

Median Charts: easy to use; shows the variation of the process. It is usually used to compare the output of several processes or various stages of the same process.

Individual Reading Chart: used for expensive measurements or when the output at any point in time remains relatively constant. Such a chart does not isolate individual steps of the process, so it can be hard to find out why there is a variation.

Mean Control Chart: shows the sample means plotted over time, to show whether the process is changing and whether it is in control. This is the type of chart we will study in this section. The chart has a center line at the target value of the process mean or at the process mean as determined by the data. Dashed lines represent control lines. These are found at the mean plus or minus three times the standard deviation divided by the square root of the number of samples. There are two cases when the process is out of control. The first is when any point falls outside the central limits, and the second is when any run of nine consecutive points falls on the same side of the center line. Figure 15.10 shows examples of two processes that are out of control (parts (a) and (b)) and one that is in control (part (c)).

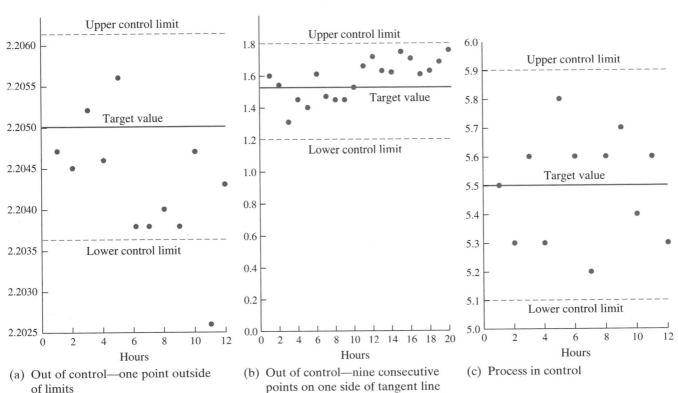

(a) Out of control—one point outside of limits

(b) Out of control—nine consecutive points on one side of tangent line

(c) Process in control

FIGURE 15.10
Three Mean Control Charts

• **EXAMPLE 1**   A manufacturer of golf balls checks a sample of 100 balls every hour. The compression of the ball is to be 90. This is the target value, and the standard deviation is ±4.5. Construct a mean control chart using the above information and the table below.

Hour	1	2	3	4	5	6	7	8	9	10	11	12
Mean	91.1	90.8	90.5	90.2	90.1	89.8	89.5	89.2	88.9	88.6	89.5	89.5

The lower and upper control limits are found by the mean ±3 standard deviations divided by the square root of the number of samples as follows:

$$90 \pm 3(4.5)/\sqrt{100} = 90 \pm 13.5/10 = 90 \pm 1.35 = 91.35 \text{ or } 88.65$$

The mean control chart is shown in Figure 15.11. The process is out of control. One point is beyond the control limits; this happened at the tenth hour. At that time, some source caused the process to go out of control. It is then up to technicians to locate the trouble and fix it.

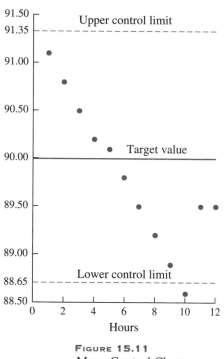

FIGURE 15.11
Mean Control Chart

Exercises   15.9

1. A certain manufacturing process has a target value of 1.20 cm and a standard deviation of ±0.15 cm. A sample of nine measurements is made each hour.

a. Draw a mean control chart using data from Illustration 1.

b. Is the process out of control? If it is, at what time does the mean control chart signal lack of control?

Hour	1	2	3	4	5	6
Mean	1.30	1.15	1.10	1.25	1.08	1.11
Hour	7	8	9	10	11	12
Mean	1.18	1.15	1.07	1.12	1.08	1.16
Hour	13	14	15	16	17	18
Mean	1.12	1.21	1.25	1.50	1.30	1.26
Hour	19	20	21	22	23	24
Mean	1.29	1.19	1.26	1.31	1.17	1.15

ILLUSTRATION 1

2. The depth of a silicon wafer is targeted at 1.0015 mm. If properly functioning, the process produces items with mean 1.015 mm and has a standard deviation of ± 0.004 mm. A sample of 16 items is measured once each hour. The sample means for the past 12 hours are given in the table below. From the data in Illustration 2, make a mean control chart and determine whether the process is in control.

Hour	1	2	3	4	5	6
Mean	1.016	1.013	1.015	1.017	1.013	1.014
Hour	7	8	9	10	11	12
Mean	1.017	1.016	1.014	1.013	1.016	1.017

ILLUSTRATION 2

3. A sporting goods manufacturer makes baseballs. The target weight of a baseball is 145.5 grams, with a standard deviation of ± 3.5 grams. A technician selects 100 balls at random per hour and records the mean weight of the samples. Illustration 3 lists the weights in grams for a 36-hour period.
   a. Make a mean control chart for the data shown.
   b. Is the process out of control? If it is, at what time does the mean control chart signal lack of control?

Hour	1	2	3	4	5	6
Weight	146.2	145.3	145.2	144.8	146.3	144.6
Hour	7	8	9	10	11	12
Weight	145.0	146.1	144.8	145.1	145.4	143.7
Hour	13	14	15	16	17	18
Weight	145.0	146.3	145.2	145.9	146.0	145.7
Hour	19	20	21	22	23	24
Weight	146.3	144.8	144.9	144.9	145.6	143.8
Hour	25	26	27	28	29	30
Weight	145.7	145.8	144.8	144.9	144.6	145.3
Hour	31	32	33	34	35	36
Weight	146.3	144.0	146.2	145.5	145.4	144.6

ILLUSTRATION 3

Chapter 15   Review

1. Find 35% of 360°.

2. Find 56.1% of 360°.

3. Draw a circle graph using the following data. In 1990–91, 26,305,999 students attended primary schools; 14,206,244 attended grades 9 to 12; and 8,175,012 attended college.

4. Draw a line graph using the data in Exercise 3.

5. In Figure 15.7 (page 462), what was the temperature at 10:00 P.M.?

For Exercises 6–8, use the data below.

A technician, using a very precise tool, measured a piece of metal to be used in a satellite. He recorded the following measurements: 7.0036 mm; 7.0035 mm; 7.0038 mm; 7.0035 mm; 7.0036 mm.

6. What is the mean measurement?
7. What is the median?
8. What is the standard deviation?
9. Given the frequency chart in Illustration 1, find **a.** the mean and **b.** the standard deviation.

For Exercises 10–12, use the data below:

A student's test and quiz scores for a quarter were recorded as follows: 72, 83, 79, 85, 91, 93, 80, 95, 82.

10. What is the mean of the scores recorded?
11. What is the median?
12. What is the standard deviation?

For Exercises 13 and 14, use the frequency chart in Illustration 2.

13. Find the mean of the data given in the frequency chart.
14. Find the standard deviation using the frequency chart.

Interval	Frequency f
10.5–21.5	4
21.5–32.5	17
32.5–43.5	10
43.5–54.5	28
54.5–65.5	13
65.5–76.5	12
76.5–87.5	9

ILLUSTRATION 1

Interval	Midpoint	Frequency
6.5–9.5	8	3
9.5–12.5	11	10
12.5–15.5	14	4
15.5–18.5	17	9
18.5–21.5	20	15
21.5–24.5	23	28
24.5–27.5	26	3
27.5–30.5	29	2

ILLUSTRATION 2

Chapter 15   Test

1. See Illustration 1. What country has the longest life expectancy?

2. See Illustration 1. What country has a life expectancy of 46.1 years?

3. Find 38% of 360°.

4. Draw a circle graph using the following data on cargo traffic between cities in the United States in billions of ton-miles: rail, 975; road, 866; air, 8.7; inland water, 435; pipeline, 587.

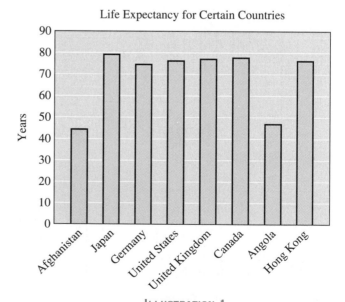

Life Expectancy for Certain Countries

ILLUSTRATION 1

5. Draw a line graph using the following data on population (in thousands) of the United States: 1940, 132,594; 1950, 152,271; 1960, 180,671; 1970, 204,879; 1980, 227,757; 1990, 249,246.

6. Using the data in Exercise 5, find the population of the United States in 1975.

7. See Illustration 2. What is the power output when the power gain is 28 db?

8. See Illustration 2. What is the power gain when the power output is 250 mW?

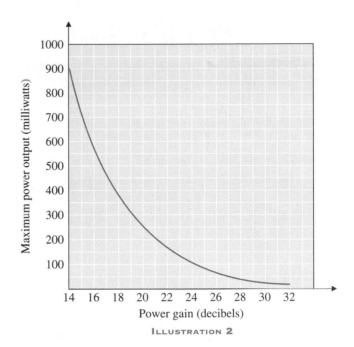

ILLUSTRATION 2

Use the following data for Exercises 9–11: 48, 54, 59, 65, 65, 65, 62, 64, 62, 66, 56, 50, 60.

9. Find the mean.   **10.** Find the median.   **11.** Find the standard deviation.

For Exercises 12–14, use the data in Illustration 3.

12. Write a frequency distribution.

13. Find the mean using grouped data.

14. Find the standard deviation.

6.0, 3.1, 0.6, 1.8, 2.1, 1.5, 4.1, 3.7, 3.3, 3.5, 2.5, 5.2, 2.5, 1.1, 3.2, 3.7, 2.7, 1.7, 4.4, 4.6, 4.0, 3.9, 2.9, 2.0, 1.9, 5.9, 2.4, 3.5, 0.9, 2.4, 0.6, 3.4, 0.5, 3.0, 3.0, 3.9, 3.3, 1.1, 3.2, 3.3, 2.5, 3.0, 3.7, 3.5, 4.2, 3.5, 1.6, 5.6, 5.2, 3.0, 3.5, 2.0, 2.6, 3.4, 3.3, 3.0, 3.0, 1.4

ILLUSTRATION 3

Cumulative Review Exercises   Chapters 1–15

1. Given the formula $R = \frac{Vt}{I}$, where $V = 32$, $t = 5$, and $I = 20$, find R.

2. Simplify: $-\frac{5}{6} - \left(-\frac{3}{5}\right)$

3. 90 kg = _____ lb

4. Read the scale.

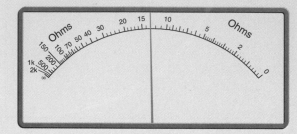

5. Simplify: $(2a^2 - 5a + 3) + (4a^2 + 3a - 1)$

6. Solve: $6 + 3(x - 2) = 24$

7. The area of a rectangle with constant width varies directly as its length. The area is 30.8 m² when the length is 12.8 m. Find the area when the length is 42.5 m.

8. Solve for y: $-5x - 3y = -8$

9. Solve: $7x - y = 4$
   $14x - 2y = 8$

10. Factor: $x^2 - 2x - 168$

11. Solve: $2x^2 - x - 8 = 0$

12. Find **a.** the area and **b.** the perimeter of the triangle in Illustration 1.

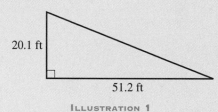

20.1 ft

51.2 ft

ILLUSTRATION 1

13. Find the volume of the frustum of the rectangular pyramid in Illustration 2.

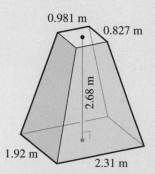

0.981 m

0.827 m

2.68 m

1.92 m

2.31 m

ILLUSTRATION 2

14. Use a calculator to find the value of tan 67.2° rounded to four significant digits.

15. If $\cos A = 0.6218$, find angle A in degrees.

From the triangle in Illustration 3, find:

16. $\angle B$

17. Side a

18. Side c

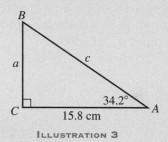

15.8 cm

34.2°

ILLUSTRATION 3

19. A roof has a rise of 8.00 ft and a span of 24.0 ft. Find its pitch and the distance from the eave to its peak rounded to the nearest inch.

20. Find the value of cos 191.13° rounded to four significant digits.

21. Draw the graph of $y = \frac{3}{2} \cos 2x$ for values of x between 0° and 180° in multiples of 15°.

22. Find the period and amplitude and draw at least two periods of the graph of the equation: $y = 2 \sin \frac{1}{2}x$

23. Given the triangle in Illustration 4, find angle C.

24. Find the mean of the following numbers: 20.2; 27.3; 35.1; 30.6; 29.6; 22.6.

25. Find the median of the data in Exercise 24.

26. Find the standard deviation of the data in Exercise 24.

27. Find the mean from the frequency distribution in Illustration 5.

28. Make a frequency distribution from the following data. The number of defective parts coming off an assembly line per eight-hour shift were as follows: 15; 12; 10; 9; 15; 22; 7; 23; 12; 8; 18; 22; 11; 30; 14; 18; 12; 20; 22; 35; 10; 8; 11; 19; 7; 23; 17; 15; 20; 16; 17; 18; 22; 15; 20; 13.

29. Find the mean using grouped data from Exercise 28.

30. Find the median of the data in Exercise 28.

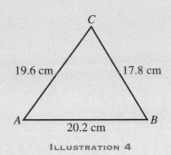

ILLUSTRATION 4

Interval	Frequency	Interval	Frequency
29.5–34.5	4	64.5–69.5	78
34.5–39.5	12	69.5–74.5	23
39.5–44.5	22	74.5–79.5	56
44.5–49.5	63	79.5–84.5	33
49.5–54.5	15	84.5–89.5	18
54.5–59.5	82	89.5–94.5	12
59.5–64.5	49		

ILLUSTRATION 5

Appendix A
Using a Calculator
with Memory

For some combinations of operations, a calculator with a memory key or keys is very helpful. A memory stores numbers for future use. Operations can then be done with numbers and the number in the memory.

Note: If your calculator has more than one memory, consult your instruction manual.

• **EXAMPLE 1** Find the value of $\dfrac{14}{\sqrt{5}} - \sqrt{\dfrac{15}{7}}$ and round the result to three significant digits.

Flowchart	Buttons Pushed	Display
Enter 14	$\boxed{1} \rightarrow \boxed{4}$	14
Push divide	$\boxed{\div}$	14
Enter 5	$\boxed{5}$	5
Push square root	$\boxed{\sqrt{}}$	2.23607
Push equals	$\boxed{=}$	6.26099
Push store or add to memory	$\boxed{\text{STO}}$ or $\boxed{\text{M+}}$	6.26099
Enter 15	$\boxed{1} \rightarrow \boxed{5}$	15

Flowchart	Buttons Pushed	Display
Push divide	÷	15
Enter 7	7	7
Push equals	=	2.14286
Push square root	$\sqrt{}$	1.46385
Push subtract from memory	+/− → SUM or M−	1.46385
Push memory recall	RCL or MR	4.79714

The result is 4.80 rounded to three significant digits.

• **EXAMPLE 2** Find the value of

$$\sqrt{\left(\frac{16}{1.35}\right)^2 + \left(\frac{1}{2\pi(60)(4 \times 10^{-5})}\right)^2}$$

rounded to three significant digits.

Flowchart	Buttons Pushed	Display
Enter 16	1 → 6	16
Push divide	÷	16
Enter 1.35	1 → . → 3 → 5	1.35
Push equals	=	11.8519
Push square	x^2	140.466
Push store or add to mcmory	STO or M+	140.466
Enter 2	2	2
Push times	×	2

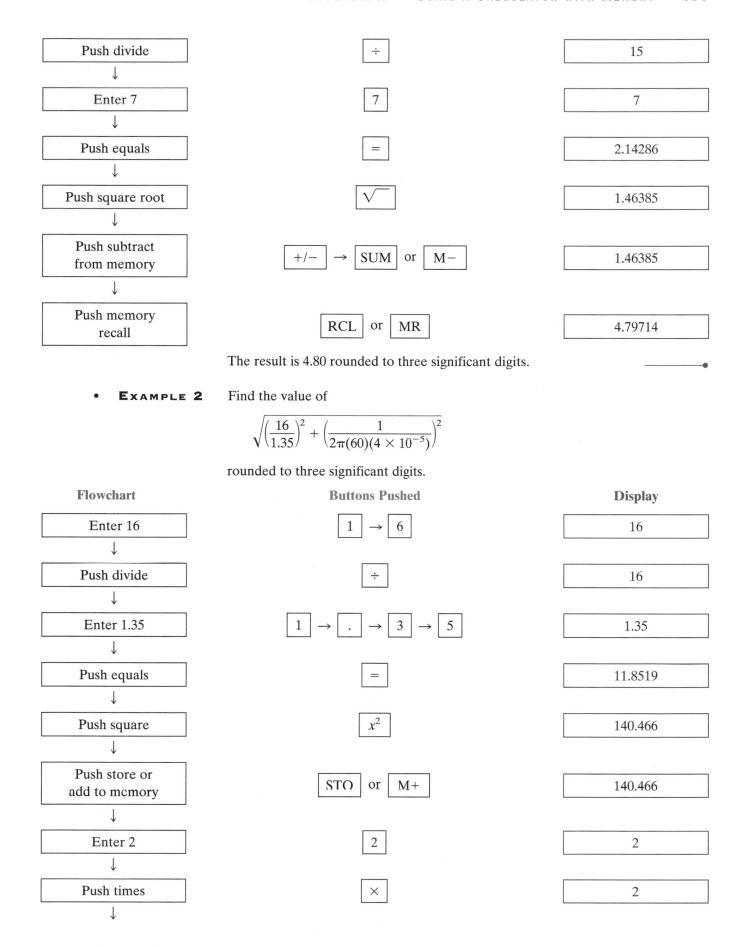

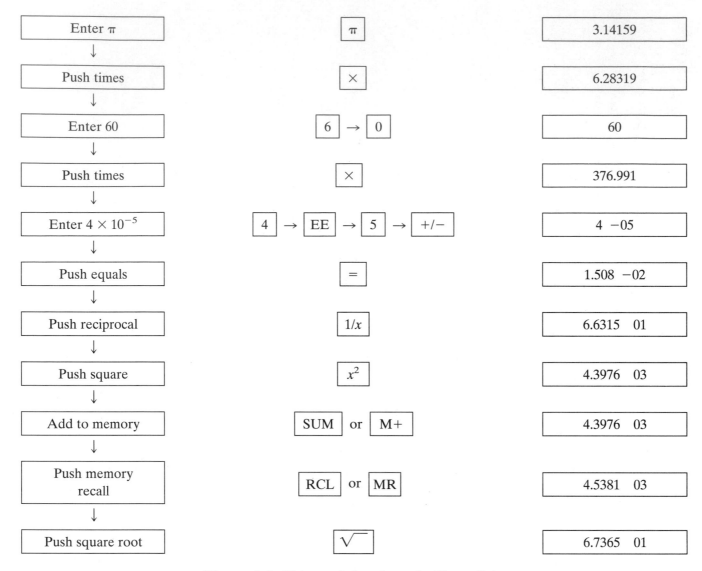

Enter π	π	3.14159
Push times	×	6.28319
Enter 60	6 → 0	60
Push times	×	376.991
Enter 4×10^{-5}	4 → EE → 5 → +/−	4 −05
Push equals	=	1.508 −02
Push reciprocal	1/x	6.6315 01
Push square	x^2	4.3976 03
Add to memory	SUM or M+	4.3976 03
Push memory recall	RCL or MR	4.5381 03
Push square root	√	6.7365 01

The result is 67.4 rounded to three significant digits.

Exercises   A

Do as indicated and round each final result to three significant digits:

1. $16\sqrt{3} + \left(\dfrac{18.1}{14}\right)^2$

2. $\dfrac{5}{\sqrt{2}} + \sqrt{\dfrac{48}{13.5}}$

3. $36^2 - \sqrt{\dfrac{19.8}{0.0246}}$

4. $\dfrac{18.3}{6\sqrt{2}} - \left(\dfrac{245}{167}\right)^2$

5. $\dfrac{49.2 + 6.8}{21.5 - 14.2}$

6. $\dfrac{29.2 - 14.2}{29.2 - 21.3}$

7. $\sqrt{\dfrac{16.8}{4.23}} + \left(\dfrac{1.45}{0.94}\right)^2$

8. $\dfrac{25.6^2}{5.2} + \left(\dfrac{41.9}{8.25}\right)^2$

9. $(13.5^2 + 18.6^2)^2 + (25.6^2 + 31.7^2)^2$

10. $(46.5^2 + 19.8^2)^2 - (44.6^2 - 23.7^2)^2$

11. $\sqrt{25.5^2 + 29.5^2} + \sqrt{36.5^2 - 27.3^2}$

12. $\sqrt{41.5^2 + 36.7^2} - \sqrt{51.7^2 - 48.7^2}$

13. $\sqrt{\dfrac{196}{3.91}} + \left(\dfrac{21.7}{0.361}\right)^2 - \dfrac{23.6}{\sqrt{2\pi}}$

14. $\dfrac{35.1}{(1.97)^2} + \left(\dfrac{25.1}{3.75}\right)^2 - \sqrt{\dfrac{14.2}{3.64}}$

15. $\sqrt{\left(\dfrac{35.6}{2.93}\right)^2 + \left(\dfrac{1}{2\pi(60)(8 \times 10^{-6})}\right)^2}$

16. $\sqrt{\left(\dfrac{855}{16.2}\right)^2 + \left(\dfrac{1}{2\pi(60)(9 \times 10^{-8})}\right)^2}$

17. $\sqrt{\left(\dfrac{90.5}{21.6}\right)^2 + \left(\dfrac{1}{2\pi(60)(1.8 \times 10^{-5})}\right)^2}$

18. $\sqrt{\left(\dfrac{185}{31.5}\right)^2 + \left(\dfrac{1}{2\pi(60)(6.5 \times 10^{-10})}\right)^2}$

19. $\dfrac{(19.2)(10.6) + (9)(15.6) - (6)(15)}{(5)(12) + (5.5)(15) + (10)(16.5)}$

20. $\dfrac{(18.6)(5.5) + (6.75)(18.5) - (9.25)(12.55)}{(5.25)(10.5) + (6.75)(18.5) + (10.5)(16.1)}$

Appendix B
Tables

TABLE 1

Formulas from Geometry
Plane Figures

In the following tables, a, b, c, and d are lengths of sides, and h is the altitude.

	Perimeter	Area

Rectangle $P = 2(b + h)$ $A = bh$

Square $P = 4b$ $A = b^2$

Parallelogram $P = 2(a + b)$ $A = bh$

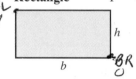

Rhombus $P = 4b$ $A = bh$

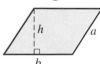

	Perimeter	Area

Trapezoid $P = a + b + c + d$ $A = \left(\dfrac{a + b}{2}\right)h$

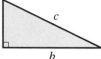

Triangle $P = a + b + c$ $A = \frac{1}{2}bh$
$$A = \sqrt{s(s - a)(s - b)(s - c)}$$
$$\text{where } s = \tfrac{1}{2}(a + b + c)$$

The sum of the measures of the angles of any triangle = 180°

Right triangle $c^2 = a^2 + b^2$ or
$$c = \sqrt{a^2 + b^2}$$

Circle   Circumference
$C = 2\pi r$
$C = \pi d$
$(d = 2r)$ $A = \pi r^2$

The sum of the measures of the central angles of a circle = 360°.

Geometric Solids

In the following table, B is the area of the base, r is the length of the radius, and h is the height.

	Volume	Lateral Surface Area
Prism	$V = Bh$	
Cylinder 	$V = \pi r^2 h$	$A = 2\pi rh$
Pyramid 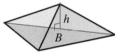	$V = \frac{1}{3}Bh$	
Cone 	$V = \frac{1}{3}\pi r^2 h$	$A = \pi rs$, where s is the slant height.
Sphere	$V = \frac{4}{3}\pi r^3$	$A = 4\pi r^2$

TABLE 2   Electrical Symbols

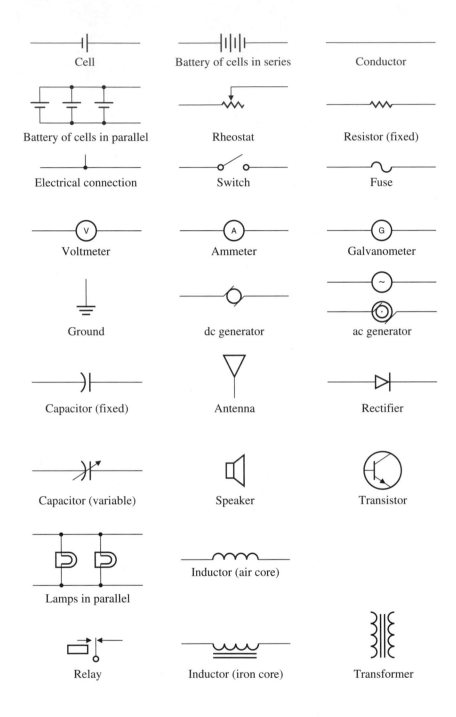

Cell	Battery of cells in series	Conductor
Battery of cells in parallel	Rheostat	Resistor (fixed)
Electrical connection	Switch	Fuse
Voltmeter	Ammeter	Galvanometer
Ground	dc generator	ac generator
Capacitor (fixed)	Antenna	Rectifier
Capacitor (variable)	Speaker	Transistor
Lamps in parallel	Inductor (air core)	
Relay	Inductor (iron core)	Transformer

Appendix C: Answers to Odd-Numbered Exercises and All Chapter Review and Cumulative Review Exercises

CHAPTER 1

EXERCISES 1.1 (page 9)

1. 3255 **3.** 2466 **5.** 1298 **7.** 1454 **9.** 795,776 **11.** 131 **13.** 5164 **15.** 26,008 **17.** 2119 **19.** 2820 **21.** 4195 Ω
23. 224 **25.** $24,715; $9,058; $33,773 **27.** I: 1925 cm³; O: 1425 cm³ **29.** 27,216 **31.** 818,802 **33.** 18,172,065
35. 26,527,800 **37.** 35,360,000 **39.** 1809 **41.** 1125 r 4 **43.** 389 **45.** 1980 r 2 **47.** 844 r 40 **49.** 496 mi **51.** 3400 cm³
53. 325 cm³ **55.** 13 km/L **57.** 4820 ft **59.** 9600 **61.** 67 in. from either corner **63.** 44 bu/acre **65. a.** 881 lb **b.** 15 lb
67. 85 bu/acre **69.** $4200 **71.** 5 A **73.** 24 V **75.** 880 oz **77.** 4

EXERCISES 1.2 (page 13)

1. 2 **3.** 10 **5.** 18 **7.** 131 **9.** 73 **11.** 23 **13.** 16 **15.** 34 **17.** 102 **19.** 137 **21.** 230 **23.** 55 **25.** 0 **27.** 4 **29.** 0
31. 13 **33.** 19 **35.** 26 **37.** 1 **39.** 102 **41.** 85

EXERCISES 1.3 (page 17)

1. 96 yd² **3.** 84 cm² **5.** 48 in² **7.** 52 in² **9.** 108 **11.** 32 gal **13. a.** $41,040 **b.** $89,072 **15.** 96 m³ **17.** 720 cm³
19. 3900 in³ **21.** 600 cm³ **23.** 69,480 lb **25.** 14,880 lb **27.** $1,250

EXERCISES 1.4 (page 22)

1. 600 **3.** 11,250 **5.** 57,376 **7.** 38,400 **9.** 8 **11.** 5017 **13.** 40 in² **15.** 810 ft² **17.** 56 m² **19.** 648 ft² **21.** 32 ft²
23. 1800 cm² **25.** 4500 cm³ **27.** 172 **29.** 16

EXERCISES 1.5 (page 25)

1. a. yes **b.** no **3. a.** yes **b.** yes **5. a.** yes **b.** no **7.** prime **9.** not prime **11.** not prime **13.** not prime **15.** yes
17. no **19.** no **21.** yes **23.** no **25.** yes **27.** yes **29.** no **31.** no **33.** yes **35.** no **37.** no **39.** yes **41.** yes **43.** yes
45. $2 \cdot 2 \cdot 5$ **47.** $2 \cdot 3 \cdot 11$ **49.** $2 \cdot 2 \cdot 3 \cdot 3$ **51.** $3 \cdot 3 \cdot 3$ **53.** $3 \cdot 17$ **55.** $2 \cdot 3 \cdot 7$ **57.** $2 \cdot 2 \cdot 2 \cdot 3 \cdot 5$ **59.** $3 \cdot 3 \cdot 19$
61. $3 \cdot 5 \cdot 7$ **63.** $2 \cdot 2 \cdot 3 \cdot 3 \cdot 7$

UNIT 1A REVIEW (page 26)

1. 241 **2.** 1795 **3.** 2,711,279 **4.** 620 **5.** 262 ft **6.** 254 bu **7.** 42 **8.** 7 **9.** 32 **10.** 499 in² **11.** 720 ft³ **12.** 180
13. 10 **14.** 300 **15.** not prime **16.** prime **17.** divisible by 3 **18.** not divisible by 5 **19.** $2 \cdot 2 \cdot 2 \cdot 5$ **20.** $3 \cdot 3 \cdot 3 \cdot 5$

EXERCISES 1.6 (page 31)

1. $\frac{3}{7}$ **3.** $\frac{6}{7}$ **5.** $\frac{3}{16}$ **7.** $\frac{1}{3}$ **9.** $\frac{4}{5}$ **11.** 1 **13.** 0 **15.** undefined **17.** $\frac{7}{8}$ **19.** $\frac{3}{4}$ **21.** $\frac{3}{4}$ **23.** $\frac{4}{5}$ **25.** $\frac{3}{10}$ **27.** $\frac{2}{3}$ **29.** $\frac{7}{8}$ **31.** $\frac{6}{7}$
33. $\frac{3}{5}$ **35.** $\frac{19}{37}$ **37.** $\frac{2}{3}$ **39.** $\frac{7}{9}$ **41.** $15\frac{3}{5}$ **43.** $9\frac{1}{3}$ **45.** $1\frac{1}{4}$ **47.** $9\frac{1}{2}$ **49.** $6\frac{1}{4}$ **51.** $\frac{23}{6}$ **53.** $\frac{17}{8}$ **55.** $\frac{23}{16}$ **57.** $\frac{55}{8}$ **59.** $\frac{53}{5}$

EXERCISES 1.7   (page 40)

1. 16   **3.** 210   **5.** 48   **7.** $\frac{5}{6}$ **9.** $\frac{5}{32}$ **11.** $\frac{11}{28}$ **13.** $\frac{29}{64}$ **15.** $\frac{7}{20}$ **17.** $\frac{13}{10}$ or $1\frac{3}{10}$ **19.** $\frac{13}{10}$ or $1\frac{3}{10}$ **21.** $\frac{19}{16}$ or $1\frac{3}{16}$ **23.** $\frac{37}{48}$
25. $\frac{13}{120}$ **27.** $\frac{67}{105}$ **29.** $\frac{43}{27}$ or $1\frac{16}{27}$ **31.** $\frac{1}{8}$ **33.** $\frac{1}{2}$ **35.** $\frac{4}{7}$ **37.** $\frac{1}{32}$ **39.** $7\frac{1}{4}$ **41.** $2\frac{5}{8}$ **43.** $4\frac{3}{4}$ **45.** $2\frac{5}{16}$ **47.** $13\frac{31}{45}$ **49.** $\frac{79}{84}$
51. $11\frac{5}{16}$ **53.** $9\frac{13}{24}$ **55.** $2237\frac{1}{4}$ ft   **57. a.** $1\frac{1}{2}$ in.   **b.** $22\frac{23}{64}$ in.   **59. a.** $\frac{1}{4}$ in.   **b.** $12\frac{5}{16}$ in.   **61.** $3\frac{1}{4}$ A   **63.** $1\frac{43}{48}$ A   **65.** $9\frac{5}{8}$ in.
67. a. $10\frac{1}{2}$ in.   **b.** $\frac{3}{4}$ in.   **69. a.** $21\frac{13}{16}$ in.   **b.** $\frac{7}{8}$ in.   **71.** $14\frac{3}{8}$ in.   **73.** $\frac{5}{64}$ in.   **75.** $1\frac{25}{32}$ in.; $1\frac{19}{32}$ in.   **77.** $3\frac{3}{4}$ in.; $10\frac{13}{16}$ in.
79. $5\frac{11}{16}$ in.   **81.** $52\frac{5}{32}$ in.

EXERCISES 1.8   (page 48)

1. 12   **3.** 9   **5.** $\frac{35}{64}$ **7.** $\frac{2}{3}$ **9.** 10   **11.** $\frac{1}{8}$ **13.** $\frac{5}{4}$ or $1\frac{1}{4}$ **15.** $\frac{2}{5}$ **17.** $\frac{18}{25}$ **19.** 18   **21.** 40   **23.** $\frac{35}{33}$ or $1\frac{2}{33}$ **25.** $\frac{88}{45}$ or $1\frac{43}{45}$
27. $\frac{9}{4}$ or $2\frac{1}{4}$ **29.** $\frac{27}{32}$ **31.** $\frac{1}{126}$ **33.** $\frac{7}{12}$ **35.** $31\frac{1}{2}$ gal   **37.** $20\frac{1}{6}$ ft   **39.** $4\frac{15}{32}$ in.   **41.** 80 bd ft   **43.** $1633\frac{1}{3}$ bd ft   **45.** $3\frac{27}{32}$ in.
47. $\frac{1}{32}$ of a set   **49.** $1\frac{7}{8}$ in.   **51.** $8\frac{5}{8}$ in.   **53.** $8\frac{2}{3}$ min   **55.** $2\frac{2}{3}$ ft³   **57.** 2750 W   **59.** 75 W   **61.** $2\frac{2}{7}$ A   **63.** $7\frac{1}{4}$ ft or 7 ft 3 in.
65. 63 lb   **67.** 40 lb   **69.** 160 bu/acre   **71.** $\frac{1}{2}$ oz   **73.** $\frac{1}{2}$ **75.** $1\frac{1}{4}$ gr   **77.** 2   **79.** 3   **81.** 4 Ω   **83.** $3\frac{1}{5}$ Ω

EXERCISES 1.9   (page 53)

1. 43   **3.** 83   **5.** 9   **7.** 10   **9.** 96   **11.** 6   **13.** 8   **15.** 5   **17.** $5\frac{1}{2}$ **19.** $3\frac{1}{2}$ **21.** $3\frac{1}{2}$ **23.** $30\frac{2}{3}$ **25.** 3520   **27.** $15\frac{5}{8}$
29. 6 ft 8 in.   **31.** 153 in.   **33.** 7 qt   **35.** $\frac{66}{125}$ Ω   **37.** $8\frac{1}{2}$ chains   **39.** $410\frac{1}{10}$ grains   **41.** 75 ft/min   **43.** 72 mi/min   **45.** $58\frac{2}{3}$ ft/s
47. 120 ft/min   **49.** 16 yd 1 ft 10 in.

UNIT 1B REVIEW   (page 54)

1. $\frac{3}{5}$ **2.** $\frac{8}{9}$ **3.** $4\frac{1}{2}$ **4.** $\frac{17}{5}$ **5.** $1\frac{1}{2}$ **6.** $2\frac{23}{24}$ **7.** $\frac{4}{15}$ **8.** $\frac{6}{13}$ **9.** $3\frac{1}{4}$ **10.** 2   **11.** $\frac{7}{8}$ **12.** $17\frac{7}{8}$ in.   **13.** $17\frac{5}{6}$ in.   **14.** $16\frac{2}{3}$ in²
15. 48 in.   **16.** 8 yd   **17.** 48 oz   **18.** 5 gal   **19.** 88 ft/s   **20.** 5 ft 8 in.

EXERCISES 1.10   (page 61)

1. Four thousandths   **3.** Five ten-thousandths   **5.** One and four hundred twenty-one hundred-thousandths   **7.** Sixty-two and three hundred eighty-four thousandths   **9.** Six and ninety-two thousandths   **11.** Two hundred forty-six and one hundred twenty-four thousandths   **13.** 5.02; $5\frac{2}{100}$ or $5\frac{1}{50}$ **15.** 71.0021; $71\frac{21}{10,000}$ **17.** 65,000; 65,000   **19.** 43.0101; $43\frac{101}{10,000}$
21. 0.375   **23.** $0.7\overline{3}$ **25.** 0.34   **27.** $1.\overline{27}$ **29.** $18.\overline{285714}$ **31.** $34.\overline{2}$ **33.** $\frac{7}{10}$ **35.** $\frac{11}{100}$ **37.** $\frac{337}{400}$ **39.** $10\frac{19}{25}$ **41.** 150.888
43. 163.204   **45.** 86.6   **47.** 15.308   **49.** 8.68   **51.** 4.862   **53.** 10.0507   **55.** a = 4.56 cm; b = 4.87 cm   **57.** 7.94 in.
59. 70.8 mm   **61.** 2.605 A   **63.** 1396.8 Ω   **65.** 0.532 in.   **67.** 0.22 in.   **69.** 1.727 in.

EXERCISES 1.11   (page 66)

1. a. 1700   **b.** 1650   **3. a.** 3100   **b.** 3130   **5. a.** 18,700   **b.** 18,680   **7. a.** 3.1   **b.** 3.142   **9. a.** 0.1   **b.** 0.057   **11. a.** 0.1   **b.** 0.070

	Hundred	Ten	Unit	Tenth	Hundredth	Thousandth
13.	600	640	636	636.2	636.18	636.183
15.	17,200	17,160	17,159	17,159.2	17,159.17	17,159.167
17.	1,543,700	1,543,680	1,543,679	—	—	—
19.	10,600	10,650	10,650	10,649.8	10,649.83	—
21.	600	650	650	649.9	649.90	649.900

23. 237,000   **25.** 0.0328   **27.** 72
29. 1,462,000   **31.** 0.0003376   **33.** 1.01

EXERCISES 1.12   (page 71)

1. 0.555   **3.** 10.5126   **5.** 15,372   **7.** 9,280,000   **9.** 30   **11.** 15   **13.** 148.12   **15.** 248.23   **17.** 3676.47   **19. a.** 37.76 m
b. 9.44 m   **21.** 80 threads   **23.** 2.95 in.   **25.** 3000 sheets   **27.** $812   **29.** 51.20 in.   **31.** 450 in³   **33.** 0.5 L   **35. a.** 0.056 in.
b. 4.535 in.   **37.** 240 gal   **39.** $132.50   **41.** 94.2 Ω   **43.** 0.288 W   **45.** 6.20 A   **47.** 136.9 Ω   **49.** 0.450 grains   **51.** 5 tablets

EXERCISES 1.13   (page 78)

1. 0.27 **3.** 0.06 **5.** 1.56 **7** 0.292 **9.** 0.087 **11.** 9.478 **13.** 0.0028 **15.** 0.00068 **17.** 0.0425 **19.** 0.00375 **21.** 54%
23. 8% **25.** 62% **27.** 217% **29.** 435% **31.** 18.5% **33.** 29.7% **35.** 519% **37.** 1.87% **39.** 0.29% **41.** 80%
43. 12.5% **45.** $16\frac{2}{3}$% **47.** $44\frac{4}{9}$% **49.** 60% **51.** 32.5% **53.** 43.75% **55.** 240% **57.** 175% **59.** $241\frac{2}{3}$% **61.** $\frac{3}{4}$ **63.** $\frac{4}{25}$
65. $\frac{3}{5}$ **67.** $\frac{93}{100}$ **69.** $\frac{11}{4}$ or $2\frac{3}{4}$ **71.** $\frac{5}{4}$ or $1\frac{1}{4}$ **73.** $\frac{43}{400}$ **75.** $\frac{107}{1000}$ **77.** $\frac{69}{400}$ **79.** $\frac{97}{600}$

EXERCISES 1.14   (page 83)

1. P = 60; R = 25%; B = 240 **3.** P = 108; R = 40%; B = 270 **5.** P = Unknown; R = 4%; B = 28,000 **7.** P = 21;
R = 60%; B = Unknown **9.** P = 2050; R = 6%; B = Unknown **11.** 10% **13.** 50% **15.** $1600 **17.** 112.8 **19.** 8.96 V
21. 440 lb, 352 lb active ingredients, 88 lb inert ingredients **23.** 850 gal, 35.7 gal butterfat **25.** 0.3 gr **27.** 25% **29.** 53.3%
31. 14.2% **33.** 11.9% **35.** $748.12 **37.** $2247.78 **39.** $3,347.79

UNIT 1C REVIEW   (page 88)

1. 1.625 **2.** $\frac{9}{20}$ **3.** 10.129 **4.** 116.935 **5.** 5.854 **6.** 25.6 ft **7.** 160.2 ft **8. a.** 45.1 **b.** 45.06 **9. a.** 45.1 **b.** 45.06
10. 0.11515 **11.** 18.85 **12.** 6 cables; 2 in. left **13.** 0.25 **14.** 72.4% **15.** 69.3 **16.** 2000 **17.** 40% **18.** $17.49

CHAPTER 1 REVIEW   (page 88)

1. 8243 **2.** 55,197 **3.** 9,178,000 **4.** 226 r 240 **5.** 3 **6.** 43 **7.** 37 **8.** 31 **9.** 340 cm^2 **10.** 30 cm^3 **11.** 10
12. 3000 **13.** No **14.** $2 \cdot 3 \cdot 3 \cdot 3$ **15.** $2 \cdot 3 \cdot 5 \cdot 11$ **16.** $\frac{9}{14}$ **17.** $\frac{5}{6}$ **18.** $4\frac{1}{6}$ **19.** $6\frac{3}{5}$ **20.** $\frac{21}{8}$ **21.** $\frac{55}{16}$ **22.** 2 **23.** $1\frac{1}{2}$
24. $\frac{103}{180}$ **25.** $14\frac{53}{84}$ **26.** $1\frac{19}{24}$ **27.** $11\frac{3}{5}$ **28.** 5 **29.** $\frac{1}{4}$ **30.** 18 **31.** $\frac{1}{16}$ **32.** $\frac{3}{8}$ **33.** $\frac{336}{25}$ or $13\frac{11}{25}$ **34.** A $3\frac{3}{8}$ in. B $5\frac{7}{16}$ in.
35. 105 **36.** 2016 **37.** 24 **38.** 63,360 yd **39.** 0.5625 **40.** $0.41\overline{6}$ **41.** $\frac{9}{20}$ **42.** $19\frac{5}{8}$ **43.** 168.278 **44.** 17.25 **45.** 68.665
46. 33.72 **47.** 3206.5 **48.** 1.9133 **49.** 3.18 **50.** 20.6 **51. a.** 200 **b.** 248.2 **c.** 250 **52. a.** 5.6 **b.** 5.65 **c.** 5.6491
53. 0.15 **54.** 0.0825 **55.** 6.5% **56.** 120% **57.** $1050 **58.** 38.1% **59.** 42.3% **60.** 48 tons

CHAPTER 2

EXERCISES 2.1   (page 96)

1. 3 **3.** 6 **5.** 4 **7.** 17 **9.** 15 **11.** 10 **13.** 7 **15.** −2 **17.** −12 **19.** 6 **21.** 12 **23.** 4 **25.** −14 **27.** −5 **29.** −12
31. −9 **33.** 4 **35.** −3 **37.** −1 **39.** −4 **41.** −6 **43.** −2 **45.** 4 **47.** 9 **49.** 3 **51.** 4 **53.** 7 **55.** −7 **57.** 5 **59.** −20
61. −2 **63.** 0 **65.** −5 **67.** −19 **69.** −12 **71.** 7 **73.** 11 **75.** 4 **77.** 4 **79.** −14

EXERCISES 2.2   (page 98)

1. −2 **3.** 11 **5.** 12 **7.** 6 **9.** 18 **11.** −6 **13.** −4 **15.** 0 **17.** 35 **19.** 14 **21.** 1 **23.** −10 **25.** −7 **27.** 15 **29.** −16
31. −10 **33.** −2 **35.** 14 **37.** 23 **39.** −15 **41.** 8 **43.** −4 **45.** 8 **47.** 2 **49.** −15 **51.** −3 **53.** −23 **55.** −4 **57.** −1
59. 2 **61.** −8 **63.** −10

EXERCISES 2.3   (page 100)

1. 24 **3.** −18 **5.** −35 **7.** 27 **9.** −72 **11.** 27 **13.** 0 **15.** 49 **17.** −300 **19.** −13 **21.** −6 **23.** 48 **25.** 21 **27.** −16
29. 54 **31.** −6 **33.** 24 **35.** 27 **37.** −63 **39.** −9 **41.** −6 **43.** 36 **45.** 30 **47.** −168 **49.** −162 **51.** 5 **53.** −9 **55.** 8
57. 2 **59.** 9 **61.** −4 **63.** 3 **65.** 17 **67.** 40 **69.** 15 **71.** 7 **73.** 4 **75.** −3 **77.** 10 **79.** 8

EXERCISES 2.4   (page 103)

1. $-\frac{3}{16}$ **3.** $\frac{1}{16}$ **5.** $-12\frac{3}{20}$ **7.** $\frac{13}{18}$ **9.** $-\frac{1}{20}$ **11.** $1\frac{1}{16}$ **13.** $5\frac{3}{4}$ **15.** $-1\frac{3}{4}$ **17.** $-\frac{1}{63}$ **19.** 6 **21.** $-\frac{9}{10}$ **23.** $\frac{7}{24}$ **25.** $-\frac{9}{20}$ **27.** $\frac{3}{8}$
29. $1\frac{1}{4}$ **31.** $3\frac{1}{4}$ **33.** $\frac{1}{20}$ **35.** $-3\frac{1}{3}$ **37.** −48 **39.** −1 **41.** $-1\frac{1}{8}$ **43.** $\frac{1}{4}$ **45.** $-\frac{1}{4}$

EXERCISES 2.5   (page 106)

1. 10^{13} **3.** 10^{12} **5.** $\frac{1}{10^{10}}$ **7.** 10^3 **9.** $\frac{1}{10^{12}}$ **11.** $\frac{1}{10^5}$ **13.** $\frac{1}{10^4}$ **15.** 10^{17} **17.** 10^6 **19.** 10^{14}

EXERCISES 2.6   (page 112)

1. 3.56×10^2 **3.** 6.348×10^2 **5.** 8.25×10^{-3} **7.** 7.4×10^0 **9.** 7.2×10^{-5} **11.** 7.1×10^5 **13.** 4.5×10^{-6} **15.** 3.4×10^{-8}
17. 6.4×10^5 **19.** 75,500 **21.** 5,310 **23.** 0.078 **25.** 0.000555 **27.** 64 **29.** 960 **31.** 5.76 **33.** 0.0000064
35. 50,000,000,000 **37.** 0.00000062 **39.** 2,500,000,000,000 **41.** 0.000000000033 **43.** 0.0048 **45.** 0.00091 **47.** 0.00037

49. 0.0613 **51.** 0.0009 **53.** 1.0009 **55.** 0.00000000998 **57.** 0.000271 **59.** 2.4×10^{-15} **61.** 3×10^{24} **63.** 1×10^{-6}
65. 1.728×10^{18} **67.** 1.1284×10^{-1} **69.** 1.11×10^{-25} **71.** 9×10^{-1} **73.** 6.67×10^{1} **75.** 7.455×10^{5} **77.** 1.1664×10^{10}
79. 9.06×10^{-11} **81.** 2.66×10^{24}

CHAPTER 2 REVIEW (page 113)

1. 5 **2.** 16 **3.** 13 **4.** 3 **5.** -8 **6.** -3 **7.** -13 **8.** -3 **9.** -11 **10.** 19 **11.** -2 **12.** 0 **13.** -19 **14.** -24 **15.** 36
16. 72 **17.** -84 **18.** 6 **19.** -6 **20.** 5 **21.** $-\frac{1}{42}$ **22.** $\frac{1}{12}$ **23.** $-3\frac{1}{8}$ **24.** $-\frac{3}{2}$ **25.** $\frac{1}{10^{5}}$ **26.** 10^{9} **27.** $\frac{1}{10^{12}}$ **28.** 1
29. 4.76×10^{5} **30.** 1.4×10^{-3} **31.** 0.0000535 **32.** 61,000,000 **33.** 0.00105 **34.** 0.06 **35.** 0.000075 **36.** 0.00183
37. 4.37×10^{-2} **38.** 2.8×10^{14} **39.** 2×10^{20} **40.** 2.025×10^{-15} **41.** 1.6×10^{37}

CHAPTER 3

EXERCISES 3.1 (page 120)

1. kilo **3.** centi **5.** milli **7.** mega **9.** h **11.** d **13.** c **15.** μ **17.** 65 mg **19.** 82 cm **21.** 36 μA **23.** 19 hL
25. 18 metres **27.** 36 kilograms **29.** 24 picoseconds **31.** 135 millilitres **33.** 45 milliamperes **35.** metre
37. ampere **39.** litre and cubic metre

EXERCISES 3.2 (page 122)

1. 1 metre **3.** 1 kilometre **5.** 1 centimetre **7.** 1000 **9.** 0.01 **11.** 0.001 **13.** 10 **15.** cm **17.** mm **19.** m **21.** km
23. mm **25.** mm **27.** km **29.** m **31.** cm **33.** mm; mm **35.** *A*: 52 mm; 5.2 cm; 0.052 m *B*: 11 mm; 1.1 cm; 0.011 m
C: 137 mm; 13.7 cm; 0.137 m *D*: 95 mm; 9.5 cm; 0.095 m *E*: 38 mm; 3.8 cm; 0.038 m *F*: 113 mm; 11.3 cm; 0.113 m
37. 52 mm; 5.2 cm; 0.052 m **39.** 79 mm; 7.9 cm; 0.079 m **41.** 65 mm; 6.5 cm; 0.065 m **43.** 102 mm; 10.2 cm; 0.102 m
45. 0.675 km **47.** 1.54 m **49.** 0.65 m **51.** 730 cm **53.** 12.5 cm **55.** Answers vary.

EXERCISES 3.3 (page 124)

1. 1 gram **3.** 1 kilogram **5.** 1 milligram **7.** 1000 **9.** 0.01 **11.** 1000 **13.** 1000 **15.** g **17.** kg **19.** g **21.** metric ton
23. mg **25.** g **27.** mg **29.** g **31.** kg **33.** metric tons **35.** 0.875 kg **37.** 85,000 mg **39.** 3600 g **41.** 0.27 g
43. 2500 kg **45.** Answers vary.

EXERCISES 3.4 (page 128)

1. 1 litre **3.** 1 cubic centimetre **5.** 1 square kilometre **7.** 1000 **9.** 1,000,000 **11.** 100 **13.** 1000 **15.** L **17.** m^2
19. cm^2 **21.** m^3 **23.** mL **25.** ha **27.** m^3 **29.** L **31.** m^2 **33.** ha **35.** cm^2 **37.** cm^2 **39.** 1.5 L **41.** 1,500,000 cm^3
43. 85 mL **45.** 0.085 km^2 **47.** 8.5 ha **49.** 500 g **51.** 0.675 ha

EXERCISES 3.5 (page 131)

1. 1 amp **3.** 1 second **5.** 1 megavolt **7.** 43 kW **9.** 17 ps **11.** 3.2 MW **13.** 450 Ω **15.** 1000 **17.** 0.000000001
19. 1,000,000 **21.** 0.000001 **23.** 350 mA **25.** 0.35 s **27.** 3 h 52 min 30 s **29.** 0.175 mF **31.** 1.5 MHz

EXERCISES 3.6 (page 133)

1. b **3.** c **5.** b **7.** c **9.** d **11.** 25 **13.** 617 **15.** 3.2 **17.** -26.7 **19.** -108.4

EXERCISES 3.7 (page 136)

1. 3.63 **3.** 15.0 **5.** 366 **7.** 66.0 **9.** 81.3 **11.** 46.6 mi **13.** 10.8 mm **15.** 30.3 L **17. a.** 38 oz **b.** 1.08 kg **19. a.** 300 ft
b. 91.4 m **21. a.** 12.7 cm **b.** 127 mm **23.** 2.51 m^2 **25.** 1260 ft^2 **27.** 116 cm^2 **29.** 1170 ft^2 **31.** 11.5 m^3
33. 279,000 mm^3 **35.** 2,380,000 cm^3 **37.** $2.25/$ft^2$; $405/frontage ft **39.** 34.4 acres **41.** 13.9 ha **43.** 0.619 acre
45. 80 acres **47.** 9397 lb/acre; 168 bu/acre **49.** 0.606 acre **51.** 363 lb/in^2 **53.** 29.1 m/s

CHAPTER 3 REVIEW (page 138)

1. milli **2.** kilo **3.** M **4.** μ **5.** 42 mL **6.** 8.3 ns **7.** 18 kilometres **8.** 350 milliamperes **9.** 50 microseconds **10.** 1 L
11. 1 MW **12.** 1 km^2 **13.** 1 m^3 **14.** 0.65 **15.** 0.75 **16.** 6100 **17.** 4.2×10^{6} **18.** 1.8×10^{7} **19.** 25,000 **20.** 25,000
21. 2.5 **22.** 6×10^{5} **23.** 250 **24.** 22.2 **25.** -13 **26.** 0 **27.** 100 **28.** 81.7 **29.** 38.4 **30.** 142 **31.** 1770 **32.** 162
33. 177 **34.** 6.07 **35.** c **36.** a **37.** d **38.** d **39.** b **40.** b **41.** a

CUMULATIVE REVIEW CHAPTERS 1–3   (page 140)

1. 72   **2.** 51 cm^2 **3.** no   **4.** $2 \cdot 3 \cdot 3 \cdot 5 \cdot 7$ **5.** $3\frac{5}{9}$ **6.** $\frac{13}{16}$ **7.** $1\frac{1}{16}$ **8.** $1\frac{7}{8}$ **9.** $\frac{1}{20}$ **10.** 83   **11. a.** 600   **b.** 615.3   **c.** 620
d. 615.288   **12.** 0.074   **13.** $3990   **14.** 662.5   **15.** 9.4%   **16.** 6.25%   **17.** 10   **18.** -432 **19.** $-2\frac{15}{16}$ **20.** $-\frac{25}{64}$
21. 3.1818×10^5 **22.** 0.00215   **23.** $\frac{1}{10^4}$ **24.** m   **25.** microsecond   **26.** 0.25   **27.** 0.02005   **28.** 30   **29.** 1050   **30.** c

CHAPTER 4

EXERCISES 4.1   (page 144)

1. 3   **3.** 4   **5.** 4   **7.** 4   **9.** 3   **11.** 4   **13.** 3   **15.** 3   **17.** 3   **19.** 4   **21.** 4   **23.** 5   **25.** 2   **27.** 6   **29.** 4   **31.** 6   **33.** 4   **35.** 2

EXERCISES 4.2   (page 148)

1. a. 0.01 A   **b.** 0.005A   **3. a.** 0.01 cm   **b.** 0.005 cm   **5. a.** 1 km   **b.** 0.5 km   **7. a.** 0.01 mi   **b.** 0.005 mi   **9. a.** 0.001 A
b. 0.0005 A   **11. a.** 0.0001 W   **b.** 0.00005 W   **13. a.** 10 Ω **b.** 5 Ω **15. a.** 1000 L   **b.** 500 L   **17. a.** 0.1 cm   **b.** 0.05 cm
19. a. 10 V   **b.** 5 V   **21. a.** 0.001 m   **b.** 0.0005 m   **23. a.** $\frac{1}{3}$ yd   **b.** $\frac{1}{6}$ yd   **25. a.** $\frac{1}{32}$ in.   **b.** $\frac{1}{64}$ in.   **27. a.** $\frac{1}{16}$ mi   **b.** $\frac{1}{32}$ mi
29. a. $\frac{1}{9}$ in^2 **b.** $\frac{1}{18}$ in^2

EXERCISES 4.3A   (page 151)

1. 27.20 mm   **3.** 63.55 mm   **5.** 8.00 mm   **7.** 115.90 mm   **9.** 71.45 mm   **11.** 34.60 mm   **13.** 68.45 mm   **15.** 5.90 mm
17. 43.60 mm   **19.** 76.10 mm

EXERCISES 4.3B   (page 155)

1. 1.362 in.   **3.** 2.695 in.   **5.** 0.234 in.   **7.** 1.715 in.   **9.** 2.997 in.   **11.** 1.071 in.   **13.** 2.503 in.   **15.** 0.317 in.   **17.** 4.563 in.
19. 2.812 in.

EXERCISES 4.4A   (page 158)

1. 4.25 mm   **3.** 3.90 mm   **5.** 1.75 mm   **7.** 7.77 mm   **9.** 5.81 mm   **11.** 10.28 mm   **13.** 7.17 mm   **15.** 8.75 mm   **17.** 6.23 mm
19. 5.42 mm

EXERCISES 4.4B   (page 161)

1. 0.237 in.   **3.** 0.314 in.   **5.** 0.147 in.   **7.** 0.820 in.   **9.** 0.502 in.   **11.** 0.200 in.   **13.** 0.321 in.   **15.** 0.170 in.   **17.** 0.658 in.
19. 0.245 in.

EXERCISES 4.5   (page 167)

1. a. 14.7   **b.** 0.017 in.   **3. a.** 16.01 mm   **b.** 0.737 mm   **5. a.** 0.0350 A   **b.** 0.00050 A   **7. a.** All have the same accuracy.
b. 0.391 cm   **9. a.** 205,000 Ω **b.** 205,000 Ω and 45,000 Ω **11. a.** 0.04 in.   **b.** 15.5 in.   **13. a.** 0.48 cm   **b.** 43.4 cm
15. a. 0.00008 A   **b.** 0.91 A   **17. a.** 0.6 m   **b.** All have same precision.   **19. a.** 500,000 Ω **b.** 500,000 Ω **21.** 18.1 m
23. 94.8 cm   **25.** 97,000 W   **27.** 840,000 V   **29.** 19 V   **31.** 45.9 cm or 459 mm   **33.** 126.4 cm   **35.** 8600 mi
37. 35 mm or 3.5 cm   **39.** 65.4 g   **41.** 0.330 in.   **43.** 26.0 mm

EXERCISES 4.6   (page 169)

1. 4400 m^2 **3.** 1,230,000 cm^2 **5.** 901 m^2 **7.** 0.13 ΩA   **9.** 7360 cm^3 **11.** 4.7×10^9 m^3 **13.** 35 A$^2\,\Omega$ **15.** 2500 in^2
17. 40 m   **19.** 340 V/A   **21.** 2.1 km/s   **23.** $3\overline{0}0$ V^2/Ω **25.** 4.0 g/cm^3 **27.** $9\overline{0}0$ ft^3 **29.** 28 m   **31.** 28.8 hp   **33.** $1\overline{0}00$ m^3
35. 270 ft^3/min

EXERCISES 4.7   (page 173)

1. 100 lb; 50 lb; 0.0357; 3.57%   **3.** 1 rpm; 0.5 rpm; 0.000571; 0.06%   **5.** 0.001 g; 0.0005 g; 0.00588; 0.59%   **7.** 1 g; 0.5 g; 0.25;
25%   **9.** 0.01 g; 0.005 g; 0.00225; 0.23%   **11.** 0.01 kg; 0.005 kg; 0.005; 0.5%   **13.** 0.001 A; 0.0005 A; 0.0122; 1.22%
15. $\frac{1}{8}$ in.; $\frac{1}{16}$ in.; 0.00526; 0.53%   **17.** 1 in.; 0.5 in.; 0.00329; 0.33%   **19.** 13.5 cm   **21.** 19.7 g

	Upper limit	Lower limit	Tolerance interval
25.	$6\frac{21}{32}$ in.	$6\frac{19}{32}$ in.	$\frac{1}{16}$ in.
27.	$3\frac{29}{64}$ in.	$3\frac{27}{64}$ in.	$\frac{1}{32}$ in.
29.	$3\frac{25}{128}$ in.	$3\frac{23}{128}$ in.	$\frac{1}{64}$ in.
31.	1.24 cm	1.14 cm	0.10 cm
33.	0.0185 A	0.0175 A	0.0010 A
35.	26,000 V	22,000 V	4000 V
37.	10.36 km	10.26 km	0.10 km

39. $53,075   **41.** $1,567,020.60

EXERCISES 4.8   (page 181)

1. 314,830 ft^3 **3.** 571,081 ft^3 **5.** 616,284 ft^3 **7.** $34.13   **9.** $153.81   **11.** 3471 kWh   **13.** 7568 kWh   **15.** +1.16 mm
17. −0.67 mm   **19.** −3.40 mm   **21.** +1.74 mm   **23.** −4.08 mm   **25.** −0.056 in.   **27.** +0.231 in.   **29.** −0.188 in.
31. +0.437 in.   **33.** 6 V   **35.** 6.4 V   **37.** 0.4 V   **39.** 1.4 V   **41.** 40 V   **43.** 230 V   **45.** 7 Ω   **47.** 12 Ω   **49.** 11 Ω   **51.** 35 Ω
53. 85 Ω   **55.** 300 Ω

CHAPTER 4 REVIEW   (page 185)

1. 3   **2.** 2   **3.** 3   **4.** 3   **5.** 2   **6.** 3   **7.** 4   **8.** 5   **9. a.** 0.01 m   **b.** 0.005 m   **10. a.** 0.1 mi   **b.** 0.05 mi   **11. a.** 100 L
b. 50 L   **12. a.** 100 V   **b.** 50 V   **13. a.** 0.01 cm   **b.** 0.005 cm   **14. a.** 10,000 V   **b.** 5000 V   **15. a.** $\frac{1}{8}$ in.   **b.** $\frac{1}{16}$ in.
16. a. $\frac{1}{16}$ mi   **b.** $\frac{1}{32}$ mi   **17.** 42.40 mm   **18.** 1.671 in.   **19.** 11.84 mm   **20.** 0.438 in.   **21. a.** 36,500 V   **b.** 9.6 V
22. a. 0.0005 A   **b.** 0.425 A   **23.** 720,000 W   **24.** $4\bar{0}0$ m   **25.** 400,000 V   **26.** 1900 cm^3 **27.** 5.88 m^2 **28.** 1.4 N/m^2
29. 130 V^2/Ω   **30.** 0.0057; 0.57%   **31.** 0.00032; 0.03%   **32.** 2200 Ω; 1800 Ω   **33.** 947,602 ft^3 **34.** −0.563 in.   **35.** 8.4 V
36. 2.6 Ω

CHAPTER 5

EXERCISES 5.1   (page 190)

1. 83   **3.** 8   **5.** 42   **7.** −6   **9.** −128   **11.** 1   **13.** −5   **15.** 3   **17.** −24   **19.** $-9\frac{3}{5}$ or $-\frac{48}{5}$ **21.** −13   **23.** $-1\frac{3}{5}$ or $-\frac{8}{5}$
25. 189   **27.** 331,776   **29.** −72   **31.** 60   **33.** 8   **35.** −78   **37.** −1   **39.** 25

EXERCISES 5.2   (page 193)

1. $a + b + c$ **3.** $a + b + c$ **5.** $a - b - c$ **7.** $x + y - z + 3$ **9.** $x - y - z - 3$ **11.** $2x + 4 + 3y + 4r$
13. $3x - 5y + 6z - 2w + 11$ **15.** $-5x - 3y - 6z + 3w + 3$ **17.** $2x + 3y - z - w + 3r - 2s - 10$
19. $-2x + 3y - z - 4w - 4r + s$ **21.** $5b$ **23.** $3x^2 + 10x$ **25.** $3m$ **27.** $a + 12b$ **29.** $6a^2 - a + 1$ **31.** $3x^2 + 3x$
33. $-1.8x$ **35.** $\frac{8}{9}x - \frac{1}{8}y$ **37.** $2x^2y - 2xy + 2y^2 - 3x^2$ **39.** $5x^3 + 3x^2y - 5y^3 + y$ **41.** 1   **43.** $3x + 4$ **45.** $5 - x$
47. $3y - 7$ **49.** $4y + 5$ **51.** 5   **53.** 6   **55.** 28   **57.** $\frac{5}{4}x - \frac{8}{3}$ **59.** $12x + 36y$ **61.** $-36x^2 + 48y^2$ **63.** $2x + 19$
65. $-8.5y - 4$ **67.** −42   **69.** $4n + 8$ **71.** $1.8x - 7$ **73.** $-6n - 2$ **75.** $-5x + 6$ **77.** $-1.05x - 8.4$

EXERCISES 5.3   (page 197)

1. Binomial   **3.** Binomial   **5.** Monomial   **7.** Trinomial   **9.** Binomial   **11.** $x^2 - x + 1$; 2nd   **13.** $7x^2 + 4x - 1$; 2nd
15. $5x^3 - 4x^2 - 2$; 3rd   **17.** $4y^3 - 6y^2 - 3y + 7$; 3rd   **19.** $-7x^5 - 4x^4 + x^3 + 2x^2 + 5x - 3$; 5th   **21.** $2a$ **23.** $2x^2 - 2x$
25. $8y^2 - 5y + 6$ **27.** $4a^3 - 3a^2 - 5a$ **29.** $7a^2 - 10a + 1$ **31.** $9x^2 - 5x$ **33.** $9a^3 + 4a^2 + 5a - 5$ **35.** $4x + 4$
37. $5x^2 + 18x - 22$ **39.** $5x^3 + 13x^2 - 8x + 7$ **41.** $a + b$ **43.** $16m - 58n + 45$ **45.** $-14a^2 - 11a - 3$ **47.** $6y^2 + 48y - 15$
49. $2x^2 + 6x + 2$ **51.** $-3x^2 + 2x + 2$ **53.** $x^3 + 3x - 3$ **55.** $8x^5 - 18x^4 + 5x^2 + 1$ **57.** $a + 3b$ **59.** $7a - 4b - 3x + 4y$
61. $x^2 - 3x + 5$ **63.** $7w^2 - 24w - 6$ **65.** $x^2 - 4x - 10$ **67.** $4x^2 - 4x + 12$ **69.** $-6x^2 + 2x - 10$

EXERCISES 5.4   (page 200)

1. $-15a$ **3.** $28a^3$ **5.** $54m^4$ **7.** $32a^8$ **9.** $-26p^2q$ **11.** $30n^3m$ **13.** $21a^4b$ **15.** $\frac{3}{8}x^3y^4$ **17.** $24a^3b^4c^3$ **19.** $\frac{3}{16}x^3y^5z^3$
21. $-371.64m^3n^2p^2$ **23.** $-255a^6b$ **25.** x^6 **27.** x^{24} **29.** $9x^8$ **31.** $-x^9$ **33.** x^{10} **35.** x^{30} **37.** $25x^6y^4$ **39.** $225m^4$
41. $15,625n^{12}$ **43.** $9x^{12}$ **45.** $8x^9y^{12}z^3$ **47.** $-32h^{15}k^{30}m^{10}$ **49.** −408   **51.** −144   **53.** 17,712   **55.** 16   **57.** −1728
59. 291,600   **61.** −1080   **63.** 324   **65.** −324   **67.** 16

EXERCISES 5.5   (page 202)

1. $4a + 24$ **3.** $-18x^2 - 12y$ **5.** $4ax^2 - 6ay + a$ **7.** $3x^3 - 2x^2 + 5x$ **9.** $6a^3 + 12a^2 - 20a$ **11.** $-12x^3 + 21x^2 + 6x$
13. $-28x^3 - 12xy + 8x^2y$ **15.** $3x^3y^2 - 3x^2y^3 + 12x^2y^2$ **17.** $-6x^3 + 36x^5 - 54x^7$ **19.** $5a^4b^2 - 5ab^5 - 5a^2b^3$
21. $\frac{28}{3}mn - 8m^2$ **23.** $16y^2z^3 - \frac{8}{35}yz^4$ **25.** $-5.2a^6 - 10a^3 - 4a$ **27.** $1334.4a^3 + 1668a^2$ **29.** $24x^4y - 16x^3y^2 + 20x^2y^3$
31. $\frac{1}{2}a^3b^3 - \frac{1}{3}a^2b^5 + \frac{5}{9}ab^6$ **33.** $-19x^2 + 8x$ **35.** $3x^2y - 2x^2y^2 + 2x^2y^3 - 7xy^3$ **37.** $x^2 + 7x + 6$ **39.** $x^2 + 5x - 14$
41. $x^2 - 13x + 40$ **43.** $3a^2 - 17a + 20$ **45.** $12a^2 - 10a - 12$ **47.** $24a^2 + 84a + 72$ **49.** $15x^2 - 4xy - 4y^2$
51. $4x^2 - 12x + 9$ **53.** $4c^2 - 25d^2$ **55.** $91m^2 + 32m - 3$ **57.** $x^8 - 2x^5 + x^2$ **59.** $10y^3 - 24y^2 - 32y + 16$
61. $24x^2 - 78x - 6y^2 - 39y$ **63.** $g^2 - 3g - h^2 + 9h - 18$ **65.** $10x^7 - 3x^6 - x^5 + 44x^4 + x^3 - x^2 + 16x - 2$

EXERCISES 5.6   (page 204)

1. $3x^2$ **3.** $\frac{3x^8}{2}$ **5.** $\frac{6}{x^2}$ **7.** $\frac{2x}{3}$ **9.** x **11.** $\frac{5}{2}$ **13.** $\frac{5a^2}{b}$ **15.** $\frac{8}{mn}$ **17.** 0 **19.** $23p^2$ **21.** -2 **23.** $\frac{36}{r^2}$ **25.** $\frac{-23x^2}{7y^2}$
27. $\frac{8}{7a^3b^2}$ **29.** $\frac{4x^2z^4}{9y}$ **31.** $2x^2 - 4x + 3$ **33.** $x^2 + x + 1$ **35.** $x - y - z$ **37.** $3a^4 - 2a^2 - a$ **39.** $b^9 - b^6 - b^3$
41. $-x^3 + x^2 - x + 4$ **43.** $12x + 6x^2y - 3$ **45.** $8x^3z - 6x^2yz^2 - 4y^2$ **47.** $4y^3 - 3y - \frac{2}{y}$ **49.** $\frac{1}{2x^2} - 3 - 2x^2$

EXERCISES 5.7   (page 207)

1. $x + 2$ **3.** $3a + 3$ r 11 **5.** $4x - 3$ r -3 **7.** $y + 2$ r -3 **9.** $3b - 4$ **11.** $6x^2 + x - 1$ **13.** $4x^2 + 7x - 15$ r 5
15. $2x^2 - 2x - 12$ **17.** $2x^2 - 16x + 32$ **19.** $3x^2 + 10x + 20$ r 34 **21.** $2x^2 + 6x + 30$ r 170 **23.** $2x^3 + 4x + 6$
25. $4x^2 - 2x + 1$ r -2 **27.** $x^3 - 2x^2 + 4x - 8$ **29.** $3x^2 - 4x + 1$

CHAPTER 5 REVIEW   (page 208)

1. a **2.** 0 **3.** 1 **4.** -2 **5.** 50 **6.** 2 **7.** 1 **8.** 9 **9.** -30 **10.** $-\frac{9}{2}$ **11.** -12 **12.** 5 **13.** $6y - 5$ **14.** $6 - 8x$
15. $10x + 27$ **16.** $3x^3 + x^2y - 3y^3 - y$ **17.** Binomial   **18.** 4 **19.** $8a^2 + 5a + 2$ **20.** $9x^3 + 4x^2 + x + 2$ **21.** $10x^2 - 7x + 4$
22. $24x^5$ **23.** $-56x^5y^3$ **24.** $27x^6$ **25.** $15a^2 + 20ab$ **26.** $-32x^2 + 8x^3 - 12x^4$ **27.** $15x^2 - 11x - 12$
28. $6x^3 - 24x^2 + 26x - 4$ **29.** $\frac{7}{x}$ **30.** $5x^2$ **31.** $4a^2 - 3a + 1$ **32.** $3x - 4$ **33.** $3x^2 - 4x + 2$

CHAPTER 6

EXERCISES 6.1   (page 215)

1. 6 **3.** 17 **5.** $10\frac{1}{2}$ **7.** 19.5 **9.** 5.2 **11.** 301 **13.** 7 **15.** 20 **17.** $4\frac{1}{14}$ or $\frac{57}{14}$ **19.** 0 **21.** 0 **23.** 16 **25.** 392 **27.** -4
29. -2 **31.** 2 **33.** -2 **35.** 32 **37.** -4 **39.** 18 **41.** $5\frac{2}{3}$ or $\frac{17}{3}$ **43.** 42 **45.** $10\frac{1}{2}$ or $\frac{21}{2}$ **47.** $-\frac{1}{3}$ **49.** -1 **51.** $\frac{77}{135}$ or 0.570
53. $1\frac{5}{7}$ or $\frac{12}{7}$ **55.** $2\frac{12}{29}$ or $\frac{70}{29}$ **57.** $-3\frac{1}{3}$ or $-\frac{10}{3}$ **59.** 25

EXERCISES 6.2   (page 217)

1. 8 **3.** 4 **5.** 4 **7.** 3 **9.** 6 **11.** 6 **13.** -5 **15.** 7 **17.** $2\frac{1}{5}$ **19.** -9 **21.** 3 **23.** 13 **25.** $\frac{3}{4}$ **27.** -1 **29.** 4

EXERCISES 6.3   (page 219)

1. 5 **3.** $\frac{2}{5}$ **5.** $-\frac{4}{3}$ or $-1\frac{1}{3}$ **7.** 6 **9.** -5 **11.** $\frac{36}{5}$ or $7\frac{1}{5}$ **13.** -5 **15.** 0 **17.** $\frac{8}{7}$ or $1\frac{1}{7}$ **19.** 9 **21.** 14 **23.** 45 **25.** 4 **27.** 22
29. -1 **31.** 10 **33.** 18 **35.** 2 **37.** -8 **39.** -2 **41.** -1 **43.** -7 **45.** $\frac{33}{4}$ or $8\frac{1}{4}$ **47.** $-\frac{4}{3}$ or $-1\frac{1}{3}$ **49.** -3 **51.** 6
53. -2 **55.** 0 **57.** $\frac{19}{24}$ **59.** $\frac{273}{44}$ or $6\frac{9}{44}$

EXERCISES 6.4   (page 224)

1. 8 **3.** 5 **5.** 3 **7.** 16 **9.** 1 **11.** 50 **13.** 6 **15.** 1 **17.** 24 **19.** 4 **21.** -56 **23.** 6 **25.** -3 **27.** $\frac{2}{3}$ **29.** $-\frac{1}{2}$ **31.** 1
33. $\frac{1}{5}$ **35.** 3 **37.** $\frac{2}{3}$ **39.** 5 **41.** $\frac{1}{7}$

EXERCISES 6.5   (page 225)

1. $x - 20$ **3.** $\frac{x}{6}$ **5.** $x + 16$ **7.** $26 - x$ **9.** $2x$ **11.** $6x + 28 = 40$ **13.** $\frac{x}{6} = 5$ **15.** $5(x + 28) = 150$ **17.** $\frac{x}{6} - 7 = 2$
19. $30 - 2x = 4$ **21.** $(x - 7)(x + 5) = 13$ **23.** $6x - 17 = 7$ **25.** $4x - 17 = 63$

EXERCISES 6.6 (page 229)

1. 10 in. **3.** 92 incandescent and 164 fluorescent bulbs **5.** $825 to John; $1650 to Maria; $2475 to Betsy **7.** 10 cm by 20 cm
9. 105 ft by 120 ft **11.** 42 ft; 42 ft; 38 ft **13.** $5\frac{1}{4}$ ft, $6\frac{3}{4}$ ft **15.** 8 @ $6.50, 12 @ $9.50 **17.** $4500 @ 6%, $3000 @ 4% **19.** 20 L
21. 320 mL of 30%, 480 mL of 80% **23.** 4 qt

EXERCISES 6.7 (page 233)

1. $\dfrac{E}{I}$ **3.** $\dfrac{F}{m}$ **5.** $\dfrac{C}{\pi}$ **7.** $\dfrac{V}{lh}$ **9.** $\dfrac{A}{2\pi r}$ **11.** $\dfrac{v^2}{2g}$ **13.** $\dfrac{Q}{I}$ **15.** vt **17.** $\dfrac{V}{I}$ **19.** $4E\pi r^2$ **21.** $\dfrac{1}{2\pi X_c C}$ **23.** $\dfrac{2A}{h}$ **25.** $\dfrac{QJ}{I^2 t}$
27. $\dfrac{5}{9}(F-32)$ or $\dfrac{5F-160}{9}$ **29.** $C_T - C_1 - C_3 - C_4$ **31.** $\dfrac{-By-C}{A}$ **33.** $\dfrac{Q_1 + PQ_1}{P}$ or $\dfrac{Q_1}{P} + Q_1$ **35.** $\dfrac{2A}{a+b}$
37. $\dfrac{l-a}{n-1}$ **39.** $\dfrac{Ft}{V_2 - V_1}$ **41.** $\dfrac{Q}{w(T_1 - T_2)}$ **43.** $\dfrac{PV}{2\pi} - 3960$

EXERCISES 6.8 (page 235)

1. a. $l = \dfrac{A}{w}$ **b.** 23.0 **3. a.** $h = \dfrac{3V}{\pi r^2}$ **b.** 20.0 **5. a.** $m = \dfrac{2E}{v^2}$ **b.** $20\overline{0}0$ **7. a.** $t = \dfrac{v_f - v_i}{a}$ **b.** 9.90 **9. a.** $h = \dfrac{v_f^2 - v_i^2}{2g}$
b. 576 **11. a.** $r_1 = \dfrac{L-2d}{\pi} - r_2$ **b.** 3.00 **13. a.** $r = \dfrac{Wv^2}{Fg}$ **b.** 1900 **15. a.** $b = \dfrac{2A}{h} - a$ **b.** 49.0 **17. a.** $d = \dfrac{2V}{lw} - D$
b. 3.00 **19. a.** $a = \dfrac{2S}{n} - l$ **b.** 16.6 **21.** 324 W **23.** 6.72 ft **25.** 12.1 m **27.** 15.0 Ω **29.** 0.43 in.

EXERCISES 6.9 (page 240)

1. 4.80 Ω **3.** 18.0 Ω **5.** 40.0 Ω **7.** 2.50 cm **9.** 44.5 cm **11.** 9.00 Ω **13.** 318 Ω **15.** 3240 Ω **17.** 6.00 μF
19. 1.91×10^{-6} F **21.** 1.74×10^{-8} F **23.** 219 Ω

CHAPTER 6 REVIEW (page 241)

1. $\frac{3}{2}$ or $1\frac{1}{2}$ **2.** -4 **3.** 57 **4.** 24 **5.** -7 **6.** -9 **7.** 8 **8.** 3 **9.** 6 **10.** $\frac{3}{5}$ **11.** 1 **12.** 6 **13.** 4 **14.** -2 **15.** 6 **16.** $\frac{1}{2}$
17. $\frac{8}{3}$ or $2\frac{2}{3}$ **18.** $-\frac{15}{2}$ or $-7\frac{1}{2}$ **19.** 26 **20.** 2 **21.** 5 **22.** 6 in. by 18 in. **23.** 4.5 L of 100%, 7.5 L of 60% **24.** $\dfrac{F}{W}$ **25.** $\dfrac{W}{P}$
26. $2L - 2A - 2B$ **27.** $\dfrac{2k}{v^2}$ **28.** $\dfrac{P_1 T_2}{P_2}$ **29.** $2v - v_f$ **30.** -144 **31.** 19.5 **32.** 37 **33.** 30.0 Ω **34.** 70.6 μF

CUMULATIVE REVIEW CHAPTERS 1–6 (page 243)

1. no **2.** $2 \cdot 2 \cdot 2 \cdot 3 \cdot 29$ **3. a.** 6 **b.** 28 **4.** $\frac{7}{13}$ **5.** $1\frac{5}{11}$ **6.** 0.0003015 **7.** 8.1% **8.** $13.93 **9.** 50,000 m^2 **10.** 2
11. 1 **12.** 5 **13. a.** 0.01 in. **b.** 0.005 in. **14.** 55.60 **15.** 0.428 in. **16. a.** 14,200 s **b.** 1.2 s **17.** 494,000 W
18. $-8x + 8y$ **19.** $2y^3 + 4y^2 + 5y - 11$ **20.** $27y^9$ **21.** $-2x^3 + 6x^2 - 8x$ **22.** $12y^4 - 16y^3 + 3y^2 + 5y - 2$
23. $20x^2 - 7xy - 6y^2$ **24.** $x^2 - 3x + 4 - \dfrac{40}{x+5}$ **25.** $\frac{7}{2}$ **26.** 56 **27.** -6 **28.** $\frac{11}{2}$ **29.** $a = 2C - b - c$ **30.** 11.1

CHAPTER 7

EXERCISES 7.1 (page 247)

1. $\frac{1}{5}$ **3.** $\frac{1}{3}$ **5.** $\frac{5}{3}$ **7.** $\frac{1}{5}$ **9.** $\frac{2}{1}$ or 2 **11.** $\frac{1}{32}$ **13.** $\frac{9}{14}$ **15.** $\frac{11}{16}$ **17.** $\frac{2}{1}$ or 2 **19.** $\frac{4}{1}$ or 4 **21.** $\frac{7}{5}$ **23.** $\frac{1}{4}$ **25.** 30 mi/gal
27. 46 gal/h **29.** 50 mi/h **31.** $\frac{3}{8}$ lb/gal **33.** $\frac{32}{5}$ or 6.4 to 1 **35.** 33 gal/min **37.** $\frac{1}{275}$ **39.** $\frac{12}{1}$ or 12 **41.** 36 lb/bu
43. 25 gal/acre **45.** 85¢/ft **47.** $42/ft^2 **49.** 50 mg/cm^3 **51.** 4 mg/cm^3 **53.** $\frac{7}{5}$ or 1.4 to 1 **55.** $\frac{11}{6}$ **57.** 50 drops/min
59. 45 drops/min **61.** $3\frac{1}{3}$ h **63.** $8\frac{1}{3}$ h

EXERCISES 7.2 (page 252)

1. a. 2, 3 **b.** 1, 6 **c.** 6 **d.** 6 **3. a.** 9, 28 **b.** 7, 36 **c.** 252 **d.** 252 **5. a.** 7, w **b.** x, z **c.** $7w$ **d.** xz **7.** yes; 30 = 30
9. no; 60 ≠ 90 **11.** yes; 12 = 12 **13.** 3 **15.** $5\frac{3}{5}$ **17.** 14 **19.** $3\frac{1}{3}$ **21.** 35 **23.** -7.5 **25.** 0.28 **27.** 0.5 **29.** 126 **31.** 20.6
33. 37.3 **35.** 38.2 **37.** 818 **39.** 9050 **41.** 21.3 **43.** 44 ft^3 **45.** $98,000 **47.** $144 **49.** $8500 **51.** 52.5 lb **53.** 108 lb
55. 1250 ft **57.** 595 turns **59.** 177 hp **61.** 21 lb **63.** 14 cm^3 **65.** 6 cm^3 **67.** 23.3% **69.** 5.9% **71. a.** 18% **b.** 117 lb
73. 19 hL **75. a.** 13.3% **b.** 33.3% **c.** 53.3% **77.** 4.5 mL **79.** 5 mL **81.** 1.5 mL **83.** 30 mL

EXERCISES 7.3   (page 258)

For all answers the allowable error is ±8 mi or ±8 km.

1. 48 mi   **3.** 60 mi   **5.** 86 mi   **7.** 104 mi   **9.** 96 mi   **11.** 527 km   **13.** 298 km   **15.** 323 km   **17.** 12.5 cm   **19.** 100 cm²
21. 1 cm × 1 cm   **23.** 6.5 cm   **25.** 6.5 cm

27.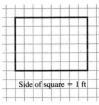

29.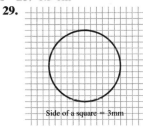

31. Yes; no   **33.** 20 : 1   **35.** 9 : 1   **37.** 68 lb   **39.** 238,500 lb
41. 81 : 1   **43.** 25 : 1   **45.** 400 lb   **47.** 400 lb   **49.** 640 lb

EXERCISES 7.4   (page 262)

1. 36 rpm   **3.** 80 rpm   **5.** 22 cm   **7.** 31 in.   **9.** 520 rpm   **11.** 50 in.   **13.** 11 in.   **15.** 10 cm; 17 cm   **17.** 160 teeth
19. 1008 rpm   **21.** 576 teeth   **23.** 96 rpm   **25.** 144 rpm   **27.** 40 teeth   **29.** 10 in.   **31.** 120 lb   **33.** 135 lb
35. $\frac{1}{2}$ ton or 1000 lb   **37.** 64 cm   **39.** $25\frac{1}{3}$ in.

CHAPTER 7 REVIEW   (page 263)

1. $\frac{1}{4}$ **2.** $\frac{3}{2}$ **3.** $\frac{2}{1}$ **4.** $\frac{11}{18}$ **5.** Yes   **6.** No   **7.** 1   **8.** 30   **9.** 24   **10.** 40   **11.** 106   **12.** 41.3   **13.** 788   **14.** 529   **15.** $187.50
16. 1250 ft   **17.** 216 h   **18.** 300 lb   **19.** Jones 58%; Hernandez 42%   **20.** 8.8%   **21.** direct   **22.** direct   **23.** inverse
24. 362.5 mi   **25.** 52.8 in.   **26.** 562.5 rpm   **27.** 75 rpm   **28.** 50 kg   **29.** 180 lb   **30.** 2 A   **31.** 108 h

CHAPTER 8

EXERCISES 8.1   (page 271)

1. $(3, 2)(8, -3)(-2, 7)$ **3.** $(2, -1)(0, 5)(-2, 11)$ **5.** $(0, -2)\left(2, -\frac{1}{2}\right)(-4, -5)$ **7.** $(5, 4)(0, 2)\left(-3, \frac{4}{5}\right)$
9. $(2, 4)(0, -5)(-4, -23)$ **11.** $(2, 10)(0, 4)(-3, -5)$ **13.** $(2, -3)(0, 7)(-4, 27)$ **15.** $(3, 10)(0, 4)(-1, 2)$
17. $(4, 14)(0, 4)(-2, -1)$ **19.** $(1, 2)(0, -\frac{5}{2})(-3, -16)$ **21.** $(2, 3)(0, 3)(-4, 3)$ **23.** $(5, 4)(5, 0)(5, -2)$ **25.** $y = \frac{6 - 2x}{3}$
27. $y = \frac{7 - x}{2}$ **29.** $y = \frac{x - 6}{2}$ **31.** $y = \frac{2x - 9}{3}$ **33.** $y = \frac{2x + 6}{3}$ **35.** $y = \frac{-2x + 15}{3}$ **37.** $A(-2, 2)$ **39.** $C(5, -1)$ **41.** $E(-4, -5)$
43. $G(0, 5)$ **45.** $I(4, 2)$ **47–65.**

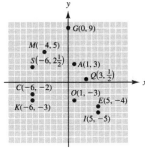

EXERCISES 8.2   (page 277)

1.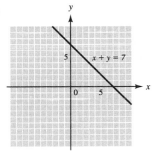

3.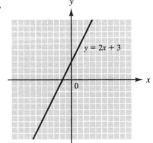

5.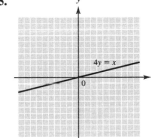

7.
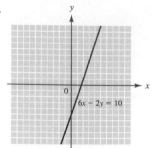
$6x - 2y = 10$

9.
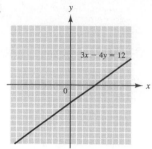
$3x - 4y = 12$

11.
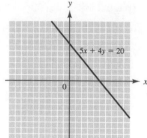
$5x + 4y = 20$

13.
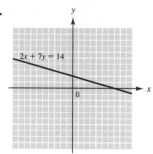
$2x + 7y = 14$

15.
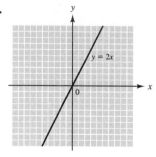
$y = 2x$

17.
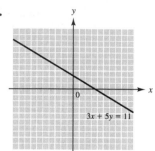
$3x + 5y = 11$

19.
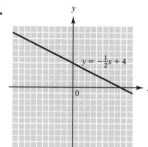
$y = -\frac{1}{2}x + 4$

21.
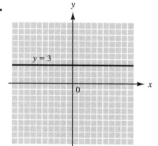
$y = 3$

23.
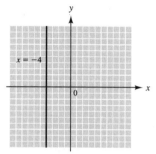
$x = -4$

25.
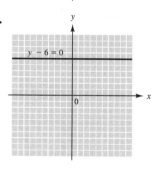
$y - 6 = 0$

27.
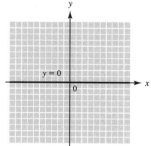
$x + 3\frac{1}{2} = 0$

29.
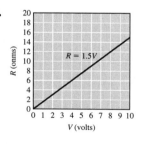
$y = 0$

	Independent	Dependent
31.	t	s
33.	V	R
35.	t	i
37.	t	v
39.	t	s

41.
$s = 5t + 10$

43.
$R = 1.5V$

45.
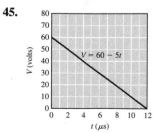
$V = 60 - 5t$

EXERCISES 8.3 (page 283)

1. 1 **3.** 6 **5.** $-\frac{5}{4}$ **7.** 0 **9.** undefined **11.** $\frac{3}{5}$ **13.** -2 **15.** $\frac{3}{7}$ **17.** 0 **19.** 6 **21.** -5 **23.** $-\frac{3}{5}$ **25.** $\frac{1}{4}$ **27.** $\frac{5}{2}$

29. undefined **31.** parallel **33.** perpendicular **35.** perpendicular **37.** neither **39.** parallel

EXERCISES 8.4 (page 288)

1.
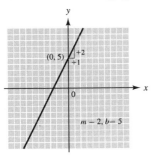

3.
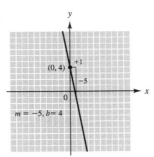

5.
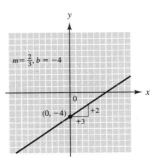

7.
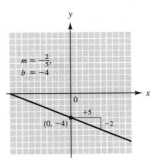

9.
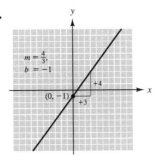

11.
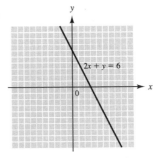

13.
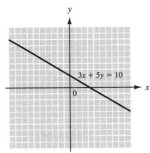

15.
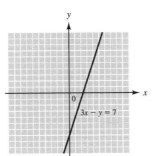

17.
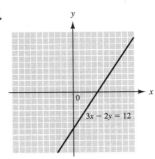

19.
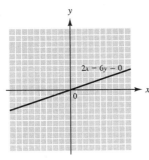

21. $y = 2x + 5$ **23.** $y = -5x + 4$ **25.** $y = \frac{2}{3}x - 4$ or $2x - 3y = 12$ **27.** $y = -\frac{6}{5}x + 3$ or $6x + 5y = 15$ **29.** $y = -\frac{3}{5}x$ or $3x + 5y = 0$

31.
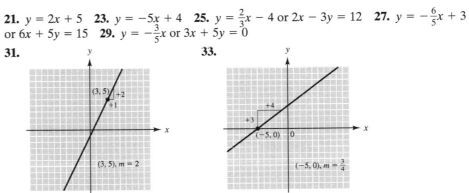

33.

35.
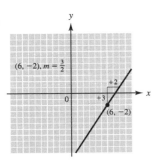

37.

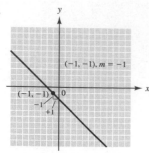

$(-1, -1), m = -1$

39.

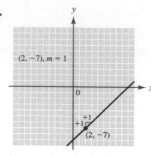

$(2, -7), m = 1$

41. $2x - y = 1$ **43.** $3x - 4y = -15$
45. $3x - 2y = 22$ **47.** $10x + 3y = -33$
49. $x + y = 2$ **51.** $2x + 3y = 13$ **53.** $2x + y = 2$
55. $x - 3y = -3$ **57.** $x - 2y = -1$
59. $x + y = 6$

CHAPTER 8 REVIEW   (page 289)

1. $(3, \frac{5}{2})(0, 4)(-4, 6)$ **2.** $(3, -2)(0, -4)(-3, -6)$ **3.** $y = -6x + 15$ **4.** $y = \frac{3x + 10}{5}$ or $y = \frac{3}{5}x + 2$ **5.** A $(3, 5)$
6. B $(-2, -6)$ **7.** C $(2, -1)$ **8.** D $(4, 0)$ **9–12.**

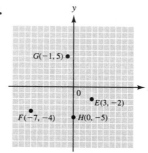

13.

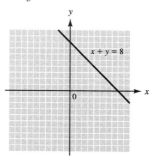

14.

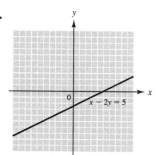

15.

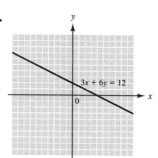

16.

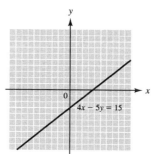

17.

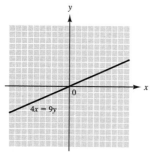

18.

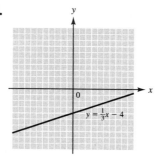

19.

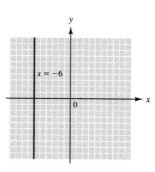

20.

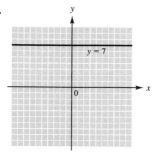

21. $\frac{9}{7}$ **22.** 1 **23.** 4 **24.** $-\frac{2}{5}$ **25.** $\frac{5}{9}$ **26.** perpendicular   **27.** neither   **28.** parallel
29. perpendicular   **30.**

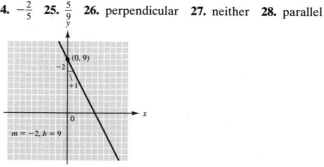

31.

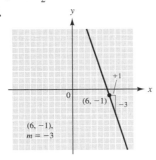

$m = -\frac{2}{3}$,
$b = -5$
0
$+3$
$(0, -5)$ -2

32.

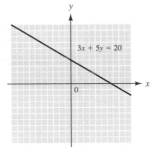

$3x + 5y = 20$
0

33.

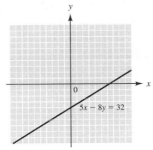

0
$5x - 8y = 32$

34. $y = -\frac{1}{2}x + 3$ or $x + 2y = 6$ **35.** $y = \frac{8}{3}x$ or $8x - 3y = 0$ **36.** $y = 0$

37.

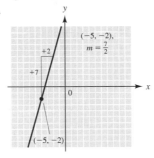

$+1$
0
$(6, -1)$ -3
$(6, -1)$,
$m = -3$

38.

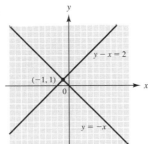

$(-5, -2)$,
$m = \frac{7}{2}$
$+2$
$+7$
0
$(-5, -2)$

39. $x + y = 6$ **40.** $x + 4y = -20$ **41.** $x - y = 5$
42. $x - 2y = 12$

CHAPTER 9

EXERCISES 9.1   (page 298)

1.

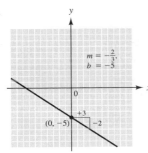

$(2, 6)$
$y = x + 4$
0
$y = 3x$

3.

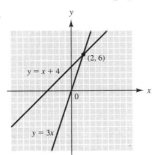

$y - x = 2$
$(-1, 1)$
0
$y = -x$

5.

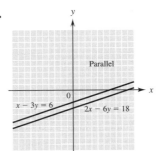

Parallel
0
$x - 3y = 6$
$2x - 6y = 18$

7.

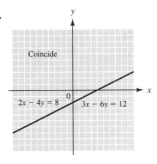

Coincide
0
$2x - 4y = 8$ $3x - 6y = 12$

9.

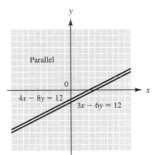

Parallel
0
$4x - 8y = 12$
$3x - 6y = 12$

11.

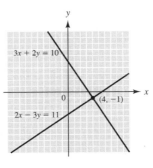

$3x + 2y = 10$
0
$(4, -1)$
$2x - 3y = 11$

13.

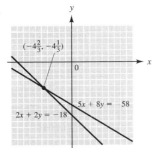

$\left(-4\frac{2}{3}, -4\frac{1}{3}\right)$
0
$5x + 8y = -58$
$2x + 2y = -18$

15.

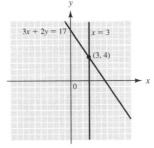

$3x + 2y = 17$ $x = 3$
$(3, 4)$
0

17.

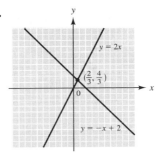

$y = 2x$
$\left(\frac{2}{3}, \frac{4}{3}\right)$
0
$y = -x + 2$

19.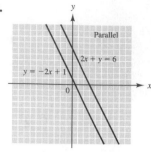

21.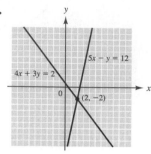

23.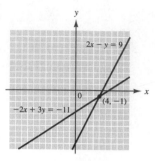

25.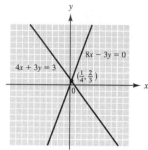

27.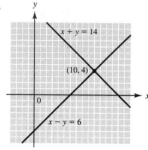

29. 4 ft³ concrete; 16 ft³ of gravel

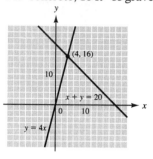

EXERCISES 9.2   (pages 302)

1. (2, 1)   **3.** (4, 2)   **5.** (−2, 7)   **7.** (3, 1)   **9.** (−5, 5)   **11.** (5, 1)   **13.** (1, − 2)   **15.** (−1, 4)   **17.** (7, −5)   **19.** (4, −1)
21. (−2, 5)   **23.** $\left(2, \frac{5}{4}\right)$ or $\left(2, 1\frac{1}{4}\right)$ **25.** (−2, 4)   **27.** coincide   **29.** (2, −1)   **31.** parallel   **33.** $\left(\frac{1}{2}, 5\right)$ **35.** coincide
37. parallel   **39.** (72, 30)   **41.** $\left(\frac{1}{4}, \frac{2}{5}\right)$ **43.** (9, −2)

EXERCISES 9.3   (pages 305)

1. 42 cm, 54 cm   **3.** $2\frac{1}{2}$ h @ 180 gal/h; $3\frac{1}{2}$ h @ 250 gal/h   **5.** 18 ft³ gravel; 4.5 ft³ cement   **7.** 30 lb @ 5%; 70 lb @ 15%
9. 2700 bu corn; 450 bu beans   **11.** 200 gal of 6%; 100 gal of 12%   **13.** 5 @ 3 V; 4 @ 4.5 V   **15.** 105 mL @ 8%; 35 mL @ 12%
17. 5 min @ 850 rpm; 9 min @ 1250 rpm   **19.** 2 h at setting 1; 3 h at setting 2   **21.** 160 L of 3% and 40 L of 8%   **23. a.** 6.8 A
b. 1.2 A   **25.** 5 h @ 140 cm³/h; 3 h @ 100 cm³/h   **27.** 31 of 2 cm³; 11 of 5 cm³   **29.** 4 one-bedroom and 9 two-bedroom

EXERCISES 9.4   (pages 309)

1. $\left(\frac{12}{5}, \frac{36}{5}\right)$ or (2.4, 7.2)   **3.** (10, 2)   **5.** (6, 6)   **7.** (4, 2)   **9.** (−3, −1)   **11.** (1, 4)   **13.** (−2, 2)   **15.** $\left(2\frac{1}{2}, -12\frac{1}{2}\right)$ **17.** (1, 3)
19. (9, −2)   **21.** 100 Ω, 450 Ω   **23.** 30 cm, 90 cm   **25.** width 40 cm; length 80 cm   **27.** 150 mA   **29.** 4 in. and 18 in.

CHAPTER 9 REVIEW   (pages 310)

1.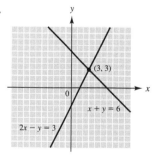

2.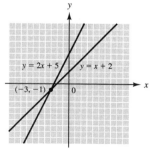

3.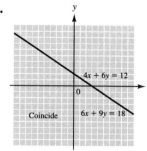

4.

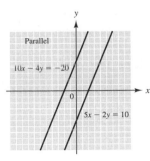

5.

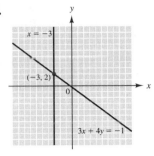

6.

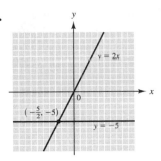

7. $(3, 4)$ **8.** $(3, 1)$ **9.** $(4, 3)$ **10.** no common solution—lines parallel   **11.** $(6, -2)$ **12.** $(3, 1)$

13. infinitely many solutions—lines coincide   **14.** $(4, 8)$ **15.** $(-1, -2)$ **16.** $\left(1\frac{9}{25}, 10\frac{11}{25}\right)$ **17.** resistor 10¢; capacitor 20¢

18. 132.5 ft by 57.5 ft   **19.** 20 mH; 70 mH   **20.** 35 ft; 55 ft

CUMULATIVE REVIEW CHAPTERS 1–9   (page 312)

1. 5   **2.** 15   **3. a.** $4\frac{3}{16}$ in.   **b.** $2\frac{7}{8}$ in.   **4.** 222.966   **5.** 1.116×10^3 **6.** 0.061 m   **7.** 5   **8.** 7.82 mm   **9.** 350 m³   **10.** 70 V

11. $7x - 11$ **12.** $-40x^4y^4$ **13.** $8x^2 - 6xy$ **14.** $-\frac{3}{5}$ **15.** $V = \frac{3s - t}{2}$ **16.** $\frac{1}{13}$ **17.** $\frac{8}{27}$ **18.** 60   **19.** 41.0   **20.** 1450

21. $102.60   **22.** 170 mi   **23.** 10 teeth   **24.** $(3, 2)\ (0, 4)\ (-3, 6)$ **25.** $y = 3x - 5$ **26.** $y = \frac{7 - 4x}{2}$

27.

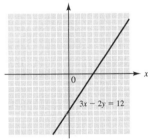

28.

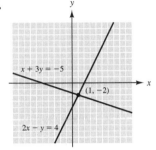

29. $(2, -4)$ **30.** graphs are parallel; no common solution

CHAPTER 10

EXERCISES 10.1   (page 316)

1. $4(a + 1)$ **3.** $b(x + y)$ **5.** $5(3b - 4)$ **7.** $x(x - 7)$ **9.** $a(a - 4)$ **11.** $4n(n - 2)$ **13.** $5x(2x + 5)$ **15.** $3r(r - 2)$
17. $4x^2(x^2 + 2x + 3)$ **19.** $9a(a - x^2)$ **21.** $10(x + y - z)$ **23.** $3(y - 2)$ **25.** $7xy(2 - xy)$ **27.** $m(12x^2 - 7)$
29. $12a(5x - 1)$ **31.** $13mn(4mn - 1)$ **33.** $2(26m^2 - 7m + 1)$ **35.** $18y^2(2 - y + 3y^2)$ **37.** $3m(2m^3 - 4m + 1)$
39. $-2x^2y^3(2 + 3y + 5y^2)$ **41.** $3abc(abc + 9a^2b^2c - 27)$ **43.** $4xz^2(x^2z^2 - 2xy^2z + 3y)$

EXERCISES 10.2   (page 318)

1. $x^2 + 7x + 10$ **3.** $6x^2 + 17x + 12$ **5.** $x^2 - 11x + 30$ **7.** $x^2 - 14x + 24$ **9.** $2x^2 + 19x + 24$ **11.** $x^2 + 4x - 12$
13. $x^2 - 19x + 90$ **15.** $x^2 - 6x - 72$ **17.** $8x^2 - 18x - 35$ **19.** $8x^2 + 6x - 35$ **21.** $14x^2 + 41x + 15$ **23.** $3x^2 - 19x - 72$
25. $6x^2 + 47x + 35$ **27.** $169x^2 - 104x + 16$ **29.** $120x^2 + 54x - 21$ **31.** $100x^2 - 100x + 21$ **33.** $4x^2 - 16x + 15$
35. $4x^2 + 4x - 15$ **37.** $6x^2 + 5x - 56$ **39.** $6x^2 + 37x + 56$ **41.** $16x^2 + 14x - 15$ **43.** $2y^2 - 11y - 21$
45. $6n^2 + 3ny - 30y^2$ **47.** $8x^2 + 26xy - 7y^2$ **49.** $\frac{1}{8}x^2 - 5x + 48$

EXERCISES 10.3   (page 321)

1. $(x + 2)(x + 4)$ **3.** $(y + 4)(y + 5)$ **5.** $3(r + 5)(r + 5)$ **7.** $(b + 5)(b + 6)$ **9.** $(x + 8)(x + 9)$ **11.** $5(a + 4)(a + 3)$
13. $(x - 4)(x - 3)$ **15.** $2(a - 7)(a - 2)$ **17.** $3(x - 7)(x - 3)$ **19.** $(w - 6)(w - 7)$ **21.** $(x - 9)(x - 10)$
23. $(t - 10)(t - 2)$ **25.** $(x + 4)(x - 2)$ **27.** $(y + 5)(y - 4)$ **29.** $(a + 8)(a - 3)$ **31.** $(c - 18)(c + 3)$ **33.** $3(x - 4)(x + 3)$
35. $(c + 6)(c - 3)$ **37.** $(y + 14)(y + 3)$ **39.** $(r - 7)(r + 5)$ **41.** $(m - 20)(m - 2)$ **43.** $(x - 15)(x + 6)$
45. $(a + 23)(a + 4)$ **47.** $2(a - 11)(a + 5)$ **49.** $(a + 25)(a + 4)$ **51.** $(y - 19)(y + 5)$ **53.** $(y - 16)(y - 2)$

55. $7(x + 2)(x - 1)$ **57.** $6(x^2 + 2x - 1)$ **59.** $(y - 7)(y - 5)$ **61.** $(a + 9)(a - 7)$ **63.** $(x + 4)(x + 14)$
65. $2(y - 15)(y - 3)$ **67.** $3x(y - 3)(y - 3)$ **69.** $(x + 15)(x + 15)$ **71.** $(x - 9)(x - 17)$ **73.** $(x + 12)(x + 16)$
75. $(x + 22)(x - 8)$ **77.** $2b(a + 6)(a - 4)$ **79.** $(y - 9)(y + 8)$

EXERCISES 10.4   (page 323)

1. $x^2 - 9$ **3.** $a^2 - 25$ **5.** $4b^2 - 121$ **7.** 9991 or $(10,000 - 9)$ **9.** $9y^4 - 196$ **11.** $r^2 - 24r + 144$ **13.** $16y^2 - 25$
15. $x^2y^2 - 8xy + 16$ **17.** $a^2b^2 + 2abd + d^2$ **19.** $z^2 - 22z + 121$ **21.** $s^2t^2 - 14st + 49$ **23.** $x^2 - y^4$ **25.** $x^2 + 10x + 25$
27. $x^2 - 49$ **29.** $x^2 - 6x + 9$ **31.** $a^2b^2 - 4$ **33.** $x^4 - 4$ **35.** $r^2 - 30r + 225$ **37.** $y^6 - 10y^3 + 25$ **39.** $100 - x^2$

EXERCISES 10.5   (page 326)

1. $(a + 4)(a + 4)$ **3.** $(b + c)(b - c)$ **5.** $(x - 2)(x - 2)$ **7.** $(2 + x)(2 - x)$ **9.** $(y + 6)(y - 6)$ **11.** $5(a + 1)(a + 1)$
13. $(1 + 9y)(1 - 9y)$ **15.** $(7 + a^2)(7 - a^2)$ **17.** $(7x + 8y)(7x - 8y)$ **19.** $(1 + xy)(1 - xy)$ **21.** $(2x - 3)(2x - 3)$
23. $(R + r)(R - r)$ **25.** $(7x + 5)(7x - 5)$ **27.** $(y - 5)(y - 5)$ **29.** $(b + 3)(b - 3)$ **31.** $(m + 11)(m + 11)$
33. $(2m + 3)(2m - 3)$ **35.** $4(x + 3)(x + 3)$ **37.** $3(3x + 1)(3x - 1)$ **39.** $a(m - 7)(m - 7)$

EXERCISES 10.6   (page 328)

1. $(5x + 2)(x - 6)$ **3.** $(2x - 3)(5x - 7)$ **5.** $(6x - 5)(2x - 3)$ **7.** $(2x + 9)(4x - 5)$ **9.** $(16x + 5)(x - 1)$
11. $4(3x + 2)(x - 2)$ **13.** $(5y + 3)(3y - 2)$ **15.** $(4m + 1)(2m - 3)$ **17.** $(7a + 1)(5a - 1)$ **19.** $(4y - 1)(4y - 1)$
21. $(3x - 7)(x + 9)$ **23.** $(4b - 1)(3b + 2)$ **25.** $(5y + 2)(3y - 4)$ **27.** $(10 + 3c)(9 - c)$ **29.** $(3x - 5)(2x - 1)$
31. $(2y^2 - 5)(y^2 + 7)$ **33.** $(2b + 13)(2b + 13)$ **35.** $(7x - 8)(2x - 5)$ **37.** $7x(2x + 5)^2$ **39.** $5a(2b + 7)(b - 5)$

CHAPTER 10 REVIEW   (page 328)

1. $c^2 - d^2$ **2.** $x^2 - 36$ **3.** $y^2 + 3y - 28$ **4.** $4x^2 - 8x - 45$ **5.** $x^2 + 5x - 24$ **6.** $x^2 - 13x + 36$ **7.** $x^2 - 6x + 9$
8. $4x^2 - 24x + 36$ **9.** $1 - 10x^2 + 25x^4$ **10.** $6(a + 1)$ **11.** $5(x - 3)$ **12.** $x(y + 2z)$ **13.** $y^2(y + 18)(y - 1)$
14. $(y - 7)(y + 1)$ **15.** $(z + 9)(z + 9)$ **16.** $(x + 8)(x + 2)$ **17.** $4(a^2 + x^2)$ **18.** $(x - 9)(x - 8)$ **19.** $(x - 9)(x - 9)$
20. $(x + 15)(x + 4)$ **21.** $(y - 1)(y - 1)$ **22.** $(x - 7)(x + 4)$ **23.** $(x - 12)(x + 8)$ **24.** $(x + 11)(x - 10)$
25. $(x + 7)(x - 7)$ **26.** $(4y + 3x)(4y - 3x)$ **27.** $(x + 12)(x - 12)$ **28.** $(5x + 9y)(5x - 9y)$ **29.** $4(x - 13)(x + 7)$
30. $5(x + 12)(x - 13)$ **31.** $(2x + 7)(x + 2)$ **32.** $(4x - 1)(3x - 4)$ **33.** $(6x + 5)(5x - 3)$ **34.** $(12x - 1)(x + 12)$
35. $2(2x - 1)(x - 1)$ **36.** $(6x + 7y)(6x - 7y)$ **37.** $2(7x + 3)(2x + 5)$ **38.** $3(5x - 7)(2x + 1)$ **39.** $4x(x + 1)(x - 1)$
40. $25(y + 2)(y - 2)$

CHAPTER 11

EXERCISES 11.1   (page 333)

1. $-4, 3$ **3.** $-5, 4$ **5.** $2, -1$ **7.** $1, -1$ **9.** $2, -2$ **11.** $-3, -2$ **13.** $7, -3$ **15.** $10, -4$ **17.** $0, 9$ **19.** $12, -9$ **21.** $-8, 2$
23. $-\frac{2}{5}, -\frac{5}{2}$ **25.** $\frac{5}{2}, -\frac{5}{2}$ **27.** $\frac{4}{3}$ **29.** $0, -3$ **31.** $\frac{1}{2}x(x + 1) = 66$; base 12 m; height 11 m **33.** length: 8 in.; width: 5 in.

EXERCISES 11.2   (page 335)

1. $a = 1, b = -7, c = 4$ **3.** $a = 3, b = 4, c = 9$ **5.** $a = -3, b = 4, c = 7$ **7.** $a = 3, b = 0, c = -14$ **9.** $-3, 2$ **11.** $-9, 1$
13. $0, -\frac{2}{5}$ **15.** $\frac{5}{4}, -\frac{7}{12}$ **17.** $1.35, -1.85$ **19.** $0, \frac{5}{3}$ **21.** $-1, \frac{3}{2}$ **23.** $-1.38, -0.121$ **25.** $-\frac{1}{4}, -1$ **27.** $2.38, -2.38$
29. $15.5, -0.453$ **31.** $4.15, -2.49$ **33.** $0.206, -2.40$ **35. a.** 4 s; 8 s **b.** 1.42 s; 10.6 s **c.** 16 s **37.** Length 7 ft, width 3 ft
39. a. 5 cm by 5 cm **b.** 7500 cm^3 **41.** 2 ft

EXERCISES 11.3   (page 341)

1.

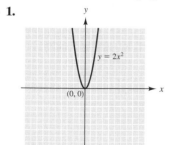

3.

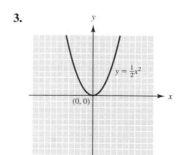

5.

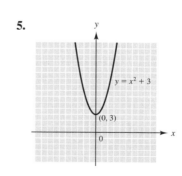

7.
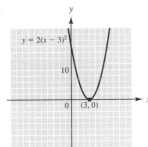

9.
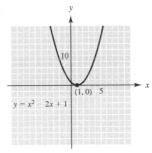

11.
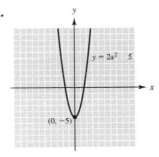

13.
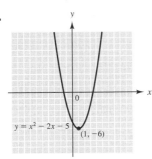

15.
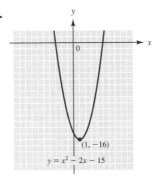

17.
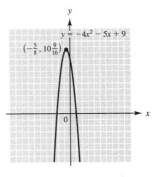

19.
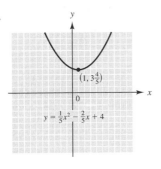

EXERCISES 11.4   (page 344)

1. $7j$ **3.** $3.74j$ **5.** $1.41j$ **7.** $7.48j$ **9.** $13j$ **11.** $5.20j$ **13.** $-j$ **15.** j **17.** $-j$ **19.** 1 **21.** -1 **23.** $-j$
25. Two rational roots   **27.** Two imaginary roots   **29.** Two irrational roots   **31.** Two imaginary roots
33. Two imaginary roots   **35.** $3 + j, 3 - j$ **37.** $7 + 2j, 7 - 2j$ **39.** $-4 + 5j, -4 - 5j$ **41.** $-0.417 + 1.08j, -0.417 - 1.08j$
43. $1 + 1.16j, 1 - 1.16j$ **45.** $-0.8 + 0.4j, -0.8 - 0.4j$ **47.** $0.2, -3$ **49.** $-0.5 + 0.866j, -0.5 - 0.866j$

CHAPTER 11 REVIEW   (page 344)

1. $a = 0$ or $b = 0$ **2.** $0, 2$ **3.** $2, -2$ **4.** $3, -2$ **5.** $0, \frac{6}{5}$ **6.** $7, -4$ **7.** $9, 5$ **8.** $6, -3$ **9.** $-\frac{8}{3}, -4$ **10.** $6, -\frac{2}{3}$
11. $0.653, -7.65$ **12.** $\frac{5}{2}, -3$ **13.** $4.45, -0.449$ **14.** $2.12, -0.786$ **15.** Length 9 ft, width 4 ft   **16. a.** $4\ \mu s, 8\ \mu s$ **b.** $6\ \mu s$
c. $2.84\ \mu s, 9.16\ \mu s$
17.
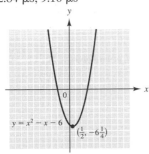

18.
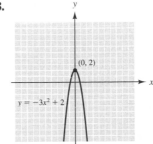

19. $6j$ **20.** $8.54j$ **21.** 1 **22.** $-j$
23. One rational root   **24.** Two imaginary roots
25. $2 + j, 2 - j$ **26.** $0.6 + 0.663j, 0.6 - 0.663j$
27. 2.5 ft by 8.5 ft   **28.** 3 in.

CHAPTER 12

EXERCISES 12.1   (page 352)

1. 47°, acute   **3.** 90°, right   **5.** 41°, acute   **7.** 156°, obtuse   **9.** Right, perpendicular   **11. a.** 1, 2; 2, 4; 3, 4; 1, 3; 5, 6; 6, 8; 7, 8; 5, 7   **b.** 1, 4; 2, 3; 5, 8; 6, 7   **13.** $\angle 2 = 57°$; $\angle 3 = 57°$; $\angle 4 = 123°$ **15.** $\angle COB = 61°$ **17.** 52°   **19. a.** Yes   **b.** Yes   **21.** 90°   **23.** 15°   **25.** 148°   **27.** 152°   **29.** Triangle   **31.** Hexagon   **33.** Quadrilateral   **35.** Heptagon

EXERCISES 12.2   (page 356)

1. $P = 60.0$ cm; $A = 225$ cm^2 **3.** $P = 32.0$ m; $A = 48$ m^2 **5.** $P = 45.0$ m; $A = 78$ m^2 **7.** $P = 89.4$ in.; $A = 411$ in^2 **9.** $P = 36.8$ cm; $A = 85$ cm^2 **11.** 24.0 cm   **13.** 11.6 mm   **15.** 5.87 ft   **17.** 13.4 ft   **19.** 8.14 m   **21.** 38.27 cm   **23.** 45.556 km   **25.** 9   **27.** 36 ft   **29.** 6 pieces   **31.** \$178,500   **33. a.** 208 in^2 **b.** 59$\overline{0}$ in^2 **35.** 40$\overline{0}$,000 acres   **37.** 7 squares   **39.** \$927.35   **41.** 250 ft^2 **43.** 568.6 ft

EXERCISES 12.3   (page 364)

1. 10.0 m   **3.** 25.0 m   **5.** 17.0 m   **7.** 13.5 cm   **9.** 1460 km   **11.** 350 ft   **13.** 59.9 in.   **15.** 33.4 in.   **17.** 2.83 in.   **19.** 16.6 in.   **21.** 116 V   **23.** 21 A   **25.** 127 Ω **27.** 2.6 Ω **29.** 158 m^2; 60.8 m   **31.** 7.80 cm^2; 18.00 cm   **33.** 395 m^2; 108.7 m   **35.** 29.9 m^2; 36.0 m   **37.** 3070 m^2; 286.7 m   **39.** 96.0 cm^2; 48.0 cm   **41.** 52 ft^2 **43.** 145 in^2 **45.** 81°   **47.** 60°   **49.** $x = 36°$, $y = 72°$

EXERCISES 12.4   (page 369)

1. 5   **3.** Yes, all angles of $\triangle ABO$ are equal to corresponding angles of $\triangle DCO$.   **5.** $5\frac{1}{3}$ **7.** 120 ft   **9.** $DE = 12.0$ m   **11.** $A = 1.33$ ft; $B = 2.67$ ft; $C = 4.00$ ft; $D = 5.33$ ft; $E = 6.67$ ft; cross-piece is 10.0 ft

EXERCISES 12.5   (page 374)

1. a. 31.4 in.   **b.** 78.5 in^2 **3. a.** 28.9 mm   **b.** 66.6 mm^2 **5. a.** 352 mi   **b.** 9890 mi^2 **7.** 171°   **9.** 43.4°   **11.** 3.00 cm   **13.** 2.60 in.   **15.** 10.0 m   **17.** 32.1 ft   **19.** 90°   **21.** 2670 ft   **23.** 13.6 in.   **25.** 1.25 in.   **27.** 1332 cm^2 **29.** 24.42 in.   **31. a.** 2.50 ft   **b.** 3.54 ft   **33.** 72°   **35.** 29°   **37.** 30°   **39.** 18,600 mi   **41.** 171°   **43.** 14°   **45. a.** 120°   **b.** 60°   **c.** 120°   **47. a.** 60°   **b.** 120°   **c.** 60°   **49. a.** 5.6 in^2 **b.** 4.13 in.   **c.** 2.26 in.   **d.** 8.4 oz

EXERCISES 12.6   (page 381)

1. 180°   **3.** 0.37 rad   **5.** 60°   **7.** $\frac{3\pi}{4}$ rad; 2.36 rad   **9.** $\frac{4\pi}{3}$ rad; 4.19 rad   **11.** $\frac{4\pi}{5}$ rad; 2.51 rad   **13.** 31.4 cm   **15.** 4.7 cm   **17.** 104 cm   **19.** 2.8 rad   **21.** 98.5°   **23.** 8.83 cm   **25.** $\frac{5}{\pi}$ or 1.59 rps   **27.** 92.2 cm   **29.** 25.8 cm   **31. a.** 26.2 m   **b.** 327 m^2 **c.** 56.4 m^2

EXERCISES 12.7   (page 385)

1. a. 44$\overline{0}$ in^2 **b.** 622 in^2 **c.** 912 in^3 **3. a.** 2490 m^2 **b.** 1230 m^2 **c.** 188 m^3 **d.** 49$\overline{0}$0 m^2 **5.** 324 ft^3 **7.** 3$\overline{0}$ in.

EXERCISES 12.8   (page 389)

1. 13,600 mm^3 **3.** 8,280,000 L   **5.** 42.6 ft   **7.** 14 : 1   **9.** 1140 ft^3 **11.** 1.79 in.   **13.** 28.3 in^3 **15. a.** 2050 mm^2 **b.** 2470 mm^2 **17.** 101,000 dm^2 **19.** 18.0 in^3 **21.** 1.38 in^2 **23.** 6.1 gal   **25.** 24.9 kg

EXERCISES 12.9   (page 394)

1. 147 in^3 **3.** 11$\overline{0}$0 m^3 **5.** 69,700 mm^3 **7.** 1440 mm^3 **9.** 312 ft^3 **11.** 1010 cm^3; 11$\overline{0}$0 cm^2 **13.** 1520 bu   **15. a.** 1690 cm^3 **b.** 672 cm^2 **17.** 37,000 ft^3, 1$\overline{0}$0 truckloads   **19.** 15.0 m^3 **21.** 108,000 ft^3; 9420 ft^2 **23.** 28.0 in.   **25.** 33.9 ft^3

EXERCISES 12.10   (page 398)

1. a. 804 m^2 **b.** 2140 m^3 **3. a.** 4120 in^2 **b.** 24,800 in^3 **5.** 114,000 m^3 **7.** 933,000 gal   **9. a.** 0.375 or $\frac{3}{8}$ **b.** 0.25 or $\frac{1}{4}$

CHAPTER 12 REVIEW   (page 399)

1. 72°   **2.** 119°   **3.** $\angle 1 = 59°$; $\angle 2 = 121°$; $\angle 3 = 59°$; $\angle 4 = 59°$; $\angle 5 = 121°$ **4.** Adjacent or supplementary   **5.** 20°   **6. a.** Quadrilateral   **b.** Pentagon   **c.** Hexagon   **d.** Triangle   **e.** Octagon   **7.** 36.00 cm; 60.0 cm^2 **8.** 360.4 m; 80$\overline{0}$0 m^2 **9.** 43.11 cm; 102 cm^2 **10.** 7.73 m   **11.** 1.44 cm   **12.** 159 m^2; 58.7 m   **13.** 796 m^2; 159.5 m   **14.** 4.73 cm^2; 10.93 cm

15. 41.9 m   **16.** 3.85 cm   **17.** 86°   **18.** 8.0 m   **19.** 423 cm²; 72.9 cm   **20.** 12.1 cm   **21.** 216°   **22.** $\frac{2\pi}{15}$ rad   **23.** 10°
24. 42.2 cm   **25.** 1.55 rad   **26.** 291 cm   **27.** 1790 m²; 2150 m²   **28.** 5760 m³   **29.** 1260 cm³   **30.** 70.4 cm³   **31.** 70.4 cm²;
95.5 cm²   **32.** 131 m³   **33. a.** 57,300 m³   **b.** 6370 m²   **34. a.** 869 m³   **b.** 44$\overline{0}$ m²   **35.** 65,900 in³; 6150 in²

CUMULATIVE REVIEW CHAPTERS 1–12   (page 403)

1. -4 **2.** $13\frac{3}{4}$ in.   **3.** b   **4. a.** 100 L   **b.** 50 L   **5.** $2x^2 - 8x - 4$ **6.** -7 **7.** 19.4   **8.** 56 lb   **9.** $y = (5x - 10)/8$ **10.** (4, 8)
11. $6x^2 + 11x - 35$ **12.** $16x^2 - 24x + 9$ **13.** $5x(x^2 - 3)$ **14.** $(x - 7)(x + 4)$ **15.** $(x + 2)(x - 2)$ **16.** $\frac{7}{2}, -3$ **17.** $\frac{3}{2}, -1$
18. 0.4, -3 **19.** $-1.72, 4.22$ **20.**

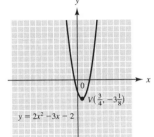

21. j **22.** $\angle 1 = \angle 2 = 82°$; $\angle 3 = \angle 5 = 98°$ **23.** 59°
24. 78.6 m, 385 m²   **25.** 76.3 cm, 265 cm²
26. 6.59 m, 1.86 m²   **27.** 45°   **28.** 191 cm², 49.0 cm
29. 7.31 cm   **30.** 987 cm³; 567 cm²

CHAPTER 13

EXERCISES 13.1   (page 409)

1. a **3.** c **5.** a **7.** B **9.** B **11.** 60.0 m   **13.** 55.2 mi   **15.** 17.7 cm   **17.** 265 ft   **19.** 35$\overline{0}$ m   **21.** 1530 km   **23.** 59,900 m
25. 0.5000   **27.** 0.5741   **29.** 0.5000   **31.** 0.7615   **33.** 2.174   **35.** 0.8686   **37.** 0.5246   **39.** 0.2536   **41.** 0.0533   **43.** 0.6104
45. 0.3929   **47.** 0.6800   **49.** 52.6°   **51.** 62.6°   **53.** 54.2°   **55.** 9.2°   **57.** 30.8°   **59.** 40.8°   **61.** 10.98°   **63.** 69.07°   **65.** 8.66°
67. 75.28°   **69.** 43.02°   **71.** 20.13°   **73.** sin A and cos B; cos A and sin B

EXERCISES 13.2   (page 412)

1. $A = 35.3°$; $B = 54.7°$ **3.** $A = 24.4°$; $B = 65.6°$ **5.** $A = 43.3°$; $B = 46.7°$ **7.** $A = 45.0°$; $B = 45.0°$ **9.** $A = 64.05°$; $B =$
$25.95°$ **11.** $A = 52.60°$; $B = 37.40°$ **13.** $A = 39.84°$; $B = 50.16°$ **15.** $A = 41.85°$; $B = 48.15°$ **17.** $A = 11.8°$; $B = 78.2°$ **19.**
$A = 31.11°$; $B = 58.89°$ **21.** $A = 15.9°$; $B = 74.1°$ **23.** $A = 64.65°$; $B = 25.35°$

EXERCISES 13.3   (page 414)

1. $b = 40.6$ m; $c = 54.7$ m   **3.** $b = 278$ km; $c = 365$ km   **5.** $a - 29.3$ cm; $b = 39.4$ cm   **7.** $a = 10.4$ cm; $c = 25.9$ cm
9. $a = 6690$ km; $c = 30,\overline{0}00$ km   **11.** $b = 54.92$ m; $c = 58.36$ m   **13.** $a = 161.7$ mi; $b = 197.9$ mi   **15.** $a = 4564$ m; $c = 13,170$ m
17. $c = 6270$ m; $b = 5950$ m   **19.** $a = 288.3$ km; $c = 889.6$ km   **21.** $a = 88.70$ m; $b = 163.0$ m   **23.** $a = 111$ cm; $b = 231$ cm

EXERCISES 13.4   (page 416)

1. $B = 39.4°$; $a = 37.9$ m; $b = 31.1$ m   **3.** $A = 48.8°$; $b = 234$ ft; $c = 355$ ft   **5.** $A = 22.6°$; $B = 67.4°$; $a = 30.0$ mi
7. $B = 37.9°$; $b = 56.1$ mm; $c = 91.2$ mm   **9.** $B = 21.2°$; $a = 36.7$ m; $b = 14.2$ m   **11.** $A = 26.05°$; $B = 63.95°$; $c = 27.33$ m
13. $B = 60.81°$; $a = 1451$ ft; $b = 2597$ ft   **15.** $A = 67.60°$; $B = 22.40°$; $c = 50.53$ m   **17.** $B = 48.9°$; $a = 319$ m; $b = 365$m
19. $A = 80.55°$; $b = 263.8$ ft; $c = 1607$ ft   **21.** $A = 44.90°$; $B = 45.10°$; $a = 268.6$ m   **23.** $A = 8.5°$; $a = 14\overline{0}0$ ft; $c = 9470$ ft

EXERCISES 13.5   (page 418)

1. 18.8 m   **3.** 5.2 m   **5.** $A = 58.2°$; 32.3 ft   **7.** 5°   **9.** 92.2 ft   **11.** 28.1°   **13.** 5.70 cm   **15.** 95.3 m   **17. a.** 430 Ω; **b.** 38°;
c. 37°; 3$\overline{0}$0 Ω   **19. a.** 53.7 V; **b.** 267 V; **c.** 236 V   **21.** 3.7529 in.   **23.** 9.5°   **25. a.** 3.46 in.; **b.** 2.04 in.   **27.** $x = 22.8$ ft,
$A = 55.2°$ **29.** $x = 4.50$ cm, $y = 3.90$ cm

CHAPTER 13 REVIEW   (page 423)

1. 29.7 m   **2.** B **3.** Hypotenuse   **4.** 43.8 m   **5.** sin A **6.** Length of side adjacent to angle A
7. $\dfrac{\text{length of side opposite angle } B}{\text{length of side adjacent to angle } B} = \dfrac{32.2 \text{ m}}{29.7 \text{ m}}$ **8.** 0.8070   **9.** 1.138   **10.** 0.4023   **11.** 45.5°   **12.** 10.4°   **13.** 65.8°   **14.** 39.2°
15. 50.8°   **16.** 7.17 km   **17.** 8.25 km   **18.** $b = 21.9$ m; $a = 18.6$ m; $A = 40.4°$ **19.** $b = 102$ m; $c = 119$ m; $B = 58.8°$
20. $A = 30.0°$; $B = 60.0°$; $b = 118$ mi   **21.** 4950 m   **22.** 5250 ft   **23.** 14.0°

CHAPTER 14

EXERCISES 14.1 (page 432)

1. 0.6820 **3.** −0.4067 **5.** −0.4352 **7.** −0.9724 **9.** 0.7701 **11.** −0.4390

13.
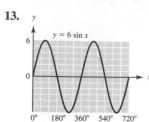

15.
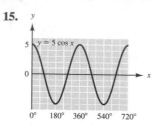

17.
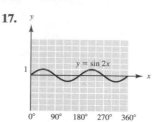

19.
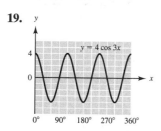

21.
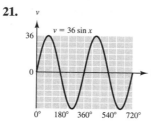

23.
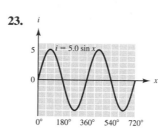

25. 25 V, −33 V **27.** 3.5 A, −3.5 A **29.** 2.0×10^{-4} s
31. 1.8 Hz **33.** 8.8×10^{9} Hz or 8.8 GHz **35.** 2.25 m/s

EXERCISES 14.2 (page 436)

1. 120°; 3
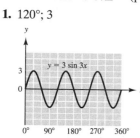

3. 60°; 8
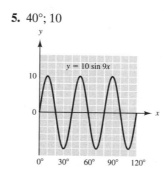

5. 40°; 10
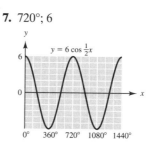

7. 720°; 6
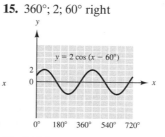

9. 540°; 3.5
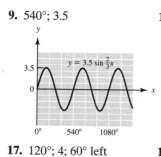

11. 144°; 4
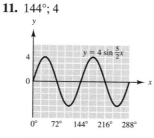

13. 360°; 1; 30° left
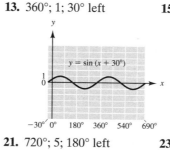

15. 360°; 2; 60° right
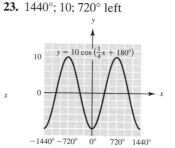

17. 120°; 4; 60° left
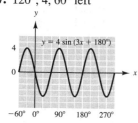

19. 90°; 10; 30° right
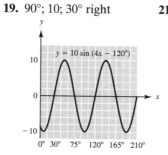

21. 720°; 5; 180° left
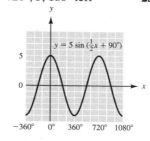

23. 1440°; 10; 720° left
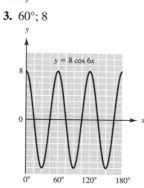

EXERCISES 14.3 (page 439)

1. $c = 24.9$ m; $B = 41.5°$; $C = 70.5°$ **3.** $a = 231$ ft; $c = 179$ ft; $C = 42.9°$ **5.** $C = 130.3°$; $a = 14.4$ cm; $b = 32.9$ cm
7. $B = 82.9°$; $b = 32.5$ m; $c = 22.4$ m **9.** $b = 79.3$ mi; $B = 145.0°$; $C = 14.9°$ **11.** $A = 74.9°$; $a = 6910$ m; $c = 5380$ m
13. $a = 26.4$ km; $A = 54.2°$; $C = 20.3°$ **15.** $c = 380$ ft; $B = 15.1°$; $C = 148.4°$ **17.** 63.5 m **19.** 7.29 km **21.** 214 ft

EXERCISES 14.4 (page 444)

1. a. One solution **b.** $B = 28.2°$; $C = 113.8°$; $c = 62.9$ m **3. a.** Two solutions **b.** $B = 28.7°$; $C = 125.7°$; $c = 517$ m and
$B = 151.3°$; $C = 3.1°$; $c = 34.4$ m **5.** No solution **7. a.** Two solutions **b.** $A = 77.0°$; $B = 31.8°$; $b = 132$ cm and
$A = 103.0°$; $B = 5.8°$; $b = 25.4$ cm **9. a.** One solution **b.** $A = 44.6°$; $C = 30.4°$; $c = 17.3$ mi **11. a.** Two solutions
b. $C = 34.3°$; $B = 114.2°$; $b = 656$ m and $C = 145.7°$; $B = 2.8°$; $b = 35.2$ m **13.** No solution **15. a.** Two solutions
b. $C = 15.3°$; $A = 156.7°$; $a = 1280$ m and $C = 164.7°$; $A = 7.3°$; $a = 412$ m **17.** 496 ft or 78.5 ft

EXERCISES 14.5 (page 449)

1. $a = 21.0$ m; $C = 69.4°$; $B = 55.6°$ **3.** $c = 476$ ft; $B = 36.9°$; $A = 28.1°$ **5.** $A = 49.6°$; $B = 60.2°$; $C = 70.2°$
7. $c = 6.72$ km; $B = 58.4°$; $A = 50.0°$ **9.** $A = 33.7°$; $B = 103.5°$; $C = 42.8°$ **11.** $A = 148.7°$; $C = 12.0°$; $b = 3070$ ft
13. $A = 30.7°$; $B = 107.3°$; $C = 42.0°$ **15.** 65.3 m **17.** $A = 40.9°$; $C = 30.5°$ **19. a.** $59.5°$ **b.** $70.5°$ **c.** 4.92 m **d.** 6.33 m
21. $114.3°$ **23.** 36 yd

CHAPTER 14 REVIEW (page 451)

1. -0.7536 **2.** -0.4970 **3.** 0.7361 **4.**

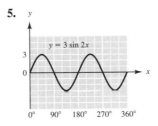

5.

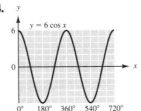

6. $120°$; 5

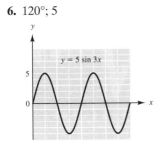

7. $90°$; 3

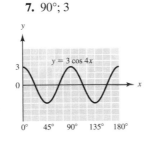

8. $360°$; 4; $60°$ left

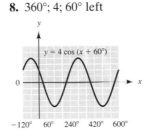

9. $180°$; 6; $90°$ right

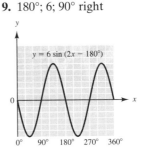

10. $c = 258$ m; $A = 42.5°$; $C = 84.8°$ **11.** $B = 43.2°$; $b = 62.2$ cm; $a = 79.6$ cm **12.** $b = 129$ m; $C = 109.0°$; $A = 53.5°$
13. $A = 66.8°$; $B = 36.2°$; $C = 77.0°$ **14.** $c = 33.4$ m; $B = 33.1°$; $C = 117.4°$ or $B = 146.9°$; $C = 3.6°$; $c = 2.36$ m
15. $c = 3010$ m; $A = 20.5°$; $C = 141.0°$ or $c = 167$ m; $A = 159.5°$; $C = 2.0°$ **16.** $A = 27.3°$; $B = 59.7°$; $C = 93.0°$
17. no solution **18. a.** $36.6°$ **b.** 52.9 m **19.** 9.40 in. **20. a.** 23.5 m **b.** 16.7 m **c.** 15.1 m **d.** 7.06 m

CHAPTER 15

EXERCISES 15.1 (page 455)

1. $1750 **3.** $780 **5.** $800 **7.** $1650 **9.** $1500 **11.** 1.5 million **13.** United States **15.** 1.8 million **17.** 13.7 million
19. 2.3 million **21.** 282 **23.** 34 **25.** 20 **27.** 203 **29.** 11.3

EXERCISES 15.2 (page 458)

1. $94°$ **3.** $55°$ **5.** $270°$ **7.** $219°$ **9.** $15°$ **11.** $328°$

13. Percentage of National Household Income by Quintile

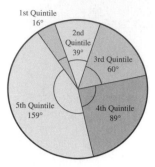

15. Industrial Technology Credit Hour Requirements

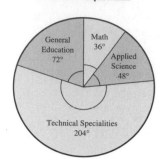

17. 1992 World Cotton Production in millions of 480-lb bales

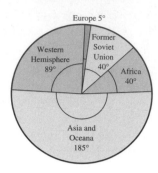

19. 1993 Federal Budget Receipts for the United States

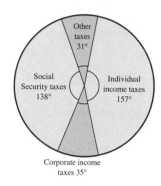

EXERCISES 15.3 (page 460)

1.

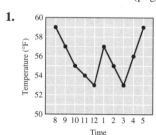

3.

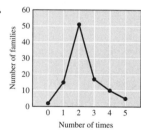

5.

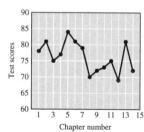

7.

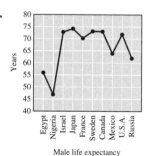

9. 29.38 in. **11.** 29.24 in. **13.** 29.40 in. **15.** 72° **17.** 70° **19.** 60%

EXERCISES 15.4 (page 464)

1. 900 mW; 14 db **3.** 85 mW **5.** 30 db and 32 db **7.** 16 db **9.** 14 db and 16 db

EXERCISES 15.5 (page 465)

1. 47.89 cm **3.** 0.2619 in. **5.** 25,770 mi **7.** 2036 km **9.** 709 lb **11.** 6911 **13.** 2020̄ **15.** 1.096 **17.** 51 **19.** 4.58

EXERCISES 15.6 (page 467)

1. 47.87 cm **3.** 0.2618 in. **5.** 25,770 mi **7.** 2025 km **9.** 781 lb **11.** 6185.5 **13.** 2023 **15.** 1.102 **17.** 53 **19.** 4.17
21. 163 **23.** 107 **25.** 23

EXERCISES 15.7 (page 471)

1. 59 **3.** 7.7 months **5.** 36 h **7.** 42 lb **9.** 11 parts **11.** 4.2 days **13.** 99 strokes **15.** 58 players

EXERCISES 15.8   (page 476)

1. 0.22 cm   **3.** 0.0003 in.   **5.** 30 mi   **7.** 33 km   **9.** 240 lb   **11.** 1848   **13.** 141   **15.** 0.060   **17.** 24   **19.** 2.38   **21.** 8.8
23. 4 months   **25.** 5.2 h   **27.** 15 lb   **29.** 4 parts   **31.** 3 days   **33.** 12 strokes   **35.** 37 players

EXERCISES 15.9   (page 478)

1. a. 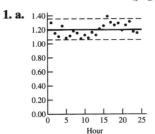   **b.** out of control; in hour 16, the value was outside of limits

3. a. 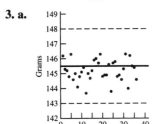   **b.** hours 5 through 17

CHAPTER 15 REVIEW   (page 479)

1. 126°   **2.** 202°   **3.**

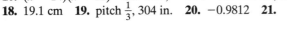

Students in 1990–91

College students 60°
Primary students 195°
Grades 9–12 students 105°

4.

5. 74° F   **6.** 7.0036 mm   **7.** 7.0036 mm
8. 0.0001 mm   **9. a.** 49.9   **b.** 18.2
10. 84.4   **11.** 83   **12.** 7.0   **13.** 19.0
14. 5.1

CUMULATIVE REVIEW CHAPTERS 1–15   (page 482)

1. 8   **2.** $-\frac{7}{30}$ **3.** 198 lb   **4.** 13 Ω   **5.** $6a^2 - 2a + 2$ **6.** 8   **7.** 102 m²   **8.** $y = \frac{8 - 5x}{3}$ **9.** graphs coincide, many solutions
10. $(x - 14)(x + 12)$ **11.** 2.27; − 1.77   **12.** 515 ft²; 126.3 ft   **13.** 6.39 m³   **14.** 2.379   **15.** 51.6°   **16.** 55.8°   **17.** 10.7 cm
18. 19.1 cm   **19.** pitch $\frac{1}{3}$, 304 in.   **20.** −0.9812   **21.** 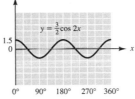   **22.**

$y = \frac{3}{2}\cos 2x$

$y = 2 \sin \frac{1}{2}x$

23. 65.2°   **24.** 27.6   **25.** 28.5   **26.** 5.0   **27.** 63

28.

Interval	Frequency	Midpoint
5.5–10.5	7	8
10.5–15.5	11	13
15.5–20.5	10	18
20.5–25.5	6	23
25.5–30.5	1	28
30.5–35.5	1	33

29. 16   **30.** 16

APPENDIX A

EXERCISES A   (page 486)

1. 29.4   **3.** 1270   **5.** 7.67   **7.** 4.37   **9.** 3,040,000   **11.** 63.2   **13.** 3610   **15.** 332   **17.** 147   **19.** 0.826

Index

Photo Credits

This constitutes an extension of the copyright page.

Chapter 1, page 1: © John Maher/Stock Boston

Chapter 2, page 93: © Suzanne Arms/The Image Works

Chapter 3, page 116: © Cary Wolinsky/Stock Boston

Chapter 4, page 141: © Billy Barnes/Stock Boston

Chapter 5, page 188: © Michael Newman/PhotoEdit

Chapter 6, page 211: © Billy Barnes/PhotoEdit

Chapter 7, page 244: © Peter Vandermark/Stock Boston

Chapter 8, page 266: © David Aronson/Stock Boston

Chapter 9, page 292: © Beringer-Dratch/The Image Works

Chapter 10, page 314: © N. R. Rowan/Stock Boston

Chapter 11, page 330: © N. R. Rowan/The Image Works

Chapter 12, page 347: © PhotoDisc, Inc.

Chapter 13, page 405: © Joseph Schuyler/Stock Boston

Chapter 14, page 426: ©Alan Carey/The Image Works

Chapter 15, page 453: © Gary Walts/The Image Works